Rheinwerk Computing

The Rheinwerk Computing series offers new and established professionals comprehensive guidance to enrich their skillsets and enhance their career prospects. Our publications are written by the leading experts in their fields. Each book is detailed and hands-on to help readers develop essential, practical skills that they can apply to their daily work.

Explore more of the Rheinwerk Computing library!

Kofler, Gebeshuber, Kloep, Neugebauer, Zingsheim, Hackner, Widl, Aigner, Kania, Scheible, Wübbeling

Hacking and Security: The Comprehensive Guide to Penetration Testing and Cybersecurity

2023, 1141 pages, paperback and e-book
www.rheinwerk-computing.com/5696

Sandip Dholakia

Modern Cryptography: The Practical Guide

2024, 425 pages, paperback and e-book
www.rheinwerk-computing.com/5856

Tobias Scheible

Hacking Hardware: The Practical Guide to Penetration Testing

2026, 644 pages, paperback and e-book
www.rheinwerk-computing.com/6181

Johannes Ernesti, Peter Kaiser

Python 3: The Comprehensive Guide (2nd Edition)

2026, approx. 1078 pp, paperback and e-book
www.rheinwerk-computing.com/6243

Sujeevan Vijayakumaran

DevOps: Frameworks, Techniques, and Tools

2025, 418 pages, paperback and e-book
www.rheinwerk-computing.com/6030

www.rheinwerk-computing.com

Haydar Yener Arici

Cyber Threat Intelligence

The Comprehensive Guide

Editors Megan Fuerst
Acquisitions Editor Hareem Shafi
Copyeditor Julie McNamee
Cover Design Abigail Damke
Photo Credits Shutterstock: 1641095257/© Dmitry Taranets; iStock 1355569095/© seamartini; Shutterstock: 674184880/© Fotomay
Layout Design Vera Brauner
Production Hannah Lane
Typesetting SatzPro, Germany
Printed and bound in the United States of America, on paper from sustainable sources

ISBN 978-1-4932-2813-3
1st edition 2026

Rheinwerk Publishing, Inc.
2 Heritage Drive, Suite 305
Quincy, MA 02171
USA
info@rheinwerk-publishing.com
+1.781.228.5070

Represented in the E. U. by:
Rheinwerk Verlag GmbH
Rheinwerkallee 4
53227 Bonn
Germany
service@rheinwerk-verlag.de
+49 (0) 228 42150-0

Library of Congress Cataloging-in-Publication Control Number: 2026018022

Contents at a Glance

Contents

Preface

To date, I've published six books, all of which met readers in Türkiye. This seventh book, published beyond the borders of Türkiye, is a natural continuation of both my professional experience and perspective that has matured over the years.

Throughout my years in the IT field, I've observed one fundamental truth: threats change, tools change, and techniques change, but one thing remains constant: uncertainty. In this uncertain environment, what truly makes a difference isn't merely technical knowledge, but the ability to interpret that knowledge and place it into context. This book was born from that need.

I didn't write this book solely to explain cyber threat intelligence (CTI). It was also written to present an approach, a way of thinking, and a culture of discipline. While working in incident response teams, threat hunting operations, and forensic investigations over the years, I encountered a recurring problem: Although there was abundant data, countless alerts, and numerous reports, context was missing. Without context, information doesn't produce decisions, and information that doesn't drive decisions isn't intelligence.

This book is the product of an effort to transform intelligence from a simple indicator of compromise (IOC) list into a living, breathing structure with real operational value.

Who This Book Is For

This book is intended for professionals and practitioners who operate in environments where cyber threats must be understood, interpreted, and acted upon. It's written for those who work at the intersection of technology, analysis, and decision making, where raw data must be transformed into meaningful intelligence.

Security analysts, threat intelligence specialists, incident responders, digital forensics investigators, security operations center (SOC) professionals, and threat hunters will find practical value in the methods and perspectives presented throughout these chapters. The book is also relevant for cybersecurity engineers and architects who want to understand how intelligence can strengthen detection capabilities, guide defensive strategies, and improve operational resilience.

Graduate students and researchers in cybersecurity, digital forensics, and information security may also benefit from this work. While the book includes practical operational insights drawn from real-world experience, it also connects those experiences with structured analytical frameworks commonly used in modern intelligence practice.

However, this book isn't written as an introductory guide to basic cybersecurity concepts. Readers are expected to possess a foundational understanding of networks, operating systems, and security operations. The discussions build on that baseline knowledge and focus on developing analytical discipline, contextual thinking, and intelligence-driven decision making.

Ultimately, this book is for individuals who recognize that cybersecurity isn't just about detecting malicious activity, but also about understanding the intent, behavior, and patterns behind it. Those who aim to move beyond simply collecting indicators to interpreting the broader threat landscape will find the perspective presented here particularly relevant.

How This Book Is Organized

CTI is often misunderstood. At times, it's confined solely to technical indicators; at other times, it's equated with strategic reporting. In reality, CTI is a holistic discipline that extends from strategic perspective to operational response, from network forensics to endpoint telemetry, and from dark web observations to threat actor behavioral mapping. Every chapter in this book was designed as a component of that integrated whole.

I began with the foundations: What is intelligence? How is it produced? Why is it fundamentally a matter of governance? I then moved beyond theoretical lifecycle diagrams and examined how intelligence should function within real operational workflows. From source reliability and open-source intelligence (OSINT) disciplines to deep and dark web collection and threat actor profiling, I focused on one central question: Does this information truly serve a purpose?

Over time, I realized an essential truth: Threat intelligence is incomplete unless it's validated. For this reason, I devoted particular attention to network-centric forensic analysis and endpoint telemetry. The true value of intelligence emerges where it intersects with reality in the field. Disruptions in network flow rhythm, silence anomalies on endpoints, and behavioral traces aren't merely technical details; they are tangible evidence of intelligence.

A significant portion of this work is also dedicated to the human factor. Threats are more than code; they reflect intent, habits, and patterns of behavior. Threat actor profiling, MITRE ATT&CK mappings, and Diamond Model analyses are therefore presented as both frameworks and tools for thinking. Technical indicators may change rapidly, but behavioral patterns evolve more slowly. That distinction provides a strategic advantage for defenders.

In the final chapters, I focused on sustainability. Modern security operations have surpassed human capacity due to the sheer volume of data. Automation and threat intelligence platforms have changed from optional to essential. However, the purpose of automation isn't speed alone—it's rhythm and consistency. Poorly designed automation scales errors. Well-structured systems transform intelligence into institutional memory.

Here's a quick glimpse of what each chapter covers in this book:

- **Chapter 1: Foundations of Cyber Threat Intelligence**
  This chapter establishes the foundational knowledge necessary to understand CTI. It introduces the origins, definitions, and strategic value of CTI in today's cybersecurity

landscape. We'll explore how intelligence is classified and how CTI supports decision-making across enterprise security operations.

- **Chapter 2: Intelligence Lifecycle in Practice**
  This chapter introduces the intelligence lifecycle as a structured process for generating and maintaining high-quality threat intelligence. The chapter walks you through each stage—planning, collection, processing, analysis, dissemination, and feedback—with an emphasis on its iterative nature in real-world operations.
- **Chapter 3: Intelligence Sources**
  This chapter provides an in-depth classification of intelligence sources used in CTI. It outlines the unique value and limitations of open-source intelligence (OSINT), human intelligence (HUMINT), and signals intelligence (SIGINT).
- **Chapter 4: Applied OSINT: Tools, Methodologies, and Operational Discipline**
  This chapter focuses on the practical application of OSINT within cyber investigations and threat intelligence gathering. It introduces a wide array of OSINT tools and techniques, covering both passive reconnaissance (e.g., search engines, metadata harvesting) and active probing (e.g., Domain Name System [DNS] zone transfers, web crawling).
- **Chapter 5: Advanced Intelligence Collection from the Deep and Dark Web**
  This chapter explores the techniques and tools for collecting intelligence from the deep web and dark web. It introduces hidden services and anonymity networks, where threat actors communicate, exchange tools, and coordinate malicious campaigns.
- **Chapter 6: Threat Actor Profiling and Behavioral Mapping**
  This chapter focuses on understanding and profiling adversaries based on their behavior, motivations, and technical patterns. Using industry frameworks such as the MITRE ATT&CK matrix and the Diamond Model, you'll learn how to structure behavioral profiles and threat models.
- **Chapter 7: Integrity, Poisoning, and Enrichment in Threat Intelligence Feeds**
  This chapter examines the reliability, integrity, and risks associated with CTI feeds. You'll explore the concept of feed poisoning—how adversaries can inject false or misleading indicators into public or commercial feeds—and the methods analysts use to detect and mitigate these risks. Additionally, the chapter covers enrichment techniques to contextualize raw indicators and transform them into actionable intelligence.
- **Chapter 8: Network-Centric Forensic Intelligence**
  This chapter delivers advanced techniques for conducting forensic analysis of network data within the context of CTI. The focus is on collecting and interpreting network traffic to uncover attacker behaviors, trace intrusion paths, and support attribution.
- **Chapter 9: Host-Based Forensic Analysis and Windows Telemetry**
  This chapter explores endpoint-level forensic intelligence, with a specific focus on Windows systems. It walks through key artifacts such as event logs, registry data, memory captures, and telemetry sources. You'll gain practical knowledge of how to acquire, analyze, and interpret host-based evidence to detect malicious behavior, uncover persistence mechanisms, and support incident response and attribution workflows.

- **Chapter 10: Integrating CTI Into Incident Response**
  This chapter illustrates how CTI supports and enhances the incident response lifecycle. From detection to containment and recovery, CTI provides context, enriches alerts, and accelerates decision-making. You'll learn how to integrate intelligence into incident response platforms, match indicators of compromise (IOCs) and indicators of attack (IOAs), and conduct threat correlation within security information and event management (SIEM) software.
- **Chapter 11: Intelligence-Driven Proactive Threat Hunting**
  This chapter explores how CTI can be operationalized to support proactive threat hunting. It discusses the limitations of purely reactive detection models and introduces a hypothesis-driven approach to identifying attacker behavior that may not trigger traditional alerts. The chapter examines how intelligence insights can guide hunting priorities, support the analysis of adversary tactics and techniques, and help analysts identify subtle behavioral patterns across enterprise environments. Through practical perspectives and real-world considerations, the chapter demonstrates how intelligence-driven hunting can reveal stealthy threats that remain invisible to conventional detection mechanisms.
- **Chapter 12: Automation and Threat Intelligence Platforms**
  The final chapter explores how automation and threat intelligence platforms (TIPs) support the operational use of CTI in modern security environments. It introduces MISP Threat Sharing (MISP) as a platform for structured intelligence sharing and discusses how standards enable scalable, machine-to-machine exchange of threat intelligence to improve analysis, correlation, and collaboration.

Acknowledgments

While writing this book, my aim wasn't just to transfer knowledge. I wanted to expand your thinking, deepen your perspective on threats, and strengthen your operational decision-making reflexes. I deliberately brought together academic frameworks and field realities, because one is incomplete without the other.

This work is also part of a personal journey. I believed that the experience and knowledge I gained in Türkiye could resonate on an international stage. Cybersecurity is universal, and threats know no borders. Knowledge, therefore, shouldn't be confined by borders either. It's my hope that this book will serve as a shared reference point for professionals across different geographies.

I'd also like to acknowledge those experiences and challenges that, at times, made my professional path more difficult. Obstacles and resistance often became unexpected sources of motivation, pushing me to think more critically, work harder, and develop greater resilience. In that sense, I'm grateful even for the moments of opposition, as they ultimately contributed to my growth and determination.

I would also like to acknowledge the challenges and obstacles that shaped my professional path. They often became sources of motivation, pushing me to think more critically and work with greater determination.

I'm grateful to the many colleagues and professionals in the cybersecurity community whose insights and discussions contributed to the ideas reflected in this work.

Finally, I'd like to thank my brother David, who has always supported me and introduced me to Python.

Conclusion

Security isn't a product, but a process. Intelligence is the compass that guides that process. When the compass functions correctly, direction isn't lost. But if the compass itself isn't questioned, flawed data can lead us toward the wrong destination. For that reason, I've embraced an approach that questions, validates the answers, and constantly seeks context.

If this book helps even one of you interpret an alert differently, model a threat actor more accurately, or make a more informed decision during an incident response, then it will have fulfilled its purpose.

Thank you for being part of this journey.

—**Haydar Yener Arıcı**

Chapter 1

Foundations of Cyber Threat Intelligence

The modern cyber battlefield is defined not only by attacks but also by how well those attacks are understood before they unfold. Cyber threat intelligence is the discipline that transforms scattered risk signals into foresight and strategic awareness. In this chapter, we begin by exploring the foundations that enable organizations to anticipate threats rather than merely react to them.

The diversity and complexity of cyber threats today clearly show that relying solely on technical solutions is no longer sufficient to stay secure. Organizations, states, and individuals must anticipate risks and prepare for them both when they come under attack and before attacks occur. This imperative has placed cyber threat intelligence (CTI) at the center of modern security strategies.

CTI is a systematic process that interprets raw data and turns it into intelligence, providing insight-driven forecasts for decision-makers. In this respect, unlike the instantaneous alerts produced by traditional security tools, CTI enables forward-looking preparation and proactive defense. In other words, CTI functions as a lens to see not both the now and the next move.

The following key characteristics serve as the building blocks of CTI:

- **An interdisciplinary approach**
  CTI isn't solely the concern of cybersecurity specialists. Concepts drawn from psychology, sociology, political science, and economics play a key role in understanding the motivations and behavior patterns of threat actors. For example, knowing only the technical methods of an attack group is often not enough; you must also analyze the group's ideological aims, financial incentives, and political context. By nature, CTI is therefore an interdisciplinary field of work.
- **Preventive and predictive role**
  Whereas traditional security approaches generally come into play after an incident has occurred, CTI aims to identify threats and anticipate risks before attacks take place. This predictive role transforms organizations into actors that both defend against attacks and develop strategies to prevent them. Consequently, CTI is one of the most important tools for moving from a reactive to a proactive approach in cybersecurity.

- **The intelligence lifecycle**
  CTI isn't a static report; it's a dynamic process that adapts to an ever-changing threat landscape. The intelligence lifecycle will be explained in detail in later sections; here, it's summarized simply to underscore that CTI is a process-based structure. This cycle makes it possible to adapt to the continually evolving threat environment. Thus, CTI is positioned as a dynamic and evolutionary process.
- **Connection to organizational decision making**
  Beyond its role as a tool used by technical security teams, CTI is also a critical decision-support mechanism for senior leadership. When investing in a new security technology, assessing supply chain risks, or preparing business-continuity plans during a crisis, CTI outputs have a direct impact. For this reason, CTI is an integral part of decision-making processes at both the operational and strategic levels.
- **Dynamic and evolutionary nature**
  The cyber threat environment isn't static; threat actors, the tools they use, and the techniques they employ are constantly changing. CTI must be continuously updated and improved to keep pace with this change. Thanks to this dynamic nature, CTI becomes a critical resource for today's threats and for future threat scenarios.

  The core idea presented in this introductory section is that CTI is a holistic decision-support mechanism. With its layered structure that extends from technical data to strategic foresight, CTI directly affects organizational security operations as well as national security, economic stability, and social order.

With these building blocks in mind, let's dive into the foundations of CTI, including its strategic importance, how it has evolved, what types of intelligence it encompasses, and more. By the end of this chapter, you'll have a solid foundation for the strategic, operational, tactical, and technical layers of intelligence that we'll examine in detail throughout this book.

1.1 What Is Cyber Threat Intelligence?

Cyber threat intelligence (CTI) is an interdisciplinary information-production and decision-support process that directly feeds organizations' strategic, operational, and tactical decisions. At its simplest, CTI is the body of knowledge made actionable for decision-makers by systematically processing, analyzing, and contextualizing raw data collected from the threat environment. The goal is to do more than detect attacks in the moment by anticipating several moves ahead and enabling a proactive defense.

In today's digital world, the threats range from state-backed advanced threat actors (also known as advanced persistent threats [APTs]) to organized cybercrime groups, hacktivist collectives, and financially motivated individual attackers. This diversity shows that technical logs generated at the time of an incident are insufficient on their own; data must be enriched with context and transformed into decision-oriented intelligence.

Three core concepts must be clearly separated to understand CTI properly:

- **Raw data**
  Unprocessed outputs such as system logs, network packets, intrusion detection system (IDS) and intrusion prevention system (IPS) alerts, endpoint agent records, and honeypot results. For example, on a firewall, "failed logon attempt from 192.168.1.15" is raw data by itself.
- **Indicator (indicator of compromise [IOC])**
  Technical findings derived from raw data that can be associated with a potential threat. Examples include IP addresses, file hashes, command-and-control (C2) domains, and phishing header patterns. Indicators are critical for detection but can be misleading without context.
- **Information**
  The meaningful whole obtained by interpreting indicators within context. For example, when it's determined that a given IP attempted a failed logon and even connected to a known APT infrastructure, the "indicator" becomes decision-grade "information."

CTI sits atop this distinction; by processing information at the strategic, operational, tactical, and technical levels, it enables you to understand, prioritize, and manage risk.

By spanning the following dimensions, CTI ensures comprehensive understanding of threats:

- **Strategic CTI**
  Guides senior leadership, for example, on the economic/geopolitical implications of APT activity targeting the energy sector.
- **Operational CTI**
  Informs the security operations center (SOC) and incident-response processes, and provides insights on campaigns, toolsets, and timelines.
- **Tactical CTI**
  Produces insight at the level of adversary tactics, techniques, and procedures (TTPs).
- **Technical CTI**
  Supplies directly actionable IOCs such as hashes, Internet Protocols (IPs), and URLs/domains.

CTI is more than technical analysis. Contributions from psychology, sociology, economics, and political science enable the behavioral patterns and operational objectives of threat actors to be understood more accurately. In this way, CTI reveals not only technical attack indicators but also the intentions and contextual objectives of adversaries, offering a multidimensional perspective. Building on this broader view, the continuity of CTI is sustained through an intelligence lifecycle that encompasses requirements definition, data collection, analysis, dissemination, and feedback. This cycle provides CTI with a dynamic, up-to-date, and adaptable structure that can be tailored to an organization's specific needs, ensuring that intelligence doesn't remain static but evolves in line with emerging challenges. This adaptability becomes even more critical in today's environment, where by 2026, the attack surface has expanded dramatically due to artificial intelligence (AI)-enabled attacks, supply chain manipulations, and sophisticated operations against critical infrastructure. Within this context, CTI shifts the defense center of gravity from reactive measures to proactive strategies, delivering technical detection, foresight, preparedness, and decision support that empower organizations to stay ahead of rapidly evolving threats.

The practical value of CTI becomes clearer when its advantages and limitations are examined together. While CTI enables organizations to anticipate potential threats, prioritize critical assets, and improve the speed and effectiveness of incident response, it also introduces operational and analytical challenges that must be carefully managed. Misleading indicators, delayed intelligence sharing, and the overwhelming volume of threat data can reduce analytical clarity and strain security teams. Table 1.1 summarizes the key benefits that CTI provides to organizations alongside the common limitations that may arise when threat intelligence is integrated into security operations.

Benefits	Limitations
■ Risk reduction: Potential attack paths and target assets become visible early; critical systems receive prioritized protection. ■ Efficient resource use: Disaggregating organization-specific threats enables efficient allocation of budget and human resources. ■ Faster incident response: Pre-modeled threat profiles shorten decision and action times. ■ Ease of compliance and audit: CTI integration is increasingly expected within frameworks such as ISO 27001, NIST Cybersecurity Framework (NIST CSF), and the EU's Digital Operational Resilience Act (DORA).	■ False flags: Adversaries may deliberately leave misleading IOCs. ■ Timing issues: Delayed sharing hinders detection and mitigation. ■ Lack of context: IOC lists alone don't deliver a durable defense. ■ Data overload: Numerous threat feeds can exceed analysis capacity; prioritization is essential. ■ Risk of misuse: Careless use of low-quality or unverified CTI feeds can trigger unnecessary alarms and waste resources.

Table 1.1: Benefits and Limitations of CTI

Internationally recognized standards and frameworks play a critical role in the effective use of CTI:

- **STIX/TAXII**
  Structured Threat Information eXpression (STIX)/Trusted Automated eXchange of Intelligence Information (TAXII) form the technical foundation for structured CTI sharing. STIX provides a standardized language for describing threat actors, indicators, campaigns, vulnerabilities, and courses of action in a machine-readable format. TAXII, in turn, functions as the secure transport mechanism that allows this structured intelligence to be exchanged automatically between organizations and platforms. Together, they enable

CTI data to move beyond static reports and become actionable intelligence that can be rapidly integrated into security tools such as security information and event management (SIEM), security orchestration, automation, and response (SOAR), and threat intelligence platforms. By establishing a common syntax and delivery channel, STIX/TAXII significantly enhance the speed, consistency, and scalability of intelligence sharing across the cybersecurity ecosystem.

- **MITRE ATT&CK**
  MITRE ATT&CK is a globally recognized knowledge base that documents real-world adversary TTPs observed in cyber operations. Rather than focusing solely on indicators such as IP addresses or file hashes, the framework emphasizes attacker behavior across the entire attack lifecycle. By mapping observed activities to the ATT&CK matrix, security teams can better understand how adversaries gain access, move laterally, escalate privileges, and achieve their objectives. This behavioral perspective enables organizations to design more resilient detection rules, conduct threat hunting, and identify defensive gaps in their security architecture. As a result, MITRE ATT&CK has become a common analytical language used by threat intelligence analysts, SOC teams, and red and blue teams worldwide.
- **ISO/IEC 27035**
  International Organization for Standardization (ISO/International Electrotechnical Commission (IEC) 27035 is an international standard that provides structured guidance for managing information security incidents within organizations. It defines processes for detecting, reporting, assessing, and responding to security incidents in a systematic and coordinated manner. In the CTI context, the standard helps organizations integrate intelligence outputs into incident response activities, enabling faster identification of threats and more informed decision-making during investigations.

These frameworks provide CTI with a common language across organizations and enable comparability. CTI is now an ecosystem practice built on sharing and collaboration. For example, consider the following sectors:

- **Finance**
  In the financial sector, CTI is commonly shared through collaborative platforms such as the Financial Services Information Sharing and Analysis Center (FS-ISAC). These networks allow banks, payment providers, and financial institutions to exchange timely information about emerging threats, attack techniques, and vulnerabilities. By sharing intelligence across the sector, organizations can detect coordinated campaigns earlier and strengthen collective defenses against financially motivated cybercrime.
- **Energy**
  In the energy sector, threat intelligence sharing is often coordinated through initiatives supported by the European Union Agency for Cybersecurity (ENISA). Sector-specific Information Sharing and Analysis Center (ISAC) structures enable energy operators, grid managers, and infrastructure providers to exchange intelligence on threats targeting

industrial control systems and critical infrastructure. This collaborative model improves situational awareness and helps organizations respond more effectively to sector-wide cyber risks.

- **Public sector**
  In the public sector, CTI is typically coordinated through national Computer Emergency Response Teams (CERTs) and international cooperation frameworks. Government agencies share threat information through national response centers and collaborative platforms established within organizations such as NATO and the EU. These structures support coordinated incident response, early warning capabilities, and the protection of critical public services against large-scale cyber threats.

This collaboration builds collective resilience against threats. The following recent cyber threat scenarios show real-world examples of CTI in action:

- **2023 MOVEit vulnerability and global data breaches**
  A critical flaw led to data exfiltration across hundreds of organizations. While raw data merely confirmed the existence of the incident, CTI analyses identified the group behind the operation, the technical method (e.g., processes based on SQL injection), and potential second-wave scenarios, thereby guiding defense.
- **AI-enabled phishing campaigns in 2024**
  The misuse of generative AI elevated text quality to a level indistinguishable from human writing. While email security captured the raw data, CTI improved detection/prevention strategies by revealing ties to prior campaigns, template-generation methods, and likely evolutionary paths.
- **APT operations against critical infrastructure in 2025**
  In advanced attacks targeting the energy sector, anomalies in network traffic were detected as raw data. CTI mapped these anomalies to the TTPs of a specific APT and illuminated possible political motivations and target expansion, accelerating strategic decisions.

CTI transforms raw data into indicators and indicators into contextualized, decision-grade information. With its layered structure, it guides both SOC operations and management-level strategic planning. Through its interdisciplinary nature, lifecycle, alignment with international standards, and culture of collaboration, CTI—as of 2026—isn't an add-on to cybersecurity but the backbone of the corporate security architecture. It illuminates today's status while making tomorrow's risks visible and rendering a proactive defense.

1.2 The Strategic Context and Importance of CTI

CTI today isn't merely a technical component; it's evaluated as a critical decision-support mechanism for national security policies, corporate strategies, and international relations.

As cyber threats have grown to directly affect economic stability, national defense, and diplomatic relations, CTI has become indispensable at the strategic level, where it fulfills three core functions:

- **Providing foresight and awareness**
  Identifies in advance the direction, origin, and likely consequences of potential threats.
- **Supporting policy and investment decisions**
  Contributes to shaping national security doctrines, critical infrastructure protection plans, and cyber defense investments.
- **Strengthening international cooperation**
  Supports intelligence sharing among countries, alliances, and institutions, thereby laying the foundation for collective defense mechanisms.

In the following sections, we'll assess CTI through a strategic lens to understand the different dimensions in which it's valuable, key considerations, the governance model, and the importance of intelligence requirement (IR) management.

1.2.1 Strategic Value Dimensions

The strategic value of CTI stems from its ability to go beyond purely technical threats. It's important to consider CTI from multiple dimensions. For example, in the area of national security, the rise of state-sponsored attacks shows that cyberspace is now an integral part of military and diplomatic strategies. CTI illuminates both the technical aspects of these attacks and their political and geopolitical backdrop. From an economic/industrial perspective, supply chain attacks and sophisticated operations targeting the energy and financial sectors make CTI a strategic tool in economic decision-making. Moreover, the protection CTI provides against industrial espionage and intellectual property theft can prevent economic losses that directly impact gross domestic product (GDP). There's also the societal dimension, where disinformation and information manipulation affect societies' sense of security and political stability. CTI plays a critical role in early detection and effective management of these threats.

In the strategic context, CTI emerges as a functional tool that supports decision-making across a wide spectrum by providing concrete outputs that guide and shape these processes:

- **Policy making**
  CTI outputs are used directly in the development of national cybersecurity strategies. Prioritization of critical sectors and identification of which threats fall into the national risk category are determined through these analyses.
- **Crisis management**
  In the event of an attack, understanding the source and intent through strategic CTI accelerates technical countermeasures as well as diplomatic and economic responses.

- **Long-term planning**
  Using CTI, trends are shown in advance, such as AI-enabled attacks or supply chain vulnerabilities, to enable defense investments to be directed accordingly.
- **Diplomacy and deterrence**
  CTI outputs enable the origin of cyberattacks to be presented as evidence in diplomatic forums. Thus, countries gain deterrence and defensive legitimacy on the international stage.

Another dimension that reinforces the strategic importance of CTI is its strong relationship with internationally recognized standards and frameworks. This connection is evident in several key initiatives:

- **NATO Cyber Defence Policy (2021)**
  CTI is positioned at the center of a common defense understanding among member states.
- **EU Network and Information Systems (NIS) 2 Directive (2023) and Digital Operational Resilience Act (DORA) (2025)**
  For the protection of critical infrastructure, the sharing and use of threat intelligence has become a legal requirement.
- **NIST CSF 2.0 (2024)**
  In the United States, institutions are explicitly advised to integrate CTI into risk management processes.

These frameworks demonstrate that CTI is a legal, political, and strategic necessity.

Looking ahead to the period between 2026 and 2030, it becomes clear that CTI will evolve into an even more critical strategic tool in the face of emerging dynamics. For example, the speed and scale of AI-based attacks will make CTI's capacity for foresight even more critical. As cryptographic standards become breakable in the post-quantum computing era, CTI will be compelled to focus on next-generation cryptanalytic threats. And by 2030, more integrated, global threat intelligence-sharing mechanisms (e.g., unified ISAC networks) are expected to form.

CTI is more than a technical security tool; it's a strategic intelligence source that directly affects economic stability, diplomatic relations, public security, and societal trust. Therefore, without strategic CTI, it's impossible for modern cyber defense policies to be effective. Moreover, thanks to its capacity for foresight, CTI both makes today's cyber threats visible and transforms defense into a proactive structure by illuminating those of tomorrow.

1.2.2 Considerations at the Strategic Level

At the strategic level, CTI does more than feed national security and corporate decision-making mechanisms; it serves as a discipline that must be addressed with its multilayered challenges and complementary dimensions. In this context, effective use of CTI requires consideration of the following critical areas:

- **Political risks of intelligence sharing**
  One of CTI's most important values is its basis in sharing and cooperation. However, sharing between countries or institutions can also create security vulnerabilities and espionage risks. The leakage of sensitive information into the wrong hands can lead to tensions in international relations. Therefore, the *trusted circles* approach is of critical importance. Structured sharing networks supported by transparency and oversight mechanisms among only trusted actors minimize political risks.
- **CTI and the legal framework**
  The place of cyber threats in international law remains a debated issue. In particular, the question of whether a cyberattack constitutes a declaration of war is a critical area of discussion in interstate relations. CTI is important in terms of the standard of proof (evidence-based intelligence). In courts, regulatory processes, or international arbitration cases, CTI outputs may carry evidentiary weight. However, to achieve this, the integrity, verifiability, and chain-of-custody security of collected data must be meticulously ensured.
- **Public-private cooperation**
  Strategic CTI isn't solely under the control of states. Because most critical infrastructures (energy, finance, healthcare, transportation) are operated by the private sector, trust-based cooperation between the state and the private sector is vital. This cooperation enhances the effectiveness of early warning mechanisms and raises national cybersecurity capacity. For example, field data provided by operators in the energy sector directly contributes to state institutions' strategic threat assessments.
- **Strategic communications dimension**
  CTI outputs are used for SOCs and in public information and crisis communications processes. Accurately conveying the attackers' intent behind an incident plays a critical role in preventing panic and disinformation. Where poorly managed strategic communications can grant psychological advantage to attackers, well-managed communications can break it. For this reason, CTI should form the foundation of communication strategies across a wide spectrum—from security teams to the media and government spokespersons.
- **Performance and measurement (metrics)**
  The success of strategic CTI should be measured not by the number of reports produced but by its tangible contribution to decision-making processes. Therefore, performance metrics (or key performance indicators [KPIs]) must be developed. For example:
  - Shortening of time-to-detect
  - Increased speed of diplomatic response
  - Accuracy of decisions in crisis management
  - Prevention rate of critical infrastructure outages

As of 2026, managing CTI without considering these additional elements creates serious gaps in the strategic security architecture.

1.2.3 Governance and Operating Models

When it comes to CTI, most discussions revolve around technical tools and data sources. Yet what truly determines success is the governance model under which this intelligence is produced, shared, and fed into decision processes. In other words, the strategic differentiators aren't "What technology do we have?" but "Who decides, based on what, and at what cadence?"

At the heart of CTI's strategic success lie the governance models that define how intelligence is produced, shared, and integrated into decision-making; therefore, examining the following approaches to governance becomes critically important:

- **Centralized model**
  There's a single decision center. All intelligence flows are collected and analyzed here, and decisions are issued from here as well. This model's advantage is discipline and speed; it ensures a single voice during crises. However, excessive centralization can deprive the organization of flexibility and lead to the loss of valuable field input.
- **Federated model**
  Different units operate their own CTI functions with a coordinating mechanism above them. This approach reflects diversity and subject-matter expertise, but it also brings the risk of misalignment and conflicting priorities. The biggest dilemma of the federated model is whose voice will prevail while everyone argues for their own truth?
- **Hub-and-spoke model**
  Here, the center (hub) provides strategic direction, and the surrounding units (spokes) contribute with their domain knowledge. This model avoids the rigidity of the centralized model and the fragmentation of the federated model to create a balanced structure between the two. This is the most preferred approach for global organizations.

One of the factors that truly defines the value of CTI is the speed at which its reports reach decision-makers, because intelligence that arrives too late may not only fail to prevent losses but could even contribute to financial damages in the millions of dollars. Reports typically follow a *routine cadence*, with presentations being made to the board on a monthly or quarterly basis. However, in *crisis cadence*with CTI, normal rhythm is suspended during an incident, creating a separate flow for rapid decisions. This distinction is what changes CTI from laboratory-grade reports into a mechanism that lets you breathe during real crises.

One of the most critical factors that determine success in CTI governance models is the clear definition of responsibility boundaries: who performs the tasks, who is accountable, whose input is sought, and who is kept informed. At this point, the RACI matrix serves as a vital compass that prevents confusion during crises and ensures that decisions reach the table in a timely and effective manner:

- **Responsible**
  The person/team doing the work.

- **Accountable**
  The one ultimately answerable for the decision.
- **Consulted**
  Stakeholders whose input is sought.
- **Informed**
  Parties notified after the decision.

If this distinction isn't made, everyone talks during a crisis but no one takes responsibility. Decisions hang in the air instead of landing on the table.

Strategic CTI governance isn't solely the responsibility of SOC analysts or the cybersecurity department; it requires the clear definition of roles for every stakeholder—from executive leadership and legal teams to communications units and beyond—to ensure that the right decisions can be made effectively during times of crisis. For example, the executive or board must ensure CTI outputs are converted into investment and policy decisions. Legal must preserve the chain of custody and ensure compliance with regulations. Communications, on the other hand, determines what message is conveyed to the public and in what language during a crisis. Poorly managed communications can create more panic than the attack itself, whereas well-managed communications can break attackers' psychological edge.

Strategic CTI governance is essentially an art of balance: constant tightrope-walking between centralized control and local flexibility, speed and accuracy, secrecy and transparency. The real value lies not so much in technology as in a properly functioning organizational architecture. Therefore, any organization seeking to make CTI successful at the strategic level must first ask, "Who manages our intelligence, by which model do we steer it, and at what cadence do we move when a crisis hits?" If the answer isn't clear, even the most advanced technologies are doomed to remain only on paper.

Case Study: Centralized Model

In 2026, a hypothetical large-scale phishing campaign targets the global network of a (fictitious) international bank called Aurora Finance. Hundreds of suspicious email reports come in from different branches.

The bank's CTI structure is entirely centralized. The main security center in London collects all data from the branches and is the sole decision-making authority.

The response process in the centralized model unfolds through the following steps:

1. Branches only transmit raw data.
2. Within 24 hours, the London center analyzes all phishing templates and identifies the fake domains used by the attackers.
3. The same day, the centralized team makes a single policy decision: "Block these domains across all branches."

The spread of the attack is prevented thanks to speed and clarity. However, because detailed analysis of field data takes time at the center, some local differences (e.g., phishing emails written in German) are overlooked.

The centralized model provides a single voice and speed during a crisis, but it risks missing local variations.

Case Study: Federated Model

In 2027, a fictitious Europe-wide energy consortium called EuroGrid is targeted by cyberattacks in different countries against its power plants, focusing on industrial control systems.

The consortium uses a federated CTI model. Each country has its own CTI unit, with a small coordinating council in Brussels.

The investigation progresses through coordinated analyses conducted by the CTI teams in different countries, each contributing findings that gradually reveal the broader structure of the attack:

- The team in Germany discovers that the attack uses a special command injection in Supervisory Control and Data Acquisition (SCADA) protocols.
- The team in Spain observes a different variant of the same attack in energy distribution software.
- The Brussels coordination team brings these findings together and reveals that the attack is actually a multilayered campaign by a single APT group.

Each country contributes its analytical findings to produce a more holistic threat picture. However, because coordination takes time, the attacker manages to cause temporary disruptions in some systems in the first days.

The federated model draws strength from diversity, but decisions may be delayed during a crisis.

Case Study: Hub-and-Spoke Model

In 2028, a fictitious multinational healthcare company called MedSecure Alliance faces ransomware attacks targeting patient records systems.

The CTI organization is designed according to the hub-and-spoke model. The center in the US (hub) provides strategic direction, and regional units (spokes) transmit their threat observations to the center.

The incident response develops through coordinated observations from regional units, with each office contributing intelligence that helps clarify how the attack is spreading across the organization:

- The Turkey office captures the first variant of the ransomware and immediately reports it to the center.
- The Canada office reports that the attack arrived not via email but through a supplier's virtual private network (VPN) access.
- The center (hub) combines all of this information, sees the cascading impact of the attack, and issues a global directive for tightened VPN access control and segmentation.

Both centralized speed and local insight are achieved. While a few regional clinics experience short interruptions, a global system collapse is prevented.

The hub-and-spoke model blends local field intelligence with centralized strategic perspective, making it the most balanced solution between flexibility and discipline.

1.2.4 Intelligence Requirement Management

In many organizations, CTI still operates on the principle of "we report what we find." However, this approach turns into a heap of randomly collected data rather than providing the information decision-makers truly need. At the strategic level, the real issue isn't "what we find" but "what we must know."

For strategic CTI to focus not on random collections of data but on what decision-makers truly need to know, it's essential to clearly define priority intelligence requirements (PIRs) and the key intelligence questions (KIQs) that support them; this structure transforms intelligence from scattered information into directed and purpose-driven outputs.

PIRs are the priority questions that the board, governments, or critical operations units absolutely want answered. PIRs guide decision-making, while KIQs are more specific subquestions that feed the PIRs. While PIRs draw the big picture, KIQs clarify what data will be collected in the field. For example, for a financial institution, here's an example of each:

- **PIR**
  Which APT groups might target the banking sector over the next six months?
- **KIQ**
  Which phishing kits have been sold on the dark web in the past three months?

This hierarchy yields directed intelligence rather than random data.

PIRs and KIQs aren't just lists of questions. They also allow the testing of strategic hypotheses. Hypotheses such as "If Group X is planning a supply chain attack, in which sectors would we see primary effects?" are validated or falsified with field data. Thus, intelligence becomes a dynamic analytic process instead of a passive flow of information.

CTI's biggest pitfall is noise from tens of thousands of IOCs, hundreds of reports, countless alerts, and so on. The question that must be asked is which decision will this information change? PIR/KIQ management transforms the chaos of information into a meaningful order. In other words, it turns intelligence into a tool that makes a tangible difference on the decision-maker's desk.

In professional CTI programs, PIR/KIQ management produces written and measurable outputs:

- **Annual/quarterly PIR list**
  Prioritized questions are aligned with strategic objectives.
- **Acceptance criteria for each PIR**
  Examples include "The APT threat list must be verified by at least three independent sources" or "A disinformation trend must be observed across multiple platforms."

These criteria guarantee the quality of intelligence.

When defining IRs, it's essential to consider ethical responsibility in addition to technical accuracy and operational effectiveness. While random data collection may lead to unnecessary privacy violations, the PIR/KIQ framework safeguards transparency and accountability at the organizational level by adhering to the principle of collecting only what is truly necessary.

Perhaps the most critical point is that the PIR list isn't just for analysts but serves as a commitment agreed upon across all management layers of the organization. PIRs are therefore the organization's collective questions.

Example

An energy company determines the following PIR/KIQs for 2026:

- **PIR**: What is the likelihood of state-sponsored attacks against our energy infrastructure in the next year?
- **KIQ 1**: Which malware families stood out in campaigns targeting the energy sector over the past six months?
- **KIQ 2**: Is there a disinformation trend on social media related to power outages?

As a result of this PIR/KIQ process, the company takes proactive measures against both technical logs and disinformation campaigns.

1.3 The Evolution of Threat Intelligence

Although the concept of threat intelligence is now at the center of cybersecurity literature, its origins date back much further to the history of military strategies and national security practices. When the term "intelligence" is mentioned, what first comes to mind are the efforts of states during wartime to learn about their adversaries' intentions, capabilities, and tactics. However, in the rapidly digitalizing world of the 21st century, threat intelligence has evolved from being solely a military tool into a multidimensional component of the global security ecosystem. We'll trace this evolution in more detail in the following sections.

1.3.1 The Theoretical Foundations of Classical Military Intelligence

The concept of intelligence isn't limited to the digital security domain of modern times; its roots stretch back thousands of years to the fundamental needs of warfare and statecraft. Throughout history, intelligence has been a strategic necessity for nations' survival, directly shaping armies' maneuvers, states' diplomatic initiatives, and societies' security.

The theoretical foundations of classical military intelligence rested on three main axes: strategic foresight, operational advantage, and undermining the adversary's capacity. These three principles are directly relatable to the building blocks of today's CTI.

These theoretical foundations have been consistently reflected in different eras of history, where intelligence shaped the destiny of states and armies alike:

- In ancient China, *The Art of War* (5th century BC) by Sun Tzu clearly highlighted the role of intelligence in military strategy. For Sun Tzu, "knowing the enemy and knowing yourself" was the most critical factor determining the outcome of war. This philosophy can be seen as one of the earliest conceptual foundations of the modern intelligence cycle (collection–analysis–assessment–dissemination).
- In the Middle Ages, the Byzantine Empire survived for centuries through a strong network of diplomacy and espionage. Intelligence reports from frontier regions guided military tactics and economic decision-making.
- During the Cold War, intelligence became a decisive factor for military superiority and the course of ideological competition. US U-2 spy planes providing aerial intelligence or the Soviet Union's successes in signals intelligence (SIGINT) were key to maintaining nuclear balance.

The foundations of classical military intelligence have directly shaped today's digital threat landscape. CTI is essentially the technological reinterpretation of these ancient principles:

- **Strategic foresight**
  As in Sun Tzu's teachings, states now develop AI-based predictive systems to anticipate rivals' intentions and capabilities.
- **Operational advantage**
  Just as Byzantium collected intelligence from its borders, today's threat hunters extract insights from the depths of networks to secure organizations' operations.
- **Undermining the adversary's capacity**
  As during the Cold War, modern threat intelligence is used to neutralize or disrupt adversaries' cyber operations through proactive defense strategies.

1.3.2 The Beginning of the Cyber Age and the APT Concept

The dawn of the cyber age marked the transformation of threat intelligence from being purely military and diplomatic into a critical security domain for states, corporations, and individuals on a global scale. One of the most defining concepts of this transformation is the *advanced persistent threat (APT)*.

An APT differs from ordinary cyberattacks in that it's long-term, organized, targeted, and often state-sponsored. In such campaigns, the attacker silently infiltrates target systems, may remain undetected for months or even years, and exfiltrates strategic data or causes operational damage at the right moment. Unlike fast and noisy attacks, APTs are characterized by deep planning, advanced techniques, and strategies for sustained access. For this reason, they have become one of the critical focal points of threat intelligence.

This evolution is best understood through landmark cases that revealed how APTs reshaped the global security landscape:

- **Moonlight Maze (1998)**
  This incident, in which the US defense systems were infiltrated for years, is considered the first large-scale APT operation.
- **Titan Rain (2003)**
  Believed to be linked to China, these attacks targeted critical institutions in the US and Europe, proving that cyber espionage had become a permanent element of interstate competition.
- **APT1 Report (2013)**
  Mandiant's report documenting operations conducted by units linked to the Chinese military brought the APT concept into both academic discourse and public attention.

As of 2026, the APT concept is no longer confined to nation-state operations. Globally, organized crime groups are conducting APT-like campaigns for financial gain. Moreover, AI-powered attack automation is making APTs more complex and unpredictable. Looking ahead to 2030, the following may become reality:

- **Autonomous APTs**
  AI-driven, self-directing attack groups are anticipated.
- **Supply chain APTs**
  Attacks like SolarWinds (2020) have shown that APTs may increasingly target systems *and* their surrounding ecosystems.
- **Quantum-era APTs**
  As quantum computers mature and classical cryptography becomes breakable, APT operations will take on a new dimension.

1.3.3 The Threat Intelligence Cycle

Threat intelligence is a structured, systematic process based on methodology. Known historically as the *intelligence cycle*, this framework has evolved from classical military doctrines to today's SOCs.

Traditionally, the military intelligence cycle comprises four stages:

1. **Collection**
   Obtaining information via espionage, surveillance, SIGINT, and so on.
2. **Processing**
   Converting collected data into usable formats.
3. **Analysis**
   Contextualizing and making sense of the data.
4. **Dissemination**
   Reporting to commanders or decision-makers.

This model was crystallized during the Cold War through human intelligence (HUMINT), signals intelligence (SIGINT), and imagery intelligence (IMINT), forming the backbone of military and diplomatic decision-making.

In the 21st century, digitalization has transformed the cycle. CTI has expanded the model by incorporating indicators of compromise (IOC), indicators of attack (IOA), and multiple layers of tactical, operational, and strategic intelligence. The stages now include the following:

- **Collection**
  Open-source intelligence (OSINT), dark web monitoring, honeypots, and malware analysis.
- **Processing**
  Data normalization, STIX/TAXII formats, and threat intelligence platforms (TIPs).
- **Analysis**
  TTP analysis via MITRE ATT&CK and threat actor profiling.
- **Dissemination**
  Automated sharing through SIEM/SOAR integrations to deliver real-time intelligence to operational security (OPSEC) teams.

The transformation of the intelligence cycle didn't occur overnight; rather, it was shaped by a series of evolutionary milestones that marked the transition from traditional military doctrines to the complex, technology-driven processes of modern CTI:

1. **Cold War intelligence cycle → early cyber era (1990s)**
   Shift from physical to digital espionage.
2. **2000s**
   Emergence of APTs introduced continuous monitoring and long-term tracking.
3. **Post-2010**
   STIX/TAXII standards enabled enterprise-level intelligence sharing.
4. **2020s**
   AI-driven automation reduced reliance on human intervention in several stages.

Looking ahead, the trajectory of the threat intelligence cycle suggests that the coming decade will be shaped by forward-leaning innovations and paradigm shifts, with the following developments expected to redefine its structure and operation:

- **Autonomous intelligence cycle**
  AI will collect, analyze, and deliver insights directly to defenses in real time.
- **Quantum-supported intelligence**
  Quantum computing will accelerate cryptography-dependent big data analysis.
- **Proactive cycle**
  The cycle will move from "threat occurs → analyze afterward" to preventing threats before they materialize.

1.3.4 Milestones Through Case Studies

The evolution of CTI is best illustrated by large-scale cyber incidents that shaped strategy, policy, and technical practices. Along with their technical implications, these attacks also had strategic and political consequences, redefining how intelligence is collected, analyzed, and shared.

Unlike classical espionage, modern cyber incidents show that intelligence has become a hybrid digital-physical domain. These events reshaped both national security policies and private-sector approaches.

This transformation can be traced through pivotal milestones—major incidents that exposed new dimensions of cyber threats and redefined the priorities and practices of threat intelligence in the modern era:

1. **Stuxnet (2010): The start of cyber-physical attacks**
   - Targeted Iran's Natanz nuclear facilities; the first attack against ICS/SCADA systems.
   - Proved that threat intelligence must address both data theft and critical infrastructure sabotage.
   - Introduced the concept of *cyber warfare* into global discourse.
2. **Titan Rain (2003): The rise of nation-state espionage**
   - China-linked espionage targeting US and European defense contractors and government agencies.
   - Demonstrated the importance of persistence and strategic targeting in CTI.
3. **APT1 Report (2013): Exposure of nation-state espionage**
   - Mandiant's report publicly documented operations by the Chinese People's Liberation Army (PLA) Unit 61398.
   - First global exposure of state-backed cyber espionage.
4. **NotPetya (2017): The digital face of economic warfare**
   - Initially targeted Ukraine, but spread worldwide, causing billions in damage.
   - Highlighted the critical role of financial and economic dimensions in CTI.

5. **SolarWinds (2020): Supply chain intelligence**
   - Compromised hundreds of US institutions and companies.
   - Showed that CTI must account for ecosystem vulnerabilities and supply chain risks.
6. **Russia–Ukraine War cyber front (2022–ongoing): Hybrid warfare**
   - Cyberattacks executed alongside conventional military operations.
   - CTI proved to be an operational asset in real-time geopolitical crises.
7. **AI-driven disinformation (2025): The societal dimension**
   - Deepfakes and AI-generated content began threatening elections and public trust.
   - Demonstrated that CTI is both technical and sociopolitical.

1.4 Types of Intelligence in Cybersecurity

Intelligence is the discipline of converting data into decisions before time overtakes it; its power lies in making strategy and technique speak at the same time. In cybersecurity, intelligence can't be handled on a single level; rather, it forms a layered structure serving different needs. Strategic, operational, tactical, and technical intelligence are interlocking parts of a cohesive whole:

- **Strategic intelligence**
  This type of intelligence plays a critical role in shaping the long-term policies of organizations and states as it reveals the future capacity of threat actors, the impact of international relations on cybersecurity, and which sectors face greater risk. For example, strategic assessments at the EU level in 2026 indicate that risk analyses related to China and Russia are influencing investment preferences. There are similar precedents: during the Cold War, strategic assessments based on SIGINT were decisive in shaping military doctrines. Today, the same approach manifests itself in national strategies targeting the security of digital infrastructures.
- **Operational intelligence**
  Unlike the strategic level, this type of intelligence focuses on shorter time horizons and concrete threat campaigns. National CERT units or organizations' SOCs use operational intelligence reports to understand which sectors are being targeted, the motivations of attacker groups, and the persistence of campaigns. In this regard, the US Cybersecurity and Infrastructure Security Agency (CISA) emphasizes transparency by publishing its reports to the public, while some Asian countries share the same information only with critical infrastructure administrators. This reflects differences in interstate intelligence-sharing cultures. Following the 2021 Colonial Pipeline attack, the US quickly issued alerts that helped prevent similar attacks in the energy sector, whereas some European companies suffered greater losses due to delays in operational intelligence sharing.

- **Tactical intelligence**
  This type of intelligence refers to the information used directly by security professionals in day-to-day operations. The attackers' TTPs are analyzed at this level. Today, the MITRE ATT&CK framework is the most widely used reference for tactical intelligence. For instance, the tactics used in the 2010 Stuxnet attack remained concealed for a long time, serving in the literature as a key case demonstrating the consequences of failing to share tactical intelligence in a timely fashion. By contrast, in a similar modern incident, rapid sharing of tactical intelligence enabled organizations to update their defenses within hours. As of 2026, this is especially evident in AI-driven phishing: threat actors use generative AI to craft convincing fraudulent emails, while security teams quickly relay these methods to organizations through tactical reports.
- **Technical intelligence**
  This is the most detailed and implementation-oriented type of intelligence that includes machine-processable indicators such as IP addresses, malware hash values, and C2 infrastructures. In the past, hash values were effective for identifying malware, but as of 2026, adversaries are developing polymorphic strains that generate different hashes on each execution. Consequently, technical intelligence can no longer rely on classical indicators alone. Today, behavioral indicators, machine-learning-supported dynamic analysis techniques, and threat-hunting processes have become inseparable components of technical intelligence. While some countries share technical indicators directly with the private sector, Europe builds shared data pools through ENISA and the Computer Security Incident Response Teams (CSIRTs) network. These differences show how international cooperation models are reflected in cybersecurity policies.

All of these types are incomplete in isolation; the real strength arises from integrating long-term strategic foresight with operational threat reporting, tactical analysis, and technical indicators. Historical cases such as Stuxnet show that failing to synchronize intelligence types can create critical gaps in the defense chain. Looking ahead, the rise of AI-enabled threats compels faster, more transparent, and more integrated use of all four intelligence types. As of 2026, the factor that determines competitive advantage for states and organizations isn't merely what data they collect, but at which level they transform that data into intelligence and how timely they act on it.

The intelligence layers discussed in this section differ in scope, as well as time horizon, analytical depth, and the audiences they support. Strategic intelligence informs long-term policy and investment decisions, whereas operational, tactical, and technical intelligence progressively focus on shorter time frames and more immediate defensive actions. Understanding these distinctions is essential for integrating threat intelligence effectively across organizational decision-making processes. Table 1.2 provides a comparative overview of these intelligence types, outlining their typical purposes, users, data sources, and operational impact within cybersecurity programs.

Type	Time Horizon	Purpose	Target Audience	Output Format	Typical Sources	Decision Impact	Example KPI
Strategic	1–5 years	Policies, investment priorities, geopolitical risks	Senior leadership, policymakers, regulators	Strategy report, scenario	Regulation, macro trends, actor ecosystems	Strategic roadmap, budget allocation	Strategy investment alignment, risk reduction score
Operational	Weeks–months	Campaign tracking, target sectors and motivations	SOC managers, CERTs, sector leaders	Campaign report, STIX/TAXII, MISP Threat Sharing	Campaign TTP/IOC clusters, sector alerts	Operational prioritization, playbook updates	Detection time, sharing latency
Tactical	Days–weeks	TTP analysis, hunting hypotheses	Threat analysts, digital forensics and incident response (DFIR) teams	ATT&CK-mapped technical note, Sigma/YARA	Sandbox, telemetry, sampling	Detection coverage, hunting success	ATT&CK coverage %, rule propagation time
Technical	Minutes–days	Real-time detection/blocking	SOC analysts, SIEM/endpoint detection and response (EDR)	IOC/IOA lists, rule sets	EDR/user and entity behavior analysis (UEBA), Domain Name System (DNS)/NetFlow, logs	Incident prevention, isolation	Mean time to detect (MTTD), mean time to remediate (MTTR), false positive/false negative rate, IOC decay time

Table 1.2: Intelligence Types in Cybersecurity

1.5 Core Concepts and Conceptual Models

CTI relies on a shared set of concepts that allow analysts, defenders, and decision-makers to interpret threats in a consistent and meaningful way. Without a common vocabulary, the signals collected from networks, threat feeds, and investigations remain fragmented and difficult to translate into actionable insight. This section therefore introduces several core terms and conceptual elements that form the analytical foundation of CTI. Together, these concepts—ranging from threat actors and behavioral patterns to technical indicators and intelligence-sharing standards—provide the framework through which cyber incidents are interpreted, communicated, and ultimately transformed into defensive action.

1.5.1 Important Terminology

In CTI, terminology forms an operational plan; misreading a term means mismanaging an attack. The terminology used in CTI may appear on the surface as another set of acronyms, but in reality, this language governs operations in the field. A misinterpreted term can lead either to underestimating the scale of an incident or, conversely, plunging an organization into unnecessary panic. Thus, these concepts aren't dictionary entries but are in fact living tools that shape the fate of operations. We shouldn't be confined by artificial order, so depending on the subject, it's sometimes more instructive to start with behavior, sometimes with the actor, and sometimes with sharing standards:

- **Threat actors**

  Recognizing a threat actor is like decoding the DNA of an attack. The motivation of a state-sponsored APT group isn't the same as that of a financially motivated cybercrime syndicate. The former is patient, collecting information over the long term; the latter seeks quick profits. An insider threat actor, however, behaves in a completely different way: a person familiar with the organization, with privileged access from within.

  Consider this example: In an attack on an energy company, the infrastructure may look ordinary at first glance. But if the attacker patiently collects data in the system for six months, this indicates a state-sponsored actor—not a typical ransomware gang. Unless the actor's intent (strategic goal), capacity (technical capability), and opportunities (ability to exploit weaknesses) are connected, the picture remains incomplete.

- **Tactics, techniques, and procedures (TTPs)**

  The most persistent signature of a threat actor is their tactics, techniques, and procedures (TTPs). A hash may change tomorrow, an IP address may vanish within minutes, but how the attacker moves within the network, what tools they use for lateral movement, and at what times they are active tend to remain constant.

  For example, in a financial institution, an attacker frequently using PowerShell with the `-enc` parameter, followed by deploying specific credential-dumping tools, creates a repeating rhythm. Detecting this rhythm provides a defensive advantage that lasts far longer than an IOC.

- **Indicators of compromise (IOCs)**

  An IOC is the observable footprint of an attack. It could be an IP address, domain name, file hash, registry key, or an unusual process chain. But IOCs must always be evaluated with their lifetimes in mind:

  - Atomic IOCs (IP, hash) are short-lived.
  - Pattern-based IOCs (command-line regex, filename template) are valuable in the medium term.
  - Behavioral IOCs often last longer than atomic or pattern-based ones, but advanced actors can eventually alter these as well.

- **STIX/TAXII**
  Intelligence loses value if it doesn't move beyond organizational boundaries. This is where STIX and TAXII come into play:
  - STIX: Defines threat information as structured objects. An IOC, a threat actor, a campaign, or a course of action can all be standardized as STIX objects.
  - TAXII: Enables secure interorganizational sharing of these objects.

  Here's a practical example: An SOC team identifies suspicious PowerShell usage and publishes it as an indicator object in STIX format. This IOC then flows via TAXII into the security platforms of other organizations. As a result, not just one institution but the entire ecosystem is strengthened simultaneously.
- **MITRE ATT&CK**
  MITRE ATT&CK is a knowledge base that documents real-world adversary behaviors in the form of TTPs. Rather than focusing only on technical indicators, the framework emphasizes how attackers operate within systems, such as privilege escalation, lateral movement, or credential dumping. By mapping observed activities to the ATT&CK matrix, analysts can identify defensive gaps and improve detection and threat hunting capabilities.
- **Cyber kill chain**
  The cyber kill chain model breaks down the progression of a cyberattack into a sequence of stages, typically beginning with reconnaissance and ending with actions on objectives. By understanding these stages—such as weaponization, delivery, exploitation, and C2—security teams can identify where defensive controls may interrupt an attack before it reaches its final objective. The model therefore provides a structured way to analyze how intrusions unfold over time.
- **Diamond Model**
  The Diamond Model is an analytical framework that focuses on the relationships between four key elements of an intrusion: the adversary, the infrastructure used, the capabilities deployed, and the victim. By examining how these elements interact, analysts can build a more complete picture of an attack campaign. This approach is particularly useful for linking individual incidents into broader threat actor activities.
- **STRIDE**
  STRIDE is a threat modeling framework used primarily in software security and system design. It categorizes potential security threats into six types: spoofing, tampering, repudiation, information disclosure, denial of service (DoS), and elevation of privilege. By systematically evaluating systems against these categories, organizations can identify design weaknesses before they are exploited by attackers.
- **Process for Attack Simulation and Threat Analysis (PASTA)**
  PASTA is a risk-driven threat modeling methodology that connects business objectives with technical threat analysis. It guides organizations through multiple stages, including defining business impact, identifying threat actors, analyzing vulnerabilities, and

simulating attack scenarios. This structured approach helps ensure that security decisions align with both technical realities and organizational risk priorities.

For instance, consider a scenario in which attackers attempt to compromise the payment infrastructure of an e-commerce platform. The MITRE ATT&CK framework can reveal which tactics and techniques were used during the intrusion, while the Diamond Model can link the attacker's infrastructure and capabilities to the targeted organization. At the same time, STRIDE may highlight the underlying software vulnerability that made the attack possible. By combining these perspectives, analysts move beyond isolated indicators and gain a layered understanding of how the attack developed.

CTI terminology is a living language that establishes communication between defense and offense. Threat actors are the subjects of this language, TTPs are the verbs, IOCs are the evidence, STIX/TAXII are the channels of transmission, and modeling frameworks are the grammar of context. If we fail to learn this language properly, our data will devolve into meaningless numbers; but when we speak it correctly, data turns into decisions, and decisions into security strategy.

1.5.2 Threat Actors and Their Motivations

The effectiveness of CTI depends on understanding who the threat actors are, what capacities they possess, and why they act. Thus, threat actors comprise a strategic element of analysis. These actors can be classified into distinct categories, as shown in Table 1.3, each defined by its resources, objectives, and methods of operation.

Nation-State Actors	Organized Crime Syndicates	Hacktivists	Insider Threats	Independent Cybercriminals and Script Kiddies
▪ State-sponsored or direc2tly state-controlled groups. ▪ Operate to protect political, military, economic, and diplomatic interests, or to weaken rival states. ▪ Have high financial resources, sophisticated infrastructure, and long-term strategies.	▪ Driven by financial gain. ▪ Ransomware, credit card theft, banking trojans, and cryptocurrency theft are the main areas. ▪ Since 2020, the ransomware as a service (RaaS) model has commercialized cybercrime.	▪ Politically or ideologically motivated. ▪ Attack governments, corporations, or social targets in protest. ▪ Effects are often symbolic but can gain significant media and public attention.	▪ Current or former employees, contractors, or third parties. ▪ Motivated by personal gain, revenge, or coercion. ▪ Often overlooked but can be among the most damaging threats.	▪ Limited technical skills; rely on ready-made tools or exploit kits. ▪ While less sophisticated than professional groups, they still expand the overall attack surface.

Table 1.3: Threat Actor Categories

Threat actors' motivations can be categorized into four main groups:

- **Political and geopolitical**
  Nation-states aiming to gain strategic advantage, influence elections, or destabilize rivals. For example, the APT28 (Fancy Bear) operation linked to Russia targeted NATO, media, and electoral systems to gather political intelligence.
- **Economic**
  Organized crime groups profiting from ransomware, fraud, or data sales. For example, DarkSide ransomware conducted the 2021 Colonial Pipeline attack, paralyzing US energy infrastructure purely for financial gain.
- **Ideological**
  Hacktivists conducting attacks based on social, religious, or environmental causes. For example, the hacker group Anonymous has launched attacks on corporations, governments, and religious groups to broadcast social and political messages.
- **Personal**
  Revenge, self-interest, or coercion, as often seen in insider threats. A well-known illustration of an insider threat is the case of Edward Snowden in 2013. Snowden, who was working as a contractor supporting the US National Security Agency (NSA), used his legitimate system access to obtain and disclose a large volume of classified documents concerning global surveillance activities. Regardless of the political debates surrounding the incident, the case clearly demonstrated a fundamental risk in cybersecurity: individuals with authorized access can circumvent technical safeguards and expose highly sensitive information.

In classical literature, the motivations of threat actors are generally classified under these four main categories. However, the boundaries between these categories have increasingly blurred beginning in 2020. Today, many cyberattacks can no longer be explained by a single motivation alone; rather, they are the result of intersecting objectives. This development made *hybrid motivations* a central theme in CTI. Here are a few examples:

- **Intersection of political and economic interests**
  One of the most evident examples of hybrid motivations is state-sponsored groups leveraging organized crime ecosystems. Such actors simultaneously target the critical infrastructure of rival states to gain geopolitical advantage, while also generating financial profit through ransomware or data theft. For instance, some Eastern European groups conduct operations aligned with their governments' foreign policy objectives while also securing direct economic gain from the same campaigns.
- **Fusion of economic gain and ideological goals**
  Hacktivist collectives are increasingly turning to cryptocurrency theft, or campaigns that begin with ideological motives but gradually evolve into profit-driven operations. In such cases, financial motivation becomes a tool to support the ideological base or to ensure the sustainability of their activities.

- **Analytical importance of hybrid motivations**
  From a CTI perspective, disregarding hybrid motivations may lead to misinterpretation of actors' behaviors. For example, a group labeled purely as economically motivated may in fact be serving a strategic political objective. Therefore, recognizing the intersections of motivations is a critical requirement in actor analysis.

Today, hybrid motivations are expected to become even more widespread. The use of proxy actors in state cyber operations, the integration of organized crime groups into geopolitical agendas, and AI-enabled attacks that produce both economic and political consequences will make multidimensional motivation analysis indispensable in CTI.

1.5.3 Threat Actors and the Intelligence Cycle

In CTI, discussion of threat actors often begins with surface-level classifications: APT groups, cybercrime syndicates, hacktivists, insider threats, and so on. These categories sound orderly, but alone they don't transform any organization's defense. Understanding an actor isn't simply labeling; it requires repositioning them at every stage of the intelligence cycle. If the actor isn't placed correctly, collected data becomes meaningless, analysis becomes contextless, and shared output becomes useless.

To grasp their true impact, threat actors must be examined in relation to each stage of the intelligence cycle, where their traces, intentions, and consequences unfold in different yet complementary ways:

1. **Collection stage**
   The threat actor is the source of clues. A C2 server's IP, alias information in domain registrations, hijacked social media accounts, or traces in dark web forums are all visible footprints. But it's not enough to record them. The critical question is "Which infrastructure habits does this actor repeat?"
2. **Analysis stage**
   The actor emerges through external traces as well as intent and motivation. The same IOC set might belong to a financially driven gang or a state-sponsored group seeking strategic intelligence. The difference lies in intent. Analysts must seek to answer the "why" when connecting the dots.
3. **Dissemination stage**
   The actor's identity determines the organization's response. If it's a ransomware gang, the priority is ensuring business continuity. If it's a state-sponsored group, the issue escalates beyond the SOC to the boardroom—even into diplomatic discussions. Thus, threat-actor analysis becomes strategically valuable when conveyed to the right audience at the right stage.

Table 1.4 breaks these stages into their actor-focused elements, key outputs, and example scenarios.

Intelligence Cycle Stage	Actor-Focused Elements	Key Outputs	Example Scenario
Collection	Actor's infrastructure preferences (C2 servers, domain registration habits, hosting providers), tools and malware used, initial traces: IP, hash, domain, social media profiles	IOC pool + early-warning signals	APT group repeatedly using the same domain registrar in the finance sector
Analysis	Actor's motivation (financial, espionage, sabotage), capacity (zero-day exploitation, supply chain abuse), operational habits (working hours, language, code signatures)	TTP maps + actor profile + contextual risk assessment	A phishing email to an energy firm resembling texts used by a known state-sponsored group
Dissemination	Who receives the report (SOC analysts, management, legal, national authorities), prioritization (APT → strategic risk, ransomware → operational risk), defense strategy aligned to actor's behavior	Strategic briefings, operational alerts, legal notifications	Detection of state-sponsored actor leads to reports escalated to board and state agencies

Table 1.4: Intelligence Cycle Interaction: Actors

The insights drawn from Table 1.4 gain their full significance when interpreted in context, as each stage of the intelligence cycle reveals a different layer of the actor's identity, intent, and impact:

- **Collection stage**
  The actor is recognized by their traces. The critical point here isn't merely capturing IOCs, but identifying infrastructure usage habits.
- **Analysis stage**
  The actor's identity begins to take shape: Why are they attacking, what capabilities do they have, how do they behave? This is where TTPs come into play.
- **Dissemination stage**
  The actor's type and intent determine to whom the information is conveyed and how. The same IOC may be a rule for the SOC, a strategic alert for management, or evidence in a legal case.

Many reports present a threat actor as a fixed identity. This is misleading. An actor isn't just a stage performer but rather a scriptwriter constantly rewriting the scene. Today, they register a C2 server in Russia; tomorrow, they repeat the same behavior using South American infrastructure. The scenery changes, but the scriptwriter's handwriting—the TTPs—remains the same. The intelligence cycle is the art of rereading this script and placing it into the correct context over and over.

The abstract principles of the intelligence cycle gain their clearest meaning when brought to life through real-world narratives. Let's consider the case of the Orion Collective—a story that shows how a threat actor's presence can be traced, analyzed, and acted upon across every stage of the cycle: One ordinary Tuesday morning, the organization's SOC noticed something unusual in the log stream. The phishing emails landing in inboxes looked unremarkable—fake invoices, misleading links, a sloppy PDF attachment. But a young analyst reviewing the code spotted a familiar marker: `orion_key`. This alias had once appeared on dark web forums from a user selling stolen credentials.

At first, no one paid much attention. After all, online fraudsters change aliases frequently. But within days, a series of attacks bearing the same signature began to appear against not only the finance teams but also the energy department's servers.

As the day progressed, analysts pieced together the puzzle. This wasn't a random phishing gang. This was the Orion Collective—a group that appeared to steal cryptocurrency but was also harvesting energy-investment documents. Their emergence unfolded like a script embedded across every stage of the intelligence cycle:

1. **Collection**
   Orion's traces were gathered-code snippets in fake emails, repeating usernames, and old forum posts. Each was a small fragment, but together they formed a mosaic.
2. **Analysis**
   The picture changed. A typical ransomware gang would only seek money. Yet Orion sometimes chose to steal data rather than cash. This contradiction revealed dual motives: financial gain alongside strategic intelligence collection. Analysts now had to ask the following: "Are these thieves or pawns in a larger game?"
3. **Dissemination**
   The matter escalated beyond SOC reports. A board briefing raised the possibility that Orion was state-sponsored. Suddenly, the issue drew the attention not of the CFO but of the foreign-relations director, because the matter now involved diplomacy as well as security.

That evening, an analyst noted that Orion's scenery kept shifting: one day a server registered in Russia, the next day hosted in South America. But behind the curtain, the handwriting never changed. In the code's margins remained the same habits, the same rigidity, the same impression. The analyst realized that the Orion Collective wasn't just an actor on stage, but a scriptwriter rewriting the play daily. Different scenery and different costumes, but the same pen.

Threat actor analysis is more than producing an ID card. Without placing the actor into the intelligence cycle, they remain only a shadow in reports. But when their traces, intentions, and outcomes are contextualized across collection, analysis, and dissemination, the lines of a report transform into a decision-shaping force at the executive table.

The real question is whether you'll stop at naming the actor, or you'll redefine them at every stage of the cycle to steer your organization's defense.

1.5.4 TTPs and Detection Engineering

To many security professionals, MITRE ATT&CK looks like a catalog: pages of tactics, techniques, and subtechniques. But reading this catalog is like standing on the shore staring at the ocean; the real task is to dive in and make it concrete through detection engineering.

TTPs describe adversary habits. *Detection engineering* is the art of converting those habits into detectable signals. Apart, TTPs remain abstract and detection engineering is directionless. Together, they yield long-lasting, behavior-based detectors that outlive IOCs.

Imagine an SOC analyst confronted with thousands of IOCs—IPs, hashes, domains. Many are outdated, and many are false positives. Eventually, the analyst drowns in noise. But if instead the analyst is handed the TTP "Credential Dumping - LSASS Access," the situation changes. This TTP can be directly converted into a detection rule:

- Windows Event ID 10 (Process Access) showing unusual access to the Local Security Authority Subsystem Service (LSASS) process.
- EDR telemetry correlating specific memory-read calls.

Now the analyst no longer wastes time on dead IOCs but instead captures the attacker's behavior.

Translating adversary behaviors into effective detection requires a disciplined workflow that turns abstract TTPs into actionable defenses through the following iterative process:

1. Form a hypothesis: "The adversary wants to access LSASS."
2. Select signals: Windows Security Event ID 10, Sysmon Event ID 4656, EDR memory-access logs.
3. Write the rule: Configure SIEM to alert on specific combinations.
4. Test: Use adversary emulation (e.g., Mimikatz in a controlled attack).
5. Refine: Reduce false positives, and set context-specific thresholds.

Repeatedly cycling through this process turns TTPs from static catalog entries into live detection strategies.

A hash may become invalid within hours as code is recompiled. But LSASS access habits, or requesting Kerberos tickets for lateral movement, don't change so easily. Detection engineering's power lies in turning such behaviors into catchable signals.

Examples

Let's consider a few examples of this TTP cycle:

- **Example 1: Ransomware behavior**
  - TTP: Deleting shadow copies before encryption.
  - Detection: Windows Event ID 524 + "vssadmin delete shadows" command-line pattern.
  - Result: Detected even if the ransomware variant changes its hash.
- **Example 2: Cloud identity abuse**
  - TTP: Unusual privilege granted to an OAuth app.
  - Detection: Microsoft Entra ID audit logs showing `Mail.ReadWrite` and `Files.ReadWrite.All` assigned simultaneously to one app.
  - Result: Behavioral alert instead of static IOC.
- **Example 3: Insider threat**
  - TTP: Abnormal data exfiltration.
  - Detection: Data loss prevention (DLP) sensors flagging large transfers outside business hours + a concurrent VPN session.
  - Result: Detects insider activity without relying solely on external IOCs.

Looking past its surface as a catalog of techniques, MITRE ATT&CK reveals itself as an architectural framework that, when mapped into detection engineering, provides three key advantages: gap analysis, structured testing, and a universal analytic language, as described here:

1. Visualizing detection gaps: Which techniques do we have rules for, which do we not?
2. Testing framework: Generating TTP-based scenarios for red-/purple-team exercises.
3. Common language: When "T1055 - Process Injection" is mentioned, analysts worldwide understand the same thing.

In academic literature, this linkage is called *TTP-to-signal mapping*. Without it, modern SOCs can't truly realize the value of threat intelligence.

1.6 Summary

This chapter has established the conceptual and strategic foundations of CTI by demonstrating that intelligence in cyberspace isn't simply the collection of technical artifacts but a structured discipline that transforms scattered signals into informed decision-making. CTI operates at the intersection of technology, strategy, and organizational governance, enabling institutions to anticipate threats, understand adversaries, and translate analytical findings into concrete defensive action.

The discussion began by clarifying the nature of CTI as a systematic process that converts raw data into contextualized knowledge through analysis, interpretation, and dissemination. Unlike traditional security monitoring, which focuses on immediate alerts, CTI introduces foresight into cybersecurity by identifying patterns, motivations, and trends before attacks fully materialize. In doing so, it shifts the defensive posture of organizations from reactive incident response toward proactive risk management.

A central theme throughout the chapter was the layered nature of intelligence. Strategic, operational, tactical, and technical intelligence form a continuum rather than isolated domains. Strategic intelligence informs long-term policy, investment priorities, and geopolitical risk assessments. Operational intelligence tracks campaigns and emerging threats affecting sectors and organizations. Tactical intelligence examines adversary behavior through TTP analysis, while technical intelligence provides machine-readable indicators that enable real-time detection. Effective cybersecurity requires the synchronization of these layers so that insights generated at one level reinforce the others.

The chapter also highlighted the importance of governance and organizational structure in making CTI effective. Intelligence must be guided by clearly defined requirements to ensure that analysis addresses the decisions that matter most. Similarly, governance models such as centralized, federated, or hub-and-spoke architectures determine how intelligence flows within institutions and how rapidly it can influence operational or strategic responses.

Another critical component examined was the evolution of threat intelligence itself. From its origins in classical military espionage to its current role in protecting digital ecosystems, intelligence practices have adapted to the realities of the cyber domain. Landmark events—from early cyber espionage campaigns to modern supply chain compromises and AI-driven disinformation—have demonstrated that along with technical systems, cyber threats now also affect economic stability, political processes, and public trust.

Understanding adversaries remains a cornerstone of CTI. Threat actors, including nation-state groups, organized cybercrime syndicates, hacktivists, insiders, and independent attackers, operate with different capabilities and motivations. These motivations are no longer strictly separable; hybrid objectives combining political influence, economic profit, and ideological messaging increasingly characterize modern campaigns. For this reason, actor analysis must be integrated into every stage of the intelligence cycle, from data collection to dissemination.

Equally important are the analytical frameworks and technical mechanisms that support CTI operations. Standards such as STIX and TAXII enable structured intelligence sharing, while frameworks such as MITRE ATT&CK, the cyber kill chain, the Diamond Model, STRIDE, and PASTA provide systematic methods for understanding adversary behavior and system vulnerabilities. When combined with detection engineering practices, these models allow organizations to convert behavioral insights into durable detection capabilities that remain effective even as indicators change.

Finally, the chapter underscored the collaborative nature of modern cyber defense. Intelligence sharing across sectors, institutions, and national boundaries has become essential in addressing threats that operate across interconnected digital ecosystems. Whether through ISAC networks, CERT coordination, or international policy frameworks, collective situational awareness strengthens the resilience of the broader cybersecurity community.

Taken together, these foundations demonstrate that CTI is no longer an auxiliary component of cybersecurity. It's a strategic capability that integrates technical analysis, organizational governance, and cross-sector collaboration. By transforming data into insight and insight into action, CTI enables organizations and states alike to anticipate risks, adapt to evolving adversaries, and build resilient security architectures in an increasingly complex digital environment.

Now, let's dive deeper into the CTI lifecycle in practice in the next chapter.

Chapter 2

Intelligence Lifecycle in Practice

The concept of cyber threat intelligence is more than a theoretical model—it's an iterative process that has a direct counterpart in the daily operations of organizations.

The intelligence lifecycle defines the systematic steps that must be followed to produce high-quality and reliable intelligence products. This framework is generally described via the following stages: planning, collection, processing, analysis, dissemination, and feedback. In practice, however, it has a far more dynamic structure in which these stages are tightly interconnected. For professionals working in security operations centers (SOCs), incident response teams, or threat-hunting activities, these stages don't represent a linear sequence of steps completed once and finished, but rather a continuously fed cycle that is reshaped according to environmental conditions. In this chapter, we'll walk through the intelligence lifecycle, which is composed of the following phases:

1. **Planning**
   One of the most common problems organizations face is the inability to correctly define their intelligence requirements (IRs). Requirements that are overly broad or poorly directed generate unnecessary noise during the collection and analysis phases that distracts the security team from their main focus. For this reason, the direction-setting step is a critical starting point that determines the quality of the entire lifecycle. At this stage, priority intelligence requirements (PIR), more specific information requirements (SIR), and essential elements of information (EEI) are clarified, thereby establishing a solid foundation for subsequent processes.

2. **Collection**
   The collection phase isn't limited to pulling data from logs, telemetry, or open sources. In addition to passive methods, it also encompasses advanced techniques such as honeypot and honeynet systems, controlled active scanning sensors, and dark web monitoring. What matters here is the diversity, reliability, and contextual value of the data obtained from different sources, as these form the basis of analytical accuracy in later stages. Note that active collection techniques may involve legal, ethical, and operational risks.

3. **Processing**
   Raw data collected on its own isn't meaningful. This data must be processed, normalized, deduplicated, assigned reliability scores, and placed into context. The processing stage ensures that analysts are presented only with a cleaned and validated dataset, reducing false positives and accelerating decision-making processes.

4. **Analysis**
   In the subsequent analysis phase, analysts don't just report specific indicators of compromise (IOCs); they also derive tactical indicators of attack (IOAs), attacker behavior patterns (tactics, techniques, and procedures [TTPs]), and campaign-level insights to understand the intent and capabilities of threat actors. Frameworks such as the Diamond Model, the MITRE ATT&CK matrix, or the kill chain provide analytical discipline to this interpretation. When analytical outputs are tailored and communicated to decision-makers at different levels, the point at which the lifecycle produces tangible value is reached.
5. **Feedback**
   The cycle doesn't end here. Feedback obtained after dissemination ensures the sustainability of the process. Input from stakeholders, evaluations of the usefulness of reports, and new requirements driven by the changing threat landscape necessitate feeding the cycle back into the planning phase. In this way, threat intelligence evolves from a reporting activity into a living mechanism that supports the continuous development of an organization's security culture. Especially in real-time environments, within SOC operations and incident response processes, this iterative structure is the most reliable way to keep pace with attackers.

The most critical characteristic of the intelligence lifecycle is that it's cyclical rather than linear, which is a point that is often overlooked. The process doesn't end after the plan, collect, analyze, and report stages; the experience and feedback gained from each cycle enhance the quality of the next one. Through this approach, threat intelligence production becomes increasingly refined, faster, and more target-focused with each iteration.

2.1 Planning and Direction Phase

Poorly defined requirements can render even the strongest analyses meaningless; correct direction, on the other hand, transforms data into strategic value. Planning and direction, the most critical step of the intelligence lifecycle, is the strategic phase that defines the framework of the entire process. If this step isn't structured correctly, no matter how technically strong the subsequent collection, processing, and analysis activities may be, the intelligence produced will fail to deliver the expected value for decision-makers.

In the following sections, we'll break down the planning phase into its key components and dive into important considerations, from relevant standards and frameworks to governance, risk tolerance, and more.

2.1.1 Components of the Planning Phase

The planning phase can be broken down into several components. Let's walk through each one:

- **Definition of IRs**
  The questions the organization seeks to answer, the risks it prioritizes, and the areas it aims to gain visibility must all be determined first. For example, for a US-based financial institution, the question of "Which emerging fraud techniques are being used to exploit online banking platforms, and how might they impact customer trust and financial stability?" may be critical, whereas for an energy company in Europe, the question "What vulnerabilities are being targeted in industrial control systems (ICS) within the energy sector, and how could these affect operational continuity?" may carry greater strategic priority. The objective here is to avoid overly general statements and to define concrete information needs that can guide operational teams.
- **Threat landscape analysis and prioritization**
  Supply chain attacks faced by US federal agencies (e.g., attempts that can be considered a continuation of incidents such as SolarWinds), or the increasing distributed denial of service (DDoS) and APT campaigns targeting critical infrastructure providers in Europe, are examples that must be directly considered during the planning process. At this point, organizations such as Cybersecurity and Infrastructure Security Agency (CISA) in the United States and European Union Agency for Cybersecurity (ENISA) in Europe provide guidance through their regularly published reports on which threats sectors should be prioritized.
- **Resource planning**
  Intelligence production is shaped by human resources, budgets, and sensor architecture. While many organizations in the US integrate MITRE ATT&CK–based cyber threat intelligence (CTI) assessments into their planning processes, financial institutions in Europe focus on diversifying intelligence sources and documenting their cyber resilience within the scope of the Digital Operational Resilience Act (DORA). As of 2026, artificial intelligence (AI) and automation have become widely used in threat intelligence collection and analysis processes; however, the proper configuration of these systems still requires the guidance of experienced analysts. Therefore, during the planning phase, analyst training programs and continuous professional development plans should be prioritized, along with the usual technology investments.
- **Assumption and bias management**
  A dimension that is frequently overlooked during planning is the management of assumptions and awareness of analytical biases. In the US, malware such as TrickBot and Emotet, which targeted the financial sector for extended periods, led organizations' threat perceptions to become narrowly focused on these actors. Today, however, double extortion ransomware groups and AI-enabled phishing techniques have come to the forefront. Similarly, in Europe, a threat landscape once primarily associated with DDoS attacks has expanded in recent years to include sophisticated attacks targeting industrial control systems. Organizations that fail to account for these changing dynamics during the planning phase risk being unprepared for future threats.

- **Threat modeling**
  Another increasingly important element is the integration of threat modeling, which is a topic we'll discuss in detail in Chapter 6. Frameworks such as MITRE ATT&CK, D3FEND, and the Diamond Model can be used both during the analysis phase and the planning stage. For example, a US healthcare organization might define the question "Which credential access techniques stand out in APT41 campaigns targeting the healthcare sector?" directly based on the ATT&CK matrix. In Europe, actors in the energy sector align ENISA reports with ATT&CK techniques to make their requirements more measurable and verifiable during the planning process.
- **Performance metric definition**
  In the US, organizations track progress using indicators such as the PIR closure rate or impact on mean time to detect (MTTD), while financial institutions in Europe emphasize metrics such as false positive reduction rates to demonstrate both operational efficiency and regulatory compliance. Including such metrics in the planning phase enables a concrete evaluation of the success of the intelligence produced.

Note

The maturity level of the organization directly affects how realistic the planning process can be. Large banks in the US with mature CTI teams prepare detailed PIR sets that prioritize threat actor TTPs, whereas smaller organizations tend to start with more general questions and develop deeper planning processes over time. In Europe, small energy distribution companies often define requirements focused primarily on regulatory compliance, while multinational energy corporations plan APT tactics in a more comprehensive manner.

- **Integration of crisis scenarios and exercises**
  In the US, Cyber Storm exercises conducted by CISA help organizations clarify their PIRs, while in Europe, EU-wide exercises such as BlueOLEx provide direct outputs regarding which threats require greater focus. In this way, planning moves beyond being a purely theoretical phase and becomes reinforced by practical field experience.
- **Time horizon planning**
  Planning shouldn't focus solely on current threats; it should also provide perspective on short-, medium-, and long-term threats. In the US, financial institutions develop scenarios addressing ransomware in the short term, AI-enabled phishing techniques in the medium term, and quantum cryptography risks in the long term. In Europe, the energy sector aims to mitigate the impact of DDoS attacks in the short term, prevent supply chain exploitations in the medium term, and prepare for sophisticated attacks targeting Internet of Things (IoT)/Operational Technology (OT) systems in the long term.
- **Use case definition**
  Finally, during the planning phase, the intelligence use cases should be clearly defined. In the US, large technology companies use produced threat intelligence in SOC operations,

product security, supply chain management, and legal departments. In Europe, particularly in the financial sector, intelligence outputs are presented to analysts, boards of directors, and regulators in regular report formats. In this way, intelligence becomes more than an operational tool and serves as an integral part of the organizational decision-support mechanism.

Throughout all of these processes, resource conflicts may arise between the priorities of different units. While the SOC team may prioritize new ransomware campaigns, the risk management team may focus more on supply chain exploitations. In the US, such conflicts are typically resolved through threat intelligence committees, whereas in Europe, guidance from regulatory authorities (e.g., the European Central Bank) plays a binding role in organizations' prioritization decisions.

As the first step of the intelligence lifecycle, planning and direction comprise the strategic compass that shapes the success of the entire process. Without clearly defined requirements, solid threat landscape analysis, realistic resource planning, measurable metrics, and clear operational objectives, the threat intelligence produced remains nothing more than a scattered collection of data for the organization. By contrast, a structured and collaborative planning process ensures that threat intelligence is produced in a form that can meaningfully support operational and strategic decision-making.

2.1.2 Deepening the Conceptual Framework: PIRs, SIRs, and EEIs

The most critical element of the planning process is the organization's ability to accurately define which questions it's truly seeking to answer. Within the intelligence lifecycle, the volume of data that can be collected is virtually unlimited; information flows continuously from logs, open sources, dark web forums, and sensors. However, how much of this data actually adds value to the organization depends on the framework within which it's collected. This is where the PIRs, SIRs, and EEIs, as mentioned earlier, come into play:

- **PIRs**

  PIRs represent the highest-level questions an organization must ask. These are generally strategic in nature and directly affect the organization's security posture. Here's an example of a PIR for a US-based financial institution: "Which ransomware groups are expected to target the U.S. banking sector in the next six months?" For a critical infrastructure energy provider in Europe, a PIR could be defined along a different axis:"Which new tactics are Russia-based APT groups adopting in their operations against energy supply chains?"

  These questions directly inform the decision-making processes of boards of directors and senior security leadership. When PIRs aren't defined correctly, they can misdirect resources across the entire lifecycle.

- **SIRs**

  SIRs are derived from PIRs and point to more targeted, operational-level information gaps. Continuing with the financial sector example, SIRs derived from the relevant PIR might include the following:

  - Which access vectors have been used in campaigns by the LockBit ransomware group targeting US banks over the past six months?
  - What are the technical characteristics (sender IP addresses, domain patterns) of the phishing emails used in these campaigns?

  At this level, the focus shifts from the strategic picture to the operational context. SIRs become practical guidance that SOC analysts and threat hunters can apply directly.

- **EEIs**

  EEIs represent the most granular level of information requirements. They define which specific details must be contained in the raw data collected from the field, for example:

  - SHA-256 hash values of phishing emails used in the LockBit campaign
  - Command-and-control (C2) IP addresses used by the attacker
  - Unique signatures of the PowerShell commands used by APT29 for credential harvesting

  EEIs are the specific elements sought directly by sensors, honeypots, log correlation mechanisms, and open-source intelligence (OSINT) collection tools.

When these three levels are considered together, the planning process moves away from random data collection efforts and becomes a measurable and verifiable methodology. PIRs set the strategic direction, SIRs fill operational gaps, and EEIs clearly define which microlevel data should be collected from the field. To illustrate, consider this analogy: In a war, a PIR asks "On which front must we achieve victory?", an SIR asks "Which bridges must we capture on that front?", and an EEI answers "What is the number and location of enemy soldiers defending the bridge?" In this sense, PIR represents vision, SIR represents operations, and EEI represents tactical execution.

Many organizations systematize this approach by using collection requirement matrices (CRM) or intelligence requirement traceability matrices (IRTM). These matrices map the relationships between IRs and the data sources used to satisfy them, allowing organizations to track how specific information needs are translated into observable indicators. An illustrative example of this relationship between PIRs, SIRs, and EEIs is presented in Table 2.1. This method ensures that intelligence planning isn't based solely on analyst experience but is also measurable and auditable.

Thus, the PIR → SIR → EEI structure provides a practical framework that supports more focused and accountable threat intelligence planning.

Level	Definition	Example Question/ Content	Area of Use
PIR	Critical information need at the strategic level. Determines the organization's security direction and investment priorities.	■ Which ransomware groups are expected to target the banking sector in the next six months?	Executive management, strategic planning, investment decisions.
SIR	More focused, operational-level information need derived from a PIR. Concentrates on specific threat actors, techniques, or campaigns.	■ Which access vectors have been used in recent LockBit campaigns targeting banks in the United States? ■ What are the technical characteristics of the phishing emails used in these campaigns?	SOC teams, threat hunters, incident response planning.
EEI	Information need at the most granular level. Defines specific indicators to be collected directly from the field.	■ SHA256 hash values of phishing emails used in the LockBit campaign ■ C2 IP addresses used by the attacker ■ Unique signatures of the PowerShell commands used by APT29 for credential harvesting	Sensors, security information and event management (SIEM), honeypots, IOC/IOA feeds.

Table 2.1: PIR → SIR → EEI Chain

2.1.3 References to Standards and Frameworks

The planning process isn't limited to addressing internal organizational needs; it must also align with global standards, regulations, and sectoral frameworks. The current state of the threat intelligence discipline clearly demonstrates that alignment with international standards is no longer just a best practice, but, in many cases, it's a legal and operational necessity.

With its 2024 update, the NIST Cybersecurity Framework (CSF) 2.0 explicitly associates threat intelligence with the Identify function (identifying assets, risks, and threats) and the Protect function (establishing protection mechanisms). This update clearly shows that, during the planning phase, threat intelligence is no longer just a technical security activity, but has become an integral part of enterprise risk management, compliance processes, and business continuity planning. For example, in the US, the risk committee of a financial institution within a planning process based on the NIST CSF must address the ransomware threat as an SOC-level security issue *and* as a business risk with direct implications for operational continuity and customer trust.

Similarly, International Organization for Standardization (ISO)/International Electrotechnical Commission (IEC) 27035, as an incident management standard, explicitly recommends defining IRs during the planning phase. This standard integrates threat intelligence with incident response processes, making it clear at the planning stage which information requirements need to be collected. For instance, a healthcare organization in Europe operating in compliance with ISO 27035 both addresses the question "How do we respond to a potential ransomware attack?" and systematically answers the question "Through which attack vectors could such attacks occur, and which indicators should we collect in advance?"

In Europe, ENISA stands out through its sector-specific guidance. ENISA's threat landscape reports, particularly those published for the energy and financial sectors, provide organizations with a direct roadmap on which threats should be prioritized during the planning phase. For example, ENISA's "Threat Landscape for Energy" report, published at the end of 2023, emphasized that energy distribution companies should treat supply chain attacks as a PIR as they entered 2025. This report has become a direct reference point for companies operating in the energy sector when defining PIRs during the planning stage.

On the US side, CISA comes to the forefront with its sectoral reports. Resources such as CISA's Known Exploited Vulnerabilities Catalog or sector-specific threat bulletins identify which vulnerabilities should be remediated with priority—particularly for critical infrastructure providers—and guide intelligence planning by indicating which threat actors should receive greater resource allocation. For example, a public energy utility in the US may, in line with CISA's guidance, treat phishing campaigns used by APT29 as a PIR-level concern.

Comparability and collaboration are also advantages of alignment with these standards and frameworks. When organizations structure their PIR/SIR definitions based on these standards, it becomes easier to speak a common language with actors in different sectors or within international collaborations. For example, while the IRs prepared by a bank in the US and the PIR list of an energy company in Europe may differ in content, both become comparable on a shared methodological foundation when grounded in the NIST CSF and ENISA reports. This significantly facilitates collaboration, particularly within cross-sector information-sharing networks such as Information Sharing and Analysis Centers (ISACs).

In short, grounding the planning phase in global frameworks such as NIST, ISO, ENISA, and CISA both simplify regulatory compliance and make organizations' intelligence activities measurable, comparable, and open to collaboration. Planning that isn't based on standards may appear to provide speed in the short term, but in the long run, it leads to an isolated, immeasurable, and unsustainable intelligence lifecycle.

To provide a clearer overview of how widely adopted standards contribute to threat intelligence planning, Table 2.2 summarizes the main focus areas of selected frameworks and their practical implications for the planning phase of the intelligence lifecycle. The table highlights how standards such as the NIST CSF, ISO/IEC 27035, ENISA Threat Landscape reports, and CISA advisories can guide organizations in defining IRs and aligning CTI activities with broader risk management and governance processes.

While these frameworks offer valuable guidance, they also differ in scope, level of detail, and practical applicability. Table 2.3 therefore presents a comparative overview of the strengths and limitations of these standards when applied to threat intelligence planning. This comparison helps illustrate how different frameworks may complement each other depending on the organization's sector, regulatory environment, and operational maturity.

Standard/Framework	Focus Area	Contribution to Planning
NIST Cybersecurity Framework (CSF) 2.0	Risk management, asset identification, protective controls	▪ Aligns PIR/SIR definitions with enterprise risks ▪ Plans threat intelligence at the SOC level and from a business continuity and governance perspective
ISO/IEC 27035 (Incident Management)	Incident detection, response, recovery	▪ Clarifies which incident types intelligence should be collected for during the planning phase ▪ Builds a bridge between incident response and CTI
ENISA threat landscape reports	Sector-specific threat landscape analyses for European energy, finance, healthcare, and related sectors	▪ Defines sectoral priorities ▪ Uses reports as direct references when creating PIRs/SIRs, ensuring alignment with European regulations
CISA Known Exploited Vulnerabilities (KEV) Catalog and sector alerts	US critical infrastructure vulnerabilities, actively exploited weaknesses	▪ Provides direct guidance on which threat actors and vulnerabilities should be prioritized during planning ▪ Helps concretize PIRs

Table 2.2: Contribution of Standards to the Planning Phase

Standard/Framework	Strengths	Weaknesses
NIST Cybersecurity Framework (CSF) 2.0	■ Integrates threat intelligence with enterprise risk management ■ Widely used in the US and recognized globally as a reference framework ■ Supports PIR definition through its Identify and Protect functions	■ Doesn't provide detailed technical guidance and remains high-level ■ Not fully aligned with European regulations, requiring additional mapping or translation
ISO/IEC 27035 (Incident Management)	■ Links incident management with IRs ■ Internationally recognized and suitable for audits ■ Establishes a methodological bridge between incident response and CTI	■ Not CTI-specific; focuses primarily on an incident management perspective ■ May offer limited flexibility in practice, as it's oriented toward formal procedures
ENISA threat landscape reports	■ Provides threat analysis tailored to European sectors (energy, finance, healthcare) ■ Directly supports compliance with regulations such as the NIS2 Directive and DORA ■ Contains concrete data for prioritizing the threat landscape	■ Published annually or semiannually, so it's not real-time information ■ Offers limited value for organizations focused on US or Asia-Pacific regions
CISA KEV Catalog and sector alerts	■ Provides directly actionable alerts for critical infrastructure in the US ■ Offers fast and practical references on exploited vulnerabilities ■ Accelerates the creation of operational PIRs/SIRs	■ Doesn't cover all threats on a global scale; US-centric ■ Focuses on tactical and operational guidance rather than strategic direction

Table 2.3: Comparative Strengths and Weaknesses of Standards

2.1.4 Analytical Biases and Bias Management

One of the most frequently overlooked elements during the planning phase of threat intelligence is the cognitive biases that analysts unconsciously bring into the process. Regardless of how advanced an organization's technology, sensors, or budget may be, decisions ultimately rely on the interpretations made by human analysts. For this reason, analytical biases can directly affect the quality of the guiding decisions made during the planning phase.

Cognitive biases can significantly influence how analysts interpret information during the planning phase of threat intelligence. These biases often operate unconsciously and may shape how threats are prioritized, which actors receive attention, and which signals are ignored. Some of the most common analytical biases encountered in threat intelligence include the following:

- **Confirmation bias**
  As one of the most common biases, this leads analysts to focus only on threat actors they already believe in or have previously validated. Analysts at a financial institution that has repeatedly been targeted by the LockBit group in the past may shape all their PIRs around this actor. However, they may overlook emerging ransomware groups such as BlackCat or Royal that are active during the same period. As a result, critical threats may remain entirely off the radar during the planning process.
- **Anchoring bias**
  Analysts place excessive weight on the first piece of information they receive and interpret subsequent data in light of these initial findings. For instance, if early indicators in an APT attack suggest a China-based threat actor, the analyst may become anchored to this initial anchor. Even if later technical evidence indicates that the attack originated from Russia, the new data may not receive sufficient consideration due to attachment to the initial assessment.
- **Survivorship bias**
  Survivorship bias poses a particularly critical risk in threat intelligence. This bias refers to focusing only on observable attacks while ignoring those that go undetected. For example, an organization may analyze and report only detected phishing campaigns. However, more sophisticated campaigns carried out by the attacker that evade detection remain invisible and therefore aren't considered during the planning phase. This leads to the threat landscape being perceived as less severe than it actually is.

Various methods can be employed to counter these biases. The devil's advocate approach is based on one team deliberately questioning the assumptions of another. For instance, when one team concludes that "APT29 isn't targeting the energy sector this quarter," another team intentionally develops a counterargument: "What if it is targeting the sector; what data might have been overlooked?" This approach helps uncover blind spots during the planning phase.

Similarly, red team/blue team exercises can be applied both in defensive testing and threat intelligence planning. The red team adopts the attacker's perspective and answers the question "Which vulnerabilities would I exploit?" whereas the blue team develops a defensive perspective by asking "Which threats should I prioritize?" This contrast helps balance cognitive biases.

Today, many organizations integrate alternative hypothesis testing into their decision-support tools. By presenting analysts with multiple scenarios rather than a single possible explanation, these tools provide warnings such as "the attack may be attributed to APT29; however, the same data also aligns with the TTPs of FIN7," preventing analysts from becoming fixated on a single judgment.

The quality of a threat intelligence team is measured not only by the technologies it uses but also by how effectively the team manages its biases. Even the most advanced SIEM or security orchestration, automation, and response (SOAR) systems can't correct a poorly directed planning phase. For this reason, analysts should keep bias awareness continuously in mind during the planning process. Regular training, scenario-based exercises, and cross-team assessments are among the most effective methods to reduce the impact of these biases.

Beyond these individual examples, cognitive biases can influence multiple aspects of the intelligence planning process, including how threats are interpreted, how priorities are established, and how analytical judgments are formed. In practice, these biases may affect the definition of IRs, the interpretation of threat indicators, and the allocation of analytical resources. Table 2.4 provides a structured overview of several common analytical biases, illustrating their definitions, practical examples, potential impact on intelligence planning, and possible mitigation approaches. By presenting these elements together, the table highlights how cognitive biases can shape threat intelligence outcomes and why structured mitigation strategies are necessary in analytical workflows.

Bias Type	Definition	Example	Impact on Planning	Mitigation Methods
Confirmation bias	Analysts focus only on data that supports their existing beliefs or previously validated assumptions.	Because the SOC team has repeatedly faced LockBit attacks in the past, it defines PIRs only around LockBit and ignores emerging threat groups such as BlackCat.	New threats or alternative attack scenarios are overlooked; planning becomes confined to a narrow scope.	Devil's advocate discussions, development of alternative hypotheses, cross-team reviews.

Table 2.4: Cognitive Biases and Their Impact on Planning

Bias Type	Definition	Example	Impact on Planning	Mitigation Methods
Anchoring bias	Analysts rely excessively on initial information, with subsequent data interpreted within that initial frame.	When an initial report attributes an attack to China, the analyst ignores later evidence pointing to Russia.	PIR/SIR definitions are misguided and resources are misallocated toward incorrect targets.	Mandatory use of multiple data sources, red team/blue team comparisons.
Survivorship bias	Analysists focus only on observable attacks while ignoring attacks that remain undetected.	The organization reports only detected phishing campaigns; sophisticated, undetected attacks remain outside the planning scope.	The threat landscape is perceived as weaker than it actually is; critical risks are excluded from planning.	Proactive threat hunting, use of honeypots/honeynets, external intelligence sharing.
Availability bias	Events that are easiest to recall or most frequently discussed are perceived as more important.	The team puts excessive focus on a ransomware group heavily covered in the media, while a quieter but more dangerous APT is ignored.	"Popular threats" are prioritized instead of actual risks.	Data-driven prioritization, statistical analysis, threat trend reports.
Groupthink	Team members avoid expressing differing opinions and conform to the majority view.	No alternative hypotheses are proposed within the analysis team; everyone aligns with the lead analyst's assessment.	Alternative scenarios aren't evaluated; planning becomes blind to risks.	Diverse teams, independent reviews, anonymous contribution mechanisms.

Table 2.4: Cognitive Biases and Their Impact on Planning (Cont.)

2.1.5 Intelligence Governance Framework

The intelligence function must be supported by technical capabilities and corporate governance principles. At this point, the concept of an intelligence governance framework comes into play. Its purpose is to make CTI processes transparent, measurable, and accountable.

An effective intelligence governance framework requires clearly defined roles, responsibilities, and accountability mechanisms within the organization. Without such structure, threat intelligence activities may remain fragmented, and the outputs produced by analysts may fail to reach decision-makers or influence organizational strategy.

At the leadership level, the *intelligence manager* or *CTI lead* is responsible for defining the strategic direction of the threat intelligence function. This role typically oversees the development of PIRs, coordinates analytical activities, and ensures that intelligence outputs support both operational and strategic objectives. In many organizations, the CTI lead also serves as a bridge between technical teams and executive leadership, communicating key threat developments to the *chief information security officer (CISO)* and, when necessary, to senior management or board-level stakeholders.

Supporting this leadership role are analysts operating at different levels of intelligence production. *Strategic analysts* focus on long-term threat trends, geopolitical developments, and adversary capabilities. *Operational analysts* examine ongoing campaigns and adversary infrastructure, while *tactical analysts* work more closely with IOCs, threat hunting, and incident response processes. Together, these roles form the analytical chain that connects high-level IRs with detailed technical observations in the field.

In addition to security teams, *business unit representatives* or liaison roles play an important part in the intelligence governance structure. These individuals communicate the needs and concerns of departments such as legal, human resources, supply chain management, and compliance. Their input ensures that IRs reflect not both technical security concerns and operational, legal, and organizational priorities.

Beyond role definitions, the governance framework must also clarify core responsibilities within the intelligence process. These responsibilities typically include defining the organization's PIRs, aligning intelligence collection and analysis activities with enterprise risk priorities, and coordinating secure information sharing with external stakeholders such as ISAC communities, Community Emergency Response Team (CERT) teams, and regulatory bodies. When these responsibilities are clearly defined, threat intelligence activities become integrated with the broader security and risk management architecture of the organization.

Finally, an effective intelligence governance structure requires clear accountability mechanisms. Organizations often measure the effectiveness of their intelligence activities through key performance indicators (KPIs), such as the rate at which PIRs are successfully addressed. Transparent documentation processes, including audit trails, help track how intelligence was collected, analyzed, and used in decision-making. Regular management reporting further ensures that CTI activities remain visible to leadership, enabling the

CISO, risk committees, and senior management to evaluate how intelligence contributes to the organization's overall security posture.

For the planning process to succeed, the hierarchical placement of the CTI team within the organization plays a critical role. A poorly positioned CTI unit may result in the intelligence it produces either not being used at all or remaining limited to the operational level. The following describes where the CTI team fits into the organization:

- **Direct reporting to the CISO**
  When the CTI team reports directly to the CISO, it gains strategic visibility. In this way, the intelligence produced is considered at the SOC level, as well as in risk management, business continuity, and board-level decision-making. For example, in large financial institutions in the US, the CTI manager presents a priority threats report to the board of directors each quarter through the CISO.
- **Integration with risk management**
  Intelligence should be treated not merely as a technical threat but as a business risk. For this reason, continuous interaction between the CTI team and the enterprise risk management function is essential:
  - In the financial sector, ransomware evolves into a risk to information security, operational continuity, and customer trust.
  - In the energy sector, APT attacks should be evaluated both from an IT perspective and within the context of supply chain risk.
- **Board-level visibility**
  For CTI outputs to be effective, they must be presented at the board level in a clear and business-oriented manner, such as "Russia-based APT29 operations in the energy sector have the potential to reduce our company's short-term revenue stream by 10%."

Through reporting, intelligence becomes an integral part of the organizational decision-support mechanism.

2.1.6 Stakeholder Analysis and Communication

In threat intelligence planning, PIRs, SIRs, and EEIs are often shaped around topics defined by technical teams. However, requirements driven solely by SOC teams, threat hunters, or forensic analysts can't ensure organizational coherence in intelligence. For a realistic and actionable intelligence structure, the active participation of different business units is essential:

- **Legal department**
  Ensures regulatory compliance (e.g., General Data Protection Regulation [GDPR], NIS2, Health Insurance Portability and Accountability Act [HIPAA], DORA). They define ethical and legal boundaries in the collection of dark web monitoring, OSINT, or honeypot data.

  Example: In a European bank, the legal department conducts a preliminary legal compliance review for every PIR involving the processing of customer data.

- **Business continuity team/operational risk team**

  Considers the impact of threat intelligence on security incidents and business continuity.

  Example: A ransomware attack disabling a call center leads both the IT team and the business continuity team to formulate PIRs.

- **Supply chain management**

  Prioritizes risks originating from suppliers, third-party service providers, and outsourced software.

  Example: In the US, an energy company redefined its PIRs after the SolarWinds incident around the question, "Which suppliers could be the primary entry points for attackers?"

- **Human resources (HR)**

  Handles insider threat risks and vulnerabilities introduced by employee turnover. It also integrates training and awareness strategies into intelligence planning.

  Example: A European telecommunications company created HR-supported SIRs specifically to measure employees' awareness levels against social engineering attacks.

In the planning process, internal stakeholders, external collaboration, and information-sharing mechanisms play a critical role.

In the US, ISACs are sector-specific sharing centers such as Financial Services ISAC (FS-ISAC), Electricity ISAC (E-ISAC), and Health ISAC (H-ISAC). They help ground PIRs in sectoral threat reports. Information Sharing and Analysis Organizations (ISAOs), on the other hand, are more flexible, sector-agnostic sharing networks that offer cost-effective collaboration opportunities, especially for smaller organizations. For example, a healthcare organization in the US used H-ISAC reports directly in its PIR definition process, translating "APT41 credential theft techniques targeting the healthcare sector" down to the SIR level.

In Europe, CERT-EU is the central CERT structure for European Union institutions, coordinating threat sharing among member states. There are also sectoral CERTs, which are dedicated to critical infrastructure sectors such as energy, finance, and transportation. For example, an energy-sector CERT established under ENISA's guidance might shape its PIR definitions directly based on sectoral threat landscape reports. The integration of internal and external stakeholders in this manner ensures that the planning process doesn't remain confined to internal priorities alone but is also aligned with the global and sectoral threat ecosystem.

Stakeholder analysis and communication ensures that threat intelligence planning aligns with organizational reality. PIR/SIR/EEI definitions must be supported by legal, operational, supply chain, and human factors. They should also be elevated to a global collaboration dimension through inputs from ISACs, ISAOs, CERT-EU, and sectoral CERTs. In this way, the planning process moves beyond a narrow technical perspective and evolves into a multi-layered, high-business-value structure.

2.1.7 Threat Intelligence Sharing Ecosystem

In the modern threat landscape, no organization can anticipate all risks on its own. Attacks are often globally distributed, simultaneous, and multilayered. A vulnerability identified in one sector can quickly affect organizations in different geographies. For this reason, threat intelligence sharing is an integral part of the planning and direction process. We'll discuss the different sharing platforms, standardizations, and more in the following sections.

Platforms for Collaboration

Sharing mechanisms in the US include the following:

- **Automated Indicator Sharing (AIS)**
  Operated by the US Department of Homeland Security (DHS) and CISA, AIS enables organizations to share threat indicators (IOC/IOA) through machine-to-machine integration. Using Structured Threat Information eXpression (STIX)/Trusted Automated eXchange of Intelligence Information (TAXII) standards, data such as IP addresses, domains, and hash values are shared in near real time. As a result, AIS transforms threat intelligence from a weekly reporting activity into an operational defense capability that can be applied within seconds.

  Example: A US healthcare organization prevented the spread of an attack by adding newly identified LockBit C2 IP addresses received via AIS to its firewall rules within minutes.

- **Joint Cyber Defense Collaborative (JCDC)**
  Established in 2021, JCDC aims to provide real-time coordination between the public and private sectors. It enables major technology companies such as Microsoft, Google, and Amazon Web Services (AWS) to collaborate directly with CISA. As a result, attacks of a scale that a single organization can't detect alone can be identified early through a collective radar system.

  Example: When the Log4j vulnerability was discovered, critical patches and threat indicators were distributed to sectors within hours—not days—through JCDC.

Collaboration platforms in Europe include the following:

- **Computer Security Incident Response Team (CSIRT) network**
  A network established under the NIS Directive that connects the CERT/CSIRT teams of EU member states. Its primary goal is to enable rapid coordination, especially during cross-border cyberattacks.

  Example: During the DDoS wave targeting critical infrastructure in Europe in 2022, technical details identified by Estonia's CERT were shared via the CSIRT network, allowing organizations in France and Germany to prepare for the same attack.

- **EU Cyber Crises Liaison Organization Network (EU-CyCLONe)**
  Focused on crisis management at the EU level, EU-CyCLONe aims to provide political and operational coordination alongside technical IOC sharing. Here, intelligence forms the basis for SOC analysts and for strategic decisions at the ministries and regulators level.

Example: During an APT attack targeting an energy company, member states used EU-CyCLONe to jointly address "Which crisis communication protocols should be activated?"

Standardized Sharing Languages and Platforms

In threat intelligence sharing, both what is shared and how it's shared are critical. When different organizations exchange information in different formats through manual processes, the effectiveness of the data is reduced. This is where standardization becomes essential. Several widely adopted standards and platforms have emerged to support structured and automated threat intelligence sharing:

- **Structured Threat Information eXpression (STIX)**
  Represents threat indicators, attack techniques, and campaign relationships through a structured data model.

  Example: Sharing an APT campaign chain-such as "credential theft → C2 communication → data exfiltration" as a visual graph via the STIX model allows analysts to derive insights much faster than from tabular IOC lists.
- **Trusted Automated eXchange of Intelligence Information (TAXII)**
  Enables secure and automated sharing of STIX data. When integrated with SOC tools, SIEM, and SOAR systems, it allows IOCs to be applied directly.

  Example: A financial institution subscribes to a TAXII feed provided by a national CERT. Newly published malicious domains and IP addresses are automatically delivered to the organization's SIEM platform. As soon as these indicators are received, detection rules are updated, and the firewall blocks connections to the identified infrastructure in near real time.
- **OpenIOC**
  An older but still-used IOC standard, particularly preferred for sharing endpoint-based threat indicators.

  Example: During a malware investigation, analysts identify several file hashes, registry keys, and suspicious process names associated with the attack. These indicators are packaged into an OpenIOC file and distributed to endpoint detection tools across the organization, enabling automated scanning of all endpoints for the same artifacts.
- **MISP Threat Sharing (MISP)**
  An open-source, collaborative intelligence-sharing platform that is compatible with STIX/TAXII and supports a broad, community-driven ecosystem.

  Example: Financial institutions in Europe regularly share phishing domains via MISP and immediately enforce them on email gateways.

Intelligence sharing that isn't based on standards results in scenarios where teams are speaking "different languages". With STIX/TAXII and MISP integration, this barrier is removed, and threat information becomes an immediate and universal language.

Strategic Added Value

The threat intelligence sharing ecosystem is more than transferring data; it's about building a collective defense architecture:

- In the US, AIS/JCDC enhance operational speed and private-sector integration.
- In Europe, the CSIRT network and EU-CyCLONe enable solidarity during multinational crises.
- STIX/TAXII and MISP standardize the technical layer, turning collaboration into concrete action.

In conclusion, intelligence planning conducted without a sharing ecosystem is like moving forward in the dark with a single flashlight. Sharing mechanisms, by contrast, illuminate the threat landscape by switching on dozens of flashlights at once.

2.1.8 Risk Tolerance and Corporate Business Objectives

One of the most critical, yet often overlooked, dimensions of the threat intelligence planning phase is its direct alignment with the organization's risk tolerance (*risk appetite*). Organizations may prioritize the same type of attack at different levels. This difference arises not only from the sector in which they operate but also from corporate business objectives, regulatory pressures, and even the strategic vision of the board of directors.

The *risk appetite framework* defines how much risk an organization is willing to accept and at which point intervention becomes necessary. This framework ensures that PIR/SIR definitions in CTI planning are aligned with the organization's business objectives:

- **If an organization has a low risk appetite**
  PIRs/SIRs are defined in a more detailed and comprehensive manner; even minor fluctuations in the threat landscape are incorporated into planning.
- **If an organization has a high risk appetite**
  CTI planning remains at a more macro level; only events that truly threaten business continuity are prioritized.

In scenarios where intelligence isn't aligned with risk appetite, CTI reports fail to resonate within the organization. This is because information that is technically accurate doesn't carry decision-making value for the board and business units unless it becomes strategically meaningful.

To demonstrate, let's examine a few sectoral risk appetite scenarios:

- **Financial sector**
  Banks and investment institutions generally have a very low risk appetite due to customer trust and regulatory compliance requirements.

  Example: In the US, a bank elevates even a small phishing campaign to the PIR level, as it may lead to reputational damage.

In Europe, due to DORA, financial institutions are required to adopt a zero-tolerance approach toward operational risks. In this context, CTI planning elevates even the slightest ransomware indicator to the strategic level.

- **Healthcare sector**

  The risk appetite of healthcare organizations is critically low due to the human life factor, although they may not have compliance mechanisms as extensive as those in finance.

  Example: In the US, under HIPAA, a hospital data breach constitutes a legal and ethical crisis. Therefore, PIR definitions directly include the question, "Which ransomware groups are targeting patient records?"

  In Europe, especially following the pandemic, the increase in APT attacks targeting the healthcare sector has further emphasized this low risk tolerance. In healthcare, threat intelligence prioritizes factors that directly affect patient safety, not just financial loss.

- **Energy sector**

  Risk appetite in the energy sector is more complex. While minor outages may be tolerated, large-scale APT attacks or supply chain exploitations escalate to the level of national security.

  Example: In Europe, an energy company may view small DDoS attacks as routine operational risks, but classify a campaign targeting the Incident Command System (ICS)/ Supervisory Control and Data Acquisition (SCADA) system as a strategic threat at the PIR level.

  In the US, the Colonial Pipeline attack fundamentally changed perceptions of risk appetite in the sector, leading even minor vulnerabilities to be viewed as critical infrastructure risks. In the energy sector, CTI planning must align with both corporate risk appetite and geopolitical risk considerations.

When threat intelligence planning isn't aligned with organizational risk appetite, two critical errors emerge:

- **Over-sensitivity**

  Low-level threats are elevated to the PIR level, resulting in resource waste and alert fatigue.

- **Over-indifference**

  Real risks are ignored, resulting in business continuity disruptions and reputational damage.

Effective planning lies in striking the right balance between these two extremes, in accordance with the organization's risk framework. Aligning CTI planning with organizational risk appetite ensures that intelligence activities remain relevant to business priorities and decision-making.

2.2 Collection: Active and Passive Techniques

The collection process functions as the "eyes" through which organizations perceive the threat landscape. Just as it's difficult for the brain to draw conclusions when the eyes don't see correctly, it's equally difficult for an intelligence team working with poorly collected data to produce meaningful results. For this reason, the reliability, integrity, source, and contextual values of every piece of collected information are of critical importance. Simply gathering large volumes of data isn't sufficient; what matters is that the data accurately reflects the threat landscape, is current, is verifiable, and highlights risks specific to the organization.

In practice, the value of the collection phase becomes evident in cybersecurity operations, where systematically collected data supports early threat detection.

Collection methods are generally categorized into two main groups: passive techniques and active techniques. We'll discuss them in detail in the following sections, but first, let's spend some time understanding why the collection phase is so important.

2.2.1 Strategic Importance of the Collection Phase

The collection phase is more than just a starting point in the intelligence lifecycle; in fact, it's the most critical building block that determines the robustness of the entire process. The quality of the data obtained at this stage directly influences decision-making processes. Simply collecting data in its raw form has no intrinsic value; what truly matters is the ability to contextualize this data and transform it into reliable insights that can guide decision-makers. For this reason, the collection phase should be regarded as a strategic investment.

The principle of "right data, right decision" is a repeatedly proven reality within the discipline of threat intelligence. If the collected data is inaccurate, incomplete, or outdated, even the most advanced analytical methods will produce flawed conclusions. For example, in the past, many organizations overlooked attacker activities by relying solely on firewall logs. When these logs weren't enriched with data from appropriate sources, it later emerged that attackers had infiltrated through backdoors and remained inside networks for months. This situation was strikingly observed in the 2013 Target data breach. The company's security systems had recorded certain signals of the attack, but because the data wasn't properly collected and correlated, these signals weren't reflected in decision-making processes. As a result, the credit card information of millions of customers was compromised.

Another notable example is the SolarWinds supply chain attack uncovered in 2020. In this case, attackers managed to infiltrate thousands of organizations by injecting malicious code into legitimate software updates. Had organizations collected only standard security logs, this attack might have gone unnoticed for a much longer period. However, some organizations were able to detect unusual communication patterns earlier because they carefully collected passive Domain Name System (DNS) records and correlated network traffic.

This example demonstrated that collection is a strategic element that directly impacts decision-making processes.

Another important consideration is the source and reliability of the data. Information obtained from unreliable or unverified sources can mislead organizations. In particular, the accuracy of information obtained from dark web forums must be questioned. In the past, some ransomware groups attempted to intimidate organizations by pretending to leak data they hadn't yet actually obtained. If such information had been incorporated into intelligence reports without verification, organizations might have taken unnecessary and misguided actions driven by panic. Therefore, source validation and data integrity checks are critical during the collection phase.

A field example that further illustrates the strategic importance of collection is the attacks carried out by APT29 against the healthcare sector. In these attacks, many organizations focused solely on collecting security incidents while overlooking user behavior. However, credential theft activities could have been detected early through careful collection of authentication and login patterns. This deficiency caused some healthcare organizations to remain compromised for months without detecting the presence of attackers within their networks.

In conclusion, the collection phase isn't a simple data accumulation task of. It's a step that shapes an organization's threat perception, provides reliable guidance to decision-makers, and generates strategic value. The key points to consider are as follows:

- **Accuracy**
  Incorrect or incomplete data leads to incorrect decisions.
- **Timeliness**
  Outdated data can obscure current threats.
- **Source reliability**
  Unverified data can mislead organizations.
- **Context**
  Data must not be evaluated in isolation, but within the broader threat landscape.

Every mistake made during the collection phase undermines the rest of the chain; conversely, a properly executed collection process enables organizations to stay one step ahead of attackers.

2.2.2 Passive Collection Techniques

Passive methods are approaches based on observing existing traces and data without establishing direct contact with attackers during the intelligence collection process. The goal of these methods is to gather clues about threat actors' activities without their awareness. In this way, security teams can obtain information without exposing their own assets or attracting the attention of attackers. The greatest advantage of passive collection lies in maintaining invisibility.

The methods used in this approach require no direct intervention by the organization. For example, information published on Internet forums or social media platforms, posts made by attackers on dark web marketplaces, passive DNS (pDNS) records, or internal log files can all be considered passive sources. Insights derived from these data sources play a critical role in identifying attackers, understanding the techniques they use, and revealing potential targets.

One of the strengths of passive collection techniques is their ability to provide a broad spectrum of data. Through these methods, it's possible to learn about threat actors' behavior without directly provoking them. For instance, pDNS solutions allow tracking of malicious domains previously used by an attacker and identifying other domains associated with the same infrastructure. Similarly, dark web monitoring tools make visible which data attackers are selling or which organizations they are targeting.

Another significant value of passive collection is that it doesn't compromise operational security (OPSEC). While attackers may notice active scanning activities, data collected through passive methods is generally derived from publicly available sources or digital traces left unknowingly by attackers. This allows security teams to expand their intelligence repository without drawing the adversary's attention.

Of course, passive collection also has limitations. Data obtained through these methods is typically retrospective; that is, it contains traces of activities attackers have already carried out or conducted in the past. Therefore, passive collection doesn't always guarantee direct prediction of future attacks. Nevertheless, understanding past behavioral patterns and deriving trends from them provides a significant advantage in anticipating potential future actions of threat actors.

So, let's take a closer look at the methods used in passive information collection:

- **Open-source intelligence (OSINT)**
  Forums, paste sites, social media platforms, and publicly shared IOCs are among passive information sources. For example, victim lists or leaked email addresses published by ransomware groups on dark web forums enable organizations to detect attack risks in advance.
- **Log and DNS analysis**
  Internal log records or pDNS solutions can reveal traces left by threat actors. For instance, an unusual DNS query may indicate covert C2 communication.
- **Dark web monitoring**
  Marketplaces on the dark web are among the most frequently used environments by cybercriminals. The sale of stolen data or malware kits may serve as an early indicator of potential attacks.

When examined in depth under three main examples, the critical role of passive collection within the intelligence lifecycle becomes clearer.

Beyond OSINT: Passive Correlation of Social Engineering Signals, Domain Registration Patterns, and WHOIS Data

OSINT is often considered limited to social media posts, forum discussions, and public IOC lists. However, in advanced passive collection activities, the correlation of social engineering signals, domain registration patterns, and Who Is (WHOIS) data provides far more valuable clues.

The objectives of this analysis include the following:

- To detect attackers' attempts to create fake identities at an early stage
- To identify domains registered for phishing campaigns at the moment of registration
- To map the infrastructure of the same attacker group by correlating recurring email addresses, phone numbers, or IP blocks found in WHOIS records

Real-World Examples

During the 2016 US elections, fake email campaigns conducted by Russia-based APT groups were identified through domain registration information. Similar email patterns used in WHOIS records linked multiple phishing domains together. This made it possible to reveal that campaigns appearing to be independent were actually associated with the same threat actor.

In 2022, phishing attacks targeting certain financial institutions were detected at an early stage thanks to similar domain registration patterns. Although the attackers registered domains under different top-level domains (TLDs; e.g., `.xyz` and `.info`), recurring templates in WHOIS records exposed the attackers' infrastructure.

Beyond the Dark Web: Telegram Groups, Discord Communities, and Closed Marketplaces

While dark web monitoring has long been a standard component of threat intelligence, attackers' communication methods have diversified. Today, forums on the Tor network, Telegram groups, Discord communities, and closed marketplaces have become some of the most critical sources for passive collection.

The primary objectives of monitoring these communication channels include the following:

- To track victim lists published by ransomware groups
- To monitor the sale of stolen data, credentials, and malware
- To identify emerging attack groups or collaborating actors at an early stage

Real-World Examples

In 2021, leaks associated with the Conti ransomware group were publicly disseminated through Telegram channels. Security researchers who passively monitored these channels were able to identify the group's attack methods and victims in advance and warn many organizations.

In 2023, the sale of patient data belonging to certain healthcare organizations was first observed not on dark web forums, but within a closed Discord community. Organizations that collected this information through passive methods were able to take countermeasures against potential attacks more quickly.

The Power of Passive DNS: Identifying Dormant Domains Linked to the Same Infrastructure

pDNS data records all domain names that have historically been associated with a given IP address, thereby revealing attackers' infrastructure relationships. This method is particularly critical for discovering so-called dormant domains. A *dormant domain* refers to a domain registered in advance by an attacker and held in reserve for future use.

pDNS analysis supports several intelligence objectives, including the following:

- To uncover different domains connected to the same C2 infrastructure
- To detect domains prepared before malicious activity begins
- To anticipate the long-term operational plans of APT groups

Real-World Examples

In 2017, the NotPetya attack was more comprehensively illuminated through pDNS analysis. It was discovered that malicious domains used by the attackers were linked to the same IP addresses as other domains that had been registered in advance. This finding confirmed that the attack had been planned to target Ukraine-based organizations.

In the APT41 campaign identified in 2020, hundreds of domains that attackers had registered in advance for use across different sectors were exposed through pDNS correlation. Although some of these dormant domains hadn't yet been used, they were blacklisted by the security community, preventing potential future attacks.

2.2.3 Active Collection Techniques

Active methods aim to directly observe the threat environment. However, this approach carries an inherent risk: attracting the attention of threat actors. Here are a few examples:

- **Honeypot and honeynet deployment**
  These systems are set up to attract attackers and serve as effective tools for observing their behavior. For example, attacks directed at a honeypot designed to resemble a misconfigured database can reveal which vulnerabilities are being targeted most frequently.
- **Active scanning**
  Network scans conducted using tools such as Nmap or Shodan enable the identification

of potential vulnerabilities. However, attackers may detect these scans and take countermeasures.

- **Protocol-based traffic collection**
  NetFlow or packet capture (PCAP)-based traffic analyses make unusual movements within an organization's network visible. For example, an endpoint transferring unusually large volumes of data may indicate an attempted data exfiltration.

Honeypot systems shouldn't be viewed in CTI as passive tools that record attackers' behavior. These systems also play a unique role in understanding attackers' motivations and target priorities. Motivation is key. For example, attackers showing intense interest in specific types of files placed within a honeypot environment may indicate whether their intent is financial gain or espionage. A ransomware group that exhibits only encryption behavior points to one objective, whereas a state-sponsored threat actor focusing on sensitive documents signals an entirely different goal. For this reason, honeypots aren't just technical observation points, but powerful analytical platforms that enable the understanding of attacker psychology and operational motivations.

Active scanning, on the other hand, plays a critical role in strengthening an organization's security infrastructure. However, improper execution of such scans can introduce serious risks. Generating excessive noise can attract attackers' attention and provoke them. For instance, frequent and intensive port scans may signal to an attacker that a security assessment is underway. As a result, the attacker may adopt more sophisticated evasion techniques or accelerate the attack, exploiting the target before the defense team has time to respond. This risk becomes far more severe in sensitive environments such as energy or healthcare systems. Therefore, precision and controlled execution are of paramount importance in active scanning activities.

PCAP and NetFlow analyses are powerful methods used to detect anomalous patterns in network traffic. These analyses aren't limited to examining raw traffic data; when supported by machine learning techniques, they become far more predictive intelligence tools. Subtle anomalies that are virtually impossible for the human eye to detect—such as low-volume but persistently recurring data flows at specific times—can be easily identified by machine learning algorithms. This enables early detection of intrusion preparations or data exfiltration attempts. Especially given the long-term stealth (living-off-the-land) techniques used by modern APT groups, integrating AI-driven anomaly detection mechanisms into PCAP/NetFlow analysis allows organizations to stay one step ahead of attackers.

Strategic Balance

Passive methods are low risk but generally produce results more slowly, whereas active methods provide richer data at the cost of increased operational risk. Organizations can achieve maximum benefit by using these two approaches in a balanced manner. For example:

- A financial institution uses passive methods to detect leaks related to its brand on the dark web.
- Simultaneously, the institution deploys honeypot systems to observe new phishing techniques used by attackers.

2.3 Processing and Initial Analysis

The data obtained during the collection phase, as is often the case, isn't always in a form that can be used directly. Within large-scale datasets, there are false positives, duplicate records, and context-free information. This is precisely why the processing phase serves as a critical filter in threat intelligence. The goal at this stage is to reduce noise and enable the analyst to reach information that is truly valuable.

In the following sections, we'll explore the key steps in the processing phase and initial analysis, resulting in a reliable output.

2.3.1 Data Normalization

Log and telemetry data collected from different sources often use different formats. For example, event logs recorded on Windows Server have their own XML-based structure. An intrusion detection system (IDS) used in the same environment—through solutions such as Snort or Suricata—typically outputs data in JavaScript Object Notation (JSON) or in a custom rule format. Intrusion prevention system (IPS) devices, on the other hand, generate logs for similar events in entirely different formats. This diversity can create significant incompatibility in threat intelligence processes. This is exactly where normalization comes into play. Through normalization, these different formats are converted into a common schema, enabling analysts to read the data as a single, unified whole.

Let's consider a practical example: Failed login attempts in Windows Server logs are recorded with different timestamps and error codes. At the same time, an IDS device detects anomalous traffic from the same IP address and records it separately in JSON format. If these two data sources aren't normalized onto a single timeline, what is actually the same attack attempt may appear as two separate incidents. However, when timestamps are standardized into a common format and shared field names (source IP, destination port, username) are mapped, the complete picture of the attack becomes clear.

Threat intelligence platforms (TIPs) are powerful tools that automate this process. For instance, platforms such as MISP or Anomaly can bring together Windows Server logs, IDS/IPS logs, and external threat intelligence feeds under a single schema. This enables an organization to evaluate millions of lines of data from different sources through a "single language." In this way, a hash value may appear as merely a file in Windows logs, while IDS records reveal malicious traffic behavior associated with the same hash. Without normalization, these two pieces of information remain disconnected; but through a shared schema, it becomes clear that the attack is occurring simultaneously at both the endpoint and network levels.

Especially in large-scale organizations, normalization is more than just convenient; it also reduces the workload of SOCs. Standardizing logs flowing from various systems enables correlation rules to function correctly and reduces false positives. When the high-volume alerts generated by IDS/IPS systems are placed into the same template as Windows Server event logs, security teams can prioritize attacks much faster. In practice, data normalization enables more effective correlation across security logs and improves early detection capabilities.

2.3.2 Tagging and Contextualization

An IP address, hash value, or domain record standing alone may appear among millions of log lines as nothing more than a number or an alphanumeric string. In such a case, the meaning an analyst can derive from that data is limited. However, when properly tagged, the same data becomes a critical link in an attack chain. For example, a seemingly ordinary IP address, when labeled as **APT29/Spear Phishing Campaign**, ceases to be a simple connection record; it becomes a prioritized indicator directly associated with a known operation of a specific threat actor—an element of critical importance in the analyst's decision-making process.

The first key consideration in the tagging process is the accuracy of context. Incorrect or hastily applied tags can misrepresent the data. For instance, if an IP address used only for spam traffic is mistakenly labeled as an APT indicator, the security team may be misled and waste resources unnecessarily. Therefore, source reliability, verification mechanisms, and cross-checks must always be applied when tagging.

Another critical point is consistency. If indicators used in different campaigns of the same threat actor are labeled under different names across different platforms, analysts can't maintain coherence in the overall picture. To solve this problem, integration among endpoint detection and response (EDR) and extended detection and response (XDR) systems and TIPs is essential. For example, an XDR solution may label suspicious file hashes collected from endpoints as **Suspicious**. But if the same hash is recorded in a TIP with the label **FIN7/Point-Of-Sale Malware**, the analyst can see with a single click that this hash is more than suspicious—it's directly linked to an organized criminal group targeting the financial sector.

EDR systems have repeatedly proven the importance of tagging in real-world environments. During the ProxyLogon vulnerabilities disclosed in 2021, EDR solutions in many organizations detected unusual web shell files. At first glance, these findings looked like mere technical anomalies. However, when TIPs tagged these IOCs as **APT Hafnium Exploit Campaign**, security teams understood that the incident wasn't a local case, but part of a state-sponsored operation. Because the correct context was provided, the severity of the event was quickly recognized and the response time was reduced dramatically.

On the XDR side, contextualization becomes stronger by unifying data collected from different security layers under a single umbrella. A domain record may appear on its own as nothing more than a query count. But when XDR correlates it with IPS logs, firewall alarms, and email gateway records and forwards it to the TIP, the same domain may match the **Cobalt Strike C2 Infrastructure** label. As a result, the analyst understands that this isn't just a frequently queried domain, but a critical threat source associated with adversary infrastructure.

Another important point here is the time factor. If tags aren't kept up to date, the analyst may focus on outdated threat campaigns and miss real attacks. For example, if a malware hash used in 2017 is still labeled years later as an **Active Threat**, incorrect prioritization can occur. For this reason, it's critical that TIPs indicate the validity period of IOCs through regular updates.

Tagging is also a powerful tool for prioritization. An SOC analyst dealing with thousands of alerts per day can't examine every record in depth. However, if IOCs are properly contextualized, an IP address labeled **APT29/Credential Harvesting**, for example, will be handled before ordinary brute-force attempts. This method ensures that resources are directed toward the most critical threats.

A case example is the SingHealth data breach in 2018. During the incident, EDR systems recorded unusual movement in certain network segments. Initially, this was considered ordinary network congestion. However, contextual tags from TIPs later showed that this traffic was associated with known C2 infrastructure used by APT41. This critical information revealed that the attack was a state-sponsored operation and accelerated the government's response process.

Tagging and contextualization aren't for "decorating" data; instead, they serve as the meaning-making process in threat intelligence. When data collected by EDR/XDR systems is combined with the tags provided by TIPs, an ordinary log line becomes a critical signal that is part of an attack. Therefore, the points to pay attention to are as follows:

- Avoid incorrect tagging and ensure verification.
- Maintain consistency; that is, the same threat must be labeled the same way across different platforms.
- Keep tags current; IOCs that have lost validity must be revised.
- Base priorities on context; indicators associated with critical actors must be elevated.

Effective threat intelligence goes beyond collecting IOCs; they must be accurately tagged and contextualized.

2.3.3 Automated Analysis of Malware Samples

Automatically analyzing malware samples obtained during collection is one of the most critical steps in threat intelligence because malware developed by attackers often attempts to bypass an organization's defense mechanisms through different variants. For this reason, manual analysis alone isn't sufficient; sandbox environments come into play here as well. A *sandbox* enables secure observation of a malware sample's impact by executing it in an isolated virtual environment.

For example, when ransomware is submitted to a sandbox, the file extensions it encrypts, the new files it creates, and the registry keys it modifies all become immediately visible. When malware such as TrickBot or Emotet is executed in sandbox environments, they trigger PowerShell commands and attempt lateral movement across the network. Such findings reveal the malware's behavior and enable the direct updating of the organization's security policies.

Integration with email security systems is also an indispensable part of this process. Today, a significant portion of malware spreads through email attachments. For example, an email containing a suspicious Word document attachment is automatically sent to a sandbox environment. The sandbox executes the document's macros and observes whether any malicious behavior occurs. If the macro attempts to connect to a known C2 server, this information is immediately shared with the email security system. In this way, the user is warned and other emails from the same campaign are automatically blocked.

Network and firewall integrations turn the results of this process into concrete defensive actions. Malicious IP addresses, domains, or hash values identified through sandbox analysis are forwarded directly to the firewall and IPS/IDS systems. For instance, if a malware sample analyzed in the sandbox is observed continuously sending beacon traffic to an IP address in China, that IP can be automatically added to the firewall blacklist, thus blocking the analyzed sample and proactively preventing all potential variants using the same infrastructure.

An incident in a European bank in 2019 clearly demonstrated the importance of this process. The bank's email security system forwarded fake invoice attachments sent to customers into the sandbox. The sandbox analysis revealed that the attachments were actually a TrickBot variant that modified specific registry keys once the attachments were opened. This information was instantly pushed to the firewall and EDR systems, stopping the attack before it spread. Without sandbox integration, this malware would've caused serious damage within the bank's internal network.

Automated analysis of malware samples accelerates an organization's defensive reflexes and makes them proactive. Sandbox environments reveal malware behavior, email security systems use this knowledge at the first point of contact, and network/firewall integrations apply the findings directly in the field. When this chain functions correctly, it forms one of the most effective defense lines against modern threats.

2.3.4 Enriching Indicators of Compromise

IOCs, on their own, are often just technical records. An IP address, a file hash, or a domain name carries limited meaning by itself. For example, it's impossible to make a decision based solely on an IP address—the address may belong to a malicious C2 server, or it may be part of an ordinary content delivery network (CDN) infrastructure. This is where IOC enrichment comes into play. Enrichment helps the analyst make faster and more accurate decisions by adding additional context and layers of evidence to raw indicators.

Online analysis and data-sharing services are among the best-known examples of this process. VirusTotal reveals how a hash value is flagged by dozens of antivirus engines and which campaigns it's associated with. Similarly, sandbox services such as Hybrid Analysis or ANY.RUN provide additional context by sharing dynamic analysis reports of files. For example, a hash that looks like `d41d8cd98f00b204e9800998ecf8427e` by itself may reveal connections to multiple malware campaigns when scanned on VirusTotal. In this way, the analyst sees a file signature and the entire attack ecosystem.

Open-source threat feeds are also powerful sources for enrichment. For example, Zeus or URLhaus feeds provided by abuse.ch associate specific IPs and domains with malicious campaigns. These IOCs can be directly transferred into the organization's firewall or IDS/IPS systems. A domain may appear as a query in pDNS records, but open-source feeds can reveal that it's part of an active phishing campaign. This helps the organization reduce false positives and prioritize threats more quickly.

Among enterprise solutions, Microsoft Defender for Endpoint is a strong example that automates IOC enrichment. For example, when a hash of an executable file detected on an endpoint is sent to Defender, the system compares it with its own threat intelligence pool. If the same hash has previously been seen in TrickBot or Emotet variants, the file is automatically tagged as a threat. In this way, the security analyst learns that the file exists and that it's a known tool used by attackers for credential theft.

Similarly, the CrowdStrike Falcon platform adds deep context to IOCs. For instance, an IP address may have been recorded as only a **Connection Attempt**. But if Falcon's threat intelligence engine indicates that the same IP was previously used as a C2 server in an APT campaign, that record becomes a high-priority alert. Without this context, traffic that appears ordinary can be correctly evaluated as part of an attack chain through enrichment. IOC enrichment enhances the value of raw indicators by providing the context needed for more effective analysis.

2.3.5 Output: A Processed and Reliable Dataset

At the end of the processing phase, the analyst now has a cleansed, verified, and contextualized information repository—not just a pile of data. This dataset minimizes false positives, removes duplicate records, and carries a high degree of integrity and reliability. As a result, the analyst can focus on attack patterns without wasting time, determine which IOC truly represents a threat, and take action faster.

Malware analysis and sandbox systems are among the strongest tools in this process. These isolated environments, where malware samples are executed automatically, provide direct evidence of attacker behavior. For example, when a suspicious Excel file sent to a bank is placed in a sandbox, it triggers macro commands and attempts to establish an external C2 connection. This output provides far more than the **Suspicious File** label; it gives the analyst a detailed behavioral profile, including which IP address is contacted, which registry keys are modified, and which files are created. Data enriched with this context is then directly added to the IOC list.

The integration of TheHive and Cortex is one of the key platforms that makes processed datasets operationally valuable. While TheHive centralizes incident management, Cortex activates different analysis engines to validate and contextualize IOCs. For example, when an IP address is entered into a case, Cortex automatically queries it across sources such as VirusTotal, AbuseIPDB, and MISP. The final output delivered to the analyst shows a suspicious IP with information about which campaigns it has been seen in, which malware it has been associated with, and in which geographic regions it has been active. This increases accuracy and prevents false positives.

TIPs such as MISP, ThreatConnect, Anomali, and ThreatQ strengthen the sharing and contextualization dimension of the processed dataset. For example, a hash uploaded to MISP may show that the same hash has been seen in other organizations, revealing the scale of an attack at the campaign level. ThreatConnect and ThreatQ provide analysts with rich context linked to threat actors' tactics, techniques, and procedures. Through these platforms, the IOC and the attack chain to which it belongs become clearly understood.

SOAR solutions (e.g., Cortex XSOAR and Splunk SOAR platforms) convert this processed dataset directly into automation. For example, a domain flagged as malicious in a sandbox can be sent as an automatic firewall rule through Cortex XSOAR. Similarly, Splunk SOAR can run an automated playbook using processed IOCs: quarantine suspicious emails, isolate the relevant endpoint in the EDR, and update firewall rules immediately. Thus, the analyst's cleansed dataset is used both for reporting and to directly activate defense mechanisms.

Real-World Example

The 2017 NotPetya attack is a striking example of how unprocessed logs can lead to disaster. While many organizations failed to detect the attack, some security teams normalized their logs, executed malware in sandboxes, and shared IOCs through MISP. These teams protected their own organizations as well as others because a processed and reliable dataset also formed a foundation for collective defense.

A processed and reliable dataset isn't just an intermediate output of threat intelligence; it's a strategic asset at the center of the defense ecosystem. Malware sandbox systems reveal behavior, TheHive and Cortex integration validate IOCs, MISP and similar platforms expand context, and SOAR solutions convert this knowledge into automated action. At the end of this chain, the analyst is left with an information pool that is verified, doesn't waste time, and is directly actionable.

At this stage of the processing workflow, several key questions help determine whether collected data has been properly prepared for effective analysis:

1. In threat intelligence, the data collected during the collection phase isn't used directly in its raw form; it contains duplicate records, context-free information, and false positives. For this reason, the processing phase serves as a critical filter. The first step, data normalization, converts logs from different sources into a common structure, making the complete picture of an attack visible. So, can you perform healthy correlation in your own systems without consolidating logs flowing in different formats into a single schema?
2. The next step, tagging and contextualization, transforms an IP address or hash that appears meaningless on its own into a powerful signal linked to known operations of threat actors. But could an incorrect or outdated tag pull you away from real threats and steer you toward the wrong resources?
3. Automated analysis of malware samples exposes attacker behavior in sandbox environments and helps update defense policies. Detecting that a macro command is secretly connecting to a C2 server offers a critical opportunity to take the right precautions. In your view, what risk would your organization face if email attachments are opened without sandbox integration?
4. The subsequent IOC enrichment process increases the value of a raw indicator. Through VirusTotal, MISP, or commercial platforms, an ordinary hash can be linked to multiple malicious campaigns. So, are you providing your analysts only raw IOCs, or are you enabling them to make reliable decisions by placing these IOCs into context?
5. At the end of all these steps, the processed and reliable dataset becomes an information pool in which false positives are minimized and the analyst can perform fast and meaningful analysis. As in the NotPetya attack, organizations that processed their logs and shared their IOCs strengthened the entire ecosystem, not only their own defense. Do you think your organization today has a cleansed, verified, and contextualized dataset, or are you still living with the risk of missing critical signals among raw data?

2.4 Interpretation and Dissemination

The real value of data collection and analysis, one of the most critical stages of the intelligence lifecycle, emerges when that data is made meaningful and shared with the right people, in the right format, at the right time, and in the right context. As an intelligence product transforms raw information into processed intelligence, it becomes a potentially powerful tool. However, the functionality of this tool remains limited unless it reaches the intended audience who will use it. If analysis results obtained through significant effort don't arrive at the decision-makers' desks or the operational teams' screens in a timely and understandable manner, even the most valuable findings become idle. This creates the institutional security problem of becoming ineffective amid an abundance of information.

Intelligence that doesn't reach the right person doesn't produce actionable decisions; a report that isn't presented in the right format doesn't meet the recipient's needs; and information that isn't delivered at the right time is often realized only after it's too late. For example, while an attack chain can be completed within minutes, an IOC report shared days later—even if technically correct—is operationally ineffective. Therefore, the value of intelligence is measured by analytical quality plus audience alignment, contextual accuracy, timely delivery, and format suitability.

As a result, this stage of the intelligence lifecycle is a strategic inflection point that directly affects an organization's security posture. Delivering analyzed data to the right decision-making bodies, on time and in an understandable way, is a critical defense mechanism that increases an organization's ability to withstand threats. Otherwise, even the most comprehensive datasets turn into forgotten digital shelfware—data piles that don't translate into action and gradually lose their value.

2.4.1 Analysis and Interpretation

At this stage, close attention is paid to what the outcomes derived from raw data actually mean. Analysis and interpretation go further than reading data to place that data into context, relate it to other data, prioritize the data, and ultimately produce outputs that can be converted into operational or strategic action. Following is an expanded explanation of the core components of this process, including methods and practical examples.

Contextualization

The first step in interpreting collected indicators is contextualization. An IOC (e.g., an IP address, domain, or file hash) provides limited information on its own. Its analytical value emerges only when it's interpreted alongside additional contextual data that clarifies how the indicator relates to the broader threat environment. Several factors should be considered when contextualizing an IOC:

- The timestamp of detection (when the indicator was observed)
- The source of the information (e.g., threat feeds, telemetry systems, or analyst reports)
- The type of targeted assets (e.g., SCADA systems, user workstations, or domain controllers)
- Geographic and infrastructure context, including Autonomous System Numbers (ASNs) or hosting providers
- The historical frequency with which the indicator has appeared in prior observations

To illustrate this process, consider an IP address that appears on a blacklist. While this may initially indicate suspicious activity, the operational priority of the indicator increases significantly when it's correlated with additional evidence. For instance, if the IP address is associated with a technique in the MITRE ATT&CK framework (e.g., C2 beaconing), linked to other suspicious IPs within the same ASN, and observed making repeated connections to a specific port during defined time windows, the analytical significance of that indicator becomes much clearer.

Various analytical tools and data sources can support contextualization. Common examples include pDNS records, WHOIS and ASN lookups, historical DNS data, sandbox analysis reports, reputation scores from platforms such as VirusTotal or AlienVault, and correlation outputs generated by SIEM systems.

Trend Analysis

Beyond individual indicators, analysts must also examine broader patterns in attacker behavior. Trend analysis focuses on identifying how threat actors' TTPs evolve over time and how these developments affect organizational risk.

Trend analysis can reveal several types of operational insights, including the following:

- Increases in the use of specific exploit kits or malware families
- Shifts in phishing techniques or social engineering approaches
- Changes in the sectors, regions, or asset types targeted by attackers

These insights are valuable at multiple levels of intelligence production. In the short term, trend analysis supports tactical decisions such as adjusting detection rules or updating firewall restrictions. Over longer time horizons, it contributes to strategic planning by identifying emerging threat patterns that may influence security investments or defensive architecture.

In practice, analysts often conduct trend analysis by applying time-series methods to historical IOC datasets. By examining how the frequency of certain indicators or techniques changes over time and correlating those changes with geography, sector, and asset classes, security teams can identify patterns that may justify proactive defensive measures.

Enrichment and Correlation

While contextualization provides background information for an indicator, enrichment and correlation extend this process by connecting the indicator to additional intelligence sources and related security events.

IOC enrichment involves augmenting raw indicators with supplementary data obtained from internal systems or external intelligence services. Typical enrichment activities include retrieving sandbox analysis results for file hashes, examining pDNS history for domains, collecting Border Gateway Protocol (BGP) or ASN information for IP addresses, or analyzing email authentication records such as DomainKeys Identified Mail (DKIM), Sender Policy Framework (SPF), and Domain-based Message Authentication, Reporting, and Conformance (DMARC).

Correlation then combines multiple telemetry sources to identify relationships between events associated with the same indicator. Analysts commonly correlate logs from endpoint detection platforms, firewalls, proxy systems, email gateways, and DNS monitoring tools. For example, if the same file hash appears both in an email attachment and on an endpoint within the network, this correlation may indicate that the attack progressed through multiple stages of the intrusion chain.

Taken together, enrichment and correlation transform isolated indicators into meaningful evidence of adversary activity.

Evaluating the Analytical Value of an IOC

When assessing the analytical value of an IOC, analysts should evaluate several key factors to determine whether the indicator provides reliable intelligence. These evaluation steps typically include the following:

1. Validating the reliability and credibility of the source
2. Adding temporal and asset-related context to the indicator
3. Enriching the indicator with historical and related intelligence data
4. Correlating the indicator with telemetry using SIEM or SOAR systems
5. Prioritizing the findings according to the appropriate reporting level (operational, tactical, or strategic)

Following these steps helps ensure that intelligence products are based on well-validated evidence rather than isolated or ambiguous signals.

Risk and Uncertainty Management

Finally, threat intelligence analysis must account for uncertainty when interpreting indicators. Not every IOC represents a confirmed attack, and failing to consider uncertainty can lead to unnecessary responses or inefficient use of resources.

When assessing IOC-related risks, analysts should therefore incorporate factors such as data reliability, likelihood of false positives, and the confidence level associated with the intelligence source. One practical approach is to classify indicators according to confidence levels. For example, organizations often categorize IOCs as follows:

- **Validated**
  Strong evidence confirms malicious activity.
- **Probable**
  There's credible but not yet confirmed threat activity.
- **Suspicious**
  Indicators require further monitoring or investigation.

Establishing these categories allows organizations to define appropriate response procedures for each confidence level, balancing automated actions with human analysis to ensure that security resources are used efficiently.

In short, the analysis and interpretation stage means transforming raw indicators into operationally meaningful, prioritizable, and actionable intelligence products by combining enrichment, correlation, trend analysis, and risk assessment. An IOC alone provides limited information; but a process that evaluates that IOC in the right context significantly strengthens your organization's detection-and-response capability and the reliability of strategic decision-making.

2.4.2 Dissemination

Even if intelligence analysis processes are carried out with great care, the outputs can't realize their true value unless they are disseminated through the right methods. Intelligence is beneficial when analysis is correct and when the results reach the right people, in the right format, with the right context, and at the right time. In other words, if dissemination—the final link of the intelligence cycle—fails, nearly all prior efforts are largely wasted. Therefore, this stage should be seen not just as a technical distribution process, but as a critical art of communication that directly affects an organization's operational resilience, decision speed, and strategic awareness.

Real-World Example

A practical scenario demonstrates why effective interpretation and dissemination are essential. In one case, a bank's SOC received a large IOC dataset from an external intelligence-sharing network. However, approximately 60% of these indicators were outdated. As a result, analysts were overwhelmed with hundreds of false alerts generated by obsolete data.

To resolve the issue, the organization implemented a lifecycle management process that filtered incoming indicators based on validation and timeliness. Only indicators confirmed to be active and relevant were distributed to operational systems. After applying this filtering process, the false alarm rate dropped dramatically-from 60% to approximately 5%.

This example illustrates a key lesson in threat intelligence operations: Without proper interpretation and structured dissemination, intelligence data can easily become operational noise rather than a meaningful contribution to organizational security.

Core Principles of Intelligence Dissemination

For threat intelligence to generate real value, analysis results must be communicated effectively to the relevant stakeholders. Along with technical accuracy, this also requires a clear understanding of how intelligence should be contextualized, structured, and delivered. Several core principles guide the dissemination of intelligence products within organizations.

The first principle is contextualization. An IOC, such as an IP address, domain name, or file hash, provides only limited insight when viewed in isolation. Its true significance emerges when it's interpreted within a broader analytical framework. For example, an IOC may become far more meaningful when mapped to a specific technique in the MITRE ATT&CK framework, associated with threat actor infrastructure through the Diamond Model, or positioned within a wider attack chain. In practical terms, a blacklisted IP address may appear to be only a minor signal on its own. However, if that same IP address is also linked to a C2 infrastructure previously used by an APT group targeting the financial sector, both the credibility and operational priority of the threat increase significantly.

A second essential principle is alignment with the intended audience. Intelligence products must be adapted to the needs, responsibilities, and technical background of the recipient. Different stakeholders require different levels of detail and different forms of interpretation. For example, operational analysts working within an SOC typically require highly technical information such as detailed IOC lists, log excerpts, and event timelines. SOC managers or incident response leads, on the other hand, benefit more from summaries describing the scope of an incident, relevant TTP patterns, and resource allocation requirements. Executive leadership usually requires an even higher-level perspective, focusing on strategic considerations such as financial impact, regulatory exposure, and potential reputational damage. For this reason, it's considered best practice to derive multiple intelligence products from the same underlying dataset, tailoring each report to operational, tactical, and strategic audiences.

Another critical principle is confidentiality and the need-to-know model. Not all intelligence should be distributed equally across an organization. Each intelligence product should therefore be classified according to its sensitivity level, for example, internal, restricted, confidential, or public. Applying the need-to-know principle ensures that stakeholders receive only the information necessary for their roles. SOC analysts may require detailed technical indicators such as file hashes or C2 IP addresses, while CFOs are more likely to require insight into the potential financial impact of a cyber incident. Careful classification helps prevent unnecessary information exposure while ensuring that each audience receives relevant and actionable intelligence.

Finally, the dissemination process must incorporate a feedback loop. Intelligence products shouldn't be treated as static reports but as outputs whose operational effectiveness must be measured. Feedback from analysts and decision-makers helps determine whether the disseminated information was useful. For instance, if only a small portion of distributed IOCs proves relevant during operational use, analysts should investigate the causes and refine their filtering or validation processes. By integrating such feedback into future cycles, organizations can gradually improve the quality and relevance of their intelligence outputs.

Dissemination Formats

Once intelligence has been interpreted and contextualized, it must be distributed using formats appropriate to the intended audience and operational environment. Organizations typically employ several complementary dissemination methods.

One common approach involves written intelligence reports. These reports are usually prepared at multiple levels of detail corresponding to strategic, tactical, and operational audiences. Regardless of the audience, each report should begin with a concise executive summary that explains the essential findings and recommended actions. This summary should clearly answer three basic questions: what happened, what actions are required, and who is responsible for responding.

In addition to written reports, organizations frequently rely on verbal briefings and coordination meetings. During active incidents or high-severity attacks, rapid decision-making is often facilitated through short briefings or "war room" meetings that bring together analysts, incident responders, and leadership. When addressing senior executives, these briefings should focus on business implications rather than technical detail, emphasizing operational impact and potential risk to the organization.

Another important dissemination mechanism involves automated alerts generated by security platforms. Real-time alerts triggered through SIEM systems enable analysts to react quickly to newly detected indicators. SOAR platforms can further accelerate response by executing automated playbooks that perform predefined defensive actions. However, the effectiveness of automated alerts depends on careful tuning; excessive false positives may lead to alert fatigue, causing analysts to overlook critical signals.

A fourth dissemination channel involves dashboards and visual analytics tools. These visualization systems allow managers to monitor threat intelligence outputs through high-level indicators such as KPI metrics, heat maps, and trend charts. When dashboards display metrics such as affected assets, IOC false-positive rates, or mean response time, they enable leadership to rapidly understand the operational situation without examining raw technical data.

Example Campaign: SPECTRUM-RED

To illustrate how intelligence dissemination differs across organizational levels, consider the hypothetical campaign SPECTRUM-RED, representing a long-term APT operation targeting the energy sector. The campaign might involve several stages typical of modern intrusion operations, including spear-phishing for initial access, the use of malicious dynamic link library (DLL) side-loading techniques, C2 communications over DNS or HTTPS beaconing, lateral movement via Remote Desktop Protocol (RDP) or Server Message Block (SMB), and encrypted data exfiltration.

In such a scenario, intelligence products are typically prepared at three different levels:

- **Operational level**
  At the operational level, intelligence is intended primarily for SOC analysts and incident response teams responsible for day-to-day detection and containment activities. These products generally include detailed technical information such as IOC lists, event timelines, and correlations between system logs and suspicious commands. Operational dissemination may occur through SIEM alerts, automated SOAR playbooks, or internal messaging systems used by security teams.
- **Tactical level**
  At the tactical level, intelligence is designed for SOC managers or incident response leads who must coordinate defensive actions and allocate resources effectively. Reports at this

level typically summarize the scope of the incident, relevant MITRE ATT&CK techniques, preliminary risk assessments, and recommended defensive measures. Tactical dissemination may occur through internal portals, scheduled security briefings, or targeted communications with management teams.

- **Strategic level**
  At the strategic level, intelligence is tailored for senior leadership and executive stakeholders. Strategic intelligence products focus less on technical indicators and more on broader organizational implications. These reports typically address potential financial impact, regulatory exposure, reputational risk, and key decision points related to crisis management. Dissemination at this level often takes the form of concise executive summaries, slide presentations, or board-level briefings.

Technical Dissemination Channels

The practical delivery of intelligence products relies on a variety of technical platforms and communication channels. Common dissemination mechanisms include the following:

- SIEM systems, which generate immediate alerts when newly identified IOCs appear in monitored environments
- SOAR platforms, which automate defensive actions such as account suspension or endpoint isolation
- TIPs that enable sharing, enrichment, and correlation of threat indicators across multiple data sources
- Secure internal portals that store full tactical or operational intelligence reports for authorized personnel
- Email or SMS notifications, typically used for concise alerts or executive summaries requiring immediate attention

Dashboard Metrics and Key Performance Indicators

To evaluate the effectiveness of intelligence dissemination, organizations often monitor a set of operational metrics through dashboards. Commonly used indicators include the following:

- **Mean time to identify (MTTI)**
  This metric measures how quickly a threat is detected after it first appears.
- **Mean time to remediate (MTTR)**
  This metric indicates the speed with which incidents are resolved.
- **IOC false-positive rate**
  This metric helps determine whether detection mechanisms generate excessive noise.
- **Number of affected critical assets**
  This metric supports prioritization during incident response.

- **Estimated volume of exfiltrated data**
  This metric is an important factor in regulatory reporting and impact assessment.

Tracking these indicators allows organizations to measure the practical effectiveness of their intelligence dissemination processes and identify areas requiring improvement.

Confidentiality and Publication Policy

Because intelligence products often contain sensitive information, organizations must establish clear confidentiality and publication policies governing their distribution. These policies typically require that reports be classified according to predefined sensitivity levels such as internal, restricted, confidential, or public.

Before sharing intelligence externally, organizations should ensure that personally identifiable information (PII) and customer data are properly masked or anonymized. Legal and communications teams should also review intelligence products intended for external distribution to avoid regulatory or reputational risks. A widely accepted guiding principle in this process is that intelligence should contain as much information as necessary, but no more detail than required for the intended audience.

Operational Workflow Checklist

To maintain consistency in dissemination processes, many organizations follow a structured workflow that includes several key steps:

1. Collected intelligence should be analyzed and contextualized, linking indicators to analytical frameworks such as MITRE ATT&CK or the Diamond Model.
2. Intelligence should be converted into actionable products, typically in the form of operational, tactical, and strategic reports.
3. The resulting intelligence products must be distributed through appropriate channels, including security platforms, internal portals, and management communication mechanisms.
4. Organizations should measure the operational impact of these products by tracking KPI metrics such as MTTI, MTTR, and false-positive rates.
5. Feedback gathered from operational teams should be integrated into subsequent intelligence cycles, enabling continuous improvement.

2.5 Feedback and the Sustainability of the Lifecycle

The CTI cycle is often perceived through a superficial lens as a linear process consisting only of the steps collect data, analyze, and report. Many new analysts assume that the job is complete once the first IOC set is published, the first CTI report is delivered, or the first dashboard is updated. Yet the nature of the cyber threat landscape clearly demonstrates how misleading this approach is. Intelligence isn't a task that ends; it's a living organism that must continuously breathe, respond instantly to environmental changes, and evolve. The reality is that in threat intelligence, the work doesn't end when a report is published; on the contrary, it begins again with feedback.

This stage isn't the last link in the chain; it's the most critical source of energy that continuously feeds the entire process and keeps the system alive. While attackers constantly change their tactics, an organization's internal risk priorities also transform over time. A threat that was critical yesterday may lose its impact today; an attack vector that seems insignificant today may become the organization's weakest link tomorrow. For this reason, the intelligence cycle must not be static, linear, or "completed"; it must be circular, iterative, and continuously self-renewing.

The feedback mechanism is the heart of this renewal. The accuracy of collected data, the operational impact of analyzed information, and the extent to which published reports contribute to decision-making can't be understood without regular, disciplined, multilayer feedback.

A remark by an analyst about a single IOC, a brief evaluation from an SOC manager regarding a report, or a strategic-level warning from a C-level executive can change the direction of the entire intelligence program because intelligence becomes real value only when it's reshaped through the eyes of the people who use it. Therefore, feedback isn't a note or a post-report formality—it actually serves as the following:

- The mechanism that keeps the intelligence program alive
- The safeguard that preserves the currency of threat perception
- The catalyst that accelerates the organization's defensive reflexes
- A mirror that improves the process and makes weak links visible
- Most importantly, the power that transforms intelligence from a data pile into a decision-support tool

If there's no feedback, threat intelligence can't go beyond keeping a retrospective record. If there's feedback, intelligence turns into an ecosystem that lives, grows, transforms, and carries the organization into proactive defense. Let's consider stakeholder feedback, specifically. Do the produced reports, alerts, or dashboards actually help the target audience? Or is there only reporting for appearance's sake?

In practice, stakeholder feedback can be grouped into several categories depending on the organizational role of the recipient.

- **Strategic feedback**
  Leadership questions the impact of reports on decision-making: "Did this report change my investment decision?"
- **Tactical feedback**
  SOC managers evaluate the contribution of reports to resource planning: "Which processes did I accelerate thanks to this information?"

- **Operational feedback**
  Analysts measure the usefulness of IOCs: "Did this alert indicate a real attack, or did it waste my time?"

Real-World Example

In a financial institution, the SOC team integrated IOC sets received from an external provider into the SIEM. In the first months, more than 500 alerts were generated daily. Analysts lost 70% of their time dealing with false positives. Once a feedback mechanism was established, the IOC provider was informed that the false positive rate is 65%. The provider improved its filtering methods, and within three months, the false positive rate dropped to 10%.

This example shows that feedback should be shared both internally and with the external ecosystem.

Threat intelligence environments evolve continuously. As attackers adopt new techniques and infrastructures, intelligence processes must adapt to remain relevant. Feedback mechanisms therefore play a critical role in ensuring that IRs, analytical methods, and reporting outputs remain aligned with real operational needs.Several key practices support this continuous adaptation:

- **Updating IRs**
  IRs defined at the beginning of a program may lose relevance as the threat landscape evolves. Organizations must therefore periodically review whether the threats they monitor still reflect the most significant risks to their operations. If analysts continue collecting indicators related to outdated threats while new attack vectors emerge, intelligence production gradually becomes disconnected from real-world risks.

Real-World Example

An energy company initially focused its intelligence program on phishing and ransomware activity. After the rise of supply-chain attacks in the sector, internal feedback revealed that these threats weren't adequately covered. As a result, the organization updated its IRs to include monitoring of software supply-chain vulnerabilities and third-party infrastructure risks.

- **Integrating new threat vectors**
  New attack techniques and technologies constantly expand the threat landscape. Intelligence teams must therefore incorporate emerging threat vectors into their analytical scope. These may include developments such as AI-assisted phishing campaigns, exploitation of cloud misconfigurations, or attacks targeting IoT devices. Monitoring these developments ensures that defensive measures evolve alongside adversary capabilities.

- **Continuous process improvement**
  Threat intelligence processes must be continuously evaluated and improved to avoid stagnation. Performance indicators such as MTTI, MTTR, and false-positive rates help organizations measure the effectiveness of their detection and response workflows. In addition, post-incident review sessions provide an opportunity to identify weaknesses and refine analytical practices.

Real-World Example

One organization launched a six-month improvement program aimed at reducing its incident response time from 3 days to 12 hours. By integrating SOAR automation, improving the quality of IOC sources, and providing additional analyst training, the organization significantly accelerated its response capabilities.

- **Embedding feedback into organizational culture**
  For feedback mechanisms to be effective, they must be integrated into the organizational culture rather than treated as a procedural formality. Analysts should feel comfortable reporting false positives or analytical mistakes, while management should actively evaluate how intelligence products influence operational and strategic decisions. Open communication channels between teams allow organizations to identify inefficiencies and improve reporting practices.

Real-World Example

A telecommunications company introduced monthly "feedback board" meetings in which security teams discussed which intelligence reports had been useful and which had not. Within two years, the organization significantly improved the efficiency and relevance of its reporting process.

- **Operational measurement and institutional learning**
  Effective feedback systems rely on structured measurement and knowledge retention. Organizations should track the operational impact of intelligence activities through defined metrics and maintain a knowledge base that preserves lessons learned from previous incidents. Maintaining institutional memory helps prevent teams from repeating the same analytical mistakes and enables new analysts to learn from past experience.
- **Automation and technology support**
  Modern threat intelligence environments increasingly rely on automated mechanisms to process feedback. Security orchestration platforms can automatically flag false positives and adjust detection logic accordingly. In addition, machine learning techniques can incorporate analyst feedback to refine detection models and reduce unnecessary alerts. These technological mechanisms ensure that lessons learned during operations are rapidly reflected in defensive capabilities.

- **External collaboration and information sharing**
  Feedback shouldn't remain confined within a single organization. Collaboration with external partners allows organizations to exchange operational insights and improve collective defense capabilities. Sector-specific information-sharing communities, national CERT or CSIRT organizations, and standardized sharing mechanisms such as STIX/TAXII enable feedback from multiple organizations to be incorporated into the intelligence lifecycle.
- **Regulatory and compliance considerations**
  In many sectors, feedback processes are also linked to regulatory frameworks and incident management standards. Frameworks such as the NIST CSF emphasize post-incident evaluation within the Respond and Recover functions. Similarly, ISO/IEC 27035 requires organizations to conduct structured "lessons learned" analyses following security incidents. These processes ensure that incident experience is translated into improved future practices.
- **Economic value of feedback**
  Effective feedback mechanisms also produce measurable economic benefits. By reducing false positives and improving the relevance of intelligence outputs, organizations can significantly decrease the amount of analyst time spent investigating irrelevant alerts. Strategic feedback can also influence investment decisions by highlighting which defensive measures provide the greatest operational value.

Real-World Example

After implementing a structured feedback process with its external IOC provider, one organization significantly reduced unnecessary alerts generated by outdated indicators. This improvement saved approximately 200 analyst hours per month, creating a substantial annual operational cost advantage.

2.6 Summary

This chapter examined how the CTI lifecycle operates as a continuous process that transforms raw data into actionable intelligence. The lifecycle begins with planning and direction, where IRs are defined and aligned with organizational risks and priorities.

During the collection phase, information is gathered from multiple sources such as OSINT, security logs, and external intelligence feeds. The processing stage then normalizes, enriches, and organizes this data so that analysts can work with reliable and structured information.

In the analysis phase, indicators are interpreted within broader threat contexts, enabling organizations to understand attacker behavior and potential risks. The resulting intelligence is delivered through the dissemination stage, where reports, alerts, and briefings are tailored to operational teams, managers, and executive leadership.

Finally, feedback mechanisms ensure continuous improvement by evaluating the usefulness of intelligence outputs and adapting processes to evolving threats. Together, these stages illustrate how the intelligence lifecycle supports informed security decisions and strengthens organizational resilience against cyber threats.

In the next chapter, we'll move on to an overview of the intelligence sources used in CTI.

Chapter 3

Intelligence Sources

Before exploring the specific sources that feed cyber threat intelligence, it's useful to briefly frame the perspective of this chapter. Intelligence rarely emerges from a single observation; it develops through the careful interpretation of signals coming from different environments. The following sections introduce the principal intelligence sources and examine how they contribute to building a more complete and actionable security picture.

The world of intelligence isn't a field where technical terms are lined up one after another; it's a deep journey that sometimes confronts us at unexpected moments of decision and often carries the stories of real people in its background. Cyber threat intelligence (CTI) forms the backbone of this journey. Although it may initially appear to be a narrow subject concerned only with security, over time, it has evolved into an almost indispensable strategic discipline for the survival of organizations. As digital systems have expanded, threats have multiplied at the same pace. Today, simply detecting an attack or glancing at logs no longer constitutes "defense." The sources that feed these threats must be examined to understand what those sources are actually trying to convey and to see how they affect an organization's risk management. Sometimes, a single detail can change the entire picture.

One of the biggest misconceptions in many organizations today is confining security solely to technical data. Yet the attacker ecosystem isn't limited to a few visible indicators on the surface; it's a vast environment spread across different platforms, sometimes concealing itself, sometimes appearing completely random. It's here that years of experience whisper the same thing: Don't believe everything you see; organizations that can read what is unseen move one step ahead.

For this reason, attempting to produce intelligence by relying on a single source is like trying to interpret a picture by looking at only half of it. Organizations need a multisource approach in which different inputs communicate with and complement one another.

The purpose of this chapter isn't to list data sources one by one. The real goal is to show how these sources are transformed into intelligence, how they shape decision-making reflexes, and where they transform an organization's security culture. Because a well-designed intelligence architecture can bring together fragmented and sometimes seemingly insignificant pieces of information to create a powerful advantage, it's somewhat like collecting scattered clues and writing a new story. It requires effort, but in the end, the picture that emerges is well worth it.

3.1 Understanding Intelligence Source Classifications

CTI essentially revolves around a simple question: "Where did I obtain this information, and why should I trust it?" Although the question itself appears straightforward, the answer is often complex. The threat landscape changes so rapidly that an attack pattern we encounter today may appear with an entirely different face tomorrow. Moreover, different disciplines are deeply intertwined. Sometimes the information is right in front of us, and other times it's hidden in a small detail tucked away in a corner.

For this reason, merely monitoring attacks can no longer be considered a defense strategy. It's essential to take a step further and examine the source of the information, the context in which it emerged, and the level of reliability it carries. This is where the foundation of an effective intelligence architecture is laid.

We need to ask ourselves a few more critical questions:

- Which source complements what?
- Which data is more valuable at which stage of analysis?
- Most importantly, is this information truly reliable?

When clear answers to these questions are established, source classification ceases to be a dry list and instead becomes a backbone that carries an organization's decision-making reflexes. Whether it's strategic planning or operational readiness, everything rests on this foundation.

In the following sections, we'll set the stage by introducing the key source types and their role in CTI. We'll also examine a comparative model and discuss a framework for source reliability.

3.1.1 The Role of Source Types in Cyber Intelligence

In CTI, when we speak of sources, what often comes to mind are structures that provide data. In reality, however, every source does more than produce information; from its position, it opens a window onto the threat landscape. The width, depth, and perspective of that window determine the fate of the intelligence obtained. The value of a source isn't measured by how much data it provides, but by how accurately, how contextually, and how timely it answers a particular question. Mature CTI structures interpret, filter, and transform data into decision-making processes, which is where their true strength emerges.

At the strategic level, the sources used seek to understand both what threat actors are doing and why. Long-term objectives, geopolitical triggers, sectoral pressure points, and global threat trends form the core material at this level. At this point, open-source intelligence (OSINT) efforts, publications from government institutions, multilateral intelligence-sharing networks, and academic analyses come into play. What matters here is access to information, and the degree to which that information is reliable, the perspective from which it's produced, and the context in which it's interpreted are decisive. Strategic intelligence looks less for an answer to "what will happen?" and more for "which direction are we heading?" And these answers sometimes change budgets, priorities, and an organization's entire security vision.

At the operational level, the picture changes. Here, rather than abstract analyses, more concrete, field-adjacent data comes to the forefront. Threat actors' infrastructures, campaign patterns, methods, modes of movement, and so on all guide security operations center (SOC) analysts, threat hunters, and incident response teams. Honeypot data, log telemetry, malware analyses, and commercial CTI reports constitute the building blocks of this layer. Working with the right sources at this level makes it possible to understand how the attacker thinks. Sometimes a small behavioral pattern becomes the key to a large campaign, and sometimes a single correlation reveals a hidden chain.

Tactical intelligence, on the other hand, races against time. Here, the issue isn't theory but reflexes. It's immediate. An anomaly coming from an endpoint, a newly observed domain, unexpected process behavior, or an unusual email subject line are all part of this layer. Tactical data answers the question "what is happening right now?" and usually requires reaction times measured in seconds or minutes. But this speed also brings risk. An out-of-context or incorrect indicator of compromise (IOC) can mislead defense teams and unnecessarily lock down systems. By contrast, properly linked and well-analyzed tactical data can stop an attack before it escalates. Sometimes, the difference is only a matter of a few minutes.

Generally, there are three main source groups:

- **Open-source intelligence (OSINT)**
  OSINT provides a broad and easily accessible pool of information; however, it contains a great deal of noise, such as a blog post, a hastily prepared report, an unverified post, and so on. Without proper enrichment, such data can mislead the analyst.
- **Human intelligence (HUMINT)**
  HUMINT, on the other hand, reveals the human side of the equation: an employee's observation, a tiny suspicion coming from within, a conversation encountered on the dark web, and so on. All of these carry high value, yet they also include human bias. For this reason, it's a source that is highly valuable but fragile.
- **Signals intelligence (SIGINT)**
  SIGINT constitutes the technical backbone of the effort with traffic flows, encrypted communications, telemetry traces, and so on. It's rigid, numerical, and reliable, but also difficult to collect and subject to numerous technical and legal constraints.

Reading these sources correctly saves organizations from collecting data at random. When trying to understand whom an APT group is targeting, HUMINT is invaluable; but when extracting the technical traces of a campaign, SIGINT comes into play. OSINT generally acts as the glue that binds these two worlds together.

A well-managed source structure does more than strengthen reports; it sharpens an organization's perception of threats. Analysts develop foresight more quickly, risk management is built on firmer foundations, and long-term decisions become more accurate. And, having the right information at a critical moment is often a power that can change the entire game. It creates a peculiar sense of confidence from knowing what you're dealing with.

For this reason, organizations that can skillfully manage their sources act swiftly in the field and take more confident steps toward the future. A strong intelligence structure detects threats and nourishes intuition about what lies ahead.

Ultimately, sources aren't just data points. Each one is a perspective, a narrative, or a trace. The maturity of CTI lies in reading these traces correctly—sometimes slowly, sometimes quickly, but always consciously.

3.1.2 Comparative Model of OSINT, HUMINT, and SIGINT

Within the CTI ecosystem, OSINT, HUMINT, and SIGINT sources may appear on the surface as three independent types of data, yet in reality, they are tightly interconnected and complementary building blocks that enable organizations to make sense of threats. Each of these sources offers distinct characteristics in terms of speed, accuracy, cost, accessibility, and risk levels. These characteristics directly influence the efficiency of intelligence processes and an organization's resilience against threats, its defensive reflexes, and the maturity of its decision-making mechanisms. For this reason, viewing these three sources as simply data providers is insufficient; each one is, in fact, a critical perspective that makes a different layer of the threat ecosystem visible.

At a high level, we can draw clear comparisons between OSINT, HUMINT, and SIGINT in terms of their advantages, limitations, and usage, as shown in Table 3.1.

Comparison Point	OSINT	HUMINT	SIGINT
Advantages	This indispensable part of early warning systems is one of the most dynamic components of modern intelligence architectures due to the discovery advantage provided by its rapidly obtainable and broad data pool.	This provides the strongest contextual insight and answers the "why" question behind an attack. Technology often answers the questions "what happened?" and "how did it happen?"; however, answers to critical questions such as "with what motivation was it carried out?" are typically obtained through HUMINT.	This is the most reliable and evidence-based type of intelligence in terms of accuracy, as it's grounded in technical infrastructure and derived from direct system signals. It carries the lowest risk of manipulation because the data comes from machines rather than humans.

Table 3.1: Intelligence Source Comparison

Comparison Point	OSINT	HUMINT	SIGINT
Limitations	Speed and scope also bring a high level of noise. Fake threat reports, manipulated data, unverified claims, and malicious information pollution make OSINT risky when used in isolation. For this reason, professional analysts never treat OSINT data as a final conclusion, but rather as a starting point that must be verified.	Due to the uncertainties inherent in human nature, HUMINT sometimes carries the risk of bias, misinterpretation, or manipulation. Therefore, although HUMINT is a high-value source, it must always be supported by other data types and subjected to cross-validation.	Collecting SIGINT is more costly and technically complex compared to OSINT. In addition, legal and ethical boundaries must be carefully observed, especially when analyzing traffic that contains personal data.
Example	A dark web post in which a ransomware group announces new infrastructure can enable SOC teams to prepare even before an attack begins.	A brief suspicious behavior report submitted by an employee can enable the early detection of a major insider threat case.	Unusual traffic directed toward the command-and-control (C2) infrastructure of an APT group can be detected at an early stage of the attack chain, preventing a major incident.

Table 3.1: Intelligence Source Comparison (Cont.)

The differing speed, accuracy, and cost profiles of these three sources are of great importance for an organization's risk management. OSINT helps explore the broad threat surface, HUMINT makes attackers' motivations visible, and SIGINT provides concrete technical evidence that enables rapid response. If an organization relies solely on OSINT, it may see external threats but miss insider threats; if it relies only on HUMINT, it may gain contextual insight but lack technical verification. Finally, if it relies only on SIGINT, it may detect technical attacks but struggle to interpret attacker intent. Therefore, a strong CTI structure doesn't view these three sources as alternatives to one another, but as complementary elements.

3.1.3 Source Reliability Framework

The ability to accurately assess the reliability of an intelligence source is one of the fundamental building blocks of modern CTI programs. This is because the threat environment isn't composed solely of technical indicators; it's filled with data that may be misleading, deliberately distorted, taken out of context, or entirely fabricated. For this reason, the reliability level of a source directly affects the accuracy of the intelligence products produced, the precision of operational decisions, and the effectiveness of organizational defense mechanisms.

Source reliability is evaluated within a multidimensional framework that includes speed, accuracy, context, historical performance, data production methodology, and verifiability. At the core of this framework are reliability classifications in which sources are graded from A to F. Class A defines sources with a strong history of accuracy, robust methodology, and consistent data production patterns, while Class F refers to sources with a high risk of manipulation, unverified origins, or a history of providing misleading information. This grading guides the risk level and verification requirements of every intelligence product.

However, reliability grading alone isn't enough. Professional CTI teams know that no single source provides absolute truth on its own. Excessive dependence on a single information source exposes intelligence processes to blind spots, false positives, and bias-driven errors. Reducing reliance on single sources is achieved through multisource validation, cross-correlation, comparison with alternative sources, and confirmation via independent datasets. For example, an IOC obtained from OSINT is expected to be verified through the same source, as well as through passive Domain Name System (DNS) records, SIGINT telemetry, sandbox analyses, or trusted commercial CTI providers. An insider threat insight obtained via HUMINT is evaluated not solely based on one individual's statement, but together with behavioral analyses, access logs, communication patterns, and awareness levels. Even abnormal traffic behavior detected through SIGINT doesn't constitute a final conclusion unless it's supported by multiple telemetry sources. This approach both reduces false positives and builds a more resilient intelligence structure against potential disinformation efforts by attackers.

Data integrity is a critical component of this process. Having access to a dataset doesn't mean that it's accurate or reliable. Attacker groups may deliberately flood OSINT platforms with false information to cover their tracks; publish fake leaks on dark web forums; or mislead analysts through social media accounts impersonating threat actors. Similarly, HUMINT data may be distorted due to personal interests, fear, misperception, or psychological factors. Technical data, on the other hand, may appear out of context due to proxy usage, virtual private network (VPN) tunnels, spoofed traffic, or misconfigurations in security products. Therefore, protecting data integrity means measuring the reliability of a source, understanding the risks of data manipulation, and developing analytical defenses against those risks.

Just as important as being aware of source manipulation risks is establishing processes that mitigate them. Ensuring that every piece of information has a theoretically verifiable framework, monitoring the historical behavior of sources, detecting conflicts between sources, and being aware of the analyst's own cognitive biases are core components of this process.

This multidimensional approach demonstrates that intelligence is more than collecting information; it's also about the art of producing and verifying knowledge. When the reliability framework is applied correctly, organizations evolve into structures that both react to threats and anticipate them to become resilient to disinformation, strong in data-driven decision-making, and equipped with advanced defensive reflexes. As a result, the accuracy of intelligence products increases, risk management decisions become more precise, and the organization becomes more resilient to threats at both the tactical and strategic levels.

3.2 Open-Source Intelligence

On the IT side, the areas of greatest interest include computer forensics, network forensics, and OSINT. When you bring all three to the same table, the picture that emerges is truly something else. When you read the trace on the disk in an incident, catch the shadow in network traffic, and combine it with open-source signals gathered from the outside world, you start to see what kind of story is actually unfolding behind a log entry. And isn't that what we ultimately want? Not just the correct result, but a perfect and reliable result.

OSINT is both the broadest domain of CTI and one of its most complex layers. It feeds on publicly available data; in essence, OSINT is the disciplined work of collecting, filtering, interpreting, and turning fragments of information that are already out there on the internet into a threat picture. Its strength comes from how easy it is to access and how virtually limitless the data is. But precisely for that reason, OSINT carries a heavy responsibility: it's simultaneously the fastest layer for producing information and the one that contains the most false, incomplete, and manipulated content. That is why reducing OSINT to just data collection is unfair; this is, in fact, the discipline of filtering, skepticism, double-checking, and placing information into context.

We'll dive into key considerations for implementing OSINT next, and then we'll explore common examples of OSINT data types in your digital landscape.

3.2.1 The Role of OSINT in Cyber Intelligence

The reason OSINT has become so critical in today's threat environment is simple: No matter how much threat actors try to hide, most of their activities leave a direct or indirect digital trace. Brief conversations on dark web forums, leak announcements on paste sites, ransomware groups' leak sites, researchers' blogs and reports, seemingly harmless code changes on GitHub, discussions on social media, technology blog posts, malware analysis reports, open DNS records and Who Is (WHOIS) database changes, digital footprint scans, and so on are each signals in their own right. At first glance, they may look scattered, unrelated, or even insignificant. But when placed side by side in the right way, they can turn into a remarkably clear picture of the attacker's intent, their preparation process, the targets they choose, and the methods they use to progress.

Example

When a ransomware group is working on a new variant, sensing the operating systems (OS) the new version focuses on, the vulnerabilities it exploits, and the sectors that may fall into its sights in the next wave only through OSINT. A short message on the dark web can be the trailer of a campaign that's "coming soon." A sudden change in a domain's WHOIS information may point to a C2 infrastructure that is being prepared. A short but technical discussion among a few security researchers on X (formerly Twitter) may indicate that an exploit not yet fully disclosed has begun circulating. When these little signals come together, OSINT becomes a "threat intuition" layer that not only describes attacks that have already happened but also helps you feel the wave that's approaching.

Still, the true value of OSINT lies not in its diversity, but in the context it builds. A single IP address flagged as malicious may not mean much on its own. But when considered together with the domains tied to that IP, infrastructure patterns used in past campaigns, and the same actor's previously preferred systems and regions, an entirely different story emerges. Suddenly it stops being just a "suspicious IP" and becomes a scenario that directly affects an organization's risk level. These scenarios help SOC analysts update their triggers and rule sets, enable incident response teams to determine which alerts to prioritize, and allow executives to increase their risk awareness in a realistic way.

One of OSINT's strengths is that it offers the opportunity to notice attacker behavior before an incident erupts. Before threat actors fully deploy an operation into the field, they often leave small traces—unintentionally. Early samples of malware appear on platforms such as VirusTotal, newly registered domains surface in passive DNS (pDNS) records, hacktivist groups intimidate target organizations through social media, and sellers of stolen credit card databases circulate signals of an impending leak on the dark web. OSINT can capture these crumbs and enable organizations to see risks that haven't yet materialized, that is, see one or two steps ahead.

Of course, producing real value from OSINT at an organizational scale isn't possible with a "let's collect everything and look later" approach. Scattered piles of data without a clear target produce, at best, noise. The first step for effective OSINT is clarifying the organization's real risks based on its sector, geography, regulations, and technology stack. In the finance sector, the primary focus may be credit card databases, crypto fraud networks, and the banking trojan ecosystem; in the energy sector, Incident Command System (ICS)/ Supervisory Control and Data Acquisition (SCADA) vulnerabilities, critical infrastructure threats, and state-sponsored APT activity come to the forefront. The same data pool can mean entirely different things for different organizations; what makes the difference is filtering that pool through the right lens.

Another critical dimension is verification. Open-source data is the most susceptible to manipulation. Threat actors may deliberately spread false information and create fake attack announcements, frame rival groups, or produce fabricated leak content to confuse security researchers. In short, sometimes part of the OSINT "stage" is written by the attacker themselves. That is why no single-source OSINT output should be confidently turned into action. Professional intelligence teams ask themselves three fundamental questions at this point:

- Who is the source of this information, where is it from, and how reliable is it?
- Do other sources also confirm the same information?
- If this information turns out to be wrong, what will the cost be for the organization?

When used correctly, OSINT is a powerful intelligence layer that provides early warning, expands threat visibility, enables contextual analysis, and makes defense proactive. When used incorrectly—scattered or aimlessly—it turns into a risky domain that produces noise, can lead analysts down the wrong trails, and causes wasted resources. That is why OSINT shouldn't be positioned as collecting data from the internet, but as an analytical and intuitive mechanism that reads the digital traces of threat actors and tries to predict their future moves. The real value emerges precisely when this perspective shifts.

3.2.2 Blogs, Security Reports, and Threat Intelligence Portals

When OSINT is mentioned, most people first think of data. Yet the real strength of this field emerges where data is processed, interpreted, and given meaning. Blog posts, technical analysis reports, and threat intelligence portals published by leading security firms in the industry come into play here.

The content produced by organizations such as Unit42, Talos, Mandiant, SecureWorks, CrowdStrike, SentinelOne, and Kaspersky describes what happened and then places how it happened, why it happened, and how could it happen again squarely on the table. In this respect, these sources provide analyses that go far beyond raw data—analyses filtered through experience and collective intelligence.

The true value of these reports lies in making the behavior of threat actors visible. Here are a few examples:

- In the analyses Mandiant publishes on APT groups, it's possible to read a threat actor's technical capabilities, operational habits, timing reflexes, and even indirect cultural traces. These reports can be accessed at *https://cloud.google.com/security/resources*.
- Talos's in-depth examinations of malware families allow analysts to follow, step by step, the evolution of a piece of malware from its earliest versions to its current variants. Talos research and reports are available at *https://blog.talosintelligence.com*.
- Secureworks Cyber Threat Unit (CTU) reports go beyond technical details, revealing attackers' economic motivations, preferred target sectors, and campaign timing. Secureworks CTU research can be found at *www.sophos.com/en-us/blog/category/threat-research*.

This documentation reflects the threat economy and attacker psychology. However, the value of this content isn't measured by how directly applicable it is. The real issue is using it within the correct context. Every report belongs to a specific geography, a specific customer profile, and a specific time frame. An IOC marked as critical in a Mandiant report may represent low risk for your organization. Or a TTP that makes waves on a global scale may not pose a real threat to your infrastructure at this time. Therefore, in a professional CTI approach, OSINT content is never accepted as "direct truth"; it's always questioned, compared, and placed into context.

In the verification process, a multisource approach plays a vital role. An IP address flagged as malicious by Talos is examined through pDNS records, compared against telemetry from other threat intelligence providers, and, if necessary, correlated with the organization's own logs and flow data. If no activity related to that IP is observed within internal systems, the risk score is naturally lowered. Similarly, a TTP set attributed to an APT group is analyzed within the MITRE ATT&CK framework, and its level of overlap with existing security controls is assessed. This process elevates OSINT from information to operational intelligence.

Another critical contribution of these blogs and reports is timeliness. In the world of attacks, everything changes rapidly. The fact that a new vulnerability has begun to be actively exploited is often first announced in a blog post or a brief analysis note. These early signals directly affect how quickly organizations can take action. Decisions such as accelerating patching processes, deploying temporary security controls, or placing certain systems under heightened monitoring are often driven by such OSINT content.

More importantly, these sources hint at both current attacks and emerging threat trends. Large campaigns rarely begin out of the blue. They are preceded by small test attacks, infrastructure setup, and probing of weak points. A Talos initial observation article may foreshadow a future threat trend even when no major attack is yet visible. A short assessment note from Mandiant may indicate an increase in an APT group's operational tempo. Being able to read these weak signals directly strengthens an analyst's intuitive capacity. At this point, OSINT reports cease to be merely technical sources and become analytical guides. Some approach attacks through a kill chain logic. Others map TTPs through the MITRE ATT&CK framework. Still others interpret risk concentration through statistical trends. This diversity enables analysts to approach the world with a multilayered perspective instead of a single window.

Of course, at the center of everything lies organizational context. While a vulnerability targeting ICS systems may constitute a red alert for an energy company, the same vulnerability may be secondary for an e-commerce business. A Linux-based rootkit may not pose an immediate risk for an environment that primarily operates on Windows. At this point, the role of CTI teams is to filter, weigh, and transform OSINT-derived data into organization-specific intelligence products.

These reports also serve as mirrors. They clearly reveal which mistakes attackers exploit, which vulnerabilities organizations repeatedly postpone addressing, and which controls fail to work effectively in practice. In this way, organizations also gain the opportunity to rethink their own security posture when monitoring for external threats. In other words, OSINT is both a window looking outward and a light shone inward.

Beyond all this, the quietest yet most powerful contribution of this content is the creation of a shared language among analysts. Over time, the way attack techniques are described, the approach to classifying TTPs, and the reflexes used in risk assessment become more standardized and consistent. This, in turn, increases team cohesion and elevates the quality of intelligence production.

OSINT sources aren't content to be read and set aside. They function as a compass and an early warning system for modern security teams, helping organizations anticipate evolving threats and adapt to changing attacker tactics.

3.2.3 Social Media and Community-Based Sources

The concept of community in threat intelligence is no longer a romantic ornament sitting on the sidelines; it's a player at the very center of the game, quietly changing the rules. In the modern OSINT ecosystem, social media and community-based platforms constitute both one of the fastest channels for data flow and one of the most slippery grounds for analysts. X, Telegram, Reddit, Discord, and similar channels make the pulse of threat actors, the first observations of researchers, community analyses, and global security trends visible almost in real time.

This speed provides serious advantages for corporate threat intelligence:

- Receiving early signals
- Seeing campaigns before they spread widely
- Entering the IOC-sharing cycle much faster

At the same time, however, this speed brings another reality with it that involves algorithms, echo chambers, manipulation, information pollution, and deliberate misdirection. Consequently, this environment is a double-edged sword: when managed correctly, it offers analysts extraordinary visibility; when used incorrectly, it can systematically distort the threat picture.

The X platform is, in practice today, the "main stage" of the threat intelligence ecosystem. Leading analysts, researchers, product teams, and independent security professionals share their initial findings largely on this platform. Early hints about zero-day vulnerabilities, indicators of emerging attack campaigns, fresh variants of malware families, and unusual activity by specific APT groups usually first appear in the X feed. A single, seemingly simple IOC share can reach thousands of analysts' screens within minutes, dramatically increasing global detection and response speed.

However, there's a critical breaking point here: Content that spreads this quickly is, by nature, not always fully verified, contextually clarified, or technically tested. Speed doesn't—and often can't—always align with academic accuracy and methodological rigor.

Telegram, Reddit, and Discord function more like back rooms. On Telegram, you encounter the following:

- Closed security channels
- Leak-sharing groups
- Internal conversations within researcher or threat actor communities

On Reddit, you'll see the following:

- Technical discussions
- Proof-of-concept (POC) shares
- Analyses within cybersecurity subforums

On Discord, you'll find long-running, relatively less visible conversations that take place within private research servers. These fragments, which may appear small or even scattered on the surface, can carry high-value signals from a threat intelligence perspective. Some Telegram groups have become spaces where ransomware groups directly discuss operational processes, announce stolen data, or publicize new communication channels. A single indicator buried in a Reddit thread may in fact be part of a much broader attack wave. A behavioral analysis shared by a niche research community on Discord may point to a critical TTP that hasn't yet appeared in any official report.

The main factor adding complexity to this picture is the fragility of the concept of "truth" on social media. Fake analyst profiles, deliberately spread false IOCs, staged discussions organized by threat actors to muddy their own traces, propaganda content, and consciously generated noise can all be used to mislead threat intelligence analysts. Therefore, every piece of data derived from social media should first be treated as a claim in professional practice and never accepted directly as an action-triggering fact.

For example, a statement shared in a Telegram channel claiming "this IP definitely belongs to this APT group" has no analytical value on its own. The analyst must do the following:

- Examine pDNS records.
- Query associated domains.
- Review WHOIS changes.
- Check for activity related to this IP in the organization's SIGINT, NetFlow, or log data.

Similarly, a vulnerability disclosure circulating on X shouldn't be entered into the corporate risk register without support from official Common Vulnerabilities and Exposures (CVE) records, vendor advisories, and reverse-engineering analyses. From an academic perspective, these steps are essentially the digital adaptation of classical *source criticism* and *multi-source verification* processes.

Organizations that succeed in this complex environment use social media platforms simultaneously for three core functions:

- Signal-collection space
- Early-warning mechanism
- Laboratory for observing behavioral patterns

As part of a successful organization in this area, effective CTI teams must do the following:

- Recognize trusted analysts and accounts.
- Know which profiles specialize in which domains.
- Isolate channels that provide high signal-to-noise ratios.
- Filter and consume with caution sources that carry a high risk of manipulation.

Here, the analyst evaluates both the content and the person producing it:

- What is their historical accuracy rate?
- In which topics have they consistently been wrong?
- In which situations have they chosen to remain silent?

This is, in effect, a reversed form of social engineering—the practice of analyzing the human behind the data.

Social media and community-based OSINT also open a window into the cultural fabric of the attacker ecosystem. Some threat actors communicate with each other on X, some develop tools on Discord, some announce operations on Telegram, and others discuss defense bypass techniques on Reddit. These behaviors contain meaningful clues about the threat economy, motivational structures, hierarchical organization, and internal group dynamics.

For example, a new ransomware group posting messages on Telegram along the lines of "we're starting a new project" shortly before announcing its leak site may be one of the earliest indicators of an upcoming attack wave. Such behaviors can be interpreted in academic literature as preoperational chatter—noise signals that precede an operation.

When all of these channels are considered together, social media–based OSINT occupies a position in modern threat intelligence architectures that is both indispensable and requires extreme caution. When analyzed correctly, the following benefits are possible:

- The first stirrings of attacker behavior can be detected.
- Trends can be identified earlier.
- Critical signals can be captured in time.
- Defensive mechanisms can be updated without waiting for classical reporting cycles.

However, the uncontrolled consumption of unverified or context-free information can result in the following negative impacts:

- Distracts the analyst's focus
- Distorts the organization's risk prioritization
- Leads to the consumption of resources against threats that don't actually exist

For this reason, the same source can become both a lifesaving radar and, when misconfigured, a noise machine that creates alarm fatigue. The real power of sources based on social media and the community often comes not from explicitly shared content, but from digital behavioral signals: an innocent-looking message posted by a threat actor, a screenshot unintentionally shared by a researcher, the writing style of an anonymous Telegram account, the appointment of a new moderator to a closed Discord channel, multiple Reddit accounts simultaneously converging on the same topics, and so on. Each of these may represent early indicators that emerge before any IOC is produced, any CVE record is published, or any official report is written.

The point where a professional analyst makes the difference is by focusing not on where the crowd is looking, but on the fracture created by a small anomaly that no one cares about. Let's consider a few practical examples using various platforms:

- **Telegram**

  Some closed Telegram groups provide good examples of this phenomenon. These groups are often not spaces for long technical discussions; short, fragmented, and relatively superficial messages circulate. Yet the signal is hidden within this simplicity. The following seemingly insignificant dynamics can be early indicators of new attack preparations or the development of new tools/payloads for an experienced OSINT analyst:

  - A sudden change in the group administrator's profile picture
  - A sudden shift in the time zones during which participants are active
  - The group going completely silent for a few days and then rapidly becoming active again

- **Reddit**

  Similar background stories can be observed on Reddit as well. Technical questions that appear innocent on the surface may actually be indirect signals of an upcoming attack, for example, one user repeatedly asking over several days why "a specific hex sequence causes crashes in shellcode," followed by another account claiming that "a new packer automatically inserts this hex sequence," and so on. When writing styles, active time windows, and previously asked questions are considered together, a strong intuitive profile may form suggesting that both accounts are operated by the same threat actor.

  For an ordinary user, this is merely a technical discussion; for a CTI analyst, it's the first draft of malware that hasn't yet been deployed.

- **X**

  Directly tracking threat actors themselves on X may not always be possible; however, following the surrounding ecosystem often yields more productive results. A sudden concentration of certain analyst accounts on the same theme, increased technical discussions just before a vulnerability disclosure, synchronized silence among accounts that track threat groups, bot accounts repeatedly amplifying similar content can all be interpreted as behavioral indicators of an approaching attack wave.

- **Discord**

  Recognizing such signals requires more than classical IOC collection practices; it demands a form of social media literacy and behavioral analytics. On Discord servers, threat actors rarely take the stage themselves; instead, the ecosystem's "proxy" players reveal themselves through their behavior. For example, the following often indicates that an organized group is training new members:

  - A sudden influx of new accounts join training channels.
  - These accounts consecutively raise topics such as "PowerShell obfuscation," "LOLBins," or "Cobalt Strike alternatives."

  Such behavioral patterns are frequently associated with campaigns that emerge in the field a few weeks later. What appears to be random chatter from the outside is, in reality, operational preparation hidden within noise.

On top of all this comes the fake news and manipulation dimension, which is perhaps the riskiest aspect of social media. Threat actors sometimes generate deliberate fake leaks to overshadow their own operations, write fabricated messages accusing rival groups, or initiate artificial debates to distract the security community.

Yet even this fake content has analytical value for a professional analyst: the target, language, and method of manipulation carry critical clues about the threat actor's real operational intent. For instance, falsely accusing a specific APT group with incorrect IOCs on X often suggests a diversion from the real attack planned in another geography or sector. Claims circulating on Telegram that "this CVE is being actively exploited" may signal that an undisclosed vulnerability is being tested in the background.

In conclusion, producing threat intelligence from social media isn't just about collecting what's visible. It's about reading the invisible, understanding the intent behind behaviors, placing small anomalies into the larger picture, and generating coherent, strategic insights amid information chaos. This process requires the following:

- Human intuition beyond classical OSINT tools
- Patience and continuity
- A behavioral analysis reflex

3.2.4 DNS Data and Passive Internet Telemetry

After setting up an IT infrastructure, many IT professionals don't examine it again as long as no problems occur. They leave it as it is—like a boiler running in the background. Security hardening usually isn't performed either.

DNS data and passive internet telemetry are often among the most powerful yet least understood building blocks of modern threat intelligence. In the following sections, we'll walk through the relevant data sources in this area, explain how DNS can be used in threat hunting, and close with a practical example.

Data Sources

Regardless of their technical capabilities, the moment threat actors connect their infrastructure to the internet, they unknowingly leave digital footprints. pDNS records, WHOIS changes, Autonomous System Number (ASN) movements, and Border Gateway Protocol (BGP) routing anomalies are the most valuable components of these invisible traces. These signals can reveal threat actors' preparation processes long before an attack campaign begins and provide organizations with the opportunity to identify risks that haven't yet become visible. For this reason, DNS-based telemetry is considered one of the most intuitive domains of threat hunting.

Let's take a look at the relevant data sources in DNS-based telemetry:

- **Passive DNS**
  pDNS data is a unique tool for understanding how attackers shape their domain infrastructure. pDNS is like a time machine that preserves the IP addresses a domain has resolved to, the infrastructure it used in the past, the regions in which it was active, and how long it remained online. For example, when a threat actor establishes a new C2 infrastructure, a domain may initially resolve to a benign IP address; however, a few days later, that IP may rapidly change and move to a malicious server. Standard DNS queries don't reveal this, but pDNS records all changes over time. This gives analysts the ability to observe the silent periods of infrastructure preparation and capture signals before an attack is launched.
- **WHOIS**
  WHOIS data is an important complementary source for understanding domain ownership structures. Although threat actors typically try to conceal their identities by using privacy services, subtle changes in WHOIS records can expose infrastructure preparation. For example, if it's known that an attacker has used a specific privacy service provider for years and a new domain is registered through the same provider, that domain may be considered suspicious even before it exhibits malicious activity. Some advanced analysts can differentiate threat actors by analyzing behavioral characteristics such as writing styles in WHOIS records, email structure conventions, registration frequency, and domain renewal rhythms. These types of microsignals create infrastructure signatures that traditional security solutions can't detect.
- **ASN**
  ASN data also plays a critical role in understanding the attack surface. Many threat actors prefer specific hosting providers, and these preferences tend to become consistent over time. For example, it's known that some APT groups frequently use the same small-scale data centers, prefer suspicious ASNs in certain regions, or distribute multiple malicious campaigns through the same ASN clusters. If an organization observes that an IP address belongs to a particular ASN, this alone doesn't necessarily indicate malicious intent; however, from a threat-hunting perspective, it can be categorized as a high contextual risk. This allows analysts to identify infrastructure that doesn't appear in IOC lists but carries behavioral risk.

- **BGP**
  BGP data provides a far more refined and advanced analysis layer. BGP routing changes can contain leaks that emerge during the infrastructure setup stages of certain attacks. For example, when a threat actor moves a C2 server to a new IP range before launching an operation, short-lived routing changes may appear in BGP. These changes may go unnoticed by an organization, but a threat intelligence team monitoring BGP telemetry can detect this activity. Similarly, some attackers may attempt BGP hijacking on specific IP blocks to buy time. These attempts aren't always successful, but even a failed attempt leaves behind a critical signal. This signal can reveal where the attacker's infrastructure may be moving, which IP blocks may be used, and in which geography activity is planned.

We'll take a closer look at how these data sources are acquired later on as part of our coverage in Chapter 4 through Chapter 7, as well as Chapter 10.

DNS-Based Threat Hunting

Rather than waiting for alerts, threat hunters form hypotheses, identify anomalies, and seek out adversarial behaviors using intelligence-derived patterns. While our in-depth exploration of threat hunting will wait until Chapter 11, in this section, we'll introduce threat hunting to demonstrate the utility of DNS data and passive telemetry.

The role of domain infrastructure changes in threat hunting may be the least known yet most powerful aspect of this topic. Threat actors usually change their infrastructure according to operational phases. A domain is first parked to appear benign, then briefly activated to manage C2 traffic, and later quickly abandoned. This pattern repeats across many attacks. Professional threat hunters can understand this model and derive an infrastructure lifecycle. For example, a domain moving to four different IP addresses within 48 hours, frequent changes in time-to-live (TTL) values, regular updates to TXT records, or infrastructure that is active only during certain hours are signals that indicate very early stages of attack preparation. These signals emerge long before they are listed as IOCs, added to antivirus engines, or reported publicly.

For this reason, DNS-based intelligence is a form of behavioral analysis that reveals threat actors' operational discipline, economic preferences, logistical constraints, and even time-management habits. How they rent C2 infrastructure, how long they keep domains, renewal times, and which providers they prefer all play a critical role in attacker profiling.

DNS data and passive internet telemetry serve as valuable early warning sources in threat hunting. When analyzed carefully, subtle infrastructure changes that appear before attacks can reveal patterns that help organizations anticipate emerging threats and strengthen proactive defense. The Windows Active Directory (AD) DNS infrastructure is one of the critical areas in internal threat hunting that most analysts tend to overlook, yet it's also one of the areas where attackers leave the most traces. AD DNS is a name resolution service that serves as the behavioral map of the internal network. Users, servers, service accounts, Kerberos ticket distribution, Lightweight Directory Access Protocol (LDAP) requests, and AD replication traffic all leave visible traces within DNS. For this reason, when advanced threat actors manage to penetrate an internal network, they initially prepare their attack by manipulating AD DNS records or altering DNS traffic patterns; however, these changes can't be detected solely through standard security logs and require very fine-grained DNS telemetry analysis.

For example, after preparing Kerberoasting or silver ticket activity, an intruding threat actor generates phantom DNS queries that appear to originate from the same IP range as the domain controllers (DCs). These queries often resemble legitimate DC names such as `dc01.internal.local.svc`; however, in pDNS, the timing of these queries doesn't align with real operational hours. While many SOC analysts focus only on Event IDs 4768 or 4771 because these authentication-related logs are commonly used to detect Kerberos anomalies and failed ticket requests, in reality, minimal TTL changes within AD DNS reveal the earliest stage of the attack. Professional analysts can recognize an attacker's DNS reconnaissance fingerprinting behavior by examining DNS logs for specific anomalies before the attacker fully establishes themselves in the environment. In practice, this includes focusing on indicators such as unusual query patterns targeting DC–like hostnames, repeated queries that generate `NXDOMAIN` responses, abnormal query timing outside normal operational hours, and subtle TTL inconsistencies within AD DNS records. Analysts should also look for repeated resolution attempts against internal service naming conventions and unexpected query sources originating from hosts that normally don't perform directory-related lookups. By correlating these signals, analysts can detect early reconnaissance activity and infer how attackers are attempting to map the internal network topology.

Another critical weakness of internal DNS is the attacker's abuse of the dynamic DNS update mechanism. In many organizations, service accounts or application servers are allowed to dynamically update their own DNS records. An advanced threat actor, from a compromised machine, can issue a record update request to add a new subdomain to internal DNS and use that domain as a pivot point for C2 communication. For example, the following behavior has been observed in real-world advanced APT operations:

- `.hidden-upd.internal.local`
- `.sysdc-cache.internal.local`
- `.krb-replica.internal.local`

These domains are named to appear harmless, but the clues are hidden in TTL inconsistencies. While a normal AD domain record uses a TTL of 3,600 seconds, records created by the attacker typically fall within the 60–120 second range. This TTL deviation is a behavioral indicator that even the most advanced APTs can't easily conceal.

DNS services positioned in the demilitarized zone (DMZ) represent the most important externally exposed manipulation surface for attackers. DMZ DNS servers usually act both as proxies to internal resources and as access points for external users to the organization's public resources. This structure provides attackers with two critical opportunities:

- Inside-to-outside DNS tunneling
- Outside-to-inside domain shadowing attacks

DNS tunneling becomes an attacker's primary escape path, especially when outbound traffic from the internal network is restricted. Many analysts focus on the size of DNS query packets; however, professional CTI analysts examine the entropy values of DNS packets. For example, domains of the following type resolved multiple times on a DMZ DNS server are actually early signals of covert C2 communication:

- `x8d92k3k1q9.internal-eval.com`
- `dkgj9821df098asdj.subdomain.control-check.net`

In normal user behavior, randomly generated high-entropy character strings are almost never seen. For this reason, entropy analysis of DNS query content is one of the most powerful yet least known methods for detecting tunneling attacks.

In outside-to-inside domain shadowing attacks, the threat actor creates a subdomain within the organization's public DNS records for which they don't legitimately hold authority. For example, `cdn.company.com` may appear completely normal; however, by exploiting zone transfer weaknesses, the attacker may actually have created the following record:

`auth.cdn.company.com` → `185.xx.xx.xx` (APT infrastructure)

This domain appears to be a CDN service and is often perceived as benign by sandbox systems. However, professional analysts trace domain shadowing through the following microsignals:

- Timestamp inconsistencies between Start of Authority (SOA) and Name Server (NS) records
- Sudden changes in the frequency of authoritative queries on DMZ DNS
- Short TTL values for the shadowed domain
- Domain only active during specific hours
- Absence of a reverse DNS record

None of these signals are automatically generated by security information and event management (SIEM) rules; this detection capability requires advanced analyst intelligence.

Going beyond these attack pathways, there are a number of analysis techniques that can be employed for DNS threat hunting:

- **Replication drift**
  Another professional-level method is AD DNS replication drift analysis. If an attacker interferes with DC replication timing—for example, by registering a C2 domain on one DC and delaying its replication to another—even millisecond-level anomalies appear in the replication delta. This is a signal that many organizations miss but advanced threat hunters can detect. Normally, AD DNS replication is consistent; a record existing on only one

DC and not on others is almost a definitive indicator of an unauthorized DNS record insertion attempt.

- **Behavioral time signatures**
  One of the strongest aspects of DNS-based threat hunting is the ability to discover behavioral time signatures that attackers leave behind unknowingly. Internal DNS and DMZ DNS servers reveal both query–response relationships and the attacker's working tempo, operational hours, testing cycles, and infrastructure setup processes. While most CTI analysts focus on the content of DNS queries, advanced analysts examine the rhythm, frequency, time patterns, and relationships between query clusters. These microsignals provide threat indications far earlier than traditional IOCs.
- **DNS reconnaissance balancing**
  For example, after infiltrating an internal network, an advanced attacker may use a reconnaissance technique known as *DNS reconnaissance balancing* before fully compromising the AD environment. This technique is identified by the attacker resolving DNS names for many systems that follow the same naming convention, but doing so in waves rather than all at once. In the first wave, workstation clusters are resolved, followed by a pause of several hours; in the second wave, patterns resembling DC names are resolved; in the third wave, file servers are resolved; and in the fourth wave, service (SRV) records related to service accounts are targeted. This rhythmic reconnaissance behavior is fully visible in pDNS and exposes the attacker's lateral movement plan before it even begins.
- **TTL variance pattern**
  Similarly, one of the most subtle signals that attackers leave behind on DMZ DNS servers is the TTL variance pattern. Under normal circumstances, TTL values within a corporate DNS zone are governed by defined policies and remain within fixed ranges, such as 60 seconds, 300 seconds, or 3,600 seconds. However, if an attacker is performing DNS tunneling through the DMZ, the C2 communication they establish is typically carried out using algorithmically generated domains (domain generation algorithm [DGA]), and the TTL values of these domain names are recalculated by automated tools for each query. As a result, TTL values may suddenly appear at nonstandard levels such as 17, 21, 19, or 25. A professional analyst can easily detect this irregularity by examining the TTL histogram of DNS queries.
- **Name resolution patterns**
  An even more advanced dimension of DNS hunting is the time-series analysis of name resolution inconsistencies. Advanced attackers keep their C2 domains active only during specific hours to avoid detection, while making the domain appear inactive at other times. For example, a C2 infrastructure may resolve between 01:00 and 06:00 UTC and return `NXDOMAIN` responses at other hours. Such time-window behavior is extremely rare in normal infrastructures. By examining `NXDOMAIN` → `NOERROR` → `NXDOMAIN` transitions on DMZ DNS servers along a time axis, an analyst can identify the attacker's operational time window. This information plays a critical role in understanding the attacker's geographic location and the working hours of the operational team.

- **SRV records**

  On the internal DNS side, an even more advanced technique involves anomalous analysis of SRV records. While attackers attempt to expand their privileges within the domain, they perform reconnaissance against Kerberos, LDAP, Common Internet File System (CIFS), or Windows Remote Management (WinRM) services. During this reconnaissance, the timing and volume of queries to records such as `_ldap._tcp.dc._msdcs.` and `_kerberos._tcp` change noticeably. Under normal conditions, these records are queried only by system processes and exhibit a consistent rhythm. However, just before an attacker manipulates Kerberos using tools such as Mimikatz, Rubeus, or custom ticketing utilities, SRV queries suddenly spike. This increase isn't visible in traditional security logs, but becomes very clear when examining internal DNS query volume graphs.

- **Quiet fallback domains**

  On the DMZ side, attacker testing of C2 infrastructure generates another critical signal: quiet fallback domains. This technique allows the attacker to fall back to backup domains if the primary domain is blocked by Domain Name System Security Extensions (DNSSEC) or reputation-based systems. Most of these backup domains appear benign; in fact, attackers often name them to closely resemble legitimate corporate domains, for example:

  - `api2.company-cdn.net`
  - `updates.company-support.org`
  - `authsync.company-ops.com`

  At first glance, these domains appear to be official subservices; only advanced analysts can determine, through DNS ASN correlations, that these domains have no actual relationship with the organization. The IP blocks hosting these domains typically belong to low-cost virtual private server (VPS) providers and are resolved on the DMZ DNS server at very low but regular volumes. Another indicator is the DNS query pattern; the domains are resolved at very low but consistent intervals by internal or DMZ DNS servers. Detecting this low-volume, regular resolution pattern reveals that the attacker is preparing infrastructure even before launching an attack.

- **Query flags**

  Another professional dimension of DNS hunting is DNS query flag behavior analysis. Advanced threat actors use custom-written DNS clients instead of standard resolvers to evade security systems. These clients often generate `QUERY` flags incorrectly. For example, individual queries with the **Recursion Desired (RD)** flag disabled are almost never seen in normal client behavior. Likewise, an attacker's custom DNS agent may generate Extension Mechanisms for DNS (EDNS(0)) parameters incorrectly or omit them entirely. These micro-anomalies expose the specialized toolset used by the attacker within the internal network; moreover, this signal can't be detected by any AV, endpoint detection and response (EDR), or intrusion detection system (IDS).

Beyond all of this, one of the most critical use cases of internal and DMZ DNS analysis is the ability to derive the attacker's lateral movement map through DNS. Before moving from one machine to another, an attacker must resolve the target system via DNS. Therefore, unusual hostname resolutions—especially cases where machines that don't follow internal naming conventions or have no legitimate business relationship suddenly resolve each other—make the attacker's intended path through the internal network visible. For example, an accounting server suddenly attempting to resolve the hostname of a machine in the software development environment is with 99% probability a reconnaissance preceding lateral movement in real-world scenarios.

When all of these signals are analyzed together (internal DNS, DMZ DNS, and global telemetry), the attacker can be detected even as they take their very first steps. This unique power of DNS is often overlooked; yet when used correctly, it becomes one of the most effective, quietest, and earliest-detecting sensors in modern cyber threat hunting.

Example Scenario: Advanced DNS Telemetry Hunting Analysis

The DNS topology shown in Figure 3.1 illustrates the DNS resolution flow of a typical corporate network:

- Internal network clients (workstations + servers)
- Internal DNS (AD-integrated)
- DMZ DNS/External Resolver
- Internet Root → Recursive flow

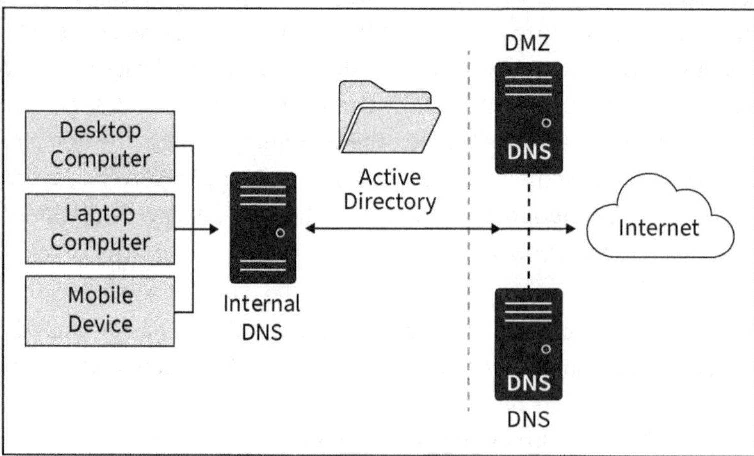

Figure 3.1: Simplified Internal DNS Architecture

Let's begin our example scenario with the attacker's initial target—that is, infiltrating the internal DNS behavior. The internal DNS zone layer shown in Figure 3.1 represents the core of the AD environment. For APT actors, this area contains critical reconnaissance information such as the following:

- Internal topology
- Hostnames of user workstations
- Server placements
- Locations of DCs
- SRV records (LDAP, Kerberos)
- Dynamically registered records distributed via Dynamic Host Configuration Protocol (DHCP)

After gaining access to the internal network, the first action an attacker takes isn't to execute commands on the endpoint, but rather to run the following commands at very low frequency and with distributed timing:

- `nslookup`
- `nltest /dsgetdc`
- `resolve-dnsname`

This behavior can easily go unnoticed under normal conditions. However, with CTI, microsignals appear as early warnings in the internal DNS behavior.

Professional CTI teams analyze internal DNS behavior of APT actors through the following four signals:

- **Unexpected hostname pattern queries**
  Before directly searching for DCs, an APT actor generates queries such as these:
  - `wsus01.internal.local`
  - `backup-sql.internal.local`
  - `fs-archive.internal.local`
  - `msk-kdc.internal.local`

  These names aren't known to ordinary users. Therefore, such queries reveal an attacker's lateral movement planning to a threat hunter.
- **TTL pattern deviation**
  Under normal conditions, DNS TTL values typically follow established patterns within the internal network infrastructure:
  - DC records: 3,600 seconds
  - Server records: 900–1,200 seconds
  - Client records: 300 seconds

  When an APT actor creates a dynamically generated fake record, the TTL typically appears as 60–120 seconds. This is the earliest detectable signal within the internal DNS layer shown earlier in Figure 3.1.
- **Nonhuman query cadence**
  Human users generate DNS queries not in burst patterns, but with random-hesitant

behavior. APT actors, by contrast, generate mathematically rhythmic queries, such as these:

- One `_ldap._tcp.dc._msdcs` query every 17 seconds
- One reverse lookup every 50 seconds

This frequency clearly indicates deliberate automation.

- **Reverse lookup anomalies**

  For external infrastructure pivoting (EIP)—a reconnaissance technique in which attackers use reverse DNS queries to identify reachable systems and map external-facing infrastructure—an APT actor may issue consecutive reverse DNS queries such as these:

  - 10.1.15.72: PTR lookup
  - 10.1.15.73: PTR lookup

  This behavior constitutes a red flag along the internal resolution path shown in Figure 3.1.

The external DNS flow depicted in Figure 3.1 represents the attacker's second objective: They want to exit the internal DNS, create a tunnel through the DMZ DNS, and connect to C2 domains on the internet. Here, APT actors leave behind distinctive professional signals:

- **Dual-stack query signal**

  APT actors deploy C2 infrastructure across dual IPv4 + IPv6 address blocks. This becomes visible on the DMZ DNS via the following comparison:

  - Normally used domains: A records
  - Attacker C2 domains: Simultaneous A + AAAA records

  This behavior is almost never normal in enterprise environments.

- **Time-sliced NXDOMAIN model**

  During APT preparation, a domain responds as follows:

  - Between 08:00 and 09:00: `NoError`
  - At other times: `NXDOMAIN`

  This behavior aims to evade sandboxes, maintain a low reputation score, and bypass DNS firewall detection. On the DMZ DNS, this temporal change is clearly visible in time-series graphs.

- **DNS tunneling via the DMZ relay**

  Tunneling activity leaves the following signals on the DMZ DNS:

  - High-entropy (e.g., Base64-like) subdomains
  - Optimal tunnel sizes between 150 and 250 bytes
  - Query frequencies following prime-number rhythms such as 12, 17, or 23 per minute (a signature APT behavior)

  These traces constitute advanced indicators of C2 activity within the DMZ DNS layer shown in Figure 3.1.

Our next step is to consider APT behavior observed in the AD DNS replication. The internal DNS zone shown in Figure 3.1 is, in fact, AD-integrated DNS. Often unknown to many analysts, advanced actors use a DNS record without allowing it to replicate to all DCs. How does this occur?

- The attacker adds a hidden domain record (e.g., `sys-upd.internal.local`) on DC1.
- The attacker deliberately abuses AD replication delay.
- The attacker conducts operations before DC2/DC3 become aware of the record.

This behavior becomes visible between the internal DNS zone and replication layer in Figure 3.1. Replication drift is one of the most definitive indicators of attacker activity.

So, how can an organization detect an APT early using everything we've covered? We recommend the following analyses for this scenario:

- **Internal DNS**
  - Unexpected SRV lookups
  - TTL deviations
  - Reverse lookup bursts
- **DMZ DNS**
  - DGA-like subdomain entropy
  - `NXDOMAIN` time-based patterns
  - Sequential AAAA queries
- **Global DNS telemetry**
  - IP pivot chains
  - ASN profiling
  - Domain shadowing

When these three analyses are combined, an APT actor can be detected purely through DNS behavior, without modifying domain records, without executing a single command, and without generating a single log entry.

This represents one of the most advanced CTI techniques available. The following list describes the essence of a horizontal DNS topology:

- A map highlighting often-overlooked signals for APT hunting
- A visualized form of a DNS-based early detection model
- A guide for reading the attack chain between internal–DMZ–internet layers
- A reference architecture for professional CTI behavioral analysis

3.3 Human Intelligence

Although CTI is often perceived as a discipline based primarily on technical signals, telemetry, and automation layers, true operational superiority is only possible through a structure capable of interpreting human behavior. HUMINT is the most critical intelligence source used to reveal intentions, motivations, decision-making processes, and the invisible relationships between actors that digital traces can't explain. For this reason, HUMINT has become a decisive factor that can change the game, especially in complex and long-term threats.

In this section, we'll address HUMINT's role within CTI, its operational contribution model, source evaluation methods, balance between reliability and risk, and tangible value it provides when combined with digital intelligence, at both the technical and strategic levels.

3.3.1 The Role of HUMINT in Cyber Intelligence

The strength of HUMINT lies in its ability (unlike data produced by technical platforms) to answer the question of who the threat actor is through behavioral patterns, operational habits, communication preferences, and levels of operational discipline, instead of through identity information. Using HUMINT-derived signals is often the only way to understand whether an APT group has entered a new campaign phase, is experiencing internal fragmentation, has shifted sponsor relationships, or is facing an internal weakness affecting its operational capability.

This is why advanced operational units, particularly those tracking state-sponsored threats, use HUMINT in a hybrid model integrated with digital intelligence flows: Technical data explains what the threat actor is doing; HUMINT explains why. The combination of these two layers advances threat anticipation by enabling predictions regarding attack timing, target selection motivations, levels of operational seriousness, and potential future vectors.

HUMINT also illuminates the background of elements that are difficult to detect in the digital domain, such as high-risk vulnerabilities, critical infrastructure targeting, the financial flows of ransomware operations, role distributions within cybercrime organizations, and vendor–broker relationships. This intelligence enables organizations to defend against attacks and develop proactive strategies by understanding the adversary's decision-making cycle.

In today's threat ecosystem, HUMINT relies on multilayered methods such as infiltrating internal communications using false identities, exchanging information with forum moderators, analyzing the behavior of threat operators, examining operational security (OPSEC) weaknesses in the social domain, and resolving trust networks between actors. Due to this structure, HUMINT is an analytical discipline capable of identifying human errors, psychological fractures, and organizational fault lines.

3.3.2 Internal Source Interviews and Internal Information Flow

While corporate CTI often focuses on external sources, the information carrying the earliest indicators of attacks is frequently found within the organization itself. Professional interviews with internal stakeholders provide a level of originality and contextual richness that no other data source can offer. For this reason, internal-source HUMINT is one of the most critical building blocks for predicting the likelihood of an attack, rather than just responding to one.

An SOC analyst's remark regarding a misclassified alert—such as "the tone of this alert has changed"—may represent a behavioral signal that automation systems are unable to analyze. When an incident response analyst states that a recurring spike in instantaneous connections on a previously isolated host is statistically abnormal, this may constitute the first human observation of an unreported lateral movement. Even a comment from the red team such as "there's unexpected resistance in this user group" can point to an insider threat layer that remains technically invisible. The value of internal interviews emerges precisely at this intersection between human intuition and technical telemetry.

Another key source of insight in corporate operations is management. A department manager's observation that "data requests have increased unusually this quarter" may normally be perceived merely as workload; however, when evaluated within a HUMINT framework, it serves as an early warning of data-focused attack preparation, internal information leakage, or a covert third-party request. Strategic-level managers often unknowingly carry critical organizational signals: team changes, role shifts, sudden behavioral changes in specific employees, unusual access requests, and quietly conducted projects. These signals may represent the first indicators of the presence of an active or potential threat actor within the organization.

The most critical domain of contextual HUMINT is *insider threat analysis*. Technical systems often fail to detect the early stages of insider threats. Human behavior, however, produces weak signals that function as early warnings. These signals aren't necessarily explicit or obvious; they are often micro-indicators that only a careful observer can interpret meaningfully:

- An increased interest in sensitive data by specific users
- The clustering of individuals within the same team who repeatedly request similar types of privileges
- Technical issue complaints raised by an employee that actually serve as a pretext for testing specific systems
- The presence of behaviors such as topic shifting, fluctuating agendas, or deliberate ambiguity in internal communications

These silent signals are inadvertently disseminated through informal communication channels within the organization. For the HUMINT analyst, the primary challenge is to render these signals readable. Technical telemetry measures events that are the result of human behavior; HUMINT identifies the intent behind those events.

Collecting internal intelligence doesn't rely solely on formal meetings. The social dynamics of organizations generate information that is less formal, yet closest to reality, such as a brief sentence overheard in a break room, a new project on which a team is concentrating, an increase in the number of specific individuals working late at night, the red team reporting unexpected difficulty, or an SOC analyst remarking to colleagues that "there's something odd about that AD group". Each of these constitutes an early warning mechanism that precedes technical logs. For a HUMINT analyst, internal informal networks are high-accuracy, low-volume information sources, much like closed forums on the dark web. The following examples illustrate typical forms of informal signals that may emerge within an organization's internal environment:

- **Source historical accuracy**
  The degree to which information previously provided by the source has been verified.
- **Contextual depth of information**
  The extent to which the signal intersects with internal operations.
- **Behavioral consistency**
  The clarity of the source's motivation, psychological disposition, and intent.

Through this approach, the organization is protected both by technical data and the observational capabilities of its personnel. Technology generates numerous alerts; HUMINT indicates which signal among them truly smells of an attack.

Ultimately, internal-source HUMINT derives meaning from what people within the organization say and what they don't say. When weak signals absent from formal reports, realistic insights carried by informal networks, and microshifts in employee behavior converge, the organization's defensive capacity extends far beyond technical capabilities. In such a structure, HUMINT is no longer just a reaction to an attack, but an invisible security layer that prevents the attack from emerging in the first place.

One of the most striking examples of this approach was observed in an anomaly detected within an advanced organization: The SOC team had noticed low-priority yet repetitive DNS queries originating from an old production server that had been quiet for weeks. While this appeared normal from a systems perspective, an IT operations employee remarked during a coffee break that "someone keeps querying that server during the night shift, as if probing a machine that isn't supposed to be active." This behavioral suspicion, not clearly visible in logs, gained meaning when evaluated through HUMINT as an indicator of a larger preparation involving the testing of privileges by an internal actor. One week later, credential chaining attempts were uncovered through the same server that had been tested. The incident began not with technical telemetry, but with informal internal observation.

Similarly, in another case not reflected in incident response team reports, a red team assessment revealed that a team member had noticed certain users approaching social engineering scenarios with an abnormally high level of caution. Although this appeared positive on the surface, when analyzed from a HUMINT perspective, it became evident that the user group had previously been in contact with an external actor and was therefore exhibiting an unusual learned reflex. The threat actor had failed to complete the phishing process but had left a behavioral trace within the team. This trace was invisible to technical systems; however, HUMINT analysis identified an incomplete social engineering attempt.

Analysis of Silent Signals

Within organizations, threat-related behavior rarely begins loudly; on the contrary, the threat actor first adapts to the organization's natural lifecycle. The true strength of HUMINT lies in its ability to read the microvibrations that disrupt this adaptation, for example:

- **A decrease in privilege revocation requests**
  In teams that are normally dynamic, the sudden cessation of such requests may indicate that an internal group has entered into a silent agreement to preserve access levels.
- **A sudden change in an employee's long-standing routine tasks**
  This is often interpreted as workload pressure; however, within HUMINT, it may be assessed as operational avoidance behavior.
- **Short, closed-door meetings between individuals in the same department who normally have no direct interaction**
  This may indicate preparation for internal information transfer.

None of these examples constitutes a direct breach; however, all of them are traces left by the threat actor during the preparation phase. For a HUMINT analyst, these traces indicate the direction and severity of risk before the technical aspects of an incident are reached.

Conducting interviews alone isn't sufficient; an internal intelligence structure must also understand the natural channels through which information flows organically. In modern organizations, information flow is examined at three levels:

- **Formal flow**
  This flow includes reports, meetings, incident tickets, and change records.
- **Operational flow**
  This flow includes team chats, shift handover notes, and technician observations.
- **Silent flow**
  This flow includes behavioral patterns, cultural changes, and disruptions in work rhythm.

Most organizations focus only on the first two flows; however, the earliest signals often emerge within the silent flow. Reading this flow requires the HUMINT analyst to understand the organization not from the outside, but as an internal observer.

For example, one technique used in advanced HUMINT practices is the *operational rhythm map*. This map outlines an organization's natural behavioral flow over a one-week period and measures disruptions within that flow. Increased file requests at unexpected hours, unrelated teams querying the same dataset, or certain employees repeatedly asking security-related questions may indicate attack preparation. Such analysis is unknown to external teams because it emerges only through long-term observation of internal organizational rhythm.

The ultimate objective of a mature internal intelligence structure isn't to search for a culprit. The goal is to interpret internal behavioral patterns to identify the following:

- When the threat actor may take action
- Which weaknesses are being targeted
- Whom the actor is attempting to manipulate internally
- Which operational backchannels are being used to leak information

This approach provides a level of foresight that can't be achieved through technical systems alone.

3.3.3 Informants, Researchers, and Dark Web Engagements

One of the sharpest edges of CTI is the capability to obtain information directly from within the threat ecosystem itself. While open-source monitoring, technical telemetry, and organizational logs provide visibility up to a certain level, understanding the true intentions, transactions, disputes, and weaknesses of threat actors often requires penetrating closed environments. At the heart of this domain lie informant relationships, coordination with independent researchers, and controlled intelligence-gathering activities conducted on the dark web.

Dark web forums, which we'll discuss in more detail in Chapter 5, may appear from the outside as a homogeneous criminal ecosystem; in reality, they are highly complex, multilayered structures with their own internal dynamics. Each forum has its own hierarchy, moderation culture, admission mechanisms, and communication rhythm. The key to conducting effective HUMINT in this environment is to observe it not through the eyes of a technical analyst, but through those of a behavioral scientist. This is because threat actors reveal themselves through their behavioral patterns.

Deciphering these behavioral layers isn't a technical action; it requires a high level of HUMINT literacy. We'll explore the different aspects of this application of HUMINT in the following sections.

The Strategic Value of HUMINT on the Dark Web

The primary objective of dark web engagements isn't to combat crime or provoke threat actors. The real goal is to expose the genuine weaknesses of threat actors by understanding the following elements within closed ecosystems:

- Power relationships
- Disputes

- Operational gaps
- Commercial motivations
- Security vulnerabilities
- Internal group fractures

Technical defenses protect systems; HUMINT, on the other hand, resolves the psychological and operational balance of the threat actor. This balance is often the most critical factor determining the direction, intensity, and timing of an attack.

Every engagement with the dark web ecosystem requires both reading the platform and interpreting the architecture of human behavior that sustains it. Threat groups may conceal technical traces, mask identities, or constantly change their infrastructure, but they can't transform their behavioral patterns at the same pace. For this reason, the most advanced HUMINT operations treat intelligence collection from closed environments not as an act of access, but as a behavioral analysis process.

At this point, the analyst's primary objective is to identify the psychological models of individuals forming the forum, the power relationships they establish with one another, their commercial priorities, and their breaking points. This analysis is conducted through the fixed parameters of human nature, not through technical tools.

Controlled Access and Secure Interaction

Accessing the dark web entails far more multidimensional risk than commonly assumed. For this reason, professional intelligence units treat access as an operational decision, not a technical event. This approach is based on three fundamental principles:

- **Consistency, not invisibility, is essential.**
  Threat actors detect inconsistent profiles more quickly than anonymous ones. Therefore, controlled personas are built on behavioral continuity.
- **The objective isn't access, but a listening position.**
  The goal isn't interaction, but observation, that is, understanding which topics actors focus on, whom they conflict with, and which vulnerabilities they discuss.
- **Technical secrecy alone is insufficient; operational anonymity is required.**
  Threat actors often analyze behavioral traces more than technical ones. A profile's writing style, timing rhythm, or word choice can all carry identity signals.

For this reason, in professional systems *metadata hygiene* is applied at the IP, message authentication code (MAC), and browser-fingerprint level, as well as at the controlling communication rhythms level.

Some of the most valuable intelligence in the modern threat landscape comes from independent sources that already have access to specific ecosystems. These sources are examined in three main categories:

- **Individuals with organic access to the dark web**
  Having been in contact with relevant communities for years, they carry contextual signals that can't be obtained externally.
- **Independent researchers and journalists**
  Some researchers, through long-term observation of specific threat groups, possess knowledge of relationship networks that organizations can't access.
- **Sector-specific informants**
  Individuals working in finance, hosting, crypto services, or logistics may notice backend movements of threat actors.

When working with these individuals, a fundamental HUMINT principle applies: Contact with a source isn't a process of extracting information, but a process of building trust. This trust is evaluated at three levels:

- **Operational trust**
  The source's ability to present the obtained information in a verifiable manner.
- **Behavioral trust**
  The clarity and understandability of the source's motivation: financial, personal, or ideological.
- **Temporal trust**
  The source's consistency in providing reliable signals over time.

Without these three elements, information—no matter how technically impressive—is worthless intelligence-wise.

Advanced OPSEC: Behavioral Privacy

Operational security (OPSEC) is the discipline of protecting sensitive activities by identifying and minimizing information that adversaries could use to infer intentions or capabilities. In intelligence and cyber operations, OPSEC is critical because even small behavioral patterns can reveal the presence, structure, or objectives of an operation. The technical dimensions of OPSEC are widely known; however, in professional operations, the most valuable component is behavioral OPSEC. This means controlling communication style, timing habits, and content-production rhythm. Here are a few examples:

- If a profile is active only on weekdays between 09:00 and 17:00, threat actors may infer a corporate operation.
- A user who employs identical terminology in every message may be flagged as a "constructed persona."
- A profile that consistently focuses on the same groups may attract suspicion due to one-dimensional behavior.

For this reason, advanced OPSEC is based not on technically hiding a profile, but on behaviorally normalizing it. This technique is used in state-level HUMINT operations, and the same logic applies in dark web intelligence: not to hide, but to remain ordinary enough to be invisible.

Real-World Example: Reading the Vulnerability of a Closed Ecosystem

Failures in behavioral OPSEC often reveal hidden operational structures that technical analysis can't detect. The following anonymized real-world example illustrates how subtle linguistic patterns can expose identities that operators attempt to conceal.

In a forum, a particular vendor appeared to be selling large datasets for months. Technical analysis was unable to determine the source of the data. However, a researcher noticed that a subtle change in the vendor's writing style—specifically, an increased frequency in the use of a certain punctuation mark—resembled the style of another vendor. HUMINT analysis ultimately revealed that the two profiles were in fact operated by the same individual, and that the vendor had manipulated forum administration to present themselves as a multi-vendor ecosystem.

No technical tool could establish this connection. However, human behavior revealed the true structure of the threat.

Decoding the Hidden Architecture of Group Behavior

Although many threat groups appear homogeneous from the outside, examination of their internal structures reveals three distinct roles:

- **Load carriers**
  Individuals who handle the technical aspects of operations, typically maintaining a low profile while demonstrating high productivity.
- **Directors**
  Actors who steer discussions, exert influence over competing vendors, and shape forum culture.
- **Disruptors**
  A user segment that instigates conflict in the background, engages in manipulation, or provokes rival groups.

Even a minor shift among these three roles indicates that forum dynamics are changing. The value of HUMINT lies in its ability to detect such shifts before an incident occurs, for example:

- If disruptors suddenly fall silent, this often indicates that behind-the-scenes negotiations have concluded.
- If directors suddenly adopt a harsher tone, an internal transfer of power within the forum may have begun.
- If load carriers begin listing similar services at reduced prices within the same time frame, intergroup market competition is intensifying.

From the outside, these signals may appear to be ordinary messaging activity. However, to a HUMINT specialist skilled in behavioral analysis, they constitute early warnings of impending operations or internal conflicts.

Real-World Example: A Role Shift Hidden Beneath Signal Noise

An anonymized case from the dark web illustrates this clearly: A particular user profile had been known for years as a midlevel vendor on a forum. Over time, however, the profile began to assume an arbitrator role in intergroup disputes without providing any explanation. This was a function typically performed by moderators. Examination of message content revealed no technical inconsistencies. HUMINT analysis, however, identified a subtle change in message timing: A profile that had previously been active during specific hours for weeks suddenly spread its activity across a much wider time range.

This signal revealed that the individual behind the profile was no longer a single person and that account usage had shifted to a divided, task-based structure. Such division of labor typically indicates the following:

- The threat actor has expanded.
- A new operational organization has been established.
- The actor seeks to assume a larger role within the forum.

This conclusion can be reached only through behavioral HUMINT; technical traces don't reveal it.

Contextualizing Data from Independent Sources

Information provided by informants, journalists, and researchers is valuable, but not sufficient on its own. For information to gain meaning, the following must be analyzed:

- The source's perspective
- The network of relationships in which the source operates
- The source's history with specific groups or vendors
- The motivation behind sharing the information

Accordingly, advanced HUMINT structures evaluate information not solely based on content but also on the behavioral motivation of the source.

For example, data conveyed by a journalist may be technically flawless; however, due to the journalist's motivation to protect their source, certain critical elements may be withheld. Conversely, a sector-specific informant, acting primarily to protect their own position, may provide more limited but significantly rawer and unfiltered information.

For this reason, high-maturity HUMINT teams assess every piece of information by asking "From which perspective can I read this most accurately?"

The Highest Level of Operational Anonymity: Distributed Behavior

Operational anonymity isn't limited to concealing a single individual; at times, the most effective method is to distribute a persona's behavioral rhythm across multiple individuals. This technique neutralizes the profile-correlation methods developed by threat actors.

In practice, this means that a single operational profile may be managed by several trained operators who follow shared communication guidelines. Each operator contributes at different times and with slight variations in writing style, timing, and interaction patterns, preventing the profile from developing a consistent behavioral signature that could be linked to one individual. For example, one operator may handle daytime discussions, another may respond during different time windows, and a third may manage longer analytical posts. As a result, behavioral analysis tools attempting to correlate language patterns, activity timing, or cognitive style struggle to attribute the profile to a single person.

This approach is used in advanced HUMINT units for the following purposes:

- To prevent the profile from exhibiting a singular cognitive signature
- To ensure natural variation in message language
- To avoid timing patterns that indicate a single geographic location
- To prevent behavioral prediction algorithms from identifying an individual

This method isn't employed for misuse, but to conceal the presence of intelligence personnel and ensure their safety.

Even threat actors conducting dark web behavioral analysis struggle to resolve such distributed rhythms, as doing so requires mimicking not a single individual, but the coordinated rhythm of a controlled team.

The Ultimate Output of Dark Web HUMINT: Mapping Pressure Points

The objective of HUMINT operations conducted on the dark web goes beyond information collection; it is to identify the ecosystem's pressure points. These pressure points represent the areas where threat actors are most vulnerable:

- Erosion of intragroup trust
- Changing commission rates among financial intermediaries
- Moderator–vendor conflicts
- Degradation within the supply chain
- Failure of new members to adapt
- Silent departure of long-standing members from forums

Such fractures are signals that emerge before major operations begin. By mapping these fractures, the HUMINT analyst can anticipate the following:

- Which group is at risk of disintegration
- Which group is preparing a new operation

- Which vendor is experiencing a loss of trust
- Which ecosystem is being driven toward conflict

This level of foresight isn't achieved through technical data, but solely through the ability to decode the intricacies of human behavior.

3.3.4 Social Engineering and the Analysis of Human Vulnerabilities

Social engineering continues to be one of the most underestimated yet most powerful attack methods within the cyber threat ecosystem. While technical vulnerabilities can be mitigated through patching, human vulnerabilities can never be fully patched. Threat actors have exploited this reality for years, increasingly bypassing technical barriers through sophisticated social engineering campaigns and targeting the weakest link in security: human behavior.

Along with being an attach technique within an advanced HUMINT approach, social engineering is an intelligence source that reveals a threat actor's motivation, maturity level, organizational structure, and behavioral patterns. A social engineering attempt often provides more significant indicators than the actor's identity itself: the language used, constructed scenario, tone of manipulation, timing rhythm, and target selection logic. All of these constitute valuable HUMINT signals, which we'll discuss in more detail in the following sections.

Social Engineering Profiles of Threat Actors

A professional HUMINT analyst classifies actors according to both their technical activities and the social engineering methods they employ. This is because each actor's manipulation style is unique and often more persistent than technical traces. There are three primary examples:

- **Procedural manipulators**
  These actors attempt to persuade the target by overwhelming them with procedures. They typically use a serious, detailed, and formal tone, relying on lengthy pseudo-policies and urgent corporate requests.
- **Emotional pressure actors**
  These profiles rush the target's decision-making process. They use a threatening tone and phrases such as "your account will be suspended" or "a critical security risk has been detected." Time pressure is the defining characteristic of these actors.
- **Trust exploiters**
  They imitate corporate relationship networks, posing as senior management or technical support units. Their temporal alignment is usually very strong; if an organization has peak workload periods, they send messages precisely during those times.

These profiles provide critical insight into an actor's operational maturity. For example, the use of procedural and corporate language may indicate an underlying organizational structure, whereas emotionally driven manipulation campaigns are typically associated with low-cost threat actors targeting broad audiences.

HUMINT Interpretation of Manipulation

While the technical details of social engineering techniques are well known, what is critical from a HUMINT perspective is the interpretation of the behavioral and operational signals behind these methods. Let's apply HUMINT to common social engineering techniques:

- **Phishing (email-based persuasion operations)**
  In a phishing email, analysis isn't limited to the URL or attachment. From a HUMINT perspective, the following questions are evaluated:
  - What does the timing of the message indicate?
  - Which internal organizational role is being targeted?
  - How familiar is the actor with the organization's terminology?
  - Which operational weakness of the organization does the target represent?

  For example, a fake accounting email sent during an organization's payroll period indicates that the actor has studied corporate cycles. This points to a threat actor conducting organization-specific reconnaissance, and not just an opportunistic attacker.
- **Vishing (voice-based persuasion and manipulation)**
  In vishing attacks, voice tone, speech frequency, pauses, and vocabulary carry significant value. HUMINT analysis places particular emphasis on the following signals:
  - Which words does the actor avoid during the conversation? (often exaggerating or incorrectly using technical terms)
  - How rehearsed is the scenario? (unnatural rhythm, memorized phrases)
  - What does the persuasion strategy rely on? (authority, fear, requesting help, urgency)

  These signals directly reflect the threat group's level of training and operational preparedness.
- **Pretexting (identity manipulation through scenario construction)**
  This method is the most complex and requires well-prepared HUMINT analysis, as pretexting reveals the attacker's ability to impersonate a persona.

  For example, when using corporate jargon indicators such as the following, this demonstrates that the attacker possesses only superficial knowledge of the organization:
  - Mispronouncing specific terms
  - Confusing department names
  - Providing unnecessary detail
  - Failing to repeat the same story consistently

  Such signals are validated through OSINT and HUMINT insights obtained internally.
- **Deepfake-supported social engineering**
  Although these methods are relatively new, they struggle to conceal behavioral signals.

Within HUMINT analysis, a deepfake video or audio recording can be identified through indicators such as these:

- Speech rhythm not aligning with natural human biology
- Emotional transitions failing to synchronize
- Inconsistencies between intonation and sentence content
- Repetitive microgestures

This analysis is conducted through technical deepfake detection tools and the integration of HUMINT expertise grounded in an understanding of human behavior.

Combining HUMINT and OSINT: Multilayered Validation

The most powerful stage of social engineering analysis occurs when HUMINT data is combined with OSINT. Through this method, a threat actor's social media traces, forum behavior, technical IP clusters, intracommunity relationships, and timing habits are correlated with HUMINT signals.

For example, if an attacker impersonating an employee repeatedly uses the word "urgent" in emails, this can be compared with writing style, tone, and word preferences analyzed through OSINT. In many cases, a pretexting attacker struggles to replicate the same linguistic patterns as the person they are impersonating.

Additionally, OSINT sources reveal critical contextual factors regarding social engineering scenarios, such as the following:

- Other organizations that were targeted previously
- Sector the threat actor is focusing on
- Economic or political period during which operations are conducted

Analyzing Inconsistencies in Human Nature

Technical systems easily detect logical errors; HUMINT, however, analyzes the inconsistencies inherent in human nature:

- Mismatch between the constructed narrative and the vocabulary used
- Vocal intonation failing to align with the emotional context
- Message timing being overly aligned with the target's routine (too perfect)
- The same attacker using similar persuasion strategies across different victims
- Changes in a profile's communication rhythm reflecting the threat actor's shift or work-cycle model

These signals don't reveal who the source of the attack is, but rather why and how the actor is operating. This, in turn, creates the capacity to anticipate the next step of the attack.

Although social engineering attacks appear on the surface as acts of communication, they are in fact among the strongest signal sources for revealing a threat actor's psychological structure, operational discipline, intragroup role, and decision-making style. For this reason, advanced HUMINT practitioners view social engineering attempts as opportunities to decode the threat actor's behavioral DNA.

The rhythm of the messages prepared by an actor, the intensity of the persuasion tools used, the complexity of the scenario, target selection, and the timing of the attack are nearly as unique as a fingerprint. This *behavioral trace* defines both the identity of the threat group and its character, which is most critical in modern intelligence.

Let's walk through the key HUMINT methods and techniques that empower you to catch these inconsistencies:

- **Micro-intent mapping**
  An important method used in advanced HUMINT operations is identifying the micro-intents behind social engineering attempts. Micro-intent refers to the microlevel objectives a threat actor embeds within a single message, for example:
  - Subtle urgency in message tone = pressure to obtain data quickly
  - Inclusion of unnecessary details = an attempt to build trust
  - Topics deliberately avoided = areas where the actor is weak
  - Exaggerated technical terminology = an attempt to exploit information asymmetry
  - Scenario changes based on the target's response speed = real-time manipulation capability

  Mapping these micro-intents clarifies the attacker's motivation: Is the actor seeking information, attempting to gain access, testing behavior, or just measuring the organization's response?

  In addition to detecting social engineering, this technique is used to predict the threat's next move.
- **Breakpoint signaling**
  One of the most revealing aspects of social engineering campaigns is the moment when the attacker's scenario breaks down. These break points provide the most valuable data about the attacker's profile.

  Examples of break-point signals include the following:
  - If the attacker's tone hardens when the target persistently asks for an answer to a specific question, the pretext isn't realistic.
  - If message volume suddenly increases, the actor has moved into a phase aimed at inducing a panic response in the target.
  - If the communication channel changes (email to phone), there's a coordination transfer within the operation, indicating multiple participants.
  - If the attacker repeats the same sentence using different wording, the script has failed and improvisation has begun; this reveals the threat actor's stress level.

  These break-point behaviors reveal the nature of the attack and the actor's organizational capacity.

Real-World Example: Silent Persuasion Pattern

In a real operational analysis (anonymized), a threat actor sent seemingly harmless emails over the course of several weeks. At first glance, there was no technical threat. However, a HUMINT specialist identified three critical silent persuasion signals within the subtext of the messages:

- Each message contained a sentence written in a "request for help" tone.
- Although not directly requested, the messages included indirect questions aimed at learning the target's system configuration.
- The timing pattern demonstrated synchronization suggesting that the actor was aware of the target's meeting schedule.

These silent signals showed that the attacker was gradually warming up the target through small steps, conducting a long-term profiling effort rather than launching a sudden attack.

Cross-validation via OSINT revealed that the attacker used the same method across different sectors, demonstrating that the actor possessed a defined psychological manipulation repertoire.

- **Character consistency test**
  Advanced HUMINT teams don't evaluate pretexting attacks solely based on content; they also measure the behavioral consistency of the character constructed by the attacker. This test technique is used in professional intelligence environments and consists of three stages:
  - Temporal consistency test: Do the response times of the fictional character align with the actual working tempo of the department being represented? (Most attackers overlook this detail.)
  - Linguistic consistency test: Does the character's writing style align with the organization's internal culture? For example, real accounting departments write concisely and directly; attackers tend to rely on lengthy explanations.
  - Informational consistency test: Do the details the fictional character claims to know align with the organization's real procedures? Even the most advanced attackers readily expose themselves at this stage.

  This method is one of the most effective HUMINT techniques for dismantling pretexting attacks at an early stage.
- **Dual-footprint correlation**
  The integration of HUMINT data with OSINT isn't limited to validation purposes; it also reveals the threat actor's *dual footprint*. This footprint represents the intersection between the attacker's behavior in closed environments and their behavior in open environments.

Example: A threat actor may use an authoritarian tone in email attacks while adopting a flexible, helpful manner on dark web forums. This inconsistency reveals the actor's operational role:

- An executor in email campaigns
- A backend designer of social engineering on the dark web

This analysis demonstrates how threat actors position themselves across different operations, which is knowledge that constitutes one of the most powerful methods for mapping a group's organizational structure.

3.4 Signals Intelligence

Within the modern threat intelligence ecosystem, signals intelligence (SIGINT) is one of the most critical intelligence layers, produced through the analysis of every signal, every packet sequence, and every electronic trace carried through the unseen veins of the digital world. HUMINT reveals the intent of the threat actor, and OSINT maps the actor's visible face, but SIGINT exposes their real-time operational activities, communication rhythms, and technical behaviors:

- HUMINT explains intent.
- OSINT presents the actor's appearance on the stage.
- SIGINT reveals the actual movements behind the scenes.

In a sense, SIGINT is a silent sensor that listens to the traces an attacker leaves behind even when they don't speak. It represents one of the most reliable stages of an attack lifecycle.

Threat actors may use social engineering propaganda to present themselves differently than they are, conceal their identity, or wear a digital mask. However, the signature they leave on the network—the Transport Layer Security (TLS) version they use, packet sequencing behavior, JA3/JA3S fingerprints, or wireless signal density—remains the most difficult aspect of their technical behavior to alter, even when they attempt to hide their identity.

For this reason, SIGINT goes well beyond data collection; it's a multidimensional discipline that analyzes the rhythm of an attack, the technical capacity of the organization, the maturity level of the operation, the geographical distribution of infrastructure, and even traffic-based behavioral changes exhibited by threat actors under stress.

In the following sections, we'll detail SIGINT's subdisciplines: packet inspection, wireless signal analysis, legal frameworks, and integration with telemetry using the most advanced methods and examples. But first, let's establish the role SIGINT plays in threat hunting and your corporate landscape.

3.4.1 The Role of SIGINT in Cyber Intelligence

Today's attack techniques often progress through invisible layers: living-off-the-land techniques, encrypted communication channels, proxy chains, VPN multiplexing, anonymized infrastructures, and Internet of Things (IoT)-based botnets. Amid all of these complex structures, the only common element that makes the threat actor visible is the signal itself. For this reason, SIGINT provides the most three-dimensional data in modern threat hunting:

- Anomalies in traffic frequency
- Behavioral patterns within encrypted communications
- Signal variations produced during malware C2 communications
- Packet timing and packet length distributions
- Anonymous signals broadcast by Wi-Fi, Bluetooth, and IoT devices
- Electromagnetic behavioral differences observed on the physical network surface

This information can reveal the threat actor's real operational model, the timing of the attack, and even whether the attacker controls the infrastructure manually or through automation.

Along with analyzing cybersecurity incidents, SIGINT provides organizations with strategic foresight. Signal behaviors often change before an attack begins, for example:

- A threat actor tests C2 infrastructure days before launching a campaign.
- A ransomware group generates atypical DNS queries during its attack preparation phase.
- An APT group sends characteristic timing packets to measure the target network's time offset.
- In wireless environments, low-power signal "heartbeats" not normally produced by IoT devices may be observed.

All of these microsignals are often overlooked within normal telemetry; however, for a SIGINT specialist, they represent the first winds of an approaching storm.

Every attack has an architecture that reveals itself through signal patterns. When combined with HUMINT and OSINT, these patterns allow us to resolve the threat actor's operational discipline, infrastructure maturity level, attack type (manual or automated), internal role distribution within the group, and logistical coordination of the attack.

For example, the JA3 fingerprint used in the encrypted traffic of a particular APT group may exhibit small, consistent variations across each campaign. Similarly, the same group creating microsecond-level timing deviations during radio-based IoT reconnaissance clearly indicates that the attack is in the preparation phase. Such indicators can be revealed only through SIGINT; no other intelligence layer can observe them.

But what about the role of SIGINT in corporate security? Every organization collects large volumes of telemetry: DNS logs, proxy records, EDR/extended detection and response (XDR) data, firewall flows, wireless access point logs, and so on. However, none of this data is meaningful on its own. SIGINT reveals the behavioral coherence embedded within all this telemetry. Logs provide the data, telemetry provides the events, and SIGINT provides the technical trace of intent.

For this reason, SIGINT feeds both the technical aspect of threat intelligence and its strategic dimension. To determine whether an organization is under attack, signatures alone are often insufficient; microdeviations in signal behavior are enough. These deviations serve as early warning mechanisms that indicate the threat temperature before an attack even occurs. SIGINT has proven a fundamental truth in the modern threat landscape: Every threat actor leaves traces in the place they believe is best concealed—within the signal.

Interpreting these traces requires both technical knowledge and an advanced analytical perspective that understands the psychology of signal behavior. This is where SIGINT derives its value in both corporate and national threat intelligence. Human behavior can be deceptive, open sources can be manipulated, and technical logs may be incomplete—but the signal doesn't lie.

3.4.2 Packet Inspection and Traffic Analysis

One of the most critical components of SIGINT is reading the content and the behavior of data transmitted over the network. Although modern threat actors attempt to reduce content visibility through encryption, tunneling, redirection, and complex C2 infrastructures, characteristics such as packet timing, size, flow, and cryptographic fingerprints make it possible to recognize the actor's technical capacity and operational discipline. For this reason, packet inspection is an intersection point between HUMINT and SIGINT that penetrates the threat actor's signal logic.

In the following sections, we examine how traffic behavior can be interpreted through different analytical lenses, including the distinction between network metadata and full packet capture (FPC), as well as the role of traffic patterns in identifying threat actor activity.

Network Metadata Versus Full Packet Capture

For many analysts, metadata and FPC may appear to carry the same intelligence value; however, in a modern SIGINT approach, these two concepts function as complementary yet distinct sensors operating from different points of view:

- **Network metadata: The behavior map**
  Metadata is used to understand the rhythm, intent, and operational purpose of traffic independently of packet content. This data typically consists of the following:
  - Source/destination IP and port
  - Packet size
  - Timestamp
  - Flow duration
  - Connection frequency

  - Communication direction
  - Protocol layers

  In modern threat intelligence, the most important characteristic of metadata is that while an attacker can conceal content, concealing the rhythm of traffic is far more difficult. For example, even before an attack begins, the following observations are strong indicators that a threat actor is in the reconnaissance phase:

  - Low-volume probing performed at regular intervals
  - Short-lived connections originating from the same subnet
  - Silent ping-like behavior recurring at specific times
  - Fixed-interval keep-alives bearing a machine signature

  Metadata doesn't show what the attacker wants to do, but how they move. For this reason, it's often more valuable than content.

- **Full packet capture (FPC): The atomic level that reveals signal identity**

  Because FPC captures the complete data stream, it can directly reveal the details of exploit chains, protocol manipulations, C2 communication formats, malware's true behavior, and unauthorized data exfiltration methods. However, the true value of FPC lies in correlating variations within packets to actor behavior, for example:

  - If a threat group changes a padding pattern every eight packets during C2 communication, this indicates the presence of a custom framework written by the operator.
  - Microsecond-level inconsistencies in packet sequencing can reveal whether the group operates its C2 server manually or through automation.
  - If the same actor uses an identical set of packet lengths across different campaigns, this demonstrates operational laziness and the development of reusable modules.

  Such signals are independent of content and can only be detected through the atomic precision of FPC.

Analysis of Encrypted Traffic

Encrypted traffic has become the cornerstone of modern attack operations. While many consider encrypted traffic to be invisible communication, advanced SIGINT analysis recognizes that encrypted traffic doesn't provide content, but its signal characteristics expose the entire architecture of the threat actor. TLS fingerprinting and JA3/JA3S techniques are at the core of this analysis:

- **TLS fingerprinting: Cryptographic behavior analysis**

  During a TLS handshake, each client and server transmits the following in a specific order:

  - Supported cipher suites
  - Extensions
  - Protocol versions

- Sequence numbers
- Key exchange methods

This order functions like a signature, revealing the actor's software stack, C2 framework, custom-developed modules, and operational standards. Even advanced threat groups can't continuously and manually alter this cryptographic signature.

- **JA3/JA3S: The silent revolution of modern threat hunting**
  JA3 (client) and JA3S (server) fingerprints generate a unique signal identity by hashing the TLS Client Hello and Server Hello structures. This does the following:
  - Identifies similar C2 infrastructures
  - Enables linking different campaigns to the same operator
  - Reveals the true nature behind proxy chains
  - Clearly distinguishes malicious traffic from legitimate traffic based on behavior

  The greatest strength of JA3 lies in the fact that while malware may attempt to mimic normal user behavior, reproducing a specific TLS fingerprint is significantly more difficult. TLS negotiation parameters tend to remain consistent across deployments, which makes them a reliable signal for identifying related infrastructure and operational tooling, for example:
  - If the JA3 fingerprint used by a ransomware group changes by only 1 byte across campaigns, this indicates that the group preserves its "operational code heritage."
  - Observing the same JA3 fingerprint on different IPs across multiple geographies reveals that the actor is using distributed infrastructure.
  - Similar JA3 fingerprints paired with different JA3S fingerprints may indicate that the threat group is using a chained reverse proxy architecture.

  This analysis provides insight into both network traffic and the actor's infrastructure design philosophy.

Real-World Example: The Link Between Cryptographic Traces and Operational Traces

In a real SIGINT analysis (anonymized), a threat actor was using VPNs, Tor, and a custom TLS wrapper to render their traffic invisible. The content was completely concealed. However, when the JA3 fingerprint and packet behavior were analyzed together, three critical signals emerged:

- Identical time intervals were observed between TLS sessions (an indicator of automation).
- The JA3 fingerprint showed an 87% similarity to previous campaigns of a group that had attacked the financial sector.
- The packet length distribution revealed that the actor was using a custom C2 library.

When these three signals were correlated, the following were accurately derived without any need for content:

- Actor's identity
- Operational model
- Level of process automation
- Tools being reused
- How the infrastructure was chained

Techniques for Packet Inspection and Traffic Analysis

In modern threat intelligence, packet inspection and traffic analysis constitute signal-based behavioral intelligence that reveals the following threat actor characteristics:

- Behavioral habits
- Operational discipline
- Infrastructure maturity level
- Current stage of the attack
- Group intent
- Campaign organization model

A threat actor may conceal content, but packet rhythm, cryptographic fingerprints, and signal behavior are the most difficult intelligence sources to hide.

Packet inspection includes much more than analyzing the visible surface of network traffic such as reading the threat actor's hidden operational model, automation quality, infrastructure maturity, and manual intervention points. At this stage, classical deep packet inspection (DPI) or superficial JA3 matching becomes insufficient. True SIGINT-level analysis focuses on the physical behavior of packets and their temporal consistencies. This approach is the most sophisticated method used to break even the most advanced concealment mechanisms employed by threat actors. The following techniques illustrate how signal-level analysis can be applied to uncover hidden operational patterns in network traffic:

- **Temporal deviation profiling (TDP)**
  This technique is based on analyzing microdeviations in packet timing intervals. Every C2 infrastructure controlled by human operators produces microsecond-level irregularities that differ from digital automation. TDP therefore identifies the following:
  - **Manual operator intervention**
    Sudden increases in inter-packet delay or disruptions in rhythm.
  - **Automation quality**
    The more flawlessly a C2 infrastructure operates, the lower the delay variance.

- **Campaign phase**
  High rhythm stability during the reconnaissance phase and rhythmic irregularities during the attack phase.

This method can reveal the threat actor's operating model solely through signal behavior without touching encrypted traffic. For example, some APT groups generate millisecond-level CPU spikes on C2 servers during manual access due to mouse movements. These subtle CPU spikes consistently affect packet timing. TDP can capture this difference. This signal proves that the attacker is using a live operator rather than full automation.

- **Packet shape fingerprinting (PSF)**
  This is a next-generation behavioral method that analyzes temporal changes in packet length distributions. Even if the signal itself doesn't change, the "shape" of packet length sequences provides critical insight into the threat actor's C2 structure. Elements identified through PSF include the following:
  - The internal mechanics of the C2 protocol
  - Payload padding patterns
  - Custom encoder–decoder behavior written by the attacker
  - The rhythm of protocol obfuscation modules

  For example, say a threat group believed it was continuously randomizing packet lengths. However, when the PSF method was applied, the randomization was found to actually be bound to a patterned cycle. This cycle repeated every 64 seconds. Based on this information, it was determined that the group was using a customized EvilSharp-based fork of a C2 framework.

  Such insights directly reveal the threat actor's toolchain, toolchain architecture, and operational capacity.

- **TLS flow personality mapping (TFPM)**
  While JA3/JA3S techniques extract cryptographic fingerprints from TLS sessions, TFPM evaluates cryptographic behavior as a form of personality analysis. Each actor exhibits distinctive personal or organizational signatures in their TLS behavior, such as the following:
  - Handshake retry rhythm
  - Fallback behavior in error conditions
  - Certificate renewal timing
  - Different extension sets observed under the same fingerprint
  - Session ticket usage ratio

  These behaviors reveal the following about the threat actor:
  - Operational discipline
  - Development team quality

- Role distribution within the C2 infrastructure
- The age of the toolset they have developed
- Consistency issues within modularized components

For example, say an APT group was using its own TLS library. However, TFPM analysis detected that the group attempted five retries instead of three in faulty handshakes. This ratio pointed to a specific retry parameter embedded within the C2 framework used by the group. The same parameter was observed in another campaign as well. As a result, two campaigns were linked to the same operator despite using different IP blocks. Such behavioral fingerprints remain valid far longer than traditional IOCs.

- **Encrypted traffic mutation analysis (ETMA)**
  This technique examines momentary variations unintentionally introduced by threat actors while manipulating encrypted traffic, for example:
  - If TLS extension ordering changes during a minor version update, the actor is updating their custom C2 library.
  - If per-second packet density spikes suddenly, multiple operators have become active simultaneously.
  - If two different JA3S fingerprints are observed within the same C2 traffic, the group is using traffic fronting or chained reverse proxies.
  - If padding patterns in packets form a three-cycle loop, the actor is using an XOR-based obfuscation module.

  This analysis exposes both the internal structure of the C2 and the group's operational organization.

Real-World Example

In this real-world case (anonymized), a threat actor used a TLS tunnel continuously for 24 hours. Everything appeared automated. However, microtiming analysis of packets revealed additional latency spikes of 5–7 milliseconds occurring only during specific hours. These hours didn't align with the following:

- Start of the workday in the actor's geographic region
- Changes in network congestion
- Anomalies observed in global automation tests

Through joint HUMINT-SIGINT analysis, the following conclusion was reached: During these hours, the operator was manually checking the connection before returning control to automation. This subtle signal proved that the C2 architecture consisted of 95% automation and 5% manual error control. This insight later played a critical role in identifying the group.

In modern threat intelligence, packet inspection and traffic analysis extend beyond technical monitoring to reveal behavioral signals embedded in network activity. By examining patterns in traffic flow, analysts can infer a threat actor's operational discipline, infrastructure design, and stage of activity, even when the underlying content remains encrypted.

3.4.3 Radio Signals and Wireless Environment Intelligence

Wireless environments are among the most overlooked yet richest SIGINT layers within modern SOC and threat intelligence operations. Data transmitted over networks can be encrypted, obfuscated, or tunneled; however, radio signals—Wi-Fi, Bluetooth, Zigbee, Bluetooth Low Energy (BLE), LoRaWAN, radio frequency identification (RFID), and other IoT protocols—can't be completely detached from the physical layer. For this reason, wireless signals provide the rawest and most difficult-to-manipulate information regarding attacker intent and device behavior.

Radio-based SIGINT is less about reading content and more about modeling the invisible electromagnetic behavior around us. Through this structure, the following elements related to a threat actor can be detected before reaching the network layer:

- Physical location
- Devices in use
- Communication rhythm
- Anomalies in signal strength
- Physical access attempts
- IoT manipulations
- Introduction of external devices

The following sections examine how wireless signal analysis can be applied in practice to detect anomalies in the radio environment and uncover early indicators of malicious activity. Each technique focuses on interpreting behavioral patterns within the wireless spectrum rather than relying solely on network layer visibility.

The Intelligence Value of Wi-Fi, Bluetooth, and IoT Signals

Wi-Fi, Bluetooth, and IoT signals provide a behavioral layer of intelligence that reveals how devices operate within a wireless environment. By analyzing these signals, analysts can identify patterns that reflect both device characteristics and user activity.

The following indicators illustrate the types of behavioral signals that can be extracted from Wi-Fi traffic during signal-level analysis:

- **Wi-Fi scanning: The electromagnetic trace of device behavior**
  Beyond just SSIDs and MAC addresses, Wi-Fi signals carry multilayered information reflecting a device's operational habits. This includes the following:
  - Probe request density
  - Packet transmission rhythm
  - Channel-hopping behavior

- Beacon frame variations
- Microsecond-level changes in power levels
- Timing sequences of management frames

The most critical insight here is that every device carries traces not of its manufacturer, but of its user's behavior. For example, consider a Wi-Fi adapter used by an APT operator that exhibited a channel-hopping pattern every 60 seconds during an attack. Normal users don't change channels at this frequency. This signal indicated that the operator was using passive reconnaissance scanners. This information later played a critical role in determining the physical location.

- **Bluetooth and BLE signals: Proximity-based threat analysis**
  Bluetooth and BLE devices continuously broadcast *advertising packets* to maintain low energy consumption. These packets represent some of the strongest signals an attacker leaves behind unknowingly. From these signals, the following can be derived:
  - Device brand/model
  - OS version
  - Behavioral rhythm
  - Power management policy
  - Device presence duration
  - User movement patterns
  - Whether multiple devices belong to the same user

  BLE devices exhibit a unique, manufacturing-originated timing jitter (microtiming deviation). This physical layer signature allows a device to be identified even if its MAC address changes. Even when threat actors use MAC randomization, this jitter profile remains unchanged. For this reason, BLE represents one of the strongest domains within SIGINT for fingerprinting anonymous devices.

- **IoT signals: Mapping invisible presence**
  IoT devices (cameras, sensors, smart plugs, beacons, Zigbee/Thread devices) generate signals such as the following:
  - Regular interval broadcasts
  - Timestamps
  - Power consumption variations
  - Firmware signatures
  - Mesh network behaviors

  Once the normal operational map of IoT signals within an organization is established, any device a threat actor attempts to introduce can be detected within seconds:
  - Unexpected beacon floods
  - Previously unseen vendor OUIs
  - Short-lived but high-power signals

- Parent-switching anomalies within mesh networks
- Abnormal density on specific channels
- Sensor activations during nighttime hours

Such indicators serve as early signals of physical intrusion attempts. For example, in a real incident, an attacker attempted to establish a shadow gateway via IoT infrastructure. However, the gateway's Zigbee timing tolerance differed by 0.3% from that of legitimate manufacturer devices. This microsignal discrepancy revealed that the device was counterfeit.

While individual wireless protocols such as Wi-Fi, Bluetooth, and IoT signals reveal device-level behavior, analyzing the broader RF environment enables analysts to understand how these signals interact across the physical network surface.

Threat Analysis of Physical Network Surfaces

The physical network surface isn't limited to cables, Ethernet ports, or access points; it also consists of an invisible map formed by electromagnetic density, signal shadows, RF echo fields, and device movement patterns. The organization's entire RF environment is mapped much like a topographic chart, capturing the following:

- Signal strength curves
- Shadows created by physical obstacles
- Relay effects
- Intersection points of Wi-Fi/BLE/Zigbee layers
- Anomalous signal accumulations
- Low-power broadcasts from covert devices

Once this map is established, any device introduced into the RF environment by a threat actor appears like a *foreign body*. During RF fingerprint mapping, analysis of low-intensity continuous carrier signals can reveal even covert micro-IoT devices used by attackers. Even if these devices don't transmit data, they generate RF thermal traces that expose their presence.

Attack attempts in the physical environment typically begin with behaviors such as these:

- Unexpected beacon generation
- Sudden high-power spikes
- Microsecond-level bursts in channel density
- Bluetooth proximity flooding
- Increases in baseline noise levels
- Spurious emissions (unintentional residual signal leakage)

These behaviors reveal the threat actor's physical location, device type, signal strength, concealment method, and stage at which the action began. For example, say a 5-second amplitude spike was observed in the organization's Wi-Fi spectrum analyzer. This power level shouldn't normally appear in the environment. Analysis of the signal model showed that the attacker was using a specialized antenna designed for Wi-Fi injection through walls.

Prior to the physical attack, there are some signals you can pick up on. Threat actors often conduct an RF reconnaissance phase before attempting network intrusion, which includes the following:

- Signal density mapping
- Device discovery
- Measuring access point power levels
- Collecting Bluetooth beacons
- Understanding the IoT mesh structure

During this phase, attackers unknowingly leave behind signals such as these:

- Periodic probe request bursts
- Unexpected channel scanning sequences
- Abnormal parent-switching behavior in Zigbee routers
- Inconsistencies in Bluetooth pairing patterns
- Gradual increases in RF noise baseline

These signals provide early warning before an attack begins. Radio signals and wireless environment intelligence demonstrate a core truth of modern threat hunting: Attackers can wear digital masks and conceal their traffic, but they can't completely erase physical signal behavior.

Wi-Fi, BLE, and IoT signals often leave identifiable behavioral traces within the RF environment. When analyzed together, these signals can reveal device activity patterns and provide early indications of physical presence or anomalous behavior.

The true strength of wireless environment intelligence lies in its ability to interpret both the existence of a signal and its behavioral patterns, identity components, physical interaction traces, and electromagnetic inconsistencies. Even if a threat actor avoids leaving traces on digital networks, the physical signal environment often exposes behavioral mistakes of which the actor is unaware. Therefore, advanced wireless SIGINT is no longer limited to Wi-Fi scanning, BLE fingerprinting, or IoT beacon tracking; it aims to identify the threat actor's physical presence, movements, and operational logic by analyzing the biomechanical effects of signals, RF reflections, electromagnetic noise correlations, and device vibration models.

To interpret these physical layer signals, analysts apply a set of specialized RF analysis techniques. Each method focuses on identifying subtle variations in signal behavior within the wireless environment. The following techniques illustrate how these patterns can be used to detect device movement, presence, and anomalous activity:

- **Signal reflection analysis**
  Methods used in signal reflection analysis examine micro-echoes created when wireless signals reflect off surrounding walls, doors, glass, and metal surfaces. These echoes provide unique data for determining whether a signal source is static or mobile and near or distant. For example, when the distribution symmetry of reflections in a Bluetooth or Wi-Fi signal begins to degrade, it can indicate that the device is being carried inside a bag or moved by an operator. In a real-world case, analysis of BLE signal reflection patterns alone made it possible to approximate which corridor of a building an attacker was traversing, revealing the threat actor's physical location even though no network connection had been established.
- **Signal noise signature analysis**
  Similarly, signal noise signature analysis examines microbehaviors at the physical layer (such as power circuitry characteristics, oscillator jitter, and harmonic leakage), even when devices aren't actively transmitting. Instead of determining the device manufacturer, these behaviors reveal production quality, circuit age, chipset type, and operating voltage levels. Even attackers employing advanced concealment techniques (MAC randomization, reduced signal strength, or intermittent transmission modes) can't alter the unique noise pattern of a device's power circuitry. In a real-world incident, an IoT camera attempted to conceal its identity using MAC randomization; however, a 0.73-microsecond jitter in its power circuit matched that of an older device from the same manufacturer, enabling detection.
- **Spectral motion mapping**
  Another critical technique in wireless SIGINT is spectral motion mapping. As people walk, devices they carry or wear generate small amplitude fluctuations in the microwave spectrum. When analyzed over time, these fluctuations allow inference of whether a device is stationary or moving, its approximate speed, where the operator pauses, and which directions they move. For example, by analyzing amplitude fluctuations of a BLE broadcast observed for only a few seconds in an organization's parking area, it was determined that the device owner was walking at approximately 1.4 m/s, paused briefly, and then continued moving. Such information is invaluable for detecting physical intrusion operations before a threat even reaches the network.
- **Electromagnetic integrity analysis**
  Electromagnetic integrity analysis focuses on identifying subtle distortions in environments where signals are expected to be pristine. Covert IoT implants, radio-based surveillance devices, modified routers, and stealthy wireless intrusion hardware often lack perfect RF isolation. Harmonic peaks, spurious emissions, RF leakage caused by improper grounding, and irregularities in power regulation are powerful indicators that expose attacker devices. In a real-world incident, an extremely weak but irregular harmonic peak at 2.45 GHz was produced by a device that was otherwise not transmitting, revealing the presence of a hidden RF implant embedded within the device.

- **IoT mesh networks**
  IoT mesh networks are also among the environments where threat actors find it most difficult to conceal their presence. Zigbee, Thread, or Z-Wave mesh networks are self-healing, and the introduction of a new device creates micropathologies measurable down to milliseconds in network topology. Sudden increases in parent-switching behavior, microsecond-level changes in routing latency, inconsistencies in Link Quality Indicator (LQI) signals, or mesh nodes abruptly rebuilding routing tables indicate attempted intrusion. In one factory, a fake device added to a Zigbee network was detected based solely on a 1.4 ms latency increase, revealing it to be a physically introduced shadow node.
- **Physical RF shadows**
  Physical RF shadows are another critical trace left unintentionally by threat actors. Every building, room, and obstacle produces specific RF shadows. When attackers attempt to hide within shadowed areas or block signals, deviations in shadow size, symmetry, reflection angles, and signal propagation times emerge. These deviations provide highly valuable intelligence signals that indicate where an operator is standing, how a device is held, and which directions the operator is moving.

When all of these techniques are combined, wireless SIGINT becomes much more than a security control; it's the deepest intelligence layer revealing a threat actor's physical presence, behavioral rhythm, device identity components, and operational intent. Digital masking, VPN chains, encryption, proxy layers, and internal network silos all leave traces in the wireless signal domain, which are the most difficult to manipulate. For this reason, wireless SIGINT offers a unique perspective in modern threat intelligence by unifying the digital and physical realms: before the attacker even sees the system, the system detects the attacker.

3.4.4 The Integrated Structure of SIGINT, Telemetry, and OSINT

In modern threat intelligence, no single data source is sufficient on its own. SIGINT provides the behavior of signals, telemetry provides the context of events, and OSINT exposes the actor's observable traces. While the information value delivered by each layer individually is significant, their true power emerges only when they are combined. This is because advanced threat groups attempt to conceal their operational traces not on a single surface but by distributing them across signals, network behavior, social traces, infrastructure relationships, and timing patterns. Multilayer integrated intelligence analysis brings all these traces together into a single picture, revealing the threat actor's*operational architecture*.

Within this structure, the SIGINT, telemetry, and OSINT triangle represents the unification of how different intelligence disciplines explain the same event at the behavioral level. As a result, not only the technical dimension of an attack becomes visible but also its strategic context, the actor's motivation, and campaign continuity.

Wireless signals, packet behaviors, and traffic rhythms expose the technical presence of the threat actor; however, understanding to whom these signals belong requires the involvement of the OSINT layer. When these two layers are combined, signals begin to correlate with the actor's traces in the open world, for example:

1. SIGINT detects the use of the same JA3 fingerprint across multiple regions.
2. Telemetry shows that the same traffic repeats at specific hours.
3. OSINT finds a similar timing rhythm in the posts of a vendor active on dark web forums during those hours.

This timing intersection provides a signal about the threat actor's operational hours. This signal can later even be used to analyze the attacker's geographic location (workday-night cycle).

Here's another possible scenario:

1. SIGINT detects an anomaly in the power consumption variation of a new node added to an IoT mesh network.
2. Telemetry records show that no authorized change was made at that time.
3. OSINT reveals that Zigbee modules have recently appeared for bulk sale in local second-hand markets.

This correlation confirms, at a very early stage, that the attacker is attempting a physical intrusion via IoT.

This *multilayer behavioral correlation* reveals the technical traces of the threat actor as well as their economic, social, and operational behaviors. SIGINT alone detects the anomaly; OSINT enables threat hunters to understand who produced that anomaly.

Advanced threat groups conduct multistage, long-running, interconnected campaigns rather than single attacks. These campaigns often use different infrastructures, encryption modules, C2 designs, pretexting methods, and geographic regions. On the surface, these may appear as independent incidents. However, the integration of SIGINT, telemetry, and OSINT can reveal that these attacks are part of a single campaign chain executed by the same actor.

Example: Advanced Integrated Analysis

Let's walk through an example to demonstrate how these three layers combine:

- **SIGINT layer**
  - The TLS traffic extension order is identical across three different incidents.
  - The BLE signal jitter observed in the Wi-Fi environment is similar.
- **Telemetry layer**
  - Anomaly times in all three incidents cluster within the same time window.
  - EDR sensors report similar application programming interface (API) call patterns across different variants of the same module.

- **OSINT layer**
  - Operational announcements by a threat group are observed on their Telegram channel at exactly these times.
  - During the same period, a specific malware loader is observed being updated on GitHub.

When these three layers are combined, the following becomes clear: The signal behavior indicates the operator's technical identity, telemetry indicates the timeline of events, and OSINT indicates the actor's social and logistical preparations.

This integrated structure proves that three separate incidents are in fact components of a single attack chain. Campaign detection becomes possible at this point.

SIGINT exposes microlevel details in signal behavior, OSINT provides context to these details, and Telemetry bridges the two by connecting them through an event sequence. For example, for an APT group, the following may not individually convey meaning:

- Minor variation in the JA3 signature
- Micro-jitter in RF signal behavior
- IoT mesh pathology
- API call sequences in telemetry
- OSINT posts about the group members' technical preferences

However, when analyzed together, they form a complete picture of the following:

- Development team's code foundations
- Timing rhythm of operational planning
- Types of devices in use
- Duration of the attack plan
- Phases of campaign objectives

This picture enables both predicting and tracking the threat actor.

Integrated analysis represents a new level of threat hunting: *signal-rhythm-telemetry flow*, in which OSINT context forms the three dimensions of the same campaign. A threat actor can erase technical traces, but concealing their operational identity is nearly impossible for the following reasons:

- Signal behavior represents technical habits.
- Telemetry represents operational routine.
- OSINT represents the social and logistical facade.

When these three converge, the threat actor's operational identity emerges (behavioral signature + technical signature + social signature). This identity provides a persistence in resolving campaign chains that classical IOCs can't offer. IOCs change within hours; operational identity remains stable for years.

The integration of SIGINT, telemetry, and OSINT reflects a core principle of modern threat intelligence: A single data layer may expose an event, but combining multiple layers reveals the underlying structure of the threat. Within this framework, isolated anomalies gain context, signal patterns scale to campaign-level insights, and OSINT findings acquire operational depth. The model described in this section is fundamentally different from the traditional SOC approach that detects attacks late. Here, the goal isn't to detect the attack, but to read the campaign and calculate the attacker's next move in advance.

3.5 Integrating and Correlating Multisource Intelligence

In modern threat intelligence, no single data source provides a complete picture. Threat actors distribute their traces across different domains: concealing intent in HUMINT signals, masking technical behavior in SIGINT layers, and limiting visibility in telemetry and OSINT environments. For this reason, meaningful intelligence emerges only when signals from multiple sources are analyzed within a unified framework. Multisource correlation therefore seeks to determine whether indicators from different disciplines describe the same operational reality and reveal what an actor is doing, why, and how.

The fusion of HUMINT, OSINT, SIGINT, and telemetry provides a level of visibility that none of these sources can offer independently. HUMINT highlights behavioral motivations, OSINT reveals public traces and relational networks, SIGINT exposes signal behavior and operational rhythm, and telemetry records concrete system activity. When combined, these layers transform isolated anomalies into coherent operational insight.

In this model, the goal isn't merely to confirm signals but to explain them across contexts. A technical indicator observed in SIGINT may correspond to historical tool usage visible in OSINT, while HUMINT signals and telemetry activity reinforce the same pattern. Through such correlation, separate indicators become part of a unified operational chain, often revealing attack preparation before it becomes visible at the network level.

Multisource fusion also requires resolving apparent contradictions between sources. Rather than representing errors, these differences typically reflect distinct perspectives of the same event: SIGINT shows technical behavior, HUMINT reveals intent, OSINT provides context, and telemetry reflects system impact. Effective analysis reconciles these signals by evaluating their timing, reliability, and context.

One important outcome of this process is the identification of a threat actor's operational identity. This identity emerges through the convergence of technical, behavioral, social, and temporal signatures. When these signatures align, analysts can recognize recurring campaigns and attribute activity with greater confidence.

Ultimately, multisource fusion enables analysts to connect fragmented signals into a coherent intelligence picture. HUMINT reveals intent, SIGINT exposes technical activity, telemetry records operational impact, and OSINT provides contextual framing. Integrating these layers allows organizations not only to detect threats, but to anticipate how they will evolve.

The practical application of this correlation model is most visible in dedicated intelligence fusion structures. In such environments, signals originating from different disciplines are analyzed together to produce a unified operational picture. The following sections examine how to implement this approach in practice.

3.5.1 Multisource Fusion Centers

At the national level, the most complex and most critical architecture for intelligence production is formed by multisource fusion centers. These centers transform fragmented data collected by different institutions—produced in different formats, at different times, and with varying levels of accuracy—into a single, holistic operational picture. The strength of a fusion center doesn't lie in the volume of data it possesses, but in the interdisciplinary logic with which it integrates that data. The national threat environment can no longer be interpreted by a single institution or a single source; cyberattacks intersect with physical threats, OSINT signals intersect with HUMINT cues, technical logs intersect with legal processes, and SIGINT findings intersect with financial analyses. Fusion centers manage this vast complexity and create a national-level threat awareness, instead of an institutional one.

Fusion centers in the United States ingest data from a broad network ranging from Department of Homeland Security (DHS) infrastructure and FBI Joint Terrorism Task Force structures to National Security Agency (NSA) SIGINT modules and local law-enforcement agencies. This model isn't unique to a single country; the same fusion logic underpins modern Community Emergency Response Team (CERT) architectures. However, a little-known fact is that fusion centers are more than data-collection entities; they are structures that build operational context architecture. The objective isn't just integrating data but revealing how the intentions embedded in data, threat chains, and actor behaviors resonate at the national level.

Let's walk through the main techniques and functions of fusion centers:

- **Relativistic context mapping**
  One of the most advanced techniques used in these centers is processing data from different sources through relativistic context maps. These maps show how an OSINT data point echoes within SIGINT, how a statement appearing in HUMINT reports is supported by telemetry flows, or how the narratives of an international threat group on social surfaces connect to technical anomalies observed in local systems as part of the same operational chain. For example, if a HUMINT signal provided by the FBI overlaps with an encrypted traffic fingerprint collected by the NSA, that signal isn't just accepted as validated information; it also elevates the assessed operational discipline level of the threat actor. Such integrated analysis is only possible within fusion centers because only these

structures possess the broad vision and technical capacity required to perform context matching across sources.

- **Synergy with national CERTs**
  National CERTs operate under a similar logic. When telemetry from the banking sector, SCADA signals from power plants, traffic anomalies at the ISP level, and OSINT reports from independent researchers are processed along the same axis, it may become evident that a very broad attack surface is being probed by a single operator. This is the strongest mechanism for proving that events which appear independent on the surface are in fact part of a single strategic campaign. For this reason, fusion centers are the core structures that reveal both technical attacks and national-scale operational motifs.
- **Managing reliability tension**
  Another lesser-known yet critical function of fusion centers is managing reliability tension between sources. At the national level, it's normal for different institutions to produce differing assessments of the same event: intelligence agencies may emphasize HUMINT signals, SIGINT units may focus on technical behavior, and local law enforcement may prioritize the physical threat dimension. Fusion centers resolve these tensions through multisource validation weighting. Under this principle, no single data point is decisive; what matters is how a data point relates to other sources. Accordingly, there's no concept of correct data in fusion centers—only data with higher contextual value.
- **Enabling holistic threat ecosystem analysis**
  Many countries use fusion centers as a shared analytical layer for terrorism, organized crime, human trafficking, financial fraud, critical infrastructure threats, and disinformation campaigns. Through this broad-scope approach, it becomes possible to understand that the actor behind a cyberattack also leaves traces in social media manipulation, exhibits suspicious financial behavior, or operates as part of an international group. No single institution can achieve this level of visibility; only a multisource fusion architecture makes it possible.

  True fusion power enables threats at the national level to be viewed as an operational ecosystem. This ecosystem brings together behavioral signatures, technical traces, geographic connections, infrastructure distributions, social-surface actor relationships, and economic motivations. As a result, a country transitions from a defense mechanism that analyzes attacks in isolation to an intelligence capability that sees the complete picture of the threat actor. Fusion centers comprise the architecture of this transformation.
- **Information circulation failure identification**
  One of the least known yet most strategic aspects of nationally operating fusion centers is their ability to identify information circulation failures across the country. Intelligence produces value when it's correctly aggregated; however, in multi-institutional structures, critical signals can be lost when data sources, context, or timing become distorted. Advanced fusion centers therefore include a mechanism known as information flow anomaly detection. This mechanism monitors interinstitutional data-sharing rhythms, reporting timelines, source density, and content diversity to identify institutional blind

spots independently of threat actors. If an institution shares less data than usual during a given period, this alone may indicate a weakness in national threat awareness. This approach is one of the rare structures that turns threat hunting not only outward, but inward as well.

- **Dynamic threat modeling**
  They model the national threat ecosystem as dynamic rather than static. These centers track both actors and the behavioral shifts over time and the evolution of operational motifs. For example, if a threat group's attack cycle intervals remain consistent for three years and then suddenly begin to shorten, fusion centers flag this as an operational stress indicator. This may suggest that the group is under financial pressure, being directed by a new sponsor, or experiencing a disruption in its international activities. Such analyses are conducted not solely through technical indicators, but through the combined examination of behavioral rhythm on OSINT surfaces, internal dynamics via HUMINT, and infrastructure changes via SIGINT.
- **Operational flow mapping**
  Moreover, advanced fusion centers don't just correlate events; they map the operational flow that threat actors create across the national security surface. This map places recurring signal patterns, activity density on dark web forums, sudden movements in financial transactions, instantaneous telemetry fluctuations in critical infrastructure, and narrative shifts on propaganda surfaces onto a single timeline. In doing so, the shape of an approaching operational wave becomes readable before the threat actor takes action. This level of visibility goes far beyond classical defense models and is only achievable within national-scale fusion centers.
- **Imitated threat filtering**
  Many people perceive fusion centers as data flow management entities; however, their most powerful capability is the ability to filter out imitated threat traces. At the national level, many threat actors deliberately plant false indicators to shift attribution toward other groups. Fusion centers determine which traces are genuine and which are deceptive by correlating SIGINT-derived technical behaviors with group-internal inconsistencies observed through HUMINT. For example, even if the infrastructure used in a particular attack appears compatible with the legacy toolchain of the APT-X group, if the API call sequences observed in telemetry don't align with that group's behavioral signature, the attack can be identified as identity imitation. This is an intelligence capability with critical national impact that smaller institutions can never achieve.
- **Operational weight modeling**
  One of the advanced capabilities of fusion centers is the evaluation of hundreds of independent data sources across the country through an operational weighting model. In this model, each data point is scored according to metrics such as timestamp, the source's historical accuracy rate, contextual alignment, degree of correlation with other signals, and its position within the threat chain. This prevents random or low-quality data from distorting analysis. Here's a critical example: Even if an OSINT data point appears highly convincing on the surface, if its timing rhythm doesn't match SIGINT

findings, fusion centers classify it as low-context data and reduce its weight. This method prevents national decision-makers from being misled by false information.

- **Synchronization break analysis**
  Another rare technique used in advanced fusion centers is synchronization break analysis, which examines the timing of traces left by a threat actor across different surfaces. If an actor's behaviors on OSINT surfaces, traffic movements in SIGINT, and systemic impacts in telemetry exhibit a natural flow, this indicates a genuine campaign. However, microsecond-level inconsistencies observed in the timeline may indicate that operations have been artificially stitched together by different groups or that one actor is imitating another. This level of detail is a signal so fine that it can only emerge in fusion centers capable of processing large-scale national data.

Beyond all of this, fusion centers illuminate the invisible intersection points within the national threat ecosystem. The connection between a cyberattack and financial manipulation may appear entirely coincidental; yet fusion centers can identify that the same cryptocurrency wallets, social media accounts, time windows, or external propaganda chains underlie both events. This level of visibility ensures that a country is prepared for attacks in general and prepared for the complex actor networks behind those attacks specifically. For this reason, fusion centers aren't the final link in the national security chain, but they are recognized as the central nervous system of national threat awareness.

3.5.2 All-Source Analysis Framework

In modern threat intelligence, the most advanced analytical model is the all-source analysis framework, which enables all sources to converge on a single plane of truth. This framework integrates the surface-level breadth of OSINT, the behavioral insights of HUMINT, the technical signal consistency of SIGINT, and the systemic reality of telemetry data into a single analytical backbone. However, the true value of this model lies not in placing these four sources side by side, but in identifying the specific operational connection points at which they intersect. A threat actor never leaves traces on a single surface: Intent is concealed within HUMINT, social traces are dispersed across OSINT, technical behavior is masked within SIGINT, and visibility is degraded at the telemetry layer. For this reason, the all-source approach is the only method capable of rendering the threat actor's true operational architecture visible.

The core philosophy of this framework is that no source is superior to another; rather, each serves to explain them, not validate them. For example, a dark web username identified on an OSINT surface may align with a group's internal communication style revealed through HUMINT; SIGINT analysis may reveal traffic timing rhythms indicating association with the same user; and telemetry may detect an identity impersonation attempt linked to that username. When these four surfaces converge, the threat actor's operational identity becomes tangible. This is where the all-source model surpasses all other analytical approaches.

The most challenging aspect of this framework is the resolution of data conflicts. By their nature, data from different sources often appears inconsistent. Intuitive insights contained in HUMINT may contradict technical findings in SIGINT; behaviors observed on OSINT surfaces may leave no trace in telemetry; and an operational rhythm visible in SIGINT may be entirely silent in OSINT. Immature analytical structures interpret these inconsistencies as contradictions. In the all-source approach, however, such conditions are evaluated as part of the actor's operational masking techniques. Data conflict is often the result of a threat actor leaving traces on one surface while erasing them on another; it's not an analytical failure, but the attacker's behavioral strategy.

The all-source analysis framework consists of the following key components:

- **Cross-layer intent resolution**
  One of the most advanced techniques used in mature all-source analysis is cross-layer intent resolution. This technique evaluates the content of each data point, the context in which it was produced, the purpose for which it was left, and the operational phase to which it belongs. An OSINT post that coincides with an internal group conflict described in HUMINT reports isn't just a public statement, but a reflection of the group's propaganda strategy. A short-lived spike in traffic observed in SIGINT, layered over a period of log silence in telemetry, indicates the use of anti-forensics techniques by the actor. Consequently, within the all-source framework, no data point is interpreted in isolation; each gains value only through its intersection with the relevant *operational windows* of other data.
- **Temporal alignment**
  Another critical component is temporal alignment. While threat actors attempt to conceal their traces, temporal consistency is the domain in which they struggle the most. The all-source model therefore aligns data from all four sources along a unified timeline to break campaign chains. For example, IPv6 traffic observed in SIGINT that coincides with infrastructure purchase dates identified through OSINT, aligns with actor motivation described in HUMINT reports, and synchronizes with API calls observed in telemetry, points to the same operational phase. Such alignment provides unique insight into the exact day, hour, infrastructure, and toolset used by the threat actor.
- **Data conflict resolution**
  Data conflict resolution techniques form the backbone of this model. The most advanced approach, known as *differential context decomposition*, assumes that conflicting data arises not from informational overload, but from contextual divergence. In this method, each data point isn't scored for absolute correctness, but for its context-consistency coefficient. HUMINT offers high contextual depth but may lack technical precision; SIGINT provides high technical accuracy but limited intent context; OSINT delivers broad coverage but with a high noise ratio; and telemetry offers definitive data but minimal context. Accordingly, in the all-source framework, data weighting is determined by the strength of the data–context relationship, not by source type.

- **Operational signature**
  In its final stage, the all-source analysis framework consolidates the traces left by the threat actor across all four layers into a single operational signature. When the technical signature (SIGINT), behavioral signature (HUMINT), surface signature (OSINT), and systemic signature (telemetry) are combined, the result is the exposure of the actor's entire campaign chain. This chain enables both the analysis of past operations and the prediction of how the actor is likely to act next. For this reason, all-source analysis is the capacity to track a threat actor using a more complex model than the actor's own.

3.5.3 Source Reliability and Prioritization Methodologies

In modern threat intelligence, one of the most critical yet least discussed issues is what value a piece of data carries within which context. This is because erroneous information, incomplete data, or deliberately misleading input is inherent to the intelligence collection process; the real skill lies in measuring the reliability of this information and assigning it the correct weight in operational decision-making processes. Source reliability and prioritization methodologies operate at this point: not all data is equal, each source carries a different weight within intelligence, and the operational value of information changes over time. For this reason, advanced intelligence structures evaluate data as a contextual signal instead of as a claim of truth.

Let's walk through some important methodologies for source reliability and prioritization:

- **Intelligence scoring**
  The primary objective of intelligence scoring systems is to generate an operational confidence value for a piece of information based on its source, timing, method of production, validation chain, and contextual alignment. This value represents usability, not absolute accuracy. For example, a HUMINT source may possess high informational depth, yet its score decreases if its historical reporting contains a margin of error. Conversely, a technical signal collected via SIGINT may not reveal operator intent, but may carry high accuracy and therefore holds strong weight within the system. OSINT data provides broad coverage, but due to its high noise ratio, its contextual score may be lower. Telemetry, while providing definitive data, lacks broad context and therefore can't be a decision driver on its own. Consequently, the true value of intelligence emerges at the point where these disparate source scores are multiplied, balanced, and correlated with one another.
- **Temporal relevance assessment**
  The timestamp is one of the most critical components of reliability methodology. Threat actors' operational behaviors are dynamic, so what was valid yesterday may be obsolete today. Therefore, information being old doesn't mean it's incorrect, but its operational value may decline. Advanced intelligence frameworks calculate this temporal effect contextually, not linearly. For example, if a C2 infrastructure used by an APT group three years ago is still emitting active signals today, the contextual value of that old information increases again. Similarly, a username mentioned on OSINT platforms may have

been dormant for years, but if a sudden DNS anomaly in telemetry correlates with that username, then dead data becomes active once more. Thus, a timestamp represents both chronological age *and* operational resonance.

- **Validation chain analysis**
  The validation chain reveals how many independent channels support a piece of information. If information is confirmed by three different sources, this doesn't necessarily mean it's reliable; if all three are derivatives of the same flawed origin, a validation illusion is created. For this reason, advanced validation chain analysis measures the number of sources and the degree of their independence. HUMINT, SIGINT, OSINT, and telemetry are independent production layers; therefore, information that resonates across all four surfaces carries extremely high operational reliability. Information validated only on the OSINT surface has low value due to noise effects; information validated only through HUMINT is more vulnerable to actor manipulation; data seen only in SIGINT lacks context; and data present only in telemetry may reflect a limited-scope event. Hence, the true strength of the validation chain lies in source diversity.
- **Context weighting**
  Context weighting represents the most abstract yet most decisive component of information valuation. Even if a data point is technically accurate, it's scored low if it doesn't align with the operational context. For instance, an OSINT report accurately describing a threat group's tactics may be technically correct; however, if the group is currently in a different strategic phase, its contextual value declines. A HUMINT-derived statement may contain a concrete threat claim, but if SIGINT doesn't support it, contextual alignment is weak. An anomaly observed in telemetry may be critical, but if it doesn't match the actor's known behavioral models, it may constitute a false positive. Thus, context weighting doesn't determine the truth of information, but its place within the operation.
- **Contextual multiplier system**
  The contextual multiplier technique is frequently used in advanced intelligence organizations, yet known by very few. In this system, each data point is evaluated through the multiplicative effect of four parameters (rather than through a single score): the source reliability coefficient, the contextual alignment coefficient, the temporal sensitivity coefficient, and the operational intersection coefficient. If any one of these values drops to 0, the operational value of the information also becomes 0. This is because data may be accurate, timely, and reliable, but if it doesn't intersect with the operational context, it influences no decision-making process. This approach enables intelligence to be evaluated not solely on accuracy, but on operational utility.

When all of these methodologies are combined, source reliability and data prioritization become an intelligence architecture and not just a ranking exercise. Through this architecture, a national security center or a corporate threat-hunting unit can clearly determine which information must be processed immediately, which should be monitored, which should be deferred, and which should be excluded entirely. Intelligence thus transitions from data chaos into a flawless filtering system that feeds decision mechanisms.

However, we can go beyond these standard methodologies and consider advanced techniques that take your prioritization architecture further:

- **Impact-weighted prioritization**

  An additional advanced stage of source reliability and prioritization processes involves evaluating intelligence by quality and by its strategic resonance coefficient. A piece of data may be reliable, yet its potential impact on an organization or a nation may be low; conversely, data with low contextual weight but high strategic resonance may become far more critical. For this reason, advanced intelligence frameworks assess each data point by both its source and the potential impact it may exert on decision-making mechanisms. This impact-weighted prioritization approach is used within the fusion centers of many countries, though rarely disclosed externally.

  In this technique, the operational value of a data point is calculated based on the increase in risk that would result if it were ignored. For example, a low-context OSINT data point may be assigned a higher impact coefficient if it indicates a potential threat chain affecting critical infrastructure. Conversely, a highly accurate SIGINT finding may be deprioritized if it affects only a low-importance system component. In this way, intelligence is classified not by accuracy alone, but also by consequences. This model is particularly valuable for countries with broad national attack surfaces.

- **Dissonance matrix**

  One of the most advanced techniques used in this process is the dissonance matrix, which helps interpret seemingly contradictory data as indicators of the threat actor's operational phase and masking strategy. For example, if HUMINT indicates that a group has launched a new campaign while SIGINT shows silence during the same period, that silence may represent a planned preparation phase rather than absence. If momentary anomalies in telemetry coincide with propaganda silence on OSINT surfaces, this may indicate that the group has transitioned to a low-noise, high-impact operational model. Dissonance isn't conflict; it's a signal of the threat actor's changing rhythm. This technique yields exceptional results in detecting phase transitions of APT groups.

- **Operational resonance analysis**

  In advanced versions of source prioritization architecture, operational resonance analysis is employed to measure whether data from different sources oscillates with the same operational rhythm. *Rhythmic resonance* unifies indicators belonging to the same campaign; *rhythmic dissonance* prevents unrelated threat surfaces from being mistakenly attributed to the same actor. In this analysis, time series, traffic bursts, acceleration of discourse on social surfaces, content density in HUMINT reports, and unexpected pattern changes in telemetry flows are examined together. If these signals progress in the same rhythm, a single actor or group is indicated. If the rhythms don't align, events are determined to belong to different operations, despite their technical similarity.

- **Source vectorization model**

  At this stage, another advanced method comes into play: the source vectorization model. In this model, each data source is transformed into a vectorial coordinate based on its accuracy coefficient, contextual weight, and which threat vectors its content indicates. For example, HUMINT represents intent, SIGINT represents infrastructure, telemetry represents system impact, and OSINT represents surface visibility. If these four vectors converge toward the same point, the operational signature becomes clear. If they don't converge, the threat surface is expanded or alternative actor models are evaluated. This technique produces powerful results in eliminating false signals, particularly in complex disinformation campaigns.

- **Time–source elasticity model**
  Another critical method employed by advanced national security centers is the time–source elasticity model. In this model, because source reliability changes over time, each source is evaluated using a dynamic confidence coefficient. For instance, a HUMINT source may demonstrate very high accuracy during a certain period, but later lose reliability due to external pressure or internal manipulation. Similarly, a SIGINT source may experience partial visibility loss following the deployment of new encryption protocols. This elastic evaluation, where sources are assessed dynamically rather than statically, is one of the defining characteristics of mature intelligence systems.

When all of these advanced methodologies are combined, intelligence organizations not only rank data, but can accurately identify the actor's behavioral model, operational rhythm, level of manipulation, and campaign phase. Ultimately, source reliability and prioritization aren't merely support functions in modern threat intelligence; they constitute an operational strategic intelligence layer that enables understanding, prediction, and the delivery of the correct response to threats.

3.6 Summary

This chapter explored how modern intelligence analysis relies on integrating multiple information sources rather than evaluating data in isolation. In complex threat environments, no single source can fully explain the behavior of a threat actor. Effective analysis therefore depends on combining HUMINT, SIGINT, OSINT, and telemetry within a unified analytical framework. Each source contributes a different perspective, and meaningful intelligence emerges when these perspectives are correlated and interpreted together.

The all-source analysis model enables analysts to identify connections between technical signals, behavioral indicators, public traces, and system activity. Through techniques such as cross-layer analysis, temporal alignment, and contextual evaluation, seemingly unrelated data points can be linked to the same operational campaign. This approach allows intelligence teams to understand individual incidents and to reveal broader operational patterns and threat actor behavior.

The chapter also emphasized the importance of evaluating the reliability and operational value of information. Methods such as intelligence scoring, validation chain analysis, temporal relevance assessment, and context weighting help analysts determine which information should guide decision-making. By applying these structured methodologies, intelligence organizations can prioritize relevant signals, reduce analytical noise, and build a clearer picture of emerging threats.

Overall, the all-source framework transforms intelligence analysis into a structured system that integrates diverse information streams, resolves data inconsistencies, and supports more accurate threat assessment and operational decision-making.

In the next chapter, we'll dig deeper into OSINT, with a focus on its practical application in CTI.

Chapter 4

Applied OSINT: Tools, Methodologies, and Operational Discipline

The open-source intelligence approach addressed in this chapter opens the door to a discipline far beyond ordinary data collection techniques. This intelligence system is built on strategic awareness, methodological integrity, and operational invisibility.

The value of open-source intelligence (OSINT) analysis isn't measured by the volume of data collected, but by the context in which that data gains meaning. Excessive data blinds the analysis process; insufficient information misleads decision-makers. Therefore, along with showing how analysts use the tools, this chapter provides a mature perspective on why data is collected, what purpose it serves, and most importantly, how it's protected once collected. The goal isn't to fill information, but to build an information architecture.

When planning an OSINT operation, the first thing analysts must do is ensure their own invisibility. The analysis method, query timing, dork structures used, research path, and even the constructed digital persona must be fully aligned. A behavior model that is not imitative yet leaves no trace is the silent backbone of real OSINT operations. If the analyst is detected, the collected information loses all meaning because the process has been sabotaged before it even begins.

This chapter also focuses on the cognitive structure of OSINT processes. While mapping a threat actor's infrastructure, unraveling a fake account network, or trying to understand the real motivations behind a domain name, the analyst's mind works like an architect completing the missing pieces of a puzzle. Technical tools are accelerators of this process; the decisive factor is the analyst's ability to perceive weak connections between multiple sources. Among hundreds of pieces, an insignificant detail may be the single key that reveals the actor's intent.

Throughout this process, information verification mechanisms are the analyst's strongest support because not every piece of data on the internet is correct. Some are deliberately misleading, some are incomplete, and others are created for manipulation purposes. This chapter presents data validation, cross-checking, source reliability grading, and timestamp consistency both as technical steps and as an analytical responsibility model. When OSINT is misdirected, it becomes one of the greatest risks to an organization; when used correctly, it becomes the strongest defense layer.

The aim of this chapter is to provide the analyst with a system of thought. The methodologies presented here go beyond data collection techniques, shedding light on how an OSINT operation should be planned, executed, sustained (including ethical and legal boundaries), and ultimately transformed into a high-value intelligence product.

4.1 Principles of Effective OSINT Collection

An effective OSINT approach is a complex system that aligns the analyst's personal intuition, operational discipline, technical competence, and intelligence model along the same axis. For this reason, before starting OSINT collection, it's necessary to think about which tools will be used and how information will gain meaning. Regardless of how advanced the technology employed may be, the factor that ultimately determines the quality of intelligence is the analyst's perspective.

A good OSINT process is shaped around a specific intelligence requirement (IR) rather than chasing random piles of data. The analyst first asks the right questions: "What do I need to know?" and "Why do I need to know it?" Any research conducted without answers to these two questions eventually turns into a self-consuming vortex—there's plenty of data, but no outcome; there's effort, but no meaning. The fundamental aim of the principles of effective OSINT collection is to eliminate this chaos and build a methodical, consistent, and purpose-oriented order.

Some OSINT operations are conducted without making any noise. An advanced analyst may sometimes analyze a threat actor's infrastructure without performing any active scanning; instead, they construct a picture solely by observing passive identifiers, timeline changes, and subtle behavioral patterns. Details that may seem incomprehensible at first glance are in fact microtraces that even an actor protected by a strong operational security (OPSEC) posture can't conceal. Being able to read these traces is one of the most critical principles of effective OSINT because information doesn't always appear openly—most of the time it's not hidden, just unnoticed.

OPSEC is another factor that makes OSINT effective. The analyst must prevent the target from becoming aware of the investigation. This isn't achieved solely by using a virtual private network (VPN) or hiding browser fingerprints; even the timing, pattern, method, and speed of queries are designed not to trigger the threat actor. The digital footprints left by a researcher can unknowingly jeopardize the entire operation. Therefore, among the principles of OSINT collection, the most overlooked yet most vital topic is managing the analyst's own invisibility.

Of course, not every piece of information collected during the OSINT process can be assumed to be true or useful. Open sources are both valuable and dangerous because they can be easily manipulated by threat actors seeking to produce false information. For this reason, source validation is a critical principle that guides the entire analysis chain. Information only gains meaning after it has been verified. Any data that isn't cross-checked is a potential trap that may lead the analyst to an incorrect conclusion. Effective OSINT advances by recognizing and avoiding these traps.

How the collected information will be used in the future is also within the scope of these principles. When preparing an intelligence report, information is contextualized, chronological relationships are established, evidence chains are created if necessary, and the data is placed within an analytical framework. A good OSINT analyst both collects data and transforms it. This transformation is what elevates OSINT from an ordinary reconnaissance activity into a strategic infrastructure capable of influencing decision-making processes at the organizational level.

In the following sections, we'll present a framework that integrates technical knowledge with methodology, methodology with ethics, and ethics with operational discipline. When applied correctly, OSINT both makes threats visible and enables the anticipation of an organization's weaknesses, competitors' strategies, and future risks.

4.1.1 Intelligence Requirement: Focused Approach

The cornerstone of an effective OSINT operation requires IR structures that define why the operation is conducted, which questions it must answer, and which decision processes it will support. Remember, the IR is the compass that determines where the analyst should focus while looking at an ocean of information. Without this compass, every piece of collected data produces only noise; a well-defined IR, on the other hand, transforms the same data into a meaningful signal. Therefore, every professional OSINT effort begins with the question "What do we need to know?" instead of "What do we know?"

At the organizational level, these requirements are structured through the creation of the priority intelligence requirements (PIRs), specific information requirements (SIRs), and essential elements of information (EEIs) chain, which we discussed in detail in Chapter 2, Section 2.1.2. When an OSINT analyst correctly establishes the relationship between these three, a collection architecture emerges that maintains the big picture without getting lost in details.

In some cases, this structure serves a technical purpose, such as uncovering the infrastructure used by a threat actor. In other cases, the objective is more abstract: predicting how risk within an organization may evolve in the future, modeling attack trends affecting a particular sector, or deciphering the motivations of a threat group. In both scenarios, OSINT conducted without an IR is like a ship without a course; it consumes time without progress and collects data but makes no sense of it.

Another critical dimension of the IR-focused approach is setting the appropriate depth of collection. An analyst shouldn't collect data at the same intensity for every target. For some targets, surface-level information may be sufficient; for others, a multilayered analysis is required, including historical DNS changes, Who Is (WHOIS) variations, social media behavior graphs, open-source code repositories, metadata remnants, and even dark web traces (refer to our exploration of these topics in Chapter 3, Section 3.2). What determines this depth isn't the analyst's personal preference, but the necessity dictated by the IR. Excessive depth generates OPSEC risks; insufficient depth results in incomplete analysis.

The delicate balance here is a point many analysts overlook: unnecessary data collection is more than a waste of time. An incorrect OPSEC step may leave unwanted records on target systems, create traces in logs, or even attract the attention of the threat actor. For this reason, the IR-focused approach determines the accuracy of intelligence production and the security of the operation itself. The analyst shouldn't expand the process without evaluating whether each piece of collected data serves the intelligence question being asked because, in OSINT, every unnecessary step amplifies invisible risks.

Another advantage of IR-focused work is analytical coherence. An OSINT analyst continuously reprocesses information over time in a cyclical manner. Changes in the target's behavior, innovations in attack infrastructure, updated code repositories, or new tactics associated with a threat actor are signals of change within the previously defined IR structure. In this way, OSINT analysis becomes a continuously evolving construct. IR provides a fixed axis along which this evolution can be interpreted.

One of the most common problems in corporate threat intelligence today is the paradox of too much data and too little insight. The IR-focused approach fundamentally resolves this paradox. From the beginning, the analyst possesses a framework that distinguishes valuable information from mere noise. When the PIR → SIR → EEI chain is correctly established, every piece of collected data is suitable for integration into the intelligence product to be produced later; no unnecessary elements are carried forward.

We'll showcase the IR-focused approach in practice through a few examples:

- **Threat actor detection via Domain Name System (DNS) change frequency**

  Threat actors often change DNS records several times before a phishing campaign actually begins. This is an operational warm-up behavior performed for testing purposes and is usually overlooked.

  In this example, our IR is "Are there possible preparatory activities targeting Bank X?" Our EEIs are as follows:

  - Candidate spoofed domain names imitating the target bank
  - DNS change rates of these domain names
  - Frequency anomalies in passive DNS (pDNS) records

  An analyst detects that a newly registered domain (e.g., `secure-banklogin[.]info`) has changed its IP address six times within 24 hours. This behavior represents an anomaly that a legitimate domain would never exhibit. Let's connect this to our PIR/SIR/EEI mapping:

  - PIR: Is there an increasing phishing risk targeting the bank?
  - SIR: Is activity involving domains impersonating the bank increasing?
  - EEI: Have DNS change anomalies been detected?

  What's our result? Because the IR is correctly defined, a minor DNS detail becomes an early warning signal of a major attack.

- **Capturing group behavior via identically worded GitHub issue descriptions**
  Many threat groups conceal their real identities within their code infrastructure, but they can't conceal their working habits.

  In this example, our IR is "Are there indicators of new attack preparations by a specific threat group?" Our EEIs are as follows:
  - Linguistic similarities in "issue description" texts on GitHub
  - Repeating writing habits across accounts that don't belong to the same user
  - Time zone alignment of commit timestamps for shared tools

  We have an unexpected finding: Across three different GitHub accounts, the same error description style appears: "Fixing minor inconsistency for better stability." This phrasing is identical to that found in the same threat group's past repositories. This represents an identity linkage that isn't overtly visible in open sources.

  Without IR, the analyst dismisses these as unrelated GitHub users. With IR, this minor linguistic alignment becomes an early indicator that the threat group is preparing a new tool.

- **How incorrect EEI definition amplifies OPSEC risk**
  When investigating a threat actor's command and control (C2) infrastructure, an analyst may perform unnecessary active scans if the IR isn't properly defined. Here's what this incorrect scenario might look like:
  - The analyst performs a port scan on the domain with no IR.
  - The threat actor detects the analyst in logs.
  - The C2 infrastructure is completely shut down or changed.

  As a result, the intelligence opportunity is lost and the operation is compromised.

  On the other hand, here's how a correct IR might look:
  - PIR: Understand whether Group A is preparing an active operation
  - SIR: Determine if the group has a new C2 infrastructure.
  - EEI: Check pDNS records, Secure Sockets Layer (SSL) fingerprints, and WHOIS changes.

  This enables intelligence collection without active scanning. The analyst remains invisible, preserving OPSEC.

- **Detecting fake profiles via temporal inconsistency analysis**
  In such cases, the analyst focuses on what the content says, when, and within which behavioral pattern it's produced to determine whether the profile reflects natural user activity or is part of a coordinated operation.

  In our example, our IR is "Is this social media account a real individual, or a state-sponsored narrative?" Our EEIs are as follows:
  - Timeline of post publication
  - Metadata of profile images
  - Writing cadence across different time zones

Let's consider an example analytical application: If a profile presents itself as residing in Turkey but displays the following warning signs, the account is likely to be artificially created or operated as part of a coordinated influence activity:

- Posts are made regularly every day between 02:15 and 05:00.
- Posting cadence aligns with Eastern European operation centers.
- Although the post language is Turkish, the keyboard error patterns are characteristic of Slavic languages.

Thanks to IR, the analyst asks the correct question: "Why is this person online at the same hours every day?"

- **Detecting attack preparation via dark web price anomalies**
  In this rarely used method, analysts examine pricing behavior within dark web marketplaces to identify abnormal shifts that may indicate preparation for a targeted cyber operation. Sudden price deviations in listings related to specific organizations can reveal changes in attacker demand, operational urgency, or the perceived strategic value of compromised data.

  In our example, the IR is "Is there preparation for a new data breach targeting the company?" The EEIs are as follows:

  - Price fluctuations of company-related accounts on the dark web
  - Update frequency of credential dumps
  - Newly created sales listings

  Let's say corporate email accounts that are normally sold for 20–30 USD suddenly rise to around 150 USD. This indicates that the data is about to be used or that a more valuable attack preparation is underway.

  Without IR, this price change appears as an insignificant fluctuation. With IR, it becomes a signal observed prior to a critical attack.

- **Identifying APT group rehearsals via SSRF Test IPs**
  Some advanced actors (advanced persistent threat [APT] groups) test server-side request forgery (SSRF) vulnerabilities using dedicated IP pools prior to an attack.

  This example uses the following IR: "Is there an APT-level attack preparation across the sector?" The EEIs are as follows:

  - Global SSRF honeypot logs
  - Unusual patterns among metadata IPs (Amazon Web Services [AWS], Digital Ocean, Google Cloud)
  - Two-second traffic patterns originating from the same IP

  Let's say a particular APT group uses the same five IP addresses across the target sector prior to an attack to test the waters. When these IPs are detected, they fit into the IR chain:

- PIR: APT threat level
- SIR: Infrastructure preparation signals
- EEI: IP pattern anomaly logs

Tip
Such detections are *pre-attack reconnaissance fingerprinting* methods known by very few people in the OSINT world. Don't forget this information!

4.1.2 Ethics, Legal Framework, and Authority Boundaries

No matter how advanced the technical capabilities of OSINT become, it must be placed within a proper ethical and legal framework. Although open-source information gathering is a broad field interpreted differently across countries, the fundamental principle that governs an analyst's behavior is always the same: Being accessible doesn't justify access. In other words, the fact that information exists on the internet doesn't mean it can be used under all circumstances. An analyst who fails to understand this distinction creates operational risk and opens the door to serious mistakes that may result in legal liability.

National and international OSINT law has become a more visible and more debated field, especially over the past decade, alongside the widespread adoption of threat intelligence activities. Regulations that include the General Data Protection Regulation (GDPR) in the EU, the California Consumer Privacy Act (CCPA) in the US, and local data protection laws in many countries impose strict limits on what information an analyst may retain, how long it may be stored, how it may be processed, and in what context it may be reported. The critical point here is that an OSINT analyst is evaluated not only as a person who collects information but also as an actor who processes personal data. For this reason, even the most well-intentioned OSINT operation can become legally problematic due to an unrecognized data processing violation.

In some cases, these boundaries are extremely sharp. For example, analyzing content from a person's public social media accounts is legal in most countries; however, feeding that data into an automated profiling engine falls under the definition of data processing and is therefore subject to legal restrictions. Similarly, while examining a domain's WHOIS information may be entirely permissible, performing active scans against a target system may be considered unauthorized information gathering activity in certain jurisdictions. For this reason, the distinction between passive and active OSINT isn't just operational but also legal. While passive OSINT is generally considered lawful in most countries, active OSINT should be carefully evaluated because it technically creates interaction with the target system.

It's critical to understand that the methods used to collect information also determine whether that information will be admissible as evidence in court. Although OSINT data can be powerful on its own, legal proceedings require the establishment of a proper chain of custody. For information to be usable in court, the accuracy of the source, timestamp, collection method, whether the data has been altered, and whether the analyst intervened in the process must all be documented. In particular, even hash verification of digitally obtained evidence is critically important in many cases. If the OSINT analyst can't maintain this chain, even the most compelling evidence of a cyberattack may lose its legal value.

The ethical framework represents a broader domain that sits above all of this legal structure. For an analyst, what is "right" isn't limited to what is merely "legal." The ethical dimension of OSINT requires awareness far beyond legal obligations. For example, a photograph shared on a public platform may theoretically be accessible to everyone; however, using that photograph in a context that could harm the person depicted is unethical. Likewise, while an employee's LinkedIn posts may be legally analyzed, drawing inferences about the employee's private life is ethically unacceptable. For this reason, professional OSINT analysts position ethical criteria one step ahead of legal requirements.

This ethical framework becomes even stronger when integrated into the IR structure that we discussed in the previous section. The more clearly the IR is defined, the easier it becomes for the analyst to adhere to the discipline of collecting only the necessary information. Every unnecessary, irrelevant, personal, or operationally unrelated piece of data creates both legal and ethical risk. Therefore, the professional OSINT approach is always based on the principle of minimal data collection: Don't collect any information you don't need. This principle serves as a safeguard from a security perspective, as well as the strongest protection mechanism in terms of the analyst's responsibility.

The distinction between passive and active OSINT forms a critical boundary at this point. Passive OSINT refers to processes in which the analyst doesn't interact with the target system in any way and only consumes existing information. Such activities are generally considered legal. Active OSINT, on the other hand, includes techniques that cause the system to generate a response, may leave traces, and in some countries are considered borderline. Techniques such as port scanning, DNS zone transfer attempts, and web application endpoint discovery may be classified in some legal frameworks not as information gathering, but as attempted unauthorized access. Professional analysts must understand this distinction and position themselves according to this boundary in daily practice.

Finally, it's important to note that the legal framework isn't globally homogeneous. An OSINT analyst must consider both the laws of a single country and data protection regulations worldwide. This is because the analyzed platform may be located in Germany, the data may be stored in the US, the threat actor may be in Russia, and the analyst conducting the operation may be in Turkey. This multilayered structure clearly demonstrates why OSINT requires a strong ethical and legal discipline.

In summary, ethical and legal frameworks aren't just restrictions; they ensure the reliability and sustainability of OSINT. An analyst must know what they can do *and* what they can't do because an unethical step can jeopardize both the analyst and the entire operation. In the following sections, we'll examine legal and ethical frameworks, using the US, EU, and UK as examples.

Frameworks in the United States

In the US, OSINT activities are regarded as a highly powerful intelligence capability due to the breadth of open sources and the richness of available data. However, this freedom is balanced by a strict legal and ethical framework. An OSINT analyst operating within US borders must comply with federal laws and state-level regulations. The US legal system provides one of the broadest freedom domains in the world regarding data usage, while simultaneously enforcing some of the strictest mechanisms for specific data categories.

In the US, OSINT isn't regulated by a legal structure explicitly defined as an OSINT law; however, many federal laws indirectly govern OSINT activities. The following laws determine how analysts may collect data, how the collected data must be stored, and under what conditions it may be reported:

- **Computer Fraud and Abuse Act (CFAA), 1986**
  The CFAA defines the most critical legal boundary for OSINT in the US. The CFAA prohibits the following:
  - Unauthorized access
  - Exceeding authorized access or unauthorized use
  - Active scans that create interaction with a system (in some cases considered risky)
  - Brute-force attempts, login attempts, or bypass efforts, even if conducted for testing purposes

  Passive OSINT (WHOIS, analysis of public social media, nonbreached data) is completely legal. Active OSINT (port scanning, directory enumeration, endpoint probing) lies in a gray area and may be interpreted as a potential attempt at unauthorized access in some states.
- **Electronic Communications Privacy Act (ECPA), 1986**
  This law prohibits the unauthorized interception or acquisition of electronic communications. An OSINT analyst may not do the following:
  - Intercept unencrypted emails
  - Monitor an ongoing communication
  - Attempt to access private messages of an account

  However, the following are fully legal:
  - Public social media posts
  - Forums
  - Public blogs
  - Search engine indexes

- **Stored Communications Act (SCA)**
  Unauthorized access to the contents of a closed platform (e.g., a private forum, Slack workspace, Discord channel) constitutes a federal crime. Even in some communities accessible via invitation links, authorization to access may be technically disputable.

The US doesn't have a centralized federal data privacy law like the GDPR; instead, it relies on sectoral and state-level regulations. Let's walk through a few data protection regulations:

- **California Consumer Privacy Act (CCPA)**
  The CCPA represents the strictest regulatory framework for data processing in the US and affects OSINT activities in the following areas:
  - How collected personal data is reported
  - The ability to disclose where an individual's data was obtained
  - Restrictions on processing data for commercial purposes

  It's important to note that CCPA applies only to California residents; however, because internet-based OSINT activities often involve this population, analysts effectively comply with this regulation.
- **Health Insurance Portability and Accountability Act (HIPAA)**
  An OSINT analyst in the US may not process any health-related personal data. Even if publicly leaked, health data is defined as a special category.
- **Family Educational Rights and Privacy Act (FERPA)**
  Student data is protected and may not be processed within the scope of OSINT.

In US law, the distinction between passive and active OSINT is drawn primarily through the CFAA and the ECPA. Decisions of district courts clearly demonstrate how this distinction is applied in practice:

- **Passive OSINT**
  Passive OSINT is generally considered fully legal in the US. This includes the following techniques:
  - Search engine operators
  - WHOIS record analysis
  - SSL/Transport Layer Security (TLS) certificate validation
  - Open profile analysis
  - pDNS
  - Metadata extraction
  - Use of satellite imagery
  - Analysis of indexed dark web content

  These methods create no interaction with the target system.

- **Active OSINT**
  Active OSINT is considered risky and is, in some cases, prohibited in the US. Most active OSINT activities in the US fall under attempted unauthorized access. Risky techniques include the following:
  - Port scanning
  - Directory enumeration
  - SSRF attempts
  - DNS zone transfer trials
  - Application endpoint probing

  It's important to note that although US courts don't explicitly define port scanning as unauthorized access, they classify it as a legal risk category because it raises suspicion of malicious intent.

In the US, there are very strict rules for OSINT-derived data to be admissible as evidence in court:

- **Authenticity**
  The source of the evidence, screenshots, log records, and timestamps must be verifiable. Hash values may be used.
- **Chain of custody**
  An OSINT analyst must be able to prove where the data was found, when, by what collection method, and that it was stored without alteration.
- **Hearsay rule**
  Information obtained from an unverified social media account isn't always admissible in court. Identity verification is required.
- **Public source doctrine**
  Evidence obtained from public sources is generally accepted in US courts. However, the legality of the collection method is a prerequisite.

Let's round out our discussion with ethics. In the US, ethical rules extend beyond legal rules. OSINT analysts shouldn't do the following:

- Make inferences about individuals' private lives.
- Produce false information or manipulative analysis.
- Collect information through social engineering.
- Report information if it could cause harm to the profile owner.
- Analyze sensitive data categories (race, religion, health).

In the US, the ethical framework is often defined by the standards of academic institutions and federal intelligence agencies. The mantra is minimum data collection, minimum harm, and maximum accuracy.

Frameworks in the European Union

Unlike the US example, OSINT activities in EU countries are evaluated within a far more centralized, written, and binding legal framework. The fundamental reason is that the EU defines personal data both as something that must be protected and as a fundamental human right. This approach directly affects the role of the OSINT analyst: the analyst is more than just a professional who collects information but is also an actor who directly interacts with essential rights and freedoms.

The primary legal text that indirectly regulates OSINT in the EU is the GDPR. The GDPR doesn't explicitly define the concept of OSINT; however, through the notions of personal data, data processing, and legitimate interest, it encompasses nearly all OSINT activities. For this reason, in the EU, an OSINT analyst is evaluated under the GDPR regardless of whether their work is conducted for intelligence, research, or security purposes.

From the GDPR perspective, the critical threshold is that the existence of publicly available information doesn't automatically legitimize the processing of that information. The fact that a person's name, photograph, or social media post is publicly accessible doesn't mean that such data can be collected, aggregated, or analyzed without limitation. An OSINT analyst must ask the legal basis question for every data point.

According to the GDPR, an OSINT analyst is, in most scenarios, assessed as either a "data controller" or a "data processor." This means that the analyst bears technical, administrative, and legal responsibilities. In particular, the following principles are directly binding for OSINT:

- **Purpose limitation**
  Collected data may only be used within the scope of a defined IR.
- **Data minimization**
  No personal data unrelated to the operation should be collected.
- **Storage limitation**
  OSINT data can't be archived indefinitely; it must be deleted once the processing purpose ceases.
- **Transparency**
  Where required, it must be possible to explain from which source the data was obtained and for what purpose it was processed.

Within this framework, advanced OSINT techniques such as automated profiling, relationship mapping, and behavioral analysis are often considered high-risk data processing activities in the EU and require additional legal justification.

In EU law, the distinction between passive and active OSINT has a narrower tolerance margin compared to the US. Even passive OSINT activities are considered data processing under the GDPR if they involve personal data. However, the general distinction is as follows:

- **Passive OSINT**
  This includes search engines, open-source reports, publicly available institutional data, and manual analysis of public social media posts.
- **Active OSINT**
  This includes any activity that creates technical interaction with the target system (scanning, probing, automation).

In most EU countries, conducting active OSINT without explicit authorization or a clear legal basis may create both ethical concerns and criminal liability. Automated data collection tools and bots, in particular, may be considered contrary to the principle of proportionality.

In EU judicial systems, OSINT data can be accepted as evidence; however, such acceptance is strictly dependent on the legality of the collection process. Open-source data obtained through unlawful methods may lose its evidentiary value regardless of how accurate its content may be. For this reason, source documentation, timestamps, integrity, and documentation of the collection method are indispensable criteria in the EU as well.

Frameworks in the United Kingdom

Although the United Kingdom has left the EU, it largely maintains the European approach in terms of OSINT and data protection. UK GDPR and the Data Protection Act 2018 are the primary texts regulating OSINT activities. The distinguishing feature of the UK approach is the emphasis placed on the concepts of proportionality and reasonable expectation.

The fact that information is publicly available doesn't always mean that analyzing that information is reasonable. Analyses that may negatively affect individuals' safety, reputation, or private lives are seriously questioned ethically, even if not always legally prohibited. In the UK, professional OSINT practices are often shaped by ethical guidelines defined by law enforcement agencies and academic institutions.

4.1.3 Operational Security

No matter how strong the OSINT operation methodology, if the analyst can't control their own visibility, the entire structure collapses. *Operational security (OPSEC)* is one of the most fundamental attributes of an analyst in the field; it often comes even before OSINT knowledge itself. Once the target becomes aware of the analyst, the data is no longer truly open source. If the target can see you, you're no longer an invisible observer—you turn into a hunter standing directly in front of the prey—and this can destroy the entire operation in a single second.

For this reason, OPSEC discipline isn't limited to identity protection alone. It's a multilayered structure in which research methods, timing, technologies used, and even the analyst's personal behavioral patterns are managed. We'll dig deeper into this multilayered architecture next and discuss different OPSEC approaches and principles.

Multilayered Architecture

A single anonymity tool is never sufficient. Regulatory structures in the US, Europe, and environments involving advanced threat actors clearly demonstrate that anonymization technologies must be used in chained layers. The multilayered architecture typically used in professional OSINT operations is as follows:

- **Physical isolation**
  The research device is completely separated from the corporate network.
- **Container sandbox**
  A separate isolated working container is used for each target.
- **Virtual private network (VPN) → Tor → proxy → browser container chain**
  The purpose of this structure is to eliminate single points of failure. If the VPN is compromised, Tor takes over; if Tor is compromised, the proxy takes over. If a container is compromised, data leakage doesn't spread across the entire device.
- **DNS over HTTPS + custom resolver**
  In this configuration, DNS queries are encrypted and routed through a dedicated resolver, preventing external observers from identifying which domains are being accessed during the research process.

The true strength of OPSEC lies in the lack of linkage between layers. There must be no logical connection left between the analyst and the research environment.

Browser Fingerprint Reduction

Modern internet infrastructure possesses mechanisms that perceive users' IP addresses, browser configurations, screen resolutions, system languages, font sets, and even device clock information as a unique profile. This fingerprint is like a digital DNA that allows systems to recognize a user.

Unless a professional analyst randomizes the fingerprint of the browser used for research, even using Tor or a VPN doesn't provide sufficient invisibility. A VPN only changes the IP address; as long as the fingerprint remains the same, the analyst may be perceived as a returning visitor. This, in turn, may cause the target to alter its behavior.

The fingerprint reduction process shouldn't rely solely on browser extensions, but should be implemented through specially prepared container-based profiles, isolated user-agent pools, and automated rotation systems. This technical approach is often overlooked in OSINT literature; however, in professional operations, it's one of the most critical OPSEC layers.

Nonattributable OSINT

The most fundamental principle of an OSINT analyst is that nothing about themselves should be visible. The nonattributable approach is more than just using pseudonymous accounts. It's a more mechanical structure in which the analyst analyzes every signal they leave on the internet, ensuring nothing remains that can be traced back.

Nonattributable OSINT is implemented at three levels:

- **Technical invisibility**
  IP masking, user-agent rotation, and fingerprint manipulation.
- **Behavioral invisibility**
  Avoiding nonhuman rhythms and avoiding the creation of suspicious patterns.
- **Purpose-oriented invisibility**
  Conducting the research through a persona independent of the target.

For example, while investigating a specific threat actor, returning to the same source every 10 minutes is an amateur move. The target understands that the observer isn't an ordinary user, but a researcher. This leads the opposing side to activate its own OPSEC mechanisms. For professional analysts, invisibility isn't about hiding identity but hiding motivation.

Behavioral OPSEC

In addition to browser fingerprints and IP addresses, threat actors can also be analyzed through user behavior. The same applies to OSINT analysts: The target has the capacity to infer that it's being analyzed based on the analyst's behavior. Behavioral OPSEC is the weakest link for most analysts.

For example, the following behaviors may trigger automated alerts on the target system:

- Checking the same domain every day at the same time
- Creating a consistent scrolling pattern for 40 minutes
- Refreshing multiple pages in a very short time
- Bot-like click speeds

More importantly, patterns that don't align with normal user behavior can attract the attention of a threat actor. For this reason, professional OSINT analysts design their timing cycles in accordance with human behavior. In some operations, *time-shifting techniques* are used: Research traffic is distributed across random hours, intervals exhibit natural variance, and no repetitive rhythm is formed.

The most advanced form of behavioral OPSEC is *noise masking*. The analyst embeds their real research traffic within normal internet traffic, making it invisible. For example, while analyzing a threat actor's blog, the analyst simultaneously visits 10 different benign news sites. This causes the traffic pattern to more closely resemble that of a real user.

More Than a Technical Skill—A Way of Thinking

Even the best OPSEC techniques are meaningless if applied with the wrong mindset. The professional analyst's mind should operate with the following reflexes:

- Does this action make me visible?
- What does this query make the target think?

- Who does this IP, this device, and this timing make me resemble?
- Is this research traceable?

Although OPSEC discipline may initially appear technical, it ultimately transforms into a psychological, behavioral, and strategic structure. The analyst positions themselves not as a data collector but as a shadow. The quieter the shadow, the more effective the OSINT.

The smallest mistake made by an OSINT analyst (e.g., an incorrect time, an incorrect IP, or an incorrect browser fingerprint) can be a trigger for some threat actors to initiate an attack. For this reason, OPSEC is critically important for ensuring the continuity of intelligence, as well as protection.

When all processes—from browser fingerprinting and traffic timing to container isolation and behavioral pattern management—work in harmony, an OSINT operation transforms into an invisible intelligence activity. True professionalism emerges at this point.

4.1.4 Chain of Custody

The value of information collected during an OSINT operation is measured by the provability of where, how, and when the data was obtained. If an analyst can't verify the origin of the information, the moment it was collected, the methods used, or the integrity of the data, then no matter how valuable that information may be, it holds no meaning within a professional intelligence approach.

The source of a piece of evidence is the first link that determines the reliability of an intelligence product. Not all information on the internet is as it appears. Some data is manipulated, some is incomplete, and some consists of traps deliberately left by threat actors to create deception. When an analyst can't verify the authenticity of a source, they not only risk reporting an incorrect conclusion but may also unknowingly contribute to a disinformation chain orchestrated by threat actors. For this reason, the identity of the source, platform on which it was found, access method, connection details, and timestamp must be systematically recorded.

The time of collection is the second critical parameter that determines the value of information. Especially in cybersecurity, information can change value within seconds. An indicator of compromise (IOC) that is valid today may be abandoned by a threat actor tomorrow. A domain that is part of a malicious infrastructure today may become entirely benign a few days later. The analyst's answer to "In what temporal context was this information valid?" directly shapes the accuracy of the intelligence. Therefore, every piece of data must be stored with an absolute timestamp in ISO 8601 format; ambiguity is unacceptable in professional analysis.

The collection method constitutes the most sensitive point of the chain of custody. How information is obtained determines both its legal validity and its admissibility in court. OSINT analysts always operate according to the lowest-risk method principle; passive OSINT techniques are always more robust than active methods. However, regardless of the method used, the analyst must keep records of the tool, script, command, or methodological step through which the data was obtained. These records allow the analyst to reproduce the operation during a future verification process and to defend the information in a provable manner.

From the perspective of evidentiary integrity, the most critical element is hash verification. OSINT data is often not a static object; it may consist of screenshots, files, metadata sets, or log fragments. If such data undergoes even the slightest modification after being recorded, the integrity of the evidence is completely compromised and it loses its admissibility in court. For this reason, every static OSINT output must be verified using strong hash algorithms such as SHA-256, and the hash values must be recorded together with timestamps. This practice ensures legal reliability and forms the cornerstone of the analyst's internal quality control.

The true strength of the chain of custody lies in combining all of these elements into a single logical sequence: source → time → method → integrity. As long as these four steps remain connected, an OSINT artifact becomes defensible. If any link in the chain is broken, the reliability of the evidence collapses instantly. For this reason, professional analysts treat every piece of data almost like a lab sample: it's isolated, labeled, sealed, and documented.

In real-world operations, the value of this mechanism becomes even clearer. OSINT data obtained to identify a threat actor, disrupt a fake social media operation, or prove an internal data breach can only be used in legal and institutional processes if the chain of custody remains intact. Otherwise, even the most critical finding remains nothing more than an allegation.

Essentially, the chain of custody isn't just a technical layer of OSINT; it safeguards the integrity of the entire intelligence process, ensuring that findings remain verifiable and defensible.

Let's walk through some concrete examples that are aligned with real OSINT operational logic and are legally, ethically, and technically sound. They include approaches and tactics that many analysts are unaware of, but they are actively applied by advanced intelligence units:

- **Change of the same evidence within three minutes**
  A code snippet used to identify a threat actor is detected on GitHub. The analyst takes a screenshot, downloads the file, verifies the hash, and saves it. Exactly three minutes later, a commit is made to the same repository. In the new commit, the following occurs:
  - One line of the original code is changed.
  - An alias is removed from the username line.
  - A line serving as a profile link is completely deleted.

If the analyst hadn't taken the hash of the first screenshot, this critical change couldn't be proven. In court or in an internal review, someone could claim that this line wasn't there. This situation is known in the literature as *transient evidence drift*. Without a chain of custody, the evidence disappears (or the accuracy of the evidence becomes disputable).

- **Detecting propaganda via hash change of an X (formerly Twitter) profile image**
  A propaganda account appears to be using the same profile picture to avoid detection. But during close analysis, the profile image looks pixel-identical, yet the hash is different. What does this mean? The attacker has re-encoded the same image to erase traces of manipulation. If the analyst had only downloaded and saved the profile image, it would be impossible to notice this change.

  Thanks to the chain of evidence, the hash of the first image, the hash of the second image, the time of the change, and the metadata on the platform are all recorded. This reveals the structural cleanup phase of the propaganda operation.

- **A dark web listing being deleted and re-uploaded 45 seconds later**
  A credential seller posts a listing on the dark web. The analyst captures a screenshot of the listing, generates a hash value to preserve evidential integrity, records the timestamp, and securely stores the file. Just 45 seconds later, the listing is deleted. Exactly 1 minute later, a similar listing appears again with a different price. There are key differences in the re-uploaded listing:
  - The username has changed.
  - A new story has been written in the "Source of Stolen Data" section.
  - The price has tripled.

  If the analyst doesn't maintain a chain of custody, this means the following: it can never be proven that the two listings were different. In fact, the first listing could be portrayed as entirely fabricated. This *volatile evidence capturing* method plays a critical role in detecting price manipulation used for disinformation by APT-level threat actors.

- **The same IP being moved across three different Autonomous System Numbers (ASNs) within 24 hours**
  Some threat groups move IP addresses between hosting providers to cover their tracks. These changes can occur so quickly that they can't be observed in normal user traffic. The analyst records the following information with a chain of evidence:
  - IP → ASN change
  - Timestamp
  - pDNS result
  - SSL certificate hash
  - Server fingerprint

  Thanks to the four-step chain, it's proven that the IP address remains under the control of the same threat group even if it's moved across three different locations. This *infrastructure lineage mapping* technique is generally known only by international intelligence units.

- **Detecting a fake news operation via the change in the number of lines in a web page's HTML**
  An article on a news site that appears harmless is manipulated by a threat actor. But the changes leave traces in both the content and in the HTML lines. The analyst records these values:
  - Hash of the HTML code
  - Total line count
  - Script blocks
  - Added invisible content
  - Timestamp

  Six hours later, the same page is updated. The line count has changed, and the hash has changed. However, the content looks the same to the naked eye. This is a critical method for detecting tracking pixels embedded into propaganda pages by APT groups.

  If a chain of custody hadn't been maintained, that operation couldn't be proven.
- **Existence verification via hash comparison of deleted Telegram messages**
  When a message is deleted in Telegram groups, it usually remains as undefined. However, if the analyst has previously recorded the message content, ID, timestamp, username, and hash value, then even after the message is deleted, the message's existence and content can be proven.

  This is a critical chain-of-custody technique, especially in financial fraud and propaganda operations. It's often assumed that Telegram leaves no linear evidence for deleted messages. That assumption is incorrect—if a chain of custody exists, proof is possible.
- **Detecting fake log generation via local time offset analysis**
  A threat actor creates fake logs to conceal their activities. However, after taking the hash of the log file, the analyst notices that although all log entries appear as UTC+2, the file creation metadata is UTC+0. This discrepancy is a common amateur mistake.

  Thanks to the chain of custody, the original log file, metadata hash, timestamp, and file modification traces are all preserved. This technique is of critical importance, especially in insider threat cases.

4.2 Passive OSINT Collection Strategies

Passive OSINT represents the most invisible yet most effective phase of cyber threat intelligence (CTI). This method refers to a process of information collection carried out by the analyst without establishing any direct contact with the target system—creating no interaction—by simply reading existing traces. From the outside, it may appear simple: Available information is taken. However, from a professional perspective, passive OSINT resembles an archaeological excavation. Information is hidden beneath layers, dispersed across relationships, faded over time, and waiting to be rediscovered. The analyst's skill lies in assembling these fragments within the correct context.

The power of passive OSINT lies in the fact that the target produces no counterreaction. When executed with disciplined OPSEC, passive OSINT typically leaves no direct interaction traces on the target system; however, absolute invisibility can't be guaranteed due to third-party logging, platform telemetry, and environmental variables. For this reason, passive OSINT is often among the lower-risk approaches for analyzing observable attacker-related traces, especially when compared to active collection methods; nevertheless, its reliability depends on source quality, platform bias, and collection context. When a threat actor believes they aren't being observed, they expose themselves more openly. This makes passive OSINT a technique that produces greater amounts of natural data compared to active reconnaissance methods, delivering high accuracy with low noise.

In the following sections, we'll address the strategic dimension of passive OSINT, the background mechanisms of the data ecosystem, and the subtle details that enable the analyst to make a difference. This includes the effective analysis of search engine operators, dork sets, metadata, digital lifecycles, and data from social media and user profiles because passive OSINT requires far more than collecting surface-level data: reading traffic flows, analyzing platform behaviors, understanding the temporal characteristics of record systems, and interpreting indirect indicators left by third-party service providers. A professional analyst must understand what the information is and how it's produced by the platform.

4.2.1 Advanced Use of Search Engine Operators

The greatest strength of passive OSINT lies in asking the right question, to the right search engine, using the right operators. Modern search engines are intelligence systems in their own right, built upon massive indexing infrastructures, complex ranking algorithms, and data normalization mechanisms. If an analyst proceeds using only simple queries without understanding these mechanisms, they are moving forward without discovering 90% of the OSINT universe. For professional and advanced OSINT analysts, search engines represent only the surface of accessible information; the real opportunities are hidden in the fine-tuning enabled by operators and in the traces of content outside the main index.

In the following sections, we'll discuss search engine operators, their value in OSINT, and important analysis techniques.

Understanding Search Engine Operators

Although major engines such as Google, Bing, Yandex (Russia-based), and Baidu (China-based) may appear similar, they don't behave in the same way. Google is the most aggressive in content filtering. Yandex can produce strong matches in certain reverse-image scenarios and may surface visually similar variants that aren't returned by other engines; performance varies by content type, region, and indexing coverage. Bing may surface a different mix of document-heavy results (e.g., PDFs/Microsoft Office files) depending on the domain and indexing patterns; cross-engine comparison is recommended for coverage. For China-hosted or China-centric ecosystems, Baidu can be a high-value complementary source due to its local indexing coverage; it's most effective when used alongside other engines. These differences produce entirely different outcomes in operator usage. For example, while the `intitle:"index of"` operator can be effective on Google, a similar technique is almost useless on Baidu because Baidu approaches security-oriented indexing with a different architectural model.

Advanced operator usage isn't just about knowing the correct operator; it's about understanding the behavioral combinations of operators with one another. When investigating a threat actor, a simple `site:` operator is often insufficient. The analyst refines the same query as follows:

- `site:domain.com -www`
- `site:domain.com inurl:admin`
- `site:domain.com ext:xml | ext:json | ext:conf`
- user `directory` remnants such as `inurl:/~username`

This construction can reveal much deeper remnants that are indexed but not obvious to users, rather than only the visible web layer. Especially in large organizations, silent indicators of vulnerability—such as developers' personal directories, test environments, or migrated but uncleaned files—are exposed only through these combinations.

Value of Search Engines in Passive OSINT

Search engines that support passive OSINT offer several key advantages. For example, reverse image search is a deceptively powerful tool. Google's accuracy threshold is high, but its algorithm is more susceptible to manipulation. Yandex processes visual similarity through regional texture and lighting pattern analysis, instead of pixels; therefore, AI-generated faces used in fake profiles can be detected much more quickly on Yandex. Tin-Eye, on the other hand, has a more limited source pool, but provides uniquely precise time-based version tracking. When a professional analyst queries the same image sequentially across these three platforms, they obtain highly valuable signals not only about visual similarity but also about geography, time, manipulation level, and the image's lifecycle across the internet.

Modern OSINT analysis must go beyond classical operators and incorporate the concept of semantic search. Traditional search engines are keyword-based; large language model (LLM)-supported semantic search is built on meaning similarity. For example, if a threat actor has formed a group named X-Squad, semantic search can capture not only this term, but all concepts within the same semantic cluster: "crew," "collective," "cell," "unit," "squadron," "shadow team," and so on.

This enables analysts to detect name variations used by threat groups attempting to conceal themselves. This technique is still not actively used by most OSINT practitioners, yet it's considered an advanced approach that provides significant advantage when applied.

Another important advantage offered by search engines is the ability to capture traces of de-indexed content. A piece of content may have been removed from a search engine, but the following traces may still remain:

- Meta descriptions in the cache
- URL patterns appearing as references on related sites
- Old links in XML sitemaps
- Content snippets appearing as small fragments in search results
- Different versions of the page in the Wayback Machine (*https://web.archive.org/*) internet archive or regional archives

Professional OSINT is an analytical approach capable of reconstructing a complete story from these fragments. For example, when a threat actor deletes a blog post, only a single sentence may remain in Google's index. However, the timestamp of that sentence, WHOIS revisions, changes in the link profile, and its relationships with other indexed pages may be sufficient to reconstruct all remaining pieces. Sometimes, even when no visible information exists, a URL structure alone can reveal the intent behind a threat.

Advanced search engine operators are also a critical tool for deep discovery, contextual analysis, and digital behavior inference. Analysts who transform querying from a simple search into a model-solving process make invisible data visible.

Search Engine Operator Analysis Techniques

The methods examined in this section encompass basic operator usage and the technical and analytical dimensions of positioning search engines as intelligence sensors. Let's walk through them:

- **Dual operator collision technique**

  Some advanced analysts report using operator-combination behaviors to surface less prominent indexed content under specific conditions. Normally, when the `intitle:` and `inurl:` operators are used together in Google, the results are narrowed. However, certain combinations temporarily place the search engine's ranking algorithm into an unstable state, causing low-priority, hidden, or normally invisible indexed content to surface.

  Let's consider the following query:

  ```
  intitle:"index of" | "directory listing" inurl:backup OR inurl:old
  ```

  This query often reveals:

  - Old directories left behind by developers
  - Backup folders
  - Archive content
  - Residual files left before server migrations

Some of this content never appears in Google's standard result lists. This approach aims to exploit occasional ranking and indexing edge cases that may cause less prominent results to appear. Outcomes can be inconsistent across time, locations, and algorithm updates.

- **Query mutation technique**
  Search engines don't evaluate different spellings of the same meaning in exactly the same way. For this reason, analysts generate mutations of the same query.

  For example, a threat group named Raven Brigade could have the following list of mutations:

  - RavenBrigade
  - Raven Brigade
  - raven-brigade
  - Raven_Brigade
  - R4ven Brigade
  - Raven Brig@de
  - Raven Brigade crew
  - Raven Brigade team

  These mutations expose subaccounts, former usernames, variants used in leaked logs, and synonymous naming across different platforms.

  Although this method is a standard practice in real CTI units, it's very rarely seen in OSINT sources.

- **Language-index divergence analysis**
  Google is optimized for English content, Yandex is optimized for Russian content, and Baidu is entirely focused on Chinese content. Bing has a broader corporate-oriented index structure, however.

  By exploiting these differences, it becomes possible to determine where a target has left traces geographically. For example, a threat actor's name may appear almost nowhere on Google, yet produce hundreds of results on Yandex. These results may include the following:

  - Discussions on Russian forums
  - Posts under `.ru` domains
  - Telegram index remnants
  - Russian translation variants

  An analyst who can recognize and evaluate this can largely confirm the actor's geographic origin.

- **Reverse image search**
  Image search answers more than just where else this photo has been used. At an advanced level, the following questions become critical:

  - When did this image first appear?
  - On which accounts was it previously used?
  - Are there changes in lighting or color structure?
  - On which sites do different versions exist?

For example, an older version of a photo found on Yandex may not appear on Google, while TinEye may display a timestamp. When the results of all three platforms are combined, the image's lifecycle on the internet can be reconstructed. This lifecycle is used for the following:

- Fake identity detection
- Identifying deepfake traces
- Revealing previous versions of propaganda accounts
- Tracking threat actor profile rotation

■ **Cache artifact exploitation**

Search engines may temporarily retain cached snapshots or snippets after content changes or removal, although retention behavior varies and isn't guaranteed. However, there's a more advanced technique: Some platforms retain short text snippets while removing the full cache page.

Analysts can access these hidden remnants using the following:

- The `cache:` command in Google
- Legacy indexers such as text uploaders
- Bing cached fragments
- Yandex snapshot viewer

The key point is that even a single word in a snippet may be enough to reconstruct a connection to a threat actor's former content. This method is especially effective when propaganda content is deleted or when threat actors attempt to hide past mistakes.

■ **Index shadowing**

Some websites block pages via `robots.txt`, but previously indexed versions aren't always completely removed. Search engines may resurface these indexes under certain conditions:

- When the query contains a very specific pattern
- When the page is referenced by another source
- When a domain migration causes resurfacing

This phenomenon is known as index shadowing and is used only by truly advanced OSINT analysts.

■ **Bing document extraction**

Search engines may differ in how they index and preview document content and metadata; therefore, running equivalent queries across multiple engines can increase discovery coverage. As a result, the following query models can produce extremely powerful results:

- `site:gov ext:pdf "internal use only"`
- `site:corp.com ext:docx "confidential"`

Bing, especially, exposes more detail in PDF metadata, including the author, last editor, creation software, version number, and date chain.

This information is surprisingly powerful and can reveal the following in OSINT:

- Internal information leaks
- Accidentally published confidential documents
- Insider threat indicators
- Threat actors' targeting strategies

- **Baidu's visual search advantage**
  Baidu's visual search operates very differently from Yandex and Google. In particular it does the following:
  - Recognizes China-region images far more effectively
  - Uses facial recognition algorithms that are trained on regionally prevalent datasets
  - Analyzes the visual origins of propaganda accounts more accurately

  For this reason, using Baidu becomes essential when investigating social media operations of China-origin APT groups.
- **Query entropy analysis**
  Even search engine autocomplete suggestions can provide intelligence about threat actors. The analyst runs the same query from different locations, at different times, and under different language settings. They then analyze changes in autocomplete behaviors, which can reveal what content is trending, how well known a threat actor is in a specific region, and in which countries disinformation campaigns are most effective. This search entropy analysis method is understood by very few in the OSINT community.

4.2.2 Using Google Dorks and Advanced Queries

Google *dorks* refer to advanced search queries that leverage the indexing structure of search engines to identify specific types of publicly accessible information. By combining specialized search operators such as `site:`, `filetype:`, `intitle:`, and `inurl:`, analysts can narrow down search results to reveal documents, configuration files, login portals, or other digital artifacts that are unintentionally exposed through search engine indexes. Rather than interacting directly with a target system, this approach relies entirely on information already collected and organized by search engines.

In OSINT practice, analysts rarely rely on a single query. Instead, they construct a structured collection of queries known as a dork set. A *dork set* is a curated group of search expressions designed to systematically explore different categories of potential exposure, such as publicly indexed documents, administrative interfaces, backup files, or misconfigured web directories.

When used responsibly, these queries enable analysts to map publicly visible aspects of a digital environment without performing intrusive actions against the target infrastructure.

For this reason, Google dorks are widely used during passive reconnaissance to identify potential information exposure while remaining within the boundaries of OSINT collection.

Let's walk through some practical examples of leveraging dork sets.

Automatic OSINT Dork Generation with Python

The queries shown in the output in Figure 4.1 represent a static OSINT dork set used to collect information from the open indexes of search engines for a domain (`haydararici.com.tr`), without performing any active scanning. Such dorks are among the most fundamental yet most effective tools used to map the attack surface during passive reconnaissance. Each dork category targets a different type of vulnerability or information residue within search engine indexes.

```
C:\users\Admin\PycharmProjects\Books>python dork_builder.py -d haydararici.com.tr
# [haydararici.com.tr] basic_visible (visibility, risk=low, phase=Recon, engines=bing,google)
#   Description: General visibility and sub-host discovery
site:haydararici.com.tr

# [haydararici.com.tr] basic_no_www (visibility, risk=low, phase=Recon, engines=bing,google)
#   Description: Find hosts other than www
site:haydararici.com.tr -www

# [haydararici.com.tr] admin_panels (admin, risk=medium, phase=ExposureMapping, engines=bing,google)
#   Description: Administration ponils and management interfaces
site:haydararici.com.tr (inurl:admin | inurl:administrator | inurl:cpanel | inurl:manage)

# [haydararici.com.tr] login_panels (ogin, risk=medium, phase=WeakneseMapping, engines=bing,google)
#   Description: Authentication / login pages
site:haydararici.com.tr (inurl:login | inurl:signin | inurl:auth)

# [haydararici.com.tr] config_files (config, risk-high, phase=ExposureMapping, engines=bing,google)
#   Description: Configuration files (xml/json/yml/conf/ini)
site:haydararici.com.tr (ext:xml OR ext:json OR ext:yml OR ext:yaml OR ext:conf OR ext:ini)

# [haydararici.com.tr] backup_dirs (dackup, risk-high, phase=ExposureMapping, engines=google)
#   Description: Backup directories and legacy content (indes of / directory listing)
site:haydararici.com.tr "drectory listing") site:haydararici.com.tr (inurl:backup OR inurl:old OR inurl:archive)

# [haydararici.com.tr] user_dirs (users, risk=medlum, phase=Profiling, engines=google,google)
#   Description: User home directory remnants (<usernans)
site:haydararici.com.tr inurl:/~

# [haydararici.com.tr] error_traces (debug, risk=medium, phase=WeaknessMapping, engines=bing,google)
#   Description: Pages containing error messages and stack traces
site:haydararici.com.tr ("PHP Notice" OR "PHP Warning" OR "Exception in thread" OR "stack trace")

# [haydararici.com.tr] indexing_keywords (sensitive_keyword, risk-high, phase=ExposureMapping, engines=bing,google)
#   Description: Broad search using sensitive keywords
#   Keyword    : confidential
site:haydararici.com.tr "confidential"

# [haydararici.com.tr] indexing_keywords (sensitive_keyword, risk-high, phase=ExposureMapping, engines=bing,google)
#   Description: Broad search using sensitive keywords
#   Keyword    : internal use only
site:haydararici.com.tr "internal use only"

# [haydararici.com.tr] indexing_keywords (sensitive_keyword, risk-high, phase=ExposureMapping, engines=bing,google)
#   Description: Broad search using sensitive keywords
#   Keyword    : password
site:haydararici.com.tr "password"

# [haydararici.com.tr] env_files (config, risk-high, phase=ExposureMapping, engines=google)
#   Description: Secret configuration files similar to .env
site:haydararici.com.tr (inurl1.env OR "APP_ANV-production" OR "DB_PASSWORD=")

# [haydararici.com.tr] git_repos (source, risk-high, phase=ExposureMapping, engines=bing,google)
#   Description: sit repository comments (.git, gitlab, bitbucket etc.)
site:haydararici.com.tr (inurl:.git OR inurl:gitlab OR inurl:bitbucket) ("HEAD" OR "refs/heads")

# [haydararici.com.tr] logs (logs, risk-high, phase=ExposureMapping, engines=google)
#   Description: Log file comments (log, access.log, error.log)
site:haydararici.com.tr (ext:log OR "access.log" OR "error.log" OR "stack traces")

# [haydararici.com.tr] test_dev (environment, risk=medium, phase=prefiling, engines=bing,google)
#   Description: URLs and directories belowwing to test / dev environments
site:haydararici.com.tr (inurl:dev OR inurl:test OR inurl:staging OR subdomain="dev.haydararici.com.tr")

# [haydararici.com.tr] ci_cd (pipeline, risk-high, phase=ExposureMapping, engines=bing,google)
#   Description: CI/CI pipeline files (.gitlab-ci.yml, jenkinsfils, azure-pipelines.yml)
site:haydararici.com.tr (".gitlab-ci.yml" OR "Jenkinskils" OR "azure-pipelines.yml")

# [haydararici.com.tr] cloud_storage (cloud, risk-high, phase=ExposureMapping, engines=google)
#   Description: Cloud bucket and file storage troces (SS, GCS, Azure BloB)
site:haydararici.com.tr ("aws_access_key_id" OR "$3.amazonows.com" OR "storage.googleapis.com" OR "blob.core.windows.net")
```

Figure 4.1: Python-Based Automatic Generation of an OSINT Dork Set for Domain Reconnaissance

The dork output presented in this section is a concrete example of the following principles:

- There's no active scanning; only data already maintained by search engines is used.
- Each dork represents a specific risk category and discovery phase.
- The dork set is designed to initiate a general, security-focused surface analysis without prior knowledge of the target's technology stack.
- The results don't mean there's a leak; they define information channels that should be reviewed.

This static dork set consolidates all critical information sources that an OSINT analyst should pay attention to for a target domain into the following five fundamental categories: visibility analysis; management and authorization surfaces; configuration and sensitive information leakage; error messages, logs, and developer artifacts; and test environments and archives.

Let's examine each dork shown in the output from the perspective of a professional security analyst:

- **`site:haydararici.com.tr`**
  This query represents the starting point of the entire OSINT process. This most basic dork is used to understand how extensively a domain is indexed by search engines:
  - How many pages are indexed?
  - Are subdirectories or old content visible?
  - Is there any content that was published unintentionally?
- **`site:haydararici.com.tr -www`**
  This focuses on discovering subdomains other than the `www` subdomain. Through this, otherwise invisible hosts used by developers or services may be revealed, such as the following:
  - `dev.example.com`
  - `api.example.com`
  - `panel.example.com`
- **`(inurl:admin | inurl:administrator | inurl:cpanel | inurl:manage)`**
  This dork is used to find management panels and control interfaces. In many organizations, forgotten or legacy panels remain accessible. These panels often contain weak passwords, outdated software, or misconfigurations.
- **`(inurl:login | inurl:signin | inurl:auth)`**
  This dork finds authentication pages. Login screens are the first contact points for attackers targeting phishing, brute-force, or credential-stuffing attacks.
- **`ext:xml OR ext:json OR ext:yml OR ext:conf OR ext:ini`**
  This dork targets configuration files that have been accidentally exposed to search engines. Such files may contain the following:

  - API keys
  - Database connection details
  - Simple Mail Transfer Protocol (SMTP) credentials
  - Application secrets
  - Deployment information

  This category is high risk, as configuration leaks often lead directly to system access.

- **intitle: "index of" OR "directory listing"**
  Directories with "index of /" enabled often contain the following:
  - Old backups
  - ZIP files
  - SQL dumps
  - Test outputs
  - Code packages

  Such open directories are among the most common sources of data leakage on the internet.

- **inurl: /~**
  Common on Apache-based servers, ~username directories historically hosted developers' test pages. These directories often contain the following:
  - Old code
  - Log files
  - Personal notes
  - Insecure test scripts

- **"PHP Warning", "PHP Notice", "Exception in thread", "stack trace"**
  This dork aims to identify error messages leaked by applications. Error messages are extremely valuable to attackers because they may expose the following:
  - File paths
  - Usernames
  - Running services
  - Version information
  - Hidden API endpoints

  This can form the first step of an attack chain.

- **"confidential", "internal use only", "password"**
  These dorks are designed to determine whether organization-owned documents have been accidentally indexed. This category typically reveals the following:
  - Reports in Word/Excel/PDF format
  - Internal correspondence
  - Personnel lists

- Policy documents
- Configuration snippets containing passwords
- GDPR (EU)/KVKK-critical documents (Turkish)

For this reason, it carries a high level of risk.

- **`(inurl:.env OR "DB_PASSWORD=")`**
  Environment variable (`.env`) files are among the most frequently leaked files in modern software stacks. They typically contain the following:
  - Database usernames and passwords
  - SMTP access details
  - JSON Web Token (JWT) secret keys
  - AWS keys
  - Redis/Memcached connection details

  A leaked `.env` file usually implies direct unauthorized access.
- **`inurl:.git OR inurl:gitlab OR inurl:bitbucket`**
  A common mistake is accidentally deploying the `.git` directory to the web root. The presence of a `.git/HEAD` file often means that the entire source code can be downloaded. Source code leakage is one of the most critical findings in corporate OSINT.
- **`ext:log OR "access.log" OR "error.log"`**
  Exposed log files may reveal the following:
  - IP addresses
  - User behavior
  - Session cookies
  - Server response codes
  - Security errors

  Such logs are among the most critical leakage types from both privacy and attack-surface perspectives.
- **`inurl:dev OR inurl:test OR inurl:staging`**
  In real-world environments, a large portion of internal leaks originate from test environments. These environments often contain the following:
  - Weaker security controls
  - Real data
  - Vulnerabilities left by developers

  This dork is used to identify nonproduction but high-risk systems.

Dynamic Domain Fingerprinting and OSINT Dork Generation with Python

This example demonstrates the automatic fingerprinting of a target domain using passive information-gathering methods, and the generation of dynamic OSINT dorks by analyzing the obtained technology, security, certificate, and infrastructure information. This approach goes beyond classical static dork lists and represents an advanced OSINT technique that understands the target's real technological profile and generates customized discovery queries tailored specifically to it.

The output shown in Figure 4.2 presents a fully passive OSINT profile obtained by running the developed `fingerprint_and_generate_dorks.py` script against the domain `xxxxx.com.tr`. The script produces two fundamental stages:

1. Domain fingerprint, that is, automatic extraction of the target's technical footprint
2. Generation of dynamic OSINT dorks based on this profile

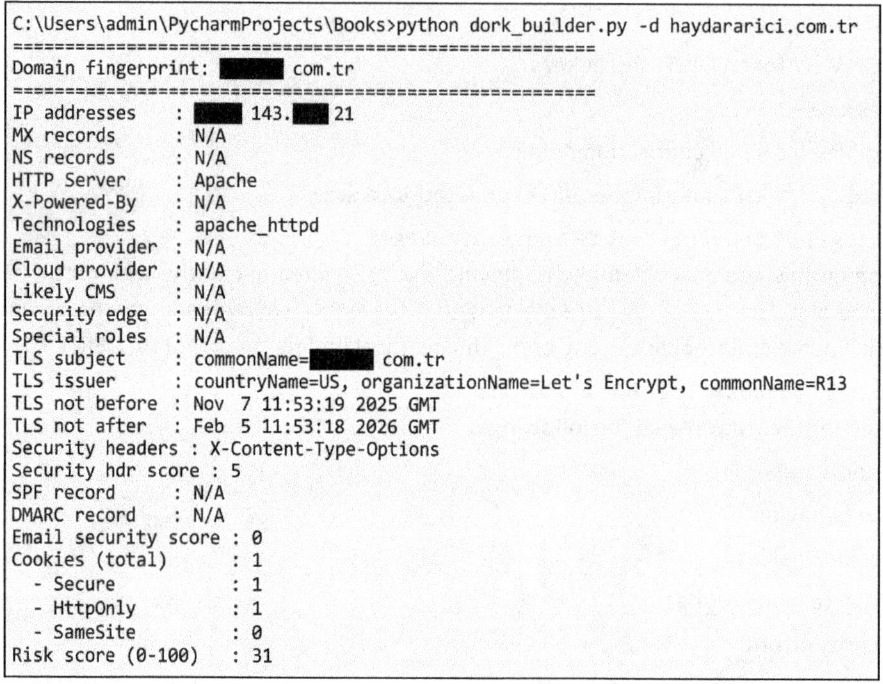

```
C:\Users\admin\PycharmProjects\Books>python dork_builder.py -d haydararici.com.tr
============================================================
Domain fingerprint: ██████ com.tr
============================================================
IP addresses     : ████ 143.███ 21
MX records       : N/A
NS records       : N/A
HTTP Server      : Apache
X-Powered-By     : N/A
Technologies     : apache_httpd
Email provider   : N/A
Cloud provider   : N/A
Likely CMS       : N/A
Security edge    : N/A
Special roles    : N/A
TLS subject      : commonName=██████ com.tr
TLS issuer       : countryName=US, organizationName=Let's Encrypt, commonName=R13
TLS not before   : Nov  7 11:53:19 2025 GMT
TLS not after    : Feb  5 11:53:18 2026 GMT
Security headers : X-Content-Type-Options
Security hdr score : 5
SPF record       : N/A
DMARC record     : N/A
Email security score : 0
Cookies (total)      : 1
  - Secure           : 1
  - HttpOnly         : 1
  - SameSite         : 0
Risk score (0-100)   : 31
```

Figure 4.2: Passive Domain Fingerprinting Output Showing Infrastructure, Security Headers, and Certificate Information

The `Domain fingerprint` section shouldn't be viewed as a single-line status summary, but as an automatically generated technical profile written for an OSINT analyst. Let's break down the components:

- **IP addresses**
  `xxxxx.com.tr` currently resolves to a single IP address. The script treats this IP information—one of the foundations of passive fingerprinting—as an entry point for further checks (e.g., alternative hostnames, reverse DNS, etc.).
- **HTTP Server/Technologies**
  Apache has been identified from HTTP headers, and the technology stack is marked as `apache_httpd`. Because no typical `X-Powered-By` header is returned, application platforms such as PHP, ASP.NET, or Node.js can't be conclusively identified at this stage.

- **`Email provider/Cloud provider/Likely CMS/Security edge/Special roles`**
  All of these fields appear as `N/A`. This doesn't mean they don't exist; rather, it means that during this fingerprinting pass, the script couldn't passively identify the following:
  - A software as a service (SaaS) email provider (Microsoft 365, Google Workspace, etc.)
  - A prominent cloud provider (AWS, Google Cloud, Cloudflare, etc.)
  - A common content management system (CMS; WordPress, Drupal, Joomla, SharePoint)
  - A security edge (web application firewall [WAF]/content delivery network [CDN])
  - A special role (e.g., Exchange Outlook Web App [OWA], SharePoint portal)

  In other words, no clear trace was found in this fingerprinting round.
- **`TLS subject/TLS issuer/TLS not before/TLS not after`**
  The certificate information indicates the use of a wildcard certificate:
  - Subject: `commonName=*.xxxxx.com.tr` is usable across subdomains such as `www.xxxxx.com.tr`, `app.xxxxx.com.tr`, and so on.
  - Issuer: Let's Encrypt R13, a free certificate authority (CA) focused on automated renewal.
  - Validity: From November 7, 2025, to February 5, 2026.

  The script records these dates without interpretation; this information is intended for analysts to map certificate lifecycle timelines, identify renewal risks, and, in some cases, correlate infrastructure changes.
- **`Security headers/Security hdr score`**
  In this example, only the `X-Content-Type-Options` header is present. The script treats this as a positive security signal and calculates a security header score of 5. This score is later used to slightly reduce the overall risk score.
- **`SPF record/DMARC record/Email security score`**
  Both `SPF` and `DMARC` fields appear as `N/A`, indicating that the script couldn't detect an Sender Policy Framework (SPF) policy (`v=spf1`) in TXT records or a Domain-based Message Authentication, Reporting, and Conformance (DMARC) policy (`v=DMARC1`) under `dmarc.xxxxx.com.tr`. As a result, the email security score remains 0, and no positive points are assigned for domain-level email protection policies.
- **`Cookies`**
  From the HTTP response, a total of one cookie was detected. This cookie has the `Secure` and `HttpOnly` flags set. The `SameSite` flag isn't present.

  The script quantifies this information; cookies marked `Secure` and `HttpOnly` are treated as positive signals that slightly reduce the risk score.
- **`Risk score (0-100)`**
  When all factors are combined, the script produces a risk score of 31/100 for this domain, which is moderate but not critical. This means there's a surface worth investigating on

this domain, but there's also an extremely exposed structure at the moment from an external perspective.

The key point is that this score serves as an automated prioritization metric. It doesn't replace the analyst's final judgment, but it saves significant time when ranking which targets should be examined first.

Now, let's interpret the selected dork categories. The following line shows which dork families are considered meaningful for this target based on the fingerprint result:

```
Selected dork categories for xxxxx.com.tr: admin, auth, backup, config,
debug, environment, logs, sensitive_keyword, users, visibility.
```

In other words, the script is indicating the following:

- **`visibility`**
  Examine overall domain visibility, subhosts, and indexed content volume.
- **`admin/auth`**
  Target web management panels and login screens.
- **`config/backup`**
  Consider scanning for accidentally indexed configuration files and `index of /` style backup or legacy content.
- **`logs/debug`**
  Investigate the possibility of leaked error messages or log files.
- **`environment`**
  Trace nonproduction environments such as test, development, or staging systems, which are often less protected.
- **`users`**
  Check for user directory remnants such as `~/username`.
- **`sensitive_keyword`**
  Search for sensitive terms such as "confidential," "internal use only," or "password" within the domain.

The critical point is that these categories aren't saying there's currently a leak. Instead, they're saying that based on this profile, these areas carry potential risk, so generate OSINT dorks accordingly.

In the following block, automatically generated dork examples are listed for each category:

- **Visibility and host discovery**
  The following queries are designed to identify the primary indexed presence of the domain and to reveal additional hosts or subdomains that may appear in search engine results:
  - `site:xxxxx.com.tr`
  - `site:xxxxx.com.tr -www`

- **Management and authentication surface**
  The following queries aim to make visible—via OSINT—the areas where an attacker would first interact, such as admin panels and login forms:
  - `site:xxxxx.com.tr (inurl:admin | inurl:administrator | inurl:cpanel | inurl:manage)`
  - `site:xxxxx.com.tr (inurl:login | inurl:signin | inurl:auth)`
- **Configuration files and `.env` residues**
  The following queries focus on identifying configuration artifacts that may have been unintentionally exposed through search engine indexing. Such files often contain application parameters, environment settings, or infrastructure references that reveal how a system is configured. During passive reconnaissance, these traces can provide valuable contextual insight into the target's technical environment:
  - `site:xxxxx.com.tr (ext:xml OR ext:json OR ext:yml OR ext:yaml OR ext:conf OR ext:ini)`
  - `site:xxxxx.com.tr (inurl:.env OR "APP_ENV=production" OR "DB_PASSWORD= ")`

  The goal here is to capture the following at the dork level:
  - XML/JSON/YAML/INI configuration files accidentally left in the web root
  - Lines directly indicating secrets, such as `DB_PASSWORD=`
- **Backup directories and `"index of"` outputs**
  The following script targets `"index of /"`-enabled backup, old, or archive directories, which often contain forgotten ZIP files, legacy code, logs, or test artifacts:

  `(intitle:"index of" OR "directory listing") site:xxxxx.com.tr (inurl:backup OR inurl:old OR inurl:archive)`
- **Logs and error messages**
  These dorks allow analysts to passively test whether the application exposes detailed error messages or unintentionally reveals stack traces or log content:
  - `site:xxxxx.com.tr (ext:log OR "access.log" OR "error.log" OR "Stack trace:")`
  - `site:xxxxx.com.tr ("PHP Notice" OR "PHP Warning" OR "Exception in thread" OR "stack trace")`
- **Sensitive keywords**
  These queries enable OSINT-level hunting for internal documents and accidentally published pages containing sensitive terms:
  - `site:xxxxx.com.tr "confidential"`
  - `site:xxxxx.com.tr "internal use only"`
  - `site:xxxxx.com.tr "password"`

  In a real CTI scenario, content found via these dorks can provide critical indicators for data classification and data leakage analysis.

- **Test/development/staging environment traces**
  Finally, this dork is used to check whether nonproduction environments—often with weaker security controls—are exposed:

```
site:xxxxx.com.tr (inurl:dev OR inurl:test OR inurl:staging OR
subdomain:"dev.xxxxx.com.tr")
```

Remember!

For a professional OSINT/CTI analyst, what matters isn't just which dorks they know, but their ability to generate them dynamically and contextually by understanding the target's real profile. The preceding output is a practical example of this approach applied to the `xxxxx.com.tr` domain.

4.2.3 Metadata Extraction

Most analysts view data as a whole; when examining a file, image, or document, they focus on the content. At a professional level, however, the real value lies in the invisible traces hidden around and beneath the content. Metadata is like the subconscious of a digital artifact: it contains the device on which it was produced, the environment in which it was edited, the changes it underwent, the paths it traveled, and often personal footprints left unintentionally. For this reason, metadata extraction is an advanced OSINT discipline that combines digital behavioral analysis with legal evidence evaluation.

In the following sections, we'll discuss the information that metadata can reveal, the different techniques to analyze it, and an in-depth practical example.

Understanding Metadata

The Exchangeable Image File Format (EXIF) data of an image can provide a wide range of information, from a camera's serial number to the shooting location, from the lens model used to ISO values, and even traces of edits applied to the photo. However, the power of professional metadata analysis doesn't lie only in reading EXIF data; it's also built on the absence of EXIF, inconsistencies within it, and signs of manipulation. Some threat actors believe they have fully cleaned EXIF data, yet even if GPS tags are removed, device models or creation software often remain. More interestingly, some platforms partially strip EXIF data while adding their own platform metadata. When an analyst can read this distinction, it becomes possible to determine whether an image has passed through Facebook, Telegram, WhatsApp, VK, or Weibo.

In PDF and Office documents, metadata is both more dangerous and equally valuable. The content of a phishing document prepared by a threat actor may not reveal an IP address, but its metadata may contain the following information:

- License owner of the software used to create it
- Username of the last editor

- Relationship between the document's creation and modification times
- Revision history
- Language pack used
- Embedded objects
- Autosave traces

This information can indicate whether the threat actor worked on a personal machine, used a corporate device, or operated in a different time zone, as well as whether correlations can be established with other previously shared content. Professional analysts use this data to verify the source of a file and to build a behavior graph to understand the actor's operational logic.

One of the most critical dimensions of metadata extraction is the editing timeline, known as the author timeline analysis. For example, the gap between a PDF's creation time and last modification time can reveal whether the actor prepared the document hastily or in a planned manner. A phishing document created within minutes usually indicates low-level attacks, whereas a document edited over hours or days signals a more sophisticated target-specific operation. This detail is a far more important signal for threat actor classification than is commonly assumed.

Detection of metadata manipulation is where advanced analysts truly make a difference. Some threat actors believe they have reset metadata, but the internal structure of the file (for example, the cross-reference tables in a PDF or the ZIP structure of an Office document) retains traces of manipulation. An image's EXIF data may be wiped, but the image-processing software used leaves an EXIF-independent signature byte pattern. These traces are characteristic of software such as Photoshop, GIMP, Lightroom, or PicsArt, and this technique is rarely mentioned in OSINT sources.

One of the most valuable metadata types in passive OSINT is chain metadata, which refers to traces left by a file as it passes through different platforms or devices. If a photo is first taken on an iPhone and later compressed on an Android device, each stage of this chain can be read from metadata. For example, Telegram's compression algorithm removes EXIF data but leaves a unique header at the beginning of the file. WhatsApp partially strips EXIF data but renames the file in a way that makes the device model inferable. VK, even when EXIF is removed, creates regional compression blocks in the image's raster pattern, making it clearly detectable to an analyst that the image passed through VK.

Such subtle indicators can reveal threat behavior: where an image came from, who uploaded it, from which device it was uploaded, through which platform it was transferred, and how many times it was compressed. Sometimes an image has no original EXIF data, yet its compression characteristics match those of typical internet platforms in a specific region. This is an extremely valuable signal, especially in propaganda and psychological operation analysis.

The story told by the content may differ from the story told by metadata; even if the content is manipulated, metadata can reveal the truth. While a file's content may appear exactly as a threat actor intends, metadata can expose who the actor is, their working discipline, the tools they use, and their operational domain. For professional analysts, metadata is therefore not just supplementary data, but the most reliable witness of an operation.

Metadata Analysis Techniques

The real power of metadata analysis lies in the technical layers of information that aren't visible on the surface. Attackers believe that by deleting metadata, they make themselves invisible; in reality, digital materials leave dozens of unnoticed side indicators. This section reveals advanced techniques known and applied by very few analysts in the modern OSINT field. Let's walk through the techniques:

- **Shadow metadata extraction**
  Even if a file's metadata is removed, some platforms embed additional metadata blocks within the file as part of their own control structures. These invisible blocks are detected through the following:
  - JPEG quantization table analysis: Each camera model uses a unique compression table; even if EXIF is removed, the Q-table reveals the device.
  - PDF XRef table pattern traces: Manipulated PDFs show abnormal offset differences in cross-reference tables.
  - Office ZIP internal structure: The `.docx`/`.xlsx` files are ZIP archives; even if `docProps` are deleted, file timestamps remain in the ZIP header.

  Even after EXIF removal, it's possible to identify the camera model and even the brand-specific algorithm of the device used.
- **Temporal drift detection**
  The timestamps of manipulated files aren't synchronized. Even if an attacker modifies EXIF data, the following often don't align:
  - File creation time
  - Revision time
  - File system timestamp
  - Application metadata time
  - Thumbnail creation time

  For example, say an image appears to have been taken in 2023 according to EXIF, but the embedded thumbnail dates to 2019. This is a mathematically provable indicator of manipulation.
- **Thumbnail forensics**
  Many devices generate a small preview image alongside the main photo. Even if an attacker deletes EXIF data, the thumbnail is often left intact. Professional analysts extract the following from thumbnails:
  - Original resolution
  - Color histograms

- Lighting inconsistencies
- Device's compression signature

A little-known technique is to identify the device model from some Android devices that compress thumbnails in reverse order.

- **Binary entropy profiling**

  A manipulated file exhibits different entropy levels than a natural file. In this context, *entropy* refers to the statistical randomness of data within a file—essentially a measure of how predictable or structured the byte patterns are. Natural files typically display relatively stable entropy patterns based on their format and compression structure. However, when content is hidden, removed, or artificially filled with random data, irregular entropy anomalies can appear within specific sections of the file. Especially after hiding, cleaning, or random filling, entropy anomalies appear. This method allows analysts to infer the following:

  - Locations of manipulated blocks
  - Editing software used by the attacker
  - Whether the file is original

  For example, entropy breaks within a PDF can reveal fake inserted page tactics.

- **Multiplatform propagation analysis**

  When a file is transferred from one platform to another, each platform applies its own unique modifications. Example signatures include the following:

  - Telegram: Compressed flag in the information header
  - WhatsApp: Date-based renaming in the filename
  - Facebook: EXIF removed but International Press Telecommunications Council (IPTC) metadata retained
  - Instagram: Unique color compression algorithm

  Using these signatures, it's possible to determine the platform chain a file has passed through. This chain reveals propaganda routes, the actor's communication model, and the authenticity of the image.

- **Scientific image residue detection**

  This advanced method is used primarily for manipulation detection. By analyzing the following, manipulated regions can be identified:

  - Light direction
  - Shadow density
  - Sharpness differences
  - JPEG artifact patterns
  - Spectrum analysis

  For example, consider a fake passport image where the JPEG compression levels of the text layer and the background layer differ. This difference is definitive proof of manipulation.

- **Document lineage tracking**
  Even if past versions of an Office or PDF document aren't directly visible in metadata, the document's lineage can be reconstructed. This professional technique includes the following:
  - Differences in embedded resource versions
  - Matching the same template with older records
  - Software version traces in the `producer` field (the metadata field that records the application used to create or export the document, such as Adobe Acrobat)
  - Inconsistencies in revision flags
  - Font version analyses

  This makes it easy to determine the following:
  - Which software and which version the threat actor used
  - How many times the document was resaved
  - Whether the same template was used in other operations
- **Metadata injection detection**
  Manipulated metadata is often easier to detect than attackers expect. Advanced detection methods include the following:
  - Timestamp inconsistency: Injected metadata doesn't align with the real device clock. For example, a user in UTC+3 uses a UTC-5 timestamp format.
  - Software version inconsistency: For example, a photo appears to be edited with Photoshop 14 according to EXIF, but the device model doesn't support Photoshop 14.
  - Byte-level alignment check: Genuine device metadata blocks are aligned; injections leave misalignment.
  - Lens model–focal length mismatch: In real captures, lens, aperture, and focal length values are consistent; in fake EXIF data, this consistency is absent.
- **Composite metadata reconstruction**
  Some files may appear to be completely sanitized. However, by combining metadata remnants from different sources, it's possible to reconstruct the capture time, file chain, platform used, editor identity, and transfer route. This technique combines the following:
  - EXIF remnants
  - IPTC fragments
  - Extensible Metadata Platform (XMP) residues
  - Device compression signatures
  - Thumbnail hashes
  - File entropy maps

  In OSINT, this is known as reconstructing the invisible layer of evidence.

Example: Shadow Metadata and Digital Footprint Analysis

Modern document formats (especially Open Office XML (OOXML)-based Word, Excel, and PowerPoint files), in addition to containing visible content, embed extensive shadow metadata related to the document's creation process, modification history, user behavior, and internal document events. This metadata layer is unnoticed by most users, yet it's extremely valuable in critical domains such as digital forensics, threat intelligence, leak analysis, and content manipulation detection.

The example analysis output shown in Figure 4.3 demonstrates an in-depth metadata examination of a Word document performed by a generated Python program called Shadow Metadata Analyzer.

```
C:\Users\admin\PycharmProjects\Books>python shadow_metadata_analyzer.py file.docx
[+] File : file.docx
[+] Type : office

=== Office OOXML ZIP Metadata Analysis ===
Is ZIP container : True
Has docProps     : True
docProps files:
  - docProps/core.xml
  - docProps/app.xml

[core.xml Properties]
  creator        : H. Yener ARICI
  lastModifiedBy : H. Yener ARICI
  revision       : 5
  created        : 2023-11-19T06:51:00Z
  modified       : 2023-11-29T13:23:00Z

[app.xml Properties]
  Template       : Normal.dotm                                         : Normal.dottm
  TotalTime      : 3                                                   : 3
  Pages          : 6                                                   : 6
  Words          : 812                                                 : 812
  Characters     : 4631                                                : 4631
  Application    : Microsoft Office Word                               : Microsoft Office Word
  Lines          : 38                                                  : 38
  Paragraphs     : 10                                                  : 10
  AppVersion     : 16.0000                                             : 16.0000

ZIP internal timeline (basic):                                      imeline (basic s):
  Earliest entry : [Content_Types].xml -> 1980-01-01T00:00:00          : [Content _Types].xml -> 1980
  Latest entry   : docProps/app.xml -> 1980-01-01T00:00:00             : docPropss/app.xml -> 1980-01
  Time span      : 0.0 seconds                                         : 0.0 seconds

[Office Structure Summary]
  Top-level folder distribution:
    - [Content_Types].xml: 1 entries
    - _rels: 1 entries
    - word: 9 entries
    - docProps: 2 entries
  Extension distribution:
    - .xml: 10 items
    - (none): 1 items
    - .png: 1 items
    - .rels: 1 items

[Office Features / Flags]
  Macros (vbaProject.bin) present : False
  Embedded OLE objects present    : False
  Embedded images present         : True
  customXml present               : False
  Signatures (_xmlsignatures)     : False
  Track Changes (TrackRevisions)  : False
  Document protection enabled     : False
  Protection type                 : None

[Timeline Analysis]
  revision_count                         : 5
  approx_words_per_minute                : 270.6666666666667
  core_timespan_seconds                  : 887520.0
  approx_seconds_per_revision            : 221880.0
  approx_revision_timestamps_iso         : ['2023-11-19T06:51:00+00:00', '2023-11-21T20:29:00+00:00', '2023-11-24T10:07:00+00:00', '2023-11-26T23:45:00+00:00', '2023-11-29T13:23:00+00:00']
  zip_earliest_minus_core_created_seconds : -1448802260.0
  zip_latest_minus_core_modified_seconds  : -1448889780.0

[Warnings / Anomalies]
  * All timestamps inside the ZIP appear to be set to year 1980 - may indicate default filesystem dates or metadata reset.
```

Figure 4.3: Advanced OOXML Metadata Forensic Analysis Revealing Document Lifecycle, Structural Artifacts, and Potential Manipulation Indicators

The tool doesn't just read the `docProps/core.xml` and `docProps/app.xml` files; it also analyzes timestamps within the ZIP structure, document architecture, hidden properties, embedded content, and potential temporal anomalies. Through this analysis, the following critical indicators can be revealed:

- Actual creation and last modification times
- User identities that edited the document
- Text volume, writing speeds, and total editing duration
- Revision count and an estimated revision timeline
- Embedded images, object linking and embedding (OLE) objects, macros, and signature structures
- Timestamp inconsistencies across all files within the ZIP container
- Inconsistencies that raise suspicion of document manipulation
- Default timestamps fixed to the year 1980 (indicative of metadata wiping)

The professional value of this output is concentrated in two key areas:

- **Forensic and nonforensic manipulation detection**
  The consistency between revision count and modified timestamps, contents of `word/settings.xml`, `trackChanges` flags, and the presence of `customXML` files generate strong signals as to whether a document was edited manually, automatically, or using an external tool.
- **Operational threat intelligence and source/identity verification**
  In cases involving leaks, breaches, threat actor operations, or insider threats, metadata can reveal—often with high accuracy—who created the document, when it was created, and through which stages it evolved.

This example represents an advanced approach that goes beyond classical metadata reading: systematically reconstructing the document's historical lifecycle, technical traces, and potential manipulation vectors.

Let's walk through the data shown in Figure 4.3:

- **File type and basic classification**
  The following code snippet verifies that the file is indeed an OOXML Office document:

  ```
  [+] File : file.docx
  [+] Type : office
  ```

  It confirms this by checking both the file extension (`.docx`) and the internal ZIP signature (`PK\x03\x04`). This verification is critical so that subsequent steps can safely parse directories such as `docProps`, `word/`, and `_rels/`.
- **`docProps` and core metadata**
  Let's move on to the next code snippet, which contains the `docProps` and core metadata:

  ```
  Is ZIP container : True
  Has docProps : True
  docProps files:
  - docProps/core.xml
  - docProps/app.xml
  ```

Here, we can see that the file is indeed a ZIP container. Both `docProps/core.xml` and `docProps/app.xml` are present:

- `core.xml`: Contains metadata fields such as the document author, last modified by information, revision count, and created/modified timestamps.
- `app.xml`: Provides document statistics, including page count, word count, and total editing time recorded by the application.

The presence of these two files indicates that Office metadata hasn't been reset yet. In other words, there's still a significant amount of forensic signal available about the document.

- **Document lifecycle**
  The next code snippet contains the `core.xml` properties:

```
[core.xml Properties]
  creator : H. Yener ARICI
  lastModifiedBy : H. Yener ARICI
  revision : 5
  created : 2025-11-19T06:51:00Z
  modified : 2025-11-29T13:23:00Z
```

  We can draw a few key conclusions:

  - `creator` = `lastModifiedBy`: The person who created the document and the user who last modified it appear to be the same. There's no indication that the document was handed off to another user and later returned. From an internal source analysis perspective, the document can be attributed to a single individual.
  - `revision` = `5`: According to Word, the document has undergone five revisions. This value is a core parameter for the script's estimated revision timeline.
  - `created` vs `modified`: The created and modified dates suggest the document was created and updated over a span of approximately 10 days and 6.5 hours. This directly answers the question: "Was this document written in one sitting, or spread across multiple days?"

- **Content volume and behavioral signals**
  Next, we arrive at the `app.xml` properties, which contain the volume and behavioral signals:

```
[app.xml Properties]
  Template : Normal.dotm
  TotalTime : 3
  Pages : 6
  Words : 812
  Characters : 4631
  Application : Microsoft Office Word
  Lines : 38
  Paragraphs : 10
  AppVersion : 16.0000
```

From this code snippet, we can infer the following:

- Total time: According to Word, there are three minutes of active editing time.
- Words: There are 812,812 words over six pages, which is roughly 135 words per page and reasonable for a Word document containing visuals and spacing.
- Approximate writing speed: Calculated by the script, this is 812 ÷ 3 ≈ 270 words per minute. This speed is high for manual typing but not impossible; it often suggests copy/paste activity or automated content generation. As mentioned earlier, values exceeding normal manual typing speeds indicate bulk insertion, automation, or pasted content.
- Template/app version: This block provides strong clues about how organic the document creation process was. `Normal.dotm` is a standard Word template, that is, not a custom corporate template. The app version is `16.0000`, which indicates the Office 2016+ family, meaning the approximate product generation is identifiable.

- **ZIP internal timeline**

  Let's move on to the next snippet, which presents something we'll call the 1980 anomaly:

```
ZIP internal timeline (basic):
  Earliest entry : [Content_Types].xml -> 1980-01-01T00:00:00
  Latest entry : docProps/app.xml -> 1980-01-01T00:00:00
  Time span : 0.0 seconds
```

  Two critical observations: 1) all internal zip file timestamps are set to `1980-01-01`, and 2) earliest and latest timestamps are identical; there's no internal time variance. Possible explanations include the following:

  - The file passed through a file system that doesn't preserve timestamps (e.g., certain embedded systems).
  - ZIP timestamps were reset or normalized. Some privacy or cleaning tools overwrite ZIP header dates with 1980; some archivers (especially older or portable ones) use this as a default.

  This is summarized by the warning:

```
All timestamps inside the ZIP appear to be set to year 1980 - may
indicate default filesystem dates or metadata reset.
```

  This warning means that low-level ZIP timestamps are unreliable, but Office core metadata is still intact. This highlights a critical forensic distinction between the following:

  - Physical file/container timestamps (ZIP headers)
  - Application-level timestamps (Office core/app XML)

  Even if the ZIP layer is wiped, Word's internal metadata still reveals the document's history.

- **Office structure summary**

  The next snippet provides the Office structure:

```
[Office Structure Summary]
  Top-level folder distribution:
  - [Content_Types].xml: 1 entries
  - _rels: 1 entries
  - word: 9 entries
  - docProps: 2 entries
  Extension distribution:
  - .xml: 10 items
  - (none): 1 items
  - .png: 1 items
  - .rels: 1 items
```

This shows the OOXML internal structure:

- Nine files under `word/` are the typical Word documents (`document.xml`, `styles.xml`, `numbering.xml`, media, etc.).
- Two files under `docProps/` are the core and app properties.
- `.png`: One item refers to at least one embedded image (reflected as `Embedded images present: True`).

This distribution helps answer questions such as the following:

- Is the document text-only, or does it include embedded objects/images/OLE?
- Does the document contain an unusually high number of `.bin` or embeddings/files that could indicate the presence of macros or OLE objects?
- Does the absence of `customXml` suggest that no advanced forms or structured document features were used?

- **Office features/flags**
  The next code snippet represents the surface of security and manipulation:

```
[Office Features / Flags]
  Macros (vbaProject.bin) present : False
  Embedded OLE objects present : False
  Embedded images present : True
  customXml present : False
  Signatures (_xmlsignatures) : False
  Track changes (trackRevisions) : False
  Document protection enabled : False
  Protection type : None
```

This block provides dense insight into attack surface and workflow:

- No macros: No direct macro-based exploitation expected
- No embedded OLE objects: No embedded Excel, Word, ActiveX, and so on
- Embedded images present: Images only
- No custom XML: Not a complex form/template-based document

- No signatures: No digital signing or timestamped signature
- Track changes disabled: Review history is not visible
- No document protection: Not read-only or password-restricted

From a CTI/digital forensics perspective, the document appears as a clean content file. There are no additional payload layers for macro-based infection. However, revision and time metadata still allow production tracking.

- **Timeline analysis**

  Our final code snippet contains all the information about the timeline:

```
[Timeline Analysis]
  revision_count : 5
  approx_words_per_minute : 270.666...
  core_timespan_seconds : 887520.0
  approx_seconds_per_revision : 221880.0
  approx_revision_timestamps_iso : [...]
  zip_earliest_minus_core_created_seconds : -1448002260.0
  zip_latest_minus_core_modified_seconds : -1448889780.0
```

  These aren't real logs, but a metadata-consistent mathematical reconstruction. This can be described as forensic approximation or heuristic reconstruction. What the script does here is extract as much derived insight as possible from available metadata:

  - `core_timespan_seconds`: Difference between modified and created: 887,520 seconds (~10.27 days).
  - `approx_seconds_per_revision`: Divides the total span by revisions: 887,520 / (5-1) = 221,880 seconds, which is roughly 2 days and 13 hours per revision.
  - `approx_revision_timestamps_iso`: Uses this interval to reconstruct an estimated revision timeline.

Note

The first revision corresponds to the document's created timestamp. The last revision corresponds to the modified timestamp recorded in the metadata. The intermediate revisions are estimated by distributing the calculated intervals evenly across the timeline.

  - `zip_earliest_minus_core_created_seconds`: Difference between ZIP's earliest timestamp (1980) and `core.created`: large negative value means ZIP timestamps clearly don't represent real dates.
  - `zip_latest_minus_core_modified_seconds`: Same conclusion for the latest timestamp.

The final result is that application-level timestamps and container-level timestamps don't align.

The Big Picture

In summary, our example output has demonstrated the following:

- The document can be attributed to a single user identity. `creator` and `lastModifiedBy` are identical.
- The document evolved over at least 10 days with five revisions. This indicates a more organic process than a one-shot copy operation.
- Word statistics show 812 words produced in three minutes. This suggests a mix of manual writing and copy–paste activity.
- No macros, OLE, signatures, or protection mechanisms are used. This is a content-focused document rather than a malicious payload container.
- ZIP timestamps are locked to 1980, but core metadata retains real dates. This means the file system or archiving layer was likely cleaned or normalized. Despite this, application-layer metadata still reveals the full story.

Shadow metadata analysis in Office documents combines core/app XML metadata with ZIP header timestamps to reconstruct the document's creation and revision timeline.

4.2.4 WHOIS, DNS, and SSL/TLS Passive Analysis

The greatest strength of passive OSINT lies in the ability to silently read infrastructure from a distance. Every domain, every IP address, every certificate, and every DNS record carries traces unintentionally left behind about who the attacker is, how they operate, which tools they use, and what they target. WHOIS data, DNS records, and SSL/TLS chains act as silent witnesses that reveal the true technical behavior model behind the mask of threat actors. This section goes beyond surface-level information easily accessible through simple queries and addresses the deepest layers of infrastructure analysis.

In the following sections, we'll explain the different data sources (WHOIS, DNS, and SSL/TLS), walk through the different techniques to analyze them, and outline an in-depth practical example.

Understanding WHOIS, DNS, and SSL/TLS Data

Although WHOIS records no longer provide as much information as they once did (largely due to privacy regulations such as GDPR and the widespread use of registrar privacy-protection services that mask registrant details), their value hasn't diminished for professional analysts. This is because WHOIS isn't merely an information repository but a behavior repository. The date a domain was registered, the frequency of registrar changes, the timing of updates, and the history of name servers all carry subtle yet critical signals about how a threat actor manages operations. For example, if most domains used by an attacker are registered around midnight, this can provide insight into the operator's time zone.

Likewise, if a threat group consistently uses a specific registrar, it may indicate that the registrar has weak oversight policies or supports anonymous payment mechanisms.

Domain lifecycle analysis doesn't simply mean looking at dates in WHOIS data; it involves mapping the entire time series. When a domain's initial registration, last update, name server changes, registrant changes, and pDNS data are analyzed together, the growth model of the threat actor's infrastructure emerges. Some actors register domains in bulk just days before a campaign, while others park domains for long periods, leaving them dormant until the right moment. Understanding these behavioral patterns is a critical analytical technique for increasing the accuracy of predicting future attacks.

Registrar footprint profiling allows analysts to read attackers' habits. Some registrars are cheap, some accept anonymous payments, and others have weak identity verification. Advanced CTI analysts observe that certain threat actors repeatedly use the same registrars and leverage this as an attribution indicator. For instance, some APT groups prefer low-cost Asian registrars, while financial fraud teams often gravitate toward specific Western registrars that offer WHOIS privacy services. Therefore, a registrar footprint provides insight beyond the source of a domain, into the threat actor's logistical preferences.

DNS analysis is where the true power of passive OSINT becomes evident. DNS records are among the strongest indicators of an attacker's operational preparation. By tracking historical versions of A, AAAA, Canonical Name (CNAME), Mail Exchange (MX), Text (TXT), and Name Server (NS) records through pDNS platforms, analysts can observe how attacker infrastructure evolves, which servers are used, which IP ranges are preferred, and what types of testing occur before an attack. For example, a domain resolving to multiple IP addresses within a short time frame is indicative of infrastructure warm-up behavior and typically signals a pre-attack preparation phase.

pDNS platforms go beyond listing DNS history to establish relationships between IP addresses, extract subdomain chains, and provide data capable of mapping an actor's entire infrastructure on a single graph. Let's take a quick look at each:

- **SecurityTrails**
  SecurityTrails is particularly useful for resolving the domain-IP-ASN relationship. It enables analysts to identify which IP addresses a domain has historically resolved to, which ASNs those IP addresses belong to, and how this infrastructure has changed over time. By correlating these elements, analysts can uncover hosting patterns, identify related infrastructure, and detect clusters of domains that may belong to the same operational network or threat actor.
- **RiskIQ**
  RiskIQ offers global passive relationship outputs for analysts combining OSINT and CTI. It aggregates large-scale internet telemetry to map relationships between domains, IP addresses, certificates, and hosting infrastructure. By analyzing historical pDNS data

and infrastructure associations, RiskIQ allows analysts to identify related assets, track infrastructure changes, and uncover hidden connections that may indicate coordinated threat activity.

- **Farsight**
  Farsight pDNS presents historical DNS resolutions as a time series, effectively turning a domain's operational periods into a chronological log.

SSL/TLS passive analysis is just as powerful as DNS analysis. An SSL certificate isn't used solely to authenticate a server's identity; it's also part of the technical signatures left behind when an attacker builds infrastructure. Information such as the CA in the certificate chain, the certificate's creation time, the key length, the list of domains in the Subject Alternative Name (SAN) field, and even minor variations introduced by the machine that generated the certificate can all be used to decode a threat actor's behavior model. Many threat groups use the same certificate-generation tools for their operational infrastructure, and the crypto signature patterns left by these tools are often unique.

Advanced OSINT analysts don't just check the validity of an SSL certificate; they analyze the entire chain. By examining the CA → intermediate → leaf certificate flow, it becomes possible to infer the country where certificates were obtained using false identities, which service providers were favored, which domains previously used the same certificate, and whether the certificate generation process was automated or manual. The appearance of the same certificate chain across different servers is one of the strongest indicators of infrastructure reuse behavior.

When data from WHOIS, DNS, and SSL/TLS is combined, passive analysis produces a vast trace map that can reveal the identity of a threat actor. Without performing any active scanning, touching any systems, or creating any risk, it's possible to reconstruct an attacker's entire operational infrastructure. What distinguishes professional CTI analysts is their ability to interpret these traces not as isolated data points, but as an interconnected behavioral graph.

WHOIS, DNS, and SSL/TLS Analysis Techniques

This section demonstrates that passive infrastructure analysis is about understanding the attacker's operational intelligence. A domain is just an address; however, when analysts examine the person who registered it, the systems that update it, the DNS records that route it, and the certificates that authenticate it all together, the attacker's entire structure becomes visible. This is the true power of passive OSINT.

Passive infrastructure artifacts such as WHOIS records, DNS resolution patterns, and certificate data often contain subtle operational signals that reveal how an attacker prepares and maintains their infrastructure. When these elements are analyzed collectively rather than in isolation via the following techniques, they allow analysts to observe patterns of activity that would otherwise remain unnoticed:

- **Domain heat signature analysis**

  Even though threat actors aim to remain anonymous, the DNS resolution intensity of their domains creates anomalies during attack preparation. Through pDNS platforms, the resolution volume of a domain can be analyzed by days and hours.

  If a domain shows any of the following a few days before an attack begins, this indicates that the actor is activating the infrastructure:

  - Sudden spike in resolution volume
  - NS change traffic
  - Time to live (TTL) reduction behavior

  This hidden technique is used to predict the attack timelines of many APT groups.

- **Registrar payment pattern analysis**

  Although WHOIS data has been reduced, some registrars leave payment method traces in metadata or residual logs. Advanced CTI analysts examine the following:

  - Which payment providers the registrar uses
  - Whether domains are billed retail or wholesale in bulk purchases
  - Whether the actor uses crypto-based payments, prepaid cards, or internal credit systems

  This information alone doesn't provide attribution; however, it reveals shopping behaviors unique to specific threat actors. For example, South Asia–based cyber fraud groups often prefer crypto-payment registrars, while low-profile spam actors use cheap registrar credits that can be activated immediately.

- **WHOIS redaction pattern analysis**

  Many domains appear protected by privacy services. However, even redaction services unintentionally emit signals to analysts.

  Companies providing privacy protection differ in the following ways:

  - UI/UX layout
  - Metadata templates
  - Registrant country formatting
  - Byte patterns of redaction text

  For example, the phrase "Redacted for Privacy" contains different Unicode sequences across registrar services. By leveraging these differences, analysts can determine which privacy service protects the domain, which services the actor used in the past, and whether other attack domains exist under the same privacy provider.

  This tactic is a professional attribution technique known by very few OSINT specialists.

- **SSL/TLS fingerprint clustering**

  Some threat actors use the same toolsets to generate certificates, for example:

  - OpenSSL default parameters
  - Traces of Lemur or mkcert tools

- Cobalt Strike panel certificates
- Let's Encrypt generation patterns

The certificate chain may reveal the following:

- Ordering of `keyUsage` fields
- SAN list formatting
- `signatureAlgorithm` variants
- Certificate validity periods (e.g., exactly 90 days → Let's Encrypt)
- Unique character strings in issuer fields

By analyzing these traces, it's often possible to infer whether an actor group likely follows a similar certificate generation pipeline; confidence increases when multiple independent signals converge:

- Automatically generates certificates with the same scripts
- Uses the same panel templates
- Follows a similar infrastructure production pipeline

- **pDNS sibling infrastructure detection**
  pDNS reveals both the historical records of a single domain and other domains previously hosted on the same IP. Using these relationships, domains can be classified as sibling infrastructure if they share the following:
  - The same IP ranges
  - The same ASN
  - Similar NS patterns
  - Overlapping registrar timestamps

  With this technique, a full campaign map can be built starting from a single phishing domain.

- **Reverse NS traversal**
  The NS record in the DNS specifies the authoritative servers responsible for resolving and managing queries for a particular domain. An NS record may look benign on the surface, but NS hostnames often reveal the following:
  - Internal IP systems of hosting providers
  - Node naming models
  - Geographic data centers

  For example,a hostname such as `ns1.panel-host-17.sgcloud.net` indicates a DNS service running on node 17 of SGCloud's Singapore data center.

  From this, analysts can infer the following:
  - Attacker's geographic distribution
  - Preferred data centers
  - Prioritized hosting providers

- **Certificate transparency (CT) log mining**

  CT logs allow many attack domains to be discovered before they are actively used. Advanced usage includes the following:

  - Email fields used in certificates
  - Unusual domains registered simultaneously
  - Other domains sharing the same certificate signing request (CSR) fingerprint
  - Domains recorded in certificates but not yet added to DNS

  This is uniquely powerful for early detection of attack campaigns.

- **DNS TTL forensics**

  TTL provides critical clues about a domain's dynamism. Advanced interpretation of TTL values can reveal the operational behavior and infrastructure strategy of the domain:

  - Low TTL: Actor plans frequent IP changes
  - TTL = 0: Infrastructure behind a reverse proxy
  - High TTL: Stable, static infrastructure (often benign)
  - Rapid TTL changes: Infrastructure tuning signal

  Anonymous proxy providers can sometimes be inferred through TTL analysis, as infrastructures that rely on proxy layers or traffic-routing services often use unusually low or dynamically changing TTL values to enable rapid backend IP rotation. Such patterns may indicate that the visible endpoint isn't the actual host but part of an intermediary proxy infrastructure.

- **NS cascade failure detection**

  During large attack campaigns, threat actors may hastily change infrastructure. This rushed change creates a DNS chain break:

  - Old NS records remain on the same IP.
  - New NS records are added before propagation completes.
  - Some regions resolve old records while others resolve new ones.

  This cascade failure indicates the urgency of the actor's operation and can be used to predict attack timing.

- **ASN behavior profiling**

  Infrastructure providers and ASNs often follow distinct operational and business practices. By examining these characteristics, analysts can evaluate how suitable a particular ASN is for malicious infrastructure and identify patterns that threat actors may prefer. Key indicators include the following:

  - Abuse handling practices
  - Server rental policies
  - Botnet hosting history
  - Pricing models
  - Know your customer (KYC; an identity verification process) requirements

If a threat actor consistently hosts domains on ASNs that are low KYC, weak in abuse response, cheap virtual private server (VPS) providers, or frequently complained about, this constitutes a threat signature. Even APT groups repeatedly exhibit such behavior.

Example: Passive OSINT Infrastructure Analysis

In this example (see Figure 4.4), we conduct a passive OSINT infrastructure analysis of the domain `XXXXX.com` using WHOIS, DNS, SSL/TLS, and ASN data.

```
C:\Users\admin\PycharmProjects\Books>python domain_passive_osint_analyzer.py -d          com
[!] CT collection error:       :om: HTTPSConnectionPool(host='crt.sh', port=443): Read timed out. (read timeout=20)
================================================================================
PASSIVE ASINT SUMMARY REPORT -                    .com
================================================================================
C:\Users\admin\PycharmProjects\Books>domain_passive_osint_analyzer.py:911: DeprecationWarning: datetime.datetime.utcnow()
  print("*Generated at: (fmt_date(dt.datetime.utcnow())) (UTC"")
Generated at: 2025-11-29 14:16:15 (UTC)
================================================================================
WHOIS SUMMARY
================================================================================
Registrar          : Hosting Concepts B.V. d/b/a Registrar.eu
Creation Date (min): 2014-05-30 18:18:17
Updated date       : 2025-05-30 05:38:11
Expiration Date    : 2027-05-30 14:18:27
Country            : NL
Status             : ClientTransferProhibited https://icann.org/epp@ClientTransferProhibited
Name Servers       : ns1.        .nl, ns1.        .nl..net, ns2.        .nl, ns3.        nl
Redacted Fields    : ['name', 'address', 'city', 'registrant_postal_code']
================================================================================
DNS SUMMARY
================================================================================
[A records]
       .com -> 3.5.54.211  (TTL=60)
[AAAA records]
       .com -> 2a05:d014:123:4601:56db:d312:356e:57a8 (TTL=28800)
[NS records]
       .com -> ns1.        nl. (TTL=28800)
       .com -> ns3.         l. (TTL=28800)
       .com -> ns2.        nl. (TTL=28800)
[TXT records]
       .com -> "abcf…string…64007c11290f467aga92" (TTL=28800)
       .com -> "google-site-verification=…                        762206676708+PFHSQ" (TTL=28800)
       .com -> "v=spf1 -alt" (TTL=28800)
       .com -> "d                              ogrye" (TTL=28800)
================================================================================
SSL / TLS PROFILE
================================================================================
Certificate #1
  Subject CN     :      .com
  Issuer         : Let's Encrypt
  Validity       : 2025-09-30 10:15:31 ->  2025-11-29 10:25:30
  SAW List       :      .com
  Signature Alg  : -
  SHA256 FP      : c251dbe89c9...full_fingerprint_string=ShA2c6b...5453b9eb05f8868b
================================================================================
CERTIFICATE TRANSPARENCY (CT) SUMMARY
================================================================================
NO CT log entries found or access failed.
================================================================================
ASN BENAVIOR SUMMARY
================================================================================
ASN 16509
  Description : AMAZON-02, US
  Country     : US
  Date        : 2017-11-29
  CIDRS       : 3.64.0.0/12
```

Figure 4.4: Passive OSINT Infrastructure Analysis of a Target Domain

The goal is to demonstrate how publicly available information can reveal the domain's registration behavior, technical infrastructure, and hosting environment.

We'll analyze these layers step-by-step to show how passive data sources can be combined to build a coherent picture of the domain's operational structure.

In the first part of the output, the `WHOIS SUMMARY` reveals the domain's basic identity card. This includes the following:

- **`Registrar: Hosting Concepts B.V. d/b/a Registrar.eu`**
  A Europe-based registrar serving corporate customers. From this, we see that the domain is managed through a professional registration operator.
- **Date information**
  The date information is provided as listed here:
  - `Creation Date (min): 2014-05-30`
  - `Updated Date: 2025-05-30`
  - `Expiration Date: 2027-05-30`

  This set of dates tells us the following:
  - The domain isn't a newly registered single-use phishing domain; it has had a lifecycle spanning many years and represents a corporate infrastructure.
  - Long-term renewal (up to 2027) shows that the organization considers the domain strategic and intends to operate it sustainably.
- **`Country: NL`**
  The registrant country being the Netherlands clearly defines the geographic and legal context of the domain.
- **`Status: ClientTransferProhibited`**
  This status is a typical security-oriented protection that prevents unauthorized transfer of the domain to another registrar. From a corporate hygiene perspective, this is a positive signal.
- **`Name Servers: ns1.xxxxx.nl, ns1.xxxxx.surf.net, ns2.xxxxx.nl, ns3.xxxxx.nl`**
  `XXXXX` is a long-established organization providing network infrastructure for academic and research institutions in the Netherlands. This indicates that the domain is likely positioned within the education/academic ecosystem.
- **`Redacted Fields: name, address, city, registrant_postal_code`**
  Personal data is masked in WHOIS. This is typical and expected behavior in terms of both GDPR compliance and WHOIS redaction pattern analysis.

Via passive OSINT, this entire section answers these questions: "Who owns this domain, how long has it existed, how corporate is it, and how stable is it?"

In the `DNS SUMMARY` section, the script collects and lists A, AAAA, NS, and TXT records in a readable format. Let's break down the different parts of this section:

- **A record:** `xxxxx.com → 3.xxx.54.211 (TTL = 60)`
  A TTL of 60 seconds suggests the use of dynamic traffic management behind the IP (e.g., a load balancer, CDN, or auto-scaling cloud infrastructure). Such short TTL values are typically used in environments where IP addresses may need to change quickly in response to traffic balancing or infrastructure scaling.
- **AAAA record:** `xxxxx.com → 2a05:d014:123:… (TTL = 28880)`
  A longer TTL on the IPv6 side reflects an operational preference, as many organizations make less aggressive changes to IPv6.
- **NS records:** `xxxxx.com → ns1.xxxxx.nl / ns2.xxxxx.nl / ns3.xxxxx.nl (TTL = 28880)`
  All name servers belonging to `XXXXX` confirm that DNS is managed through an academic network operator.
- **TXT records**
  Google site verification, SPF, and likely DMARC/DomainKeys Identified Mail (DKIM)-related configurations are visible.

  The SPF record (`v=spf1 …`) shows that the organization restricts email infrastructure to authorized IPs, which is a passive signal of good email security hygiene.

Overall, the DNS layer shows that the domain isn't a hastily deployed setup on a single VPS, but rather a corporate, well-structured infrastructure with a dedicated DNS provider, separate IPv4/IPv6 addressing, and configured email security.

In the `SSL / TLS PROFILE` section, the script directly retrieves the certificate from `xxxxx.com:443` and extracts the following information:

- **Subject CN:** `xxxxx.com`
  The Common Name (CN) identifies the primary domain for which the certificate was issued, confirming that the certificate is intended to authenticate this specific hostname.
- **Issuer:** `Let's Encrypt`
  This field indicates the CA that issued the certificate, revealing which trust provider validated the domain and signed the certificate chain.
- **Validity:** `2025-09-30 10:15:31 → 2025-12-29 10:25:30`
  This is a classic ~90-day Let's Encrypt validity period. Short certificate lifetimes indicate the use of an automated renewal pipeline.
- **SAN List:** `xxxxx.com`
  The SAN list specifies all hostnames covered by the certificate. In this case, the certificate is limited to the primary domain rather than multiple subdomains or related services.
- **Signature Alg**
  A signature algorithm consistent with Let's Encrypt's current profile indicates that the certificate was issued using widely accepted modern cryptographic standards and is compatible with the standard validation practices of contemporary web infrastructure.

- `SHA256 FP`
  A unique certificate fingerprint represents the cryptographic hash of the certificate, allowing analysts to precisely identify the certificate and correlate it with other domains or infrastructure using the same certificate instance.

From this, the following interpretations can be made:

- The organization uses an automated, free, modern CA (Let's Encrypt) rather than a commercial Extended Validation (EV) or Organization Validation (OV) certificate. This model is commonly seen in DevOps-oriented environments with strong automation.
- The certificate validity dates show that the domain is actively and currently used, not an abandoned" or forgotten asset.
- When scanning multiple domains in the future, domains sharing the same fingerprint template can be grouped using the SSL/TLS fingerprint clustering technique and linked to the same infrastructure or pipeline.

In this example, the `CERTIFICATE TRANSPARENCY (CT) SUMMARY` section shows the following line:

```
No CT log entries found or access failed.
```

The terminal warning logs a timeout (read timeout) error from the `crt.sh` side. This doesn't imply that the infrastructure is weak or misconfigured. It simply indicates that CT log access was unavailable under these conditions, so subdomain mining and certificate history analysis couldn't be performed.

In the final section, `ASN BEHAVIOR SUMMARY` resolves the IPs from the A records at the ASN level:

```
ASN: 16509
Description: AMAZON-02, US
Country: US
CIDR: 3.64.0.0/12 (example range)
```

This shows the following:

- `xxxxx.com` is hosted on the `AMAZON` infrastructure on the IPv4 side.
- While the name servers are operated by `XXXXX` in the Netherlands, the application layer is hosted on AWS, indicating a hybrid model consisting of academic DNS plus a cloud application layer. In this model, the organization maintains control of its authoritative DNS infrastructure through its own institutional network provider, while the web application itself is deployed on a scalable cloud platform. This separation allows the domain management layer and the application hosting environment to operate independently, combining institutional network governance with the flexibility of cloud-based services.

ASN information is used both for legal/jurisdictional analysis and to correlate with internal threat intelligence databases to answer questions such as the following: "Has this AS historically been associated with a high volume of malicious activity?

4.2.5 Social Media and Open User Profiles

Beyond serving as digital identities, social media profiles in modern OSINT are massive data bodies that conceal the behavioral patterns of threat actors, propaganda networks, cyber fraud organizations, and groups conducting anonymous operations. The true power of these profiles lies not in their content, but in their timing, consistency, and behavioral traces. While previous sections of passive OSINT focused on infrastructure and technical records, social media analysis targets the digital reflections of human behavior. For this reason, an analyst must develop awareness beyond technical knowledge to social structures, cultural patterns, and the psychology of digital interaction.

A social media profile may appear real on the surface because it has a photo, friends, and posts. However, a professional analyst knows that creating a fake identity is easy; what is difficult is sustaining that identity and maintaining continuity that resembles natural human behavior. Therefore, the primary goal of profile verification techniques isn't just to examine whether the information is true, but to resolve the digital lifecycle of the profile to understand who the actor behind it is. These outputs are of critical value both for uncovering social engineering attacks and for exposing state-sponsored information operations and organized fraud teams.

The profile verification techniques used by professional OSINT analysts go far beyond a simple reverse image search step. The authenticity of a profile is examined across several layers, which we'll discuss in the following sections.

Existence Continuity

Real digital identities accumulate traces over time; these traces may appear in a forgotten comment, an old username, or interactions beneath a photo uploaded years ago. Fake profiles can't replicate this natural continuity because the individual or team creating them can't simulate years of organic behavior required to produce the digital residue of a real life.

For this reason, the first thing an analyst examines isn't what the account says, but how long it has existed and how it has lived. The age of a digital presence is one of the strongest identity verification layers in social media OSINT. An analyst pays particular attention to the following areas:

- **Username evolution**
  The use of similar username variations by the same person across different platforms over the years is a sign of continuity. People typically carry the same nickname for many years; fake profiles do not.

- **Cross-platform behavioral traces**
  The same writing habits, spelling mistakes, and word choices are digital fingerprints of a real individual. Fake profiles either lack this behavioral consistency or copy it artificially.
- **Fragments of past content**
  Old comments, forgotten photos, accounts followed years ago, likes no one remembers, and so on are inevitable residues a person leaves on the internet. Fake profiles can't generate these residues because they have no past.

A real user typically has a natural digital lifecycle spanning 6–12 years. Fake profiles, on the other hand, are usually a few weeks or months old, or they are short-lived identities tied to a specific operational period.

Existence continuity isn't just a step in verifying the authenticity of an account; it's also the clearest way to understand whether the entity behind the profile is an individual, a team, or a fully automated bot system.

> **Remember!**
> Real people live, change, and leave traces. Fake profiles just appear.

Visual Lifecycle (Image Origin Tracking)

A profile picture may look new, but when the following techniques are used, the image's circulation history is revealed:

- **Yandex image matching**
  Yandex's reverse image search engine compares the image against a large visual index and often detects visually similar images even when they have been resized, slightly edited, or re-cropped. This allows analysts to identify earlier appearances of the same image across different platforms.
- **TinEye timeline analysis**
  TinEye focuses on identifying the earliest indexed occurrence of an image and provides a chronological view of where and when the image has appeared online. This helps analysts determine the approximate origin and historical spread of the image.
- **EXIF remnants**
  EXIF metadata embedded in the image file may contain technical details such as camera model, software used for editing, timestamps, and sometimes location-related data. Even partial remnants of this metadata can provide clues about how and when the image was produced or modified.

Photos of real individuals typically appear in different years, on different devices, and in different social contexts. In fake profiles, the photo is usually derived from a single source.

Social Connection Depth (Context Network Mapping)

One of the most powerful ways to assess the authenticity of a social media account is to examine its connection architecture. Even in digital environments, humans don't live in isolation; they form relationships, become part of friendship networks shaped over years, and exist within communities aligned with their interests. The connections of a real profile are natural, layered, and continuous—just like real-life relationships.

Fake profiles attempt to imitate this architecture but can't reproduce genuine connection depth because organic social bonds require time, shared history, reciprocal interaction, and naturally evolving context. When evaluating social connection depth, an analyst should examine the following:

- **Reciprocal interaction relationships**
  In real profiles, comments, likes, and messaging behaviors are multidirectional. People don't just follow; they talk, argue, laugh, and react. Bot or operational profiles exhibit one-way interaction: they follow many but receive little response.
- **Long-term digital relationship traces**
  Old school friends, close circles followed for years, and familiar accounts appearing in past posts are all proof of a real person's social history. Such long-lived ties are rare in fake profiles.
- **Natural interest clusters**
  The content followed by real users naturally reflects their hobbies, cultural background, expertise, and social environment. Fake profiles' interests are confined to narrow, repetitive clusters: politics, agenda manipulation, repeated themes, or specific operational targets.

Fake profiles often reveal themselves through the following network structures:

- **Artificial bot rings following one another**
  Account clusters are created simultaneously using the same language model and sharing the same hashtag sets.
- **Accounts producing uniform behavior**
  Identical sentence structures, identical emoji styles, and identical posting rhythms are classic indicators of copied persona generation techniques.
- **Low bidirectional interaction ratios**
  Fake profiles follow many but are followed by few. They post frequently but receive little feedback because they lack organic bonds.

Social connection depth isn't just a metric in digital identity analysis; it's the most stable structural element for determining whether an account is run by a real person or managed for operational purposes.

Timeline Consistency Analysis

A profile's posting stream, like geological layers, carries accumulated behavioral traces over time. Consistency analysis is the rhythm of the content.

To identify these patterns, analysts examine several behavioral indicators within the timeline. These indicators help distinguish natural human activity from coordinated or automated account behavior:

- **Chronological anomaly detection**
  In real accounts, things like sleep cycles, working hours, and cultural routines appear naturally. In fake accounts, you'll pick up on uniform interaction rhythms across a 24-hour period or activity concentrated in a single time zone. You might also observe bot-like repetitive posts at the same times every day.
- **Content evolution analysis**
  Real individuals change over time: style, interests, language, and content evolve. Fake accounts are generally static; no interest naturally develops.
- **Activity waves (behavior waves)**
  Real user behavior contains natural waves that increase during social periods, decrease during intense work periods, and have unusual peaks on special days. Fake accounts can't replicate this rhythm. For this reason, timeline wave analysis is a technique frequently used by professional OSINT teams.

Bot/Honeypot Profile Detection

In addition to serving as propaganda, fake accounts may be used as honeypots to trap OSINT analysts. To determine whether a profile is a bot or a trap account, you can use the following approaches:

- **Language biometrics**
  Real individuals make spelling mistakes, have personal orthographic preferences, and exhibit diverse sentence structures. Bot or honeypot accounts, on the other hand, display near-perfect correctness, repetitive template sentences, and language monotony.
- **Interaction spectrum**
  Bot/honeypot accounts produce many posts but receive few comments. They also might receive comments mostly from their own network, and they are generally followed by low-quality accounts.

  This interaction ratio can be easily detected through mathematical modeling. This interaction ratio can be detected through quantitative analysis of engagement patterns, such as comparing the number of posts to the volume of replies, likes, and reposts over time. By examining these ratios across multiple time intervals, analysts can identify abnormal interaction structures that often indicate coordinated amplification or automated activity rather than organic user engagement.

- **Visual monotony test**
  Honeypot accounts often use highly attractive, extremely clear, high-resolution, uniform photos. These images typically have cleaned EXIF data, professional lighting textures, and artificial (AI-produced) facial symmetry. They can therefore be detected using symmetry anomaly detection, a technique that examines whether interaction patterns follow unusually balanced or repetitive structures. In coordinated or automated accounts, actions such as posting, liking, and resharing often occur in mechanically consistent intervals or proportions, creating symmetrical behavioral patterns that rarely appear in natural human activity.
- **Relationship circle void**
  Real users typically have interaction networks that extend far beyond their direct contacts, forming multiple layers of connections. In contrast, fake accounts often show little or no interaction beyond their immediate contacts and a small secondary circle, creating a visible gap in the broader network structure.

  This relationship desert pattern is one of the most distinctive indicators of bot networks. The advanced analyses in this section transform social media profiles from simple digital business cards into complex structures at the intersection of behavior, time, context, and signal sciences.

4.3 Active OSINT Techniques

If passive OSINT requires reading the silent traces of the digital world, active OSINT requires gently knocking on the doors of that world, measuring system reflexes, and recording the target's unexpected behaviors. Active OSINT, as the name implies, involves interaction, but this interaction isn't an aggressive approach that constitutes a direct attack. Within a professional framework, active OSINT respects the boundaries of the target and employs only permissible levels of information-gathering tactics. Therefore, active OSINT shouldn't be confused with attack techniques; it should be regarded as the controlled and methodical reconnaissance face of intelligence.

The strength of active OSINT lies in observing how a target system behaves when it receives controlled queries. While passive techniques analyze traces left in the past, active techniques reveal how infrastructure responds in the present. This makes active OSINT particularly useful for monitoring changes in digital infrastructure and identifying signals that appear during the preparatory phases of cyber operations.

Unlike passive methods, active OSINT involves limited interaction with target systems. For this reason, analysts must collect information carefully without triggering defensive mechanisms or drawing attention to the investigation. Poorly configured scanners, excessive probing, or poorly timed queries can quickly alert administrators and disrupt the intelligence process. Professional analysts therefore treat active OSINT as a controlled behavioral observation rather than an aggressive scanning activity.

Through techniques such as DNS queries, port analysis, HTTP response observation, and banner collection, analysts can gradually reconstruct how a system is configured and maintained. These signals help reveal both the technical architecture of an infrastructure and the operational discipline behind it.

Because active OSINT involves direct interaction with external systems, it also requires strict OPSEC. Every query leaves a potential trace. When applied carefully and within legal boundaries, however, active OSINT becomes one of the most effective methods for understanding the operational behavior of digital infrastructure.

The techniques covered in this section go far beyond superficial reconnaissance. They explain how weak signals are read, how target system fingerprints are extracted, how back-end architectures are resolved through passive-aggressive traces, how actor behaviors are predicted, and how the risks left behind by active queries are minimized.

4.3.1 Port and Service Discovery

One of the most critical stages of active OSINT is port and service discovery because this is where the first contact with the external surface of the target system is established. This contact seeks to understand how the system presents itself, which services it exposes to the outside world, what kind of security habits it has, and which behavioral patterns it exhibits. For a professional analyst, port discovery both determines which doors are open and produces strong signals about who opened those doors and how they think.

Among the tools used for port scanning, Nmap is still one of the best options; however, the key point that must not be forgotten is that Nmap is only a tool. The real value lies in the intent behind the command sequence, the manner in which the scan is conducted, and the analytical discipline applied in interpreting the results. Therefore, in advanced Nmap usage, the goal isn't to speed up the scan, but to collect the most comprehensive signals possible without triggering the target's behavior.

In the following sections, we'll explain how to use Nmap for port scanning. We'll then explain some techniques for detecting attackers or honeypot providers, and walk through a detailed behavioral analysis case study.

Advanced Nmap Command Sets

Using Nmap at a superficial level is easy: you define a target, add a few basic parameters, and wait for the results. Professional-level port discovery, however, requires far more than that. True active OSINT doesn't rely on brute-force scans; it's based on understanding the target's behavior through the signals it reveals on its own. To achieve this, scanning ceases to be a mere information-gathering action and becomes the art of tracking without disturbing the target system by using the quietest possible contact.

The primary objective of advanced Nmap usage is to operate with seemingly innocuous yet effective queries, at a pace that won't trigger firewalls and at an intensity that intrusion detection system (IDS)/intrusion prevention system (IPS) systems won't classify as suspicious traffic. In other words, port scanning shouldn't be conducted like a noisy attacker forcing doors open, but as a silent observer trying to understand whether windows left open at night are the result of the wind or of a deliberate choice.

Professional analysts use Nmap as an intelligence sensor to understand the target's network architecture, service layout, routing logic, and security discipline. For this reason, the purpose of advanced command sequences isn't to scan all ports at once, but rather to collect the signals that the target reveals at its weakest level of awareness.

Special parameters used for silent scans, timing profiles, and low-interaction SYN packets make it possible to capture moments when the target system is least reactive. Most network administrators configure IDS rules based on unusual traffic volume; therefore, a steady, low-paced, and randomized scan allows you to read the entire surface without triggering alarms. A common technique used by many professionals is simply to reduce the scan speed as much as possible, randomize packet distribution, and lower detectability to a level that blends into the natural background noise of network devices.

At this level, Nmap becomes a discovery mechanism that reveals the target's timing habits, latency responses, filtering layers, load balancer behavior, and even the system administrator's configuration style. This is because, in advanced scans, the real information isn't hidden in the responses you receive, but in the responses you don't receive: in delays, packet drops, and inconsistent protocol behaviors.

Advanced Nmap usage doesn't even fully correspond to the word "scanning." This process is a practice of tracking that is carried out without disturbing the target system's own rhythm, requiring restraint, discipline, and technical intelligence.

To conduct this type of controlled observation, analysts rely on several scanning approaches designed to minimize noise while still revealing useful behavioral signals:

- **Low-noise SYN scans**

  One of the most commonly used techniques for this purpose is the low-noise SYN scan, which allows analysts to probe services while keeping network interaction at a minimal and controlled level.

  The following command demonstrates a controlled SYN scan configuration designed to minimize detection while collecting basic port-state information from the target system:

  ```
  nmap -sS -T2 --defeat-rst-ratelimit --max-retries 2 <target>
  ```

  This approach does the following:

  - Prevents aggressive firewall behavior
  - Performs silent observation without exceeding the target system's RST rate limits
  - Avoids creating a recognizable scan signature

- **Timing-based fingerprinting**

  By analyzing the target system's response delays (latency jitter), it's possible to reveal the following:
  - Virtualization layer
  - Use of proxies
  - Load balancing
  - Bandwidth throttling

  This analysis produces behavioral OSINT signals rather than purely technical OSINT material.

  The following command performs operating system (OS) detection and Transmission Control Protocol/Internet Protocol (TCP/IP) stack fingerprinting while applying controlled timing parameters to observe subtle response delays and network behavior patterns:

  ```
  nmap -O --osscan-guess --max-retries 2 --max-rtt-timeout 500ms -T3
  <target>
  ```

- **Protocol-focused scans**

  Another approach focuses on identifying services and protocol behavior by examining how applications respond during controlled version and banner queries.

  The following command performs a targeted service version scan, combining detailed version detection with banner and SSL certificate inspection to reveal service identity and configuration signals:

  ```
  nmap -sV --version-intensity 9 --script banner,ssl-cert <target>
  ```

  This approach does the following:
  - Verifies the service banner
  - Checks whether the reported version number is genuine
  - Examines the SSL certificate in a passive-aggressive manner
  - Reveals the use of fake banners if present

  Banner verification plays a critical role especially in honeypot detection. We'll discuss this further in the next section.

Service Fingerprint Comparison

Every service produces a unique fingerprint that has several key components:

- Response latency
- TCP flag sequences
- Banner format
- Whitespace structure of version strings

- Error messages
- TLS handshake patterns

Attackers or honeypot providers often attempt to mimic these signatures. However, a real service fingerprint and a fake service fingerprint are never identical.

A professional OSINT analyst uses the following comparison techniques:

- **Passive versus active fingerprint comparison**
  Fingerprint information obtained from passive sources (Shodan, Censys, FOFA, ZoomEye) is compared with the fingerprint obtained through active scanning (i.e., the Nmap-based fingerprinting techniques discussed in the previous section). If Apache appears in passive records, but Nginx appears in active scanning, this indicates a honeypot transition, a reverse proxy, or a newly activated service.
- **Banner microstructure analysis**
  The internal structure of the banner is analyzed, including the layout of spaces, hyphens, line breaks, and UTF-8 traces, for example, the line structure of a real OpenSSH banner:

  ```
  SSH-2.0-OpenSSH_7.6p1 Ubuntu-4ubuntu0.7
  ```

  In honeypot versions, the following are commonly observed:
  - Encoding errors
  - Extra spaces
  - Missing OS information
  - Incorrect punctuation
  - Inconsistencies between version number and distribution
- **Error message fingerprinting**
  Real services produce consistent error messages; honeypots usually return generalized error messages. Here are a couple of examples:
  - In a genuine Apache deployment, a malformed or incomplete request typically triggers a structured HTTP `400 Bad Request` response generated by the server's error-handling mechanism. This response may include a reference identifier or request ID that is logged internally by the server or associated monitoring systems. Such identifiers allow administrators to trace the request within server logs, correlate it with error events, and diagnose abnormal client behavior. The presence of these structured error elements is often a strong indicator of a real service stack rather than a simplified or emulated service environment.
  - In honeypot environments, error responses are often intentionally simplified and lack the structured details normally produced by genuine services. Messages such as `Invalid Request` typically appear as generic placeholders that don't include standard HTTP error formatting, server-specific diagnostics, or request identifiers. This occurs because many honeypot frameworks emulate service behavior rather than running the

original software stack. As a result, their responses often appear overly generic, inconsistent with real server implementations, or lacking the contextual metadata that production systems normally generate for logging and troubleshooting.

This difference is one of the fastest ways of detection.

Honeypot and Decoy Service Detection

As the use of honeypots has become widespread in modern networks, the task of the active OSINT analyst is no longer just to identify open ports, but to distinguish which ones are real and which are fake. Although the purpose of fake services is to mislead attackers, they can also lead intelligence analysts to incorrect conclusions.

Advanced signals for honeypot detection include the following:

- **Deterministic response patterns**
  Real services experience delays under load, change behavior based on system resources, and show fluctuating connection times. Honeypots, on the other hand, always respond with the same delay, and show no statistical variance. This is one of the most reliable detection methods.
- **Port combination mismatch**
  In real systems, certain port combinations naturally appear together. For example:
  - If port 22/SSH is open, ports 80 or 443 are usually also open.
  - If Active Directory (AD) is present, ports 88, 389, and 445 are naturally found.

  Honeypots often operate with a single port or have fake port combinations, such as the following:
  - Only 22 open
  - Only 445 open
  - Port 21 open but port 20 closed
  - Port 80 present but no favicon behavior

  This is a clear indicator of a honeypot.
- **Fail-closed versus fail-open behavior**
  When a real system encounters an error, it closes the connection (fail-closed). A honeypot often keeps the connection open (fail-open) because it wants to record the interaction.
- **TLS parameter inconsistencies**
  TLS handshakes generated by honeypots often can be characterized in the following ways:
  - Contain weak cipher suite lists
  - Have an empty SAN field
  - Include incomplete or incorrect CA chains
  - Have abnormally short certificate validity periods

  This almost certainly confirms that the service is fake.

- **Application layer inertia test**
  Real servers produce different error messages depending on the request type. In honeypots, all errors are returned using a single generic template. This test produces consistent and successful results that can be used with confidence in practical applications.

The most sensitive aspect of active OSINT is understanding which services a target offers and under which conditions those services change their behavior. The fact that ports are statically open by itself doesn't produce meaningful intelligence; system administrators often leave ports seemingly open, while in reality those ports respond differently depending on connection types, geographic locations, or specific user-agent values. For this reason, advanced port discovery must analyze two questions: "Is it open or closed?" and "To whom and how is it open?" These subtle differences in the target system's behavior reveal the logic of its security policies and the habits of its administrators.

Just as important as whether a connection is accepted or rejected are the small delays that occur during the connection process. Professional analysts examine the service response and the latency variations, jitter ranges, and microsecond-level deviations during packet responses. These minor differences often indicate the presence of load balancers, reverse proxy architectures behind the service, or traffic passing through a security layer. For example, some systems return different TTL values to requests coming with specific user-agent strings; this is a typical signal of passive-aggressive security filtering. When an advanced analyst notices such a detail, they understand both the port and the security logic operating behind that port.

The same principle applies to honeypot analysis. The biggest weakness of fake services is that over time they produce an unnatural rhythm. A real service infrastructure behaves like a living organism whose response time changes with traffic intensity. A honeypot, on the other hand, is bound to a fixed template, and when observed over long periods, that template consistently oscillates around the same values. Therefore, advanced detection doesn't focus solely on packet content; it examines microbehavioral consistency in services kept under continuous observation. Most honeypots maintain similar latency ranges even over long periods, whereas real systems unintentionally produce small variations due to factors such as CPU load, RAM usage, and disk I/O behavior.

In some environments, honeypots create rhythmic inconsistencies on specific ports and across the entire IP block. Professional analysts compare the results of low-intensity probing across other IPs within the same block—not just the target IP—to determine whether that block belongs to a production network or to a security research laboratory. In a real data center, response consistency across IPs within the same block degrades under load; honeypot clusters, however, often display artificial symmetry because they are typically controlled by a single management panel.

One of the least-known techniques in advanced port discovery is examining the direction and characteristics of the very first packet returned in response to a connection attempt. Some honeypot systems generate RST packets with a broader pattern than production systems or lack the microscopic variations left by real TCP/IP stacks in RST flows. For an experienced analyst, this difference constitutes an easily distinguishable signature. The differences between real Linux systems and emulated TCP stacks can be clearly identified using packet crafting tools, allowing the authenticity of the system to be verified.

Another important signal during active OSINT operations is port banner timing. In some services, banners are returned immediately; in others, banners arrive with a delay even after the connection is established. These timing differences depend on both the service type and the server-side throttling mechanisms. A honeypot typically sends the banner with the same delay every time, whereas a production server may vary banner timing by milliseconds due to traffic load or the state of the CPU scheduler. This variation is one of the most subtle yet most consistent indicators for determining whether a system is real.

In advanced analyses, an insight even more valuable than whether a port is open emerges: why the port is open. Some ports remain open temporarily for developer testing, while others are the result of forgotten configurations. If passive sources confirm that a port has behaved consistently for a long time, this indicates that the service is intended for production use. However, a newly appeared port that doesn't show up in passive sources and responds only to specific user-agent values typically indicates attacker infrastructure in preparation for an operation. Such temporary port behaviors are especially common in C2 infrastructures.

Examples: Active OSINT Behavioral Analysis

In a Windows Server 2022 infrastructure, different roles project different behavioral layers to the outside world. These behaviors produce strong signals for understanding elements such as reflex, latency, banner consistency, port combination logic, and security policy traces, which are the core focus of active OSINT. The following examples represent professional insights that an organization can obtain while analyzing its own servers:

- **Active Directory Domain Services (AD DS)**

  On a system where the AD DS role is enabled on Windows Server 2022, the expected behavior is to expose the following ports together:

  - TCP 88 (Kerberos)
  - TCP/UDP 389 (Lightweight Directory Access Protocol [LDAP])
  - TCP 445 (Server Message Block [SMB])
  - TCP 135 (RPC Endpoint Mapper)
  - TCP 636 (Lightweight Directory Access Protocol Secure [LDAPS])

  You might find the following on an internal AD DS server:

  - Port 389 is open but 88 is closed: A configuration anomaly in the Kerberos service
  - Port 445 is not open: A potential issue in group policy replication
  - The LDAPS port (636) is open but the certificate has not been renewed: A behavioral signal of weakness in the security policy

These aren't indicators of an attack, but of organizational visibility analysis. Changes in the behavioral surface of AD DS can point to misconfigurations in Windows security policies.

- **Windows Server 2022 DNS role**

  On a Windows server with the DNS role enabled, active OSINT analysis (for internal testing purposes) yields the following signals:

  - If TTL values aren't stable: DNS round-robin or load-balancing behavior
  - If the Start of Authority (SOA) record is older than expected: An indication of insufficient administrative maintenance
  - If response latency varies by user-agent: DNS query filtering is in effect

  These differences aren't attack techniques, but advanced intelligence signals about the operational quality of the DNS role. For example, when certain security packages are installed, Windows DNS servers alter `NXDOMAIN` response timings by microseconds; this difference can indicate the timing of changes in DNS security policies.

- **Web Server (Internet Information Services [IIS]) role**

  Behavior exposed externally by the IIS role on Windows Server 2022 includes the following:

  - The presence of the "Server: Microsoft-IIS/10.0" banner
  - Consistency of language and format in HTTP 500/404 error codes
  - Generate Random Extensions And Sustain Extensibility (GREASE) value behavior during the TLS handshake: The presence and handling of GREASE values during the TLS negotiation can reveal how strictly the server adheres to modern TLS standards and whether the underlying TLS stack properly supports extension tolerance, which in turn provides subtle signals about the server's software implementation and update maturity.
  - `HTTP/2 SETTINGS` frame parameters: The configuration and values within the `HTTP/2 SETTINGS` frame can reveal how the server is tuned for connection management, flow control, and concurrency limits, offering insight into the server's performance configuration and the operational policies applied by the hosting environment.

  An advanced active OSINT analyst (internally) may observe the following:

  - The same server showing different HTTP response latencies at different times, which provides insight into application pool recovery behavior
  - TLS certificate renewal frequency, which produces signals about the patching rhythm of operational teams
  - Different error messages returned by IIS modules to different URLs, which reveals how complex the server's module structure is

  This is high-value information for internal analysts assessing the security posture of IIS.

- **Remote Desktop Services (RDS)**
  On servers where RDS is enabled, the following elements form active OSINT signals:
  - The behavioral rhythm of TCP 3389
  - TLS negotiation patterns
  - Configuration of the RDP protocol's pre-authentication banner
  - Whether Network Level Authentication (NLA) is enforced

  From a professional internal analysis, the following conclusions may be drawn:
  - Port 3389 responding differently based on geography is a firewall rule adaptation.
  - NLA not being mandatory is a weakness in authentication policies.
  - Disrupted banner consistency is an asynchrony in RDS Gateway or farm configurations.

  These findings are measurable organizational weakness insights derived from the externally visible behavior of the RDS role.
- **Windows Server 2022 file services**
  The SMB protocol (TCP 445) provides very strong signals to active OSINT analysts. During SMB dialect negotiation, the following can be observed:
  - Is SMBv1 disabled?
  - Is SMBv2 handshake latency stable?
  - Are Distributed File System (DFS) namespace responses consistent?
  - Is SMB signing mandatory?

  Expected behavior on Windows Server 2022:
  - SMBv1 disabled
  - SMB signing enabled
  - SMB compression controlled

  During internal testing, the following are possible:
  - If the dialect list is excessively broad, incompatible devices are integrated into the network.
  - If SMB signing behavior is inconsistent, issues in Group Policy Object (GPO) replication are possible.
  - If DFS responses are delayed, network instability is possible between the domain controller and clients.

  None of these are attack methods; all are behavioral visibility analyses.
- **Windows Server 2022 Hyper-V**
  Hyper-V infrastructure doesn't directly expose ports externally. However, its behavior can be interpreted through the following:
  - External port distribution of virtual machines (VMs)
  - MAC address patterns

  - Overlay network latency
  - TTL effects of the virtual switch Network Address Translation (NAT) model

  In organizational active OSINT analysis, the following may be observed:

  - Very similar response jitter among VMs on the same physical host: Increased load
  - Certain VMs consistently showing low TTL values: Configuration differences in NAT or routing layers on the host
  - Changes in port behavior over time: Traces of snapshot restoration, cloning, or migration

  This information is used solely to assess the operational health of the infrastructure.

- **Windows Server 2022 Certificate Services - CA Behavioral Signatures**
  An Active Directory Certificate Services (AD CS) role doesn't directly expose ports, but produces signals through TLS behavior:

  - Response times of the CA issuer URL during TLS handshakes
  - Certificate revocation list (CRL) update intervals
  - Response rhythm of the online responder

  You might observe the following:

  - CRL distribution is slow: AD replication delays
  - The CA certificate chain is outdated: Weak security maintenance rhythm
  - Online Certificate Status Protocol (OCSP) response times are variable: Infrastructure load imbalance

  These are entirely professional internal audit signals.

4.3.2 DNS Enumeration

One of the most critical stages of active OSINT is DNS enumeration because the DNS structure silently contains the most fundamental road map of an organization's digital assets. How a domain is structured, how consistently different record types are maintained, how zones are segmented, and how authoritative DNS servers behave all make visible a wide spectrum ranging from the system administrator's level of transparency to operational discipline. DNS enumeration is therefore one of the most refined ways of reading an organization's externally reflected architectural characteristics.

For an advanced analyst, DNS is the point where passive OSINT and active OSINT intersect. Most DNS records are publicly accessible; however, understanding how these records are organized, which subdomains are created according to what logic, and during which periods zone structures change requires reading the microsignals produced by active queries. Small delays in DNS server response times, inconsistencies that occur when specific record types are queried, or TTL values changing at different times often carry signals of background operations, maintenance cycles, or security changes.

In the following sections, we'll discuss DNS enumeration techniques and walk through a detailed example.

DNS Enumeration Techniques

The first well-known step of DNS enumeration is attempting a zone transfer; however, a professional analyst doesn't treat zone transfer as an end goal. A zone transfer attempt is performed solely to understand how the authoritative DNS responds. Some DNS servers explicitly reject AXFR requests, some respond with high latency, and others remain completely silent. This behavioral difference forms a strong intuition about the target's security policy. A clear and fast rejection indicates a conscious configuration; silent dropping may indicate that the security layer applies passive blocking. Most importantly, even if a zone transfer is unsuccessful, the truly valuable aspect for the analyst is the behavioral trace left by DNS in response to the request.

Another way to understand DNS structure is to read the visibility model of subdomains. However, an advanced OSINT approach doesn't operate with a crude brute-force mindset here. A professional analyst narrows the scope instead of expanding the wordlist because the target organization's culture, IT teams' naming habits, and the platforms in use determine how subdomains are generated. In organizational subdomain structures, environments are generally predictable: names such as `test`, `dev`, `api`, `auth`, `sts`, `internal`, `gateway` indicate specific disciplines. Therefore, subdomain discovery isn't random trial and error, but a statistical inference process aimed at decoding the target's behavioral patterns.

Sudden changes in subdomain configurations may be silent indicators of new applications being deployed or attackers preparing infrastructure. Additionally, if TTL values differ at the subdomain level, this may indicate that the infrastructure behind a load balancer has been updated, the service distribution model has changed, or temporary redirection is being performed at the DNS level. Such subtle behavioral differences are extremely valuable for weak-signal analysis in OSINT.

Misconfiguration analysis of DNS Security Extensions (DNSSEC) also occupies an important place in active OSINT because although DNSSEC adds security, it produces the opposite effect and leaks operational errors outward when misconfigured. An analyst evaluates not merely whether DNSSEC is enabled, but how it's implemented. Inconsistencies in Delegation Signer (DS) records, signature validity periods being shorter or longer than expected, date discrepancies in Resource Record Signature (RRSIG) records, or breaks in the chain of trust reveal both technical issues and deficiencies in administrative discipline. When DNSSEC is misconfigured, it produces behavioral weakness signals, which are highly valuable from an OSINT perspective.

When signature periods in DNSSEC configurations are predictable, it becomes possible to make intuitive inferences about maintenance cycles. For example, if an organization's signature renewal period is longer than normal, it may indicate delays at the administrative or operational level. Conversely, very rapid and irregular signature renewals may indicate panic fixes or urgent configuration changes within the system.

None of these stages constitute a technical attack; all are intelligence extractions aimed at understanding how an organization manages its own DNS structure.

The most important aspect of DNS enumeration is that, for an analyst, the DNS server's response behavior is just as valuable as the DNS records themselves. Different response times to the same query at different times of day reveal how backend system traffic fluctuates; temporary inconsistencies in certain record types indicate maintenance windows or moments of configuration change. Therefore, DNS enumeration is never a one-time operation; its accuracy is strengthened only through behavioral observations conducted over a time window.

DNS is the backbone of infrastructure, and every step taken on the backbone sends unintentional signals to the outside world. The professional OSINT analyst's task is to read these signals, filter meaningful behavioral motifs from technical noise, and decode the operational logic behind every small change on the DNS surface.

Examples: Advanced DNS Behavioral Analysis

Let's walk through some examples of behavioral analysis:

- **TTL waveform behavior in the Windows Server 2022 DNS role**

  Most people think of TTL simply as a duration value; however, for an advanced analyst, the daily rhythm of TTL changes is an indicator of operational discipline.

  On Windows Server 2022 DNS, certain records (especially SRV records) produce microsecond-level TTL shifts when GPO or AD replication is delayed. This shift may be observed as 300 → 299.8 → 299.9. If the waveform pattern is highly irregular, it indicates clock synchronization issues (Network Time Protocol [NTP] drift) between domain controllers.

  This signal isn't an indicator of attack; it's an external reflection of timing and replication quality at the management layer.

- **SOA serial pattern behavior on Linux (BIND 9) servers**

  BIND 9 administrators typically update serial numbers manually or automatically. However, for an advanced analyst, even the pattern of serial increments carries information. BIND 9 is one of the most widely used open-source DNS server implementations, commonly deployed to manage authoritative and recursive DNS services in enterprise and internet infrastructure environments.

  When the serial number follows the YYYYMMDDnn format, its increment pattern usually fits one of the following models:

  - Linear rhythm: 2025022801 → 2025022802 → regular production environment
  - Broken rhythm: 2025022801 → 2025031501→ maintenance period or long configuration freeze

- Jump rhythm: 2025022801 → 2025022801 → 2025022802→ indication that the zone file is being overwritten by two different nodes

These signals aren't attacks, but operational stability diagnostic data.

- **Negative cache duration behavior in Windows server 2022 DNS**
  Windows DNS servers cache nonexistent records (`NXDOMAIN`) for a certain period.

  If the `NXDOMAIN` TTL for the same subdomain changes at different times (e.g., 900 → 300 → 900), this indicates one of two things:

  - The DNS server is returning cached responses from different AD sites.
  - The IT team is performing temporary redirection tests.

  This is an important weak signal in active OSINT for detecting operational activity on the DNS surface.

- **Resolving subdomain culture via case sensitivity anomalies on Linux systems**
  Although subdomain names are case-insensitive in Linux DNS environments, user habits are reflected in case patterns.

  If the query latencies for `login.api.example.com` and `Login.API.example.com` aren't the same, this usually indicates that different nodes behind a reverse proxy are being hit. This behavior reveals the administrator's load distribution algorithm (round-robin, IP-hash, weight-based).

  This information provides infrastructure visibility only.

- **Clock drift in SRV records on Windows Server 2022 DNS**
  TTL fluctuations in SRV records (especially `_kerberos` and `_ldap`) leak the load status of domain controllers, for example:

  - When CPU usage on a DC reaches 90%, SRV response latency increases by 2–3 ms for short periods.
  - If multiple DCs exist in the same region, this increase is reflected only on a specific node.
  - When analysts capture this difference, they can determine which DC is under load.

  This is an entirely safe, internal audit-focused OSINT extraction.

- **Wildcard behavior test on Linux DNS (BIND 9, PowerDNS)**
  Wildcard records are typically configured as `*.example.com`. However, many administrators don't fully understand how wildcard behavior works.

  In wildcard records, BIND 9 applies closest encloser behavior instead of returning `NXDOMAIN`. If wildcard resolution introduces a 2–3 ms delay, this indicates that the wildcard record triggers additional lookups on the processor. Through this signal, the following internal dynamics can be read:

  - DNS recursion policy
  - Wildcard logic
  - Zone structure

- **Extracting maintenance schedules from DNSSEC signature cycles in Windows Server 2022 DNS**
  DNSSEC RRSIG records contain fields such as inception, expiration, and key tag. If an organization's DNSSEC signature duration is normally 7 days but drops to 2 days for a period, this usually results from re-signing after an urgent patch. Conversely, if the signature duration increases from 7 days to 14 days once, this may indicate the following:
  - Heavily loaded CA server
  - Delays in AD replication
  - Staff shortages in DNS management

  DNSSEC isn't designed to conceal these signals; therefore, behavioral analysis becomes one of the richest data sources.
- **Detecting background changes via resolver behavior on Linux systems**
  Linux systems typically rely on local DNS resolver services such as `systemd-resolved`, `unbound`, or `dnsmasq` to handle domain name resolution, cache DNS responses, and manage how queries are forwarded to upstream name servers. The behavior of these resolvers produces strong indicators of how administrators organize network architecture.

  On a Linux system using `systemd-resolved`, if the response times for the same query differ by 3–5 ms when issued 2 seconds apart, this typically indicates an IPv6 fallback attempt. This means the administrator believes IPv6 is enabled, but the configuration is invalid.
- **Authority chain drift analysis in hybrid Windows + Linux environments**
  Many organizations use hybrid Windows DNS + Linux BIND 9 environments. If Windows DNS responds in 52 ms while BIND 9 responds in 47 ms for the same record, and this difference is consistently in the same direction, it indicates that BIND 9 isn't authoritative. If the difference fluctuates (52 → 49 → 51 → 46), this means neither DNS server is authoritative and recursion is in effect. Through this analysis, it becomes clear which node in a hybrid DNS infrastructure is truly authoritative.

4.3.3 Web Discovery and Scanning

The web surface is the most complex, dynamic, and signal-producing area that many organizations expose to the outside world. While only a limited amount of data can be obtained about an IP address, a web application is constantly in motion: it gets updated, old code is removed, new endpoints are added, WAF policies are tightened, and the content delivery network changes. For this reason, web discovery is both one of the richest disciplines of active OSINT and one that must be conducted with the greatest care.

A professional analyst doesn't start web discovery by asking which directories exist because the real issue is the ability to read the application's reflexes, design habits, and backend logic. A web application leaves behavioral traces just like a human. Error codes, redirect logic, comment lines inside JavaScript files, the lifetime of cookies, and the version-naming style of static content all carry direct clues about the discipline of the teams running the application, the development cycle, and the security mindset.

In the following sections, we'll dive into web discovery, including the various steps, tools, and considerations to keep in mind.

Understanding Web Discovery

One of the first and most innocently perceived steps of web discovery is examining the `robots.txt` and `sitemap.xml` files. These files are often incorrectly framed as attackers' hidden endpoint search tools; whereas from a professional OSINT perspective, these two files are the narrative map of an organization's internal architecture. Directories blocked in `robots.txt` indicate which areas they don't want to be exposed to bot traffic; the sitemap structure shows how orderly or fragmented the architecture is. The presence of paths such as `/internal` or `/beta` in `robots.txt` isn't a risk, but a signal of how the organization manages its application lifecycle. A sitemap that is consistently versioned indicates the maturity of software development teams. While examining this area, an analyst focuses not so much on the content of the files, but on how they were created, what is missing, and which unexpected directories aren't present.

One of the most decisive structures on the security surface of modern web applications is WAF deployments. The goal of WAF fingerprinting isn't to bypass the WAF, but to derive the WAF's behavioral profile because the strictness, aggressiveness, or looseness of a WAF shows the risk level at which the application operates and the security priorities by which it's managed. For example, some WAFs return the same-pattern error for malformed requests, while others produce different error messages for different user agents. These differences are strong indicators that reveal the security provider in use (F5, Cloudflare, FortiWeb, ModSecurity, etc.).

For an analyst, the real value is how the WAF reacts:

- How fast do error messages return?
- How strict is the rate-limit policy?
- Does behavior change when you send the same request back-to-back?
- Which HTTP methods are more tolerant?

These signals reveal the application's security maturity like a visual map. One of the most valuable sources of information in a web application is JavaScript files. However, the goal here isn't to review code or search for hidden endpoints. A professional analyst looks at JavaScript files through a behavioral lens:

- Is the file naming convention consistent?
- Is the versioning logic coherent?
- What do comment lines inside the file say about developer culture?

- Are API calls centralized or scattered?
- Is the code minified, or not?

The answers to these questions reflect the application's architectural maturity. Unminified JavaScript files (JavaScript files that haven't been minified or compressed for production use) aren't a risk from a CTI perspective; they are just an indicator that developers' deployment process isn't fully settled. The structure of API endpoint calls (`fetch`, `axios`, legacy `XMLHttpRequest` [XHR], etc.) shows whether the application is modern or accumulating technical debt.

JavaScript files also carry signals of internal team order. If every file has a different coding style, it indicates the absence of linting or code-style standards; if all files share the same style, it signals a strong continuous integration/continuous delivery (CI/CD) process.

A deeper branch of web discovery is behavioral analysis of the directory structure. The purpose of tools such as Gobuster or dirsearch isn't to find directories; it's to understand the rhythm of the responses given to those directories. For the same directory, if one request returns 403, and another request returns 404, then this is a strong signal showing instability in the application's backend routing logic.

In addition, in some applications, directory errors return in JSON format, while in others, they return in HTML template format. This difference may indicate that the application is running on multiple frameworks or undergoing a gradual modernization process. For example, if part of the application exhibits .NET behaviors while other parts exhibit classic ASP.NET behaviors, the organization's transition to a newer architecture may still be incomplete. In practice, this difference is often reflected in response formats: modern ASP.NET–based services frequently return JSON-structured error responses typical of API-driven architectures, whereas classic ASP.NET applications more commonly produce HTML template error pages generated by the web server or framework.

An analyst also looks at whether the same directory request has different response times at different hours? If so, this is usually an external reflection of operational activity such as the following:

- CDN cache flush
- Deployment pipeline
- Background task executions

When web discovery is done correctly, there's no need to search for any hidden area inside the application because the application's own behavior already reflects management culture, development discipline, and security maturity outward. For a professional active OSINT analyst, what matters is the character of the responses: how fast they return, how they produce errors, and when they redirect—not what they hide, but how they hide it.

Web Discovery Tools

The tools used during web discovery are often introduced as directory-finding or file-listing tools. However, for a professional analyst, the value of these tools doesn't lie in how they generate queries, but in understanding how the target system responds to those queries. For example, Gobuster's rapid requests, dirsearch's different user-agent profiles, curl's simplicity in reading raw HTTP traffic, Wget's content-processing behavior, or the network traces produced by a browser's DevTools panel aren't just tests but also sensors that measure the application's behavioral surface.

For an expert analyst, a tool's real capability is hidden not in what it produces, but in what it fails to produce. For example, when Gobuster can't find any directory, this empty result can provide valuable information about the target server's error handling, rate-limit policy, connection acceptance rhythm, or whether traffic passes through a proxy chain. When the same wordlist is run with different speed profiles, it becomes clear whether the target behaves consistently or has a microservice arrangement that behaves differently under load. Thus, the tool stops being merely a directory scanner and becomes a behavioral probe that measures backend architectural reflexes.

Basic HTTP clients such as curl and Wget play a much deeper role in professional OSINT than is commonly assumed. The TCP-level handshake time of a request made with curl can reveal the stability of the server's TLS configuration or the behavioral consistency of reverse proxy layers. Some systems show millisecond-level differences depending on different user-agent chains; these differences often reveal the type of backend routing mechanism or the stage at which WAF processes traffic—details that are unnoticed on the developer side but highly meaningful for an intelligence analyst. Such small signs create a subtle link between the speed an application appears to have on the surface and the architecture it actually runs on.

The power of tools used in JavaScript analysis emerges not in deciphering the code's content, but in observing the code's organizational structure. Signals such as whether a JavaScript file is minified, the presence of source map files, whether network requests are centralized or modular, and whether framework behavior is uniform are projections not only of the analysis tool but also of the application's entire lifecycle. Through these tools, the analyst can easily read whether the application is built on a modern foundation or a legacy base, whether all teams use the same coding style, or whether it's a fragmented structure shaped by different hands.

Tools used for WAF fingerprinting (e.g., `wafw00f`) are often described as aiming to identify the type of WAF; however, the true value of these tools is capturing rhythmic changes in WAF behavior. Even a simple probe that sends the same HTTP request every five seconds can reveal the WAF's reflexes under load, its throttling model, latency increases shown toward the same source address, or the variance of traffic passing through a CDN. Therefore, tools are used not to *classify* correctly, but to *read behavior* correctly.

Browser-based tools (Chrome DevTools, Firefox Inspector, etc.) are the least used but most valuable part of web discovery. Response chain durations observed in the **Network** panel, cross-origin resource sharing (CORS) policy consistency, preload/preconnect behavior, static file versioning logic, or the rhythm of 304 response cycles produce extremely strong signals about the application's performance management and operational discipline. Especially in large-scale organizations, even the millisecond-level delays revealed by DevTools make it possible to understand the latency chain between backend services.

None of these tools are powerful on their own; the real power is an analyst's ability to convert all signals produced and not produced by these tools into a behavioral model. If the web surface is stable, tools also produce stable outputs; if there's instability on the surface, a late deployment, a faulty routing rule, or an unplanned maintenance process, small but measurable fractures appear in the tools' logs. The professional OSINT analyst extracts meaning precisely at this point by looking at the behavior of the data, not at the data found.

Important Considerations

Web discovery and scanning processes may technically appear as innocent as "I'm only sending an HTTP request", but when applied with the wrong mindset, they can turn into a structure that both pushes legal boundaries and distorts intelligence quality. Therefore, a professional analyst must clarify their mental framework and boundaries long before their toolset. There are some basic principles that everyone looking at a web surface should know, but analysts working at an advanced level must also be mindful of additional, more layered dimensions that are often not discussed.

Let's examine these key principles and dimensions:

- **No unauthorized testing**
  First of all, web discovery should never, under any circumstances, mean unauthorized testing. Statements like "I'm only checking directories" or "I'm only looking at headers" aren't a defense from a legal perspective. The analyst must continually ask themselves if they have the authority to send this request and if they truly understand the potential impact of this request. Even in internal tests, web discovery conducted without clearly defined scope, timing, and method creates both operational risk and legal gray zones. The professional approach is to design every step like approved laboratory behavior; that is, plan in advance what you do, why you do it, and for how long.
- **Minimize noise visibility**
  Another point that requires attention is the visibility of the noise produced. Most requests sent during web discovery end up in logs; they leave traces in WAF, reverse proxy, CDN, application logs, and SIEM layers. These traces don't just produce a technical "recon was performed" signal; they can cause behavior changes on the target side. That is, the analyst inadvertently contaminates their own intelligence: the system administrator tightens the WAF rule, application developers rush a patch based on the pattern in the logs, or a new rule is defined on the CDN side. In this case, the analyst is no longer observing the natural system but the reactive system triggered by the analyst's presence. This is a rarely discussed but critical truth at an advanced level: Web discovery doesn't

leave the observed system unchanged; if done incorrectly, it changes the system and corrupts the purity of intelligence.

- **Ensure appropriate data collection**
  Both the scope and depth of data collected during web discovery are also sensitive matters. To perform behavioral analysis, there's often no need for personal data, account-level details, or session content. Nevertheless, if the analyst begins collecting more information than necessary simply because the tools at hand support it, they may inadvertently drift toward the edge of data protection laws (KVKK, GDPR, etc.). The professional approach is to place the question "How much do I truly need to produce intelligence?" at the center, instead of "What can I collect technically?". Collecting unnecessary deep logs, HTML dumps, JavaScript content, or response bodies creates unnecessary burdens both in storage and legal responsibility.
- **Avoid single-cause bias**
  In addition, while reading behavioral signals obtained during web discovery, the analyst's greatest enemy is single-cause bias. The same HTTP 500 error can indicate a poorly coded endpoint, temporary maintenance work, a misconfigured WAF rule, or a CDN-induced problem. A TTL change may indicate either attacker preparation or a completely benign migration process. Therefore, an advanced analyst consciously keeps at least two or three alternative explanations on the table for every observation and doesn't jump to dramatic conclusions from a single signal. Web discovery isn't a process of "seeking proof" but a process of narrowing by eliminating hypotheses.
- **Don't blindly trust the tool**
  Analysts must also avoid exaggerating tool behavior and treating the tool as "reality itself." Gobuster, dirsearch, curl, DevTools, and WAF fingerprinting tools look for patterns defined for them; they often ignore signals that fall outside those patterns or live at the edges. For the analyst, this means asking "What did the tool show me?" is as important as "What did the tool not show me?" Empty results, incomplete lists, continuously repeating timeouts, logs cutting off at certain points, and so on are at least as meaningful as the data that is found. The analyst who blindly believes the world seen through the tool is reality usually misses the most critical behavioral motifs.
- **Avoid high-stress timing**
  Timing is also a high-level factor to be careful about. Conducting web discovery during periods such as production hours, campaign periods, launch processes, payroll cutoff days, or times of heavy application traffic is problematic both ethically and technically. At these moments, the system is already under maximum stress; additional requests can unintentionally amplify performance issues or cause incident response teams to trigger false alarms. An advanced analyst, if possible, conducts tests within controlled time windows, and if internal, even coordinates with relevant teams. Even with an OSINT perspective, hitting while the target system is already strained isn't a professional approach.
- **Ensure anonymity**
  On the OPSEC side, the analyst must hide their own identity and also avoid putting their

organization, customer, and team at risk. The IP blocks, domains, test user agents, and even email addresses used for analysis can leave traces that may later create legal or organizational responsibility. A detail that isn't taken seriously (for example, the real company name appearing in the WHOIS record of a domain used for testing) creates unnecessary visibility about who these tests belong to in the target system. This can lay the groundwork for undesirable misunderstandings later. The professional approach is built both technically managing test identity and on keeping managerial and legal risks minimal.

- **Record and share information responsibly**
  Another advanced-level point to consider is how findings are recorded and shared. Behavioral signals obtained during web discovery, screenshots, request-response examples, header dumps, or JavaScript analysis notes—when detached from the proper context, when they fall into the wrong hands, or when they are shared without authorization—can turn into unnecessary exposure of both corporate secrets and internal processes. Therefore, the output storage environment, access permissions, naming conventions, and reporting language are at least as important as the discovery itself. Correct information recorded in the wrong way is no longer intelligence, but risk.
- **Preserve ethics**
  Finally, preserving the ethical perspective while conducting web discovery is more valuable than sharp technical vision. Every weak structure, every inconsistency, every messy JavaScript file, or every faulty WAF rule you see isn't automatically a usable vulnerability for you. The professional analyst, at a certain point, starts thinking not about what can be done, but about what should be done: Who benefits from this information, and who is harmed by it? Is it more correct to share it or to keep it limited? Do I report it, or do I store it only for trend analysis after anonymizing it? The answers given to these questions reveal a professional's maturity more than their technical level.

4.4 OSINT Data Structuring and Storage

In OSINT processes, the most difficult part isn't collecting information; the real challenge is organizing the collected data in a way that prevents it from generating meaning in the wrong context. In the modern threat intelligence world, the volume of data is already infinite: social media, DNS changes, the web surface, open-source repositories, credential leaks, dark web rumors, and SSL renewal cycles are all signals in constant motion, changing every second. The task of the OSINT analyst is both to select the signals that generate meaning among all of these and keep under control how that meaning changes over time. This is only possible with the right data structure, storage model, and contextualization logic.

Data isn't information until it's processed; likewise, unstructured data isn't OSINT—it's random digital noise. For this reason, a professional OSINT effort must place as much emphasis on processing, correlation, and validation stages as on collection. It's precisely at this point that the correct design of the data structure turns into the analyst's most powerful invisible weapon: consistency. When consistency is lost, even the most accurate signal obtained can turn into a misleading impression.

Within an OSINT architecture, data storage is a strategic decision that determines how intelligence will be processed. An incorrect storage structure renders even the strongest information ineffective in the analyst's hands. For example, failing to convert raw HTML outputs into JSON format, logs without timestamps losing value in correlation processes, or storing social media data without relational linkage are invisible traps that lead future analysis to faulty conclusions. Therefore, data structure design isn't a technical detail, but the intelligence fabric itself.

The most critical element in OSINT data is the time factor. Every piece of data represents the architecture at the moment it was collected, yet that architecture may change within minutes. For this reason, an analyst must both store the data and place it within a temporal context; in other words, to record the flow of the data as well. Although each of these bits of information—what DNS records were a week ago, when an SSL certificate was renewed, during which time zone a subdomain emerged, or on which day a GitHub commit was pushed—isn't necessarily valuable on its own, when brought together over time, they form a powerful map that reveals the silent behavioral patterns of threat actors or internal operations.

The structuring models discussed in this section will address how intelligence should be stored holistically, how temporal context should be preserved, how false positives should be filtered out, how data should be optimized for reuse, and, most importantly, how it should be organized in a way that reduces the analyst's cognitive load. A properly structured OSINT database forms the foundation of both today's analysis and future investigations.

Remember!

An OSINT data structure isn't a warehouse; it's a fluid, living, constantly updating analytical ecosystem. The architecture of this ecosystem reflects both the analyst's technical skills and perspective, which adds value to data.

4.4.1 Technical Normalization of Intelligence Data

In the OSINT world, collecting data is often easy, but making it consistent, comparable, and suitable for automated processing is the most difficult step. This is because information obtained from the digital ecosystem often has no analytical value in its raw form due to format inconsistencies across sources. A WHOIS output for an IP address may be in JSON format, the pDNS information for the same IP may come in comma-separated values (CSV) format, social media data may be in HTML structure, and commit information obtained from open-source repositories may be entirely free text. As a result, a large portion of intelligence value is lost due to unnormalized data.

The primary purpose of technical normalization is to reduce the analyst's context-building cost by fitting data into a single standardized mold. This process isn't just about unifying formats; it's also about identifying and standardizing data, associating it with time, and making it meaningful for automation. Normalizing OSINT data is less like editing a text and more like disassembling information molecules and reconstructing them.

In the intelligence ecosystem, this reconstruction begins especially with the correct use of IOC formats. Standards such as Structured Threat Information eXpression (STIX)/Trusted Automated eXchange of Intelligence Information (TAXII), OpenIOC, and JavaScript Object Notation for Linked Data (JSON-LD) are used for IOC sharing and to make OSINT data machine-readable. Because STIX can define both a technical object and a behavioral model as a threat, it preserves the time dimension of OSINT. For example, along with the registration date of a domain, its pDNS history, SSL change cycles, previously associated IPs, and actor relationships can be expressed within a single STIX structure. In this way, the analyst can derive answers to "What story does the domain tell?"

JSON-LD, on the other hand, is one of the cleanest ways to establish relational context in OSINT data. Relationships such as an IP's association with an organization, the time zone in which a GitHub user makes commits, or whether a social media account exhibits similar behavior across different platforms can be transformed into semantic nodes within JSON-LD. This approach does more than store data; it also makes the data's own relational map visible.

One of the most critical elements of normalization is data quality because data normalized with incorrect or incomplete context contaminates all subsequent analysis layers. Therefore, in professional OSINT architectures, data is both collected and measured. Three core metrics form the backbone of data quality in intelligence operations:

- **Confidence score**
  The analyst's level of confidence in the accuracy of a piece of data. This isn't a subjective rating; it's calculated based on criteria such as the source's historical accuracy rate, the data's consistency with independent sources, and signal-behavior coherence. If the confidence score is low, the data is stored but not allowed to drive operational decisions.
- **Relevance score**
  Not every data point serves every IR. An IP address may be critical for one campaign, while completely irrelevant in another context. The relevance score therefore measures the data's distance to the IR. The greatest enemy of OSINT isn't information, but an excess of irrelevant information.
- **Freshness**
  The digital threat ecosystem is time dependent. A domain may have been active a week ago, but be completely nonfunctional today. Similarly, IPs in an actor's C2 infrastructure may change within hours. The freshness metric measures the currency of data and prevents outdated information from misleading analysis. Just like food, intelligence data also has an expiration date.

When these three metrics work together, data becomes decision-capable information. Normalization may appear to be a technical process, but in reality, it's the heart of intelligence because the accuracy of intelligence depends on the correctness of the data structure even before analytical power.

High-maturity OSINT teams treat the normalization process as a discipline of accuracy, rather than as a simple pipeline. Every IOC, every timestamp, every metadata field passes through a quality control filter before entering the data warehouse. Thanks to this discipline, data coming from hundreds of sources begins to produce consistent meaning within a single centralized memory. There are several approaches that achieve this:

- **Unique data entities**
  Along with presenting data in the correct format, a frequently overlooked dimension of technical normalization is preserving its unique identity. OSINT data often consists of highly similar signals; hundreds of similar IOCs may emerge with very small variations. An IP address appearing under different names across sources, a domain being recorded together with its subdomains, or the same username appearing across different platforms with changing contexts can all cause data collision during normalization. Therefore, advanced OSINT storage architectures are built on processing each data point not merely as a string, but as a unique entity.
- **Entity resolution**
  At this stage, the most critical structure is the identity resolution process, known as entity resolution. Determining whether the same data appearing in different forms across different places is related to a single threat is the upper layer of normalization. It's a common situation for a domain to arrive from six different APIs in six different formats, with the IP WHOIS, pDNS, SSL logs, and OSINT collection tools all using different JSON schemas. When these aren't normalized, data volume increases but value remains constant; when resolution is applied, data volume decreases, but meaning density increases exponentially. For this reason, the goal in professional OSINT architectures isn't to grow data piles, but to ensure data integrity by reducing the different faces of the same threat into a single centralized entity.
- **Temporal consistency**
  An IOC is a snapshot that's part of a constantly changing ecosystem. However, many analysts store normalized data in a time-independent manner, which makes it difficult to track the evolving infrastructures of threat actors. In a professional approach, each IOC is stored not only as a singular object but as a behavior curve spread over time. Unless TTL changes of domains, IP rotation periods, SSL renewal cycles, and the emergence and disappearance cycles of subdomains are plotted along a time axis, it becomes impossible to understand the actor's operational tempo.
- **Data gravity analysis**
  This concept measures the density of other data forming around a particular data type. For example, if a domain frequently appears in passive sources, the IP records associated

with it are considered "heavier"; conversely, a newly created data point that produces many relationships indicates that the threat actor is actively building infrastructure. Data gravity provides normalization with a clean format *and* makes visible the data's pull; i.e., its position from an intelligence perspective.

- **Data aging**
  Although freshness is often described as currency, in advanced OSINT architectures, aging isn't solely time-based. A piece of data may have been collected 24 hours ago, but if all associated IOCs have changed, it's considered aged data. Conversely, SSL certificate metadata collected six months ago may still be considered fresh if it remains valid and aligns with actor behavior. This approach demonstrates that intelligence operates with contextual time, not absolute time.
- **Integration maturity**
  A data structure may be STIX-compliant, but if it can't establish connections with other systems, its analytical value decreases. Similarly, an IOC may be properly processed in JSON format, but if it's not integrated with enrichment layers, it produces only a partial signal. Therefore, the true success of normalization is measured by the degree to which data is compatible with the following:
  - Correlation layers
  - Threat hunting dashboards
  - Automated alerting systems
  - Timeline analysis engines

  The purpose of normalization is to make data mobile, not just orderly.
- **Context-weighting model**
  One of the most advanced normalization approaches is the context-weighting model. In this technique, each piece of data is assigned an accuracy or freshness score *and* a context value. For example, an IP address's association with an actor's infrastructure is normalized relative to the same IP's visibility in general internet traffic. If an IP is used by millions of people, its context value decreases; if it appears only within the actor's infrastructure, its context value increases. In this way, data begins to generate meaning rather than noise.
- **Bias filtering**
  Finally, the most critical but least discussed dimension of normalization is filtering the analyst's own biases. A piece of data may have been collected on time, correctly formatted, and perfectly linked through JSON-LD relationships, but if the analyst unconsciously projects their own assumptions onto the data, the normalization process may be technically correct but analytically flawed. For this reason, advanced OSINT teams add an *analyst-independent validation layer* as the final step of normalization. In this layer, data is reread by detaching it from the context in which it was collected, and the question of what the data truly says is reevaluated. This way, data gains an objective reality independent of the analyst's intent.

4.4.2 Correlation Architecture

The true power of OSINT data doesn't arise from the value it carries on its own, but from the density of relationships it establishes with other data. Normalization gives order to data; correlation gives intelligence to data. Because no IOC has meaning on its own, a domain is incomplete without the IPs it's connected to, an IP is anonymous unless it's linked with WHOIS records, and an SSL certificate is merely a piece of text unless it's placed on a time axis.

Correlation architecture is the framework that systematically connects all of these independent elements and makes the longitudinal fabric of the threat structure visible. For a professional analyst, correlation is the extraction of behavioral texture through data matching.

Let's dig into the layers of this architecture:

- **Domain → IP → WHOIS → SSL → ASN chain**
  The first layer of correlation is the longitudinal processing of the domain to ASN chain. This chain may appear to many as a simple data flow; however, in advanced architecture, it's examined both horizontally and by vertical depth. For example, the same domain may first be seen on a shared hosting IP, then on a cloud provider, and later on a private ASN. The timing of these changes often reveals the operational maturity of the threat actor. Some abandon infrastructure quickly, some remain in the same IP block for a long time, and some repeat certificate renewals with minute-level precision. These patterns gain meaning not only through the connection of data but through the movement of those connections over time.

  Some threat actors initially keep their WHOIS information hidden, and then later mistakenly update it with a public email address. Some have their certificates renewed automatically, while others renew manually. All of these microbehavioral differences cease to be isolated signals the moment they intersect within the correlation architecture and turn into an operational signature. Therefore, correlation doesn't work to connect IOCs, but to read the intent behind the IOCs.

- **Time-series correlation**
  Time-series correlation is a critical component of this architecture because threats often reveal themselves in a chronological order. When the time between a domain's registration and its first DNS resolution, the gap between SSL certificate creation and pDNS visibility, the time zones in which subdomains emerge, and the intervals at which ASN changes occur are all placed on a timeline, they reveal the operational speed and agility level of the threat actor. Some threat groups operate with a *cold operation model* (i.e., they do nothing for a long time, then suddenly activate all infrastructure at once), and others follow a *hot operation model* (i.e., gradual, measured, advancing in small steps). These differences can only be understood through systematic time analysis.

- **Multisource cross-validation and correlation**
  The most advanced step of correlation architecture is multisource cross-validation. Information that isn't validated by a single pDNS source is considered fragile from an

intelligence perspective. If the same domain appears at different times on Censys, in a different context in the RiskIQ database, and with a completely different IP on SecurityTrails, this reinforces the accuracy of the data and reveals where the threat infrastructure leaks signals. Inconsistencies between sources sometimes indicate infrastructure transitions by the threat actor, sometimes DNS propagation delays, and sometimes that manually made changes haven't been fully reflected across systems. The analyst reads these inconsistencies both for accuracy and for the character of the threat operation.

Once information from multiple sources has been validated, the next analytical step is correlation. At this stage, the goal isn't simply to confirm that the same data appears in different places, but to understand the context in which those relationships are recorded. For example, an IP address may appear in several data sources, yet be linked to an SSL certificate in only one of them. In such cases, what matters to the analyst isn't the number of occurrences, but the context: which system recorded the relationship, why it recorded it, and what the absence of that record in other sources may indicate. Professional correlation begins at this point by assigning meaning to the data itself and also to the absence of data.

In advanced correlation architectures, relationships aren't treated as simple links; they are also weighted according to their analytical strength. Some connections are inherently stronger than others. For instance, a domain-IP relationship generally represents a strong technical association, while a WHOIS-IP relationship may provide weaker contextual evidence, and an SSL-domain linkage often reflects a more consistent operational connection. By applying such weighting, analysts can distinguish which data points represent primary evidence and which serve only as supporting indicators. In this way, correlation evolves from a basic graph of connections into a structured prioritization mechanism.

For this reason, correlation architecture represents one of the most analytically demanding layers of OSINT. Data collection is largely mechanical, and normalization is systematic, but correlation is the stage where analytical judgment becomes essential. Correctly identifying and interpreting relationships does more than produce a map of indicators; it helps reveal the operational structure and strategic behavior of the threat actor behind them.

- **Contextual deviation analysis**
  One of the often unseen but most critical layers of correlation architecture is contextual deviation analysis. Two IOCs may appear to belong to the same chain, but if their contexts don't align, they aren't actually related. These deviations are microdistortions left behind as threat actors expand their infrastructure like an electromagnetic field. For example, if a new domain is registered in a time zone that doesn't align with the actor's previous infrastructure behavior, either the operation has been delegated to a different subteam or part of the infrastructure has been rented while another part has been stolen. The correlation system evaluates these deviations as indicators of operational change.

- **Modeling negative relationships**
  Another advanced dimension of correlation is the modeling of negative relationships. Most analysts examine only existing connections; however, the absence of a connection often produces a stronger signal. For example, if an SSL certificate has been signed by the same CA for six years, but a new domain of the same actor uses a different CA, this may indicate that the actor is experimenting with a new operational model. Such negative relationships are the gaps that don't appear in the correlation graph but are completed in the analyst's mind. Along with connecting nodes, professional correlation also interprets gaps.
- **Asymmetric relationship flow**
  Another level used in understanding network structures is asymmetric relationship flow. Some IOCs accumulate a large amount of data linked to them, yet generate few backward connections. For example, an IP receiving heavy traffic may be used by many domains; however, if the SSL certificates, WHOIS changes, and ASN transitions associated with that IP are weak, this density usually reflects the natural load distribution of a service provider, not a real operation. The purpose of correlation architecture is to detect such asymmetries and prevent reading noise as threat.
- **Infrastructure resonance analysis**
  One of the advanced layers of correlation is infrastructure resonance analysis. Some IOC clusters appear at the same frequency over time: certificate renewals on the same day, DNS updates at the same minutes, or increases in pDNS records during the same weeks. This resonance reflects the automation level and operational discipline of threat actors. If automation is strong, resonance is regular; if manual work dominates, resonance is irregular, fluctuating, and intuitive. By reading this rhythm, the analyst can infer the actor's working style, which is the most abstract yet highest value–producing dimension of correlation architecture.
- **Scale impact analysis**
  A complex but valuable aspect of time-series correlation is scale impact analysis. Even a small infrastructure expansion can evolve into a large attack campaign over time. Some actors first register a single domain, make no moves for days, create dozens of subdomains overnight, and then shortly thereafter begin IP rotation. If this scale growth occurs suddenly, the correlation architecture marks it as a hot phase. In this way, the analyst generates advanced early warning even before the operation has begun. This mechanism interprets data connections, as well as the speed, intensity, and growth vector of those connections.
- **Semantic overlap analysis**
  One of the advanced stages of multisource validation is semantic overlap analysis. IOCs observed in different sources are evaluated in both a technical and content context. For example, while a pDNS record provides technical information, the appearance of the same IOC in a code repository history may indicate that the threat actor supports its

infrastructure with open-source code. This thematic overlap reinforces the technical correlation and the narrative coherence of the threat story.

One of the biggest mistakes in correlation architecture is assuming that the analyst's confidence level will automatically increase as the relationship network grows. In reality, some large graphs are false networks formed by linking low-quality data. The upper layer of professional correlation is therefore false density analysis. If relationships are very broad but signals are very weak, this structure usually reflects data contamination, not a threat paradigm. To identify the correct relationship, the analyst focuses on the center of gravity of the relationships instead of the size of the network.

Finally, perhaps the most important aspect of correlation is its ability to generate forward-looking predictions. Correlation connects the past and produces forecasts about future infrastructure movements. If an actor rotated IPs at specific intervals in previous operations, renewed certificates through the same CA, or consistently made WHOIS changes in the same time zone, these recurring behavioral patterns form a solid basis for predicting future movements. In this way, correlation ceases to be detective work and becomes an operational prediction system.

4.4.3 Data Storage Options

For a correlation architecture to function well, it depends on both correctly associating the data and storing the data because OSINT data isn't static; it's an organism that continuously expands, deepens, updates, and at times completely loses its validity. For this reason, storage isn't the "back room" of the OSINT architecture; it's the spinal system of operational intelligence. A wrong storage choice makes even flawless correlation processes fragile. Therefore, the professional approach evaluates storage technologies in terms of technical capacity and their alignment with the logic of the intelligence flow:

- **Elastic**
  One of the most frequently used components in this ecosystem is the Elastic stack. Elastic is a search engine that provides the hybrid structure of time-series plus text-based search that is ideal for capturing the rhythm of change in OSINT data. The WHOIS changes, SSL certificate cycles, IP rotations in C2 infrastructure, and social media posting dates can be read on Elastic through both chronological and content-based queries. However, Elastic's real value is that it allows the analyst to notice instant anomalies in large-volume data. If a domain that wasn't seen yesterday shows persistently recurring DNS behavior today, Elastic's aggregations structure distinguishes this from the normal data flow and makes it visible. For this reason, Elastic turns into the early-warning radar of the OSINT architecture.
- **MISP Threat Sharing (MISP)**
  MISP is the standardized memory of threat sharing. In the OSINT world, MISP's role is often described as an IOC repository, but for an advanced analyst, MISP is much more

than that. MISP's strongest aspect is its ability to add context to IOCs: threat actor attribution notes, attack techniques, references, source reliability, correlated events, and so on. When normalized OSINT data is transferred into MISP, it's not just stored; the narrative fabric of intelligence is preserved as well. Thanks to STIX/TAXII compatibility, MISP enables automated data exchange with external systems while ensuring that the analyst doesn't lose context. Thus, internal and external data flows integrate on a single centralized memory.

- **Neo4j**
  For graph-based data needs, Neo4j comes into play. Correlation architectures, especially when the domain → IP → WHOIS → SSL → ASN chains exceed a certain threshold, become unreadable with traditional table structures. At this point, Neo4j stores data as graph nodes and edges, reflecting the analyst's mental model directly into the database. The infrastructure resonance of a threat actor, the clustering of similar SSL signatures, and the temporal distribution of domains sharing the same WHOIS email field become visible at a glance on a graph. Neo4j's real strength lies in combining the analyst's intuitive relationship-building ability with machine logic. While reading the graph, the analyst sees both relationships and the movement of relationships.
- **TimescaleDB**
  TimescaleDB is OSINT's silent hero, especially in architectures where time-series correlation is sensitive. The SSL renewal time, DNS TTL rhythm, the hours at which subdomains appear, AS rotation intervals, and the timing of botnet command signals comprise the biological rhythm of a threat infrastructure. This rhythm can only be preserved with high-resolution time data. TimescaleDB combines PostgreSQL's maturity with time-series optimizations, preserving the OSINT team's operational memory at millisecond level. This level of temporal integrity is critical, especially in deriving actors' automation levels.

Data storage isn't limited only to which platform is chosen; it's also related to how backup and chain-of-custody architecture is designed. OSINT data often begins to carry legal or audit value once it exceeds operational value. Therefore, data loss is more than a technical deficiency; it means the loss of intelligence integrity. In professional OSINT architectures, a timestamp, hash verification, and an immutability policy are applied to every piece of data. Thus, when needed, data gains the quality of evidence that can be verified back to its source. This architecture strengthens the analyst's hand in incident investigations and threat attribution studies.

Another critical dimension is the corporate access matrix. By nature, OSINT data contains layered confidentiality levels; some data should be visible only to analysts, some only to managers, and some only to technical teams. This access control can't be limited only to database authorization; it's also part of the matrix that determines at what stage and by whom the data is processed, which correlation outputs are shared with whom, and which data is kept in long-term archives. Corporate access architecture turns OSINT data into a force multiplier while also minimizing data leakage risk.

An advanced dimension of storage architecture is data lifecycle management. OSINT data isn't static; it's collected, enriched, correlated, reinterpreted, and eventually either archived or deleted. When this natural cycle isn't managed well, databases quickly reach meaningless sizes and turn into a "data landfill" that tires the analyst instead of producing value. For this reason, high-maturity organizations define a lifespan for each data type: DNS records may be short-lived, SSL data carries meaning in the medium term, and WHOIS data generally produces long-term value. Thus, storage takes on the role of strategic memory management.

Within this lifecycle, one of the most critical stages is managing data decay. Keeping data that has lost its meaning over time in the system creates noise in the correlation layer. For example, an old IP record may have long since exited the threat infrastructure, but if it remains in the database, it may cause the analyst to evaluate new relationships in the wrong context. Therefore, advanced storage solutions calculate data decay by timestamp and contextual activity criteria. If a piece of data hasn't produced any relationship with any other IOC for a long time, the system automatically assigns it the status of weakened data. This approach keeps OSINT memory broad but simultaneously precise and sharp.

Another advanced dimension of the storage ecosystem is multitier data retention policies. High-sensitivity data (such as WHOIS changes that may constitute potential attribution evidence or time series related to the operational rhythm of specific actors) is stored only in highly secure environments. In contrast, high-volume pDNS records that constitute operational noise can be kept in lower-cost storage environments. Thus, the organization achieves a balanced architecture in terms of security, cost, and performance. This hierarchical storage strategy ensures data is stored according to its value.

In addition, regardless of the storage option, the immutability layer plays an important role in advanced OSINT operations. Preventing certain data from being manipulated later is critical for the integrity of investigated incidents. Therefore, hash verification alone isn't sufficient; the dataset must be stored as an immutable block within itself. Modern approaches generally use append-only architectures or write once read many (WORM) principles. In this way, transparency is preserved in operations and the chain of trust required for legal processes is ensured.

How integration is established among storage systems is at least as important as the system itself. Platforms such as Elastic, MISP, Neo4j, and TimescaleDB can operate independently, but in mature organizations, they behave as if they are a single system through multilayer data federation. In this federation structure, each system operates in its area of expertise: Elastic provides fast search, Neo4j visualizes complex relationships, TimescaleDB ensures temporal accuracy, and MISP preserves contextual information. The greatest advantage of this integration is that the analyst can access data through a single window and obtain a natural information flow while building links across data layers.

Another advanced topic is the management of access traces. The corporate access matrix defines user permissions and records which analyst interacts with which data. These trace records increase accountability and show whether there was data manipulation or incorrect correlation in future operational reviews. This is especially critical for teams dealing with global threat actors because in high-volume data flows, analyst errors can often settle into the system unnoticed. Regular auditing of access traces is the insurance of OSINT memory against analyst impact.

At an advanced stage of data storage, organizations apply a risk-based data protection strategy. In this approach, data is stored according to its technical importance and its strategic risk level. For example, initial observations about a new threat actor, even if not yet validated, are preserved long-term due to their high strategic value. In contrast, automated scanning data that constitutes noise is cleaned after a certain period. In this way, storage architecture doesn't focus only on expansion; it makes active choices to reduce counterintelligence impact.

The common goal of these components is to transform OSINT data from a storage cluster into the organization's collective memory. The better data is stored, the sharper the reading of future threat signals becomes. The storage system serves as both the archive of the operational mind and its computation space. Analysts remember the past thanks to storage; they predict the future by reading the rhythm of this memory.

4.4.4 Example: Automatic Normalization and Scoring of the DNS-WHOIS-SSL Chain for a Single Domain

Figure 4.5 shows the complete workflow produced by the `osint_framework.py` script while analyzing the domain `haydararici.com.tr`. This illustrates the operational workflow of the framework during the analysis of a single domain. It demonstrates how raw infrastructure data obtained from DNS, WHOIS, and SSL sources is systematically processed within the framework. The collected data is normalized, evaluated through scoring logic, and recorded in the evidence log to maintain analytical traceability. At the same time, supporting metrics such as data aging and data gravity are calculated to provide additional context for the reliability and relevance of the collected indicators.

```
C:\Users\admin\PycharmProjects\Books>python osint_framework.py ******.com.tr
2025-11-30 17:44:50,423 [INFO] osint_framework :: Target domain: ******.com.tr
2025-11-30 17:45:05,485 [INFO] osint_framework :: Processing raw event ******.com.tr
2025-11-30 17:45:05,487 [INFO] osint_framework :: Normalized 1 IOC(s) from event c29ab939-9dcc-4516-b276-c57ddd630e from dns_live
2025-11-30 17:45:06,175 [INFO] osint_framework :: Processing raw event f786631b-bab1-4077-8781-e2ca890665411 from whois
2025-11-30 17:45:06,176 [INFO] osint_framework :: Normalized 1 IOC(s) from event f786631b-bab1-4077-07651-e2ca890665411
2025-11-30 17:45:06,267 [INFO] osint_framework :: Processing raw event 9db69875-940c-435c-99b6a-5196113ce20b from ssl
2025-11-30 17:45:06,269 [INFO] osint_framework :: Normalized 2 IOC(s) from event 9db69875-940c-435c-98b6a-5196113ce20b
2025-11-30 17:45:06,273 [INFO] osint_framework :: Processed IOC(s):
2025-11-30 17:45:06,273 [INFO] osint_framework :: IOC ioc_domain score={'confidence': 1.0, 'relevance': 0.0, 'freshness': 1.0}
2025-11-30 17:45:06,273 [INFO] osint_framework :: IOC 56b44a97b65a4f7a53b9e2e53e2ed6af88b16db83621222cfbf9d28****** (ssl_cert)
'subject': ((('commonName', '******.com.tr'),),),
'issuer': ((('country', 'US'),), (('organizationName', 'Let's Encrypt'),)),
'subjectAltName': (('DNS', '*.******.com.tr'), ('DNS', '******.com.tr')),
'age_days': 0.0,
'aging_status': 'fresh'
C:\Users\admin\PycharmProjects\Books>
```

Figure 4.5: DNS-WHOIS-SSL Normalization and Scoring Workflow

Let's examine what happens line by line:

- **Command and target domain**
  At the top, the following command is visible: `python osint_framework.py haydararici.com.tr`.
  The framework then records the target: `Target domain: haydararici.com.tr`. This indicates that all subsequent events and IOCs are processed within the context of a single target domain.
- **Live DNS query event (`dns_live RawEvent`)**
  This stage begins with the framework issuing a live DNS query against the target domain. The purpose is to obtain real-time infrastructure data directly from the DNS resolution process rather than relying on previously indexed sources. The resulting DNS response is then captured as an event and passed into the analysis pipeline for further processing.
  - **`Processing raw event ... from dns_live`**
    This line shows that the script performs a live DNS query via the `DNSClient` and processes the result as a `RawEvent` object.
  - **`Normalized 1 IOC(s) from event ...`**
    At least one IOC has been extracted from the DNS response (e.g., the domain name or an IP address). The `NormalizationEngine` converts this raw DNS response into a uniform `NormalizedIOC` format, as described in Chapter 6.
- **WHOIS event (`whois RawEvent`)**
  In the next stage, the framework queries the WHOIS infrastructure associated with the target domain. This step aims to retrieve registration-level metadata that describes the administrative and organizational context of the domain. Unlike DNS data, which reflects operational infrastructure, WHOIS records provide information about registration authority, jurisdiction, and management attributes. Once retrieved, this information is introduced into the analysis pipeline as an event and prepared for normalization and IOC extraction:
  - **`Processing raw event ... from whois`**
    Here, the `WhoisClient` has connected to the WHOIS server for `haydararici.com.tr`, and the returned results have entered the system as a `RawEvent`.
  - **`Normalized 1 IOC(s) from event ...`**
    At least one IOC has been produced from the WHOIS data (in this case, the domain and WHOIS-related attributes). Attributes such as registrar, country, and organization have been added, enabling the domain to be used later in contextual analyses.
- **SSL certificate event (`ssl RawEvent`)**
  In this stage, the framework retrieves the SSL/TLS certificate presented by the target domain during the TLS handshake. The certificate metadata is then captured as an event and prepared for normalization and IOC extraction within the analysis pipeline:

- **Processing raw event ... from ssl**
  This line shows that the SSLInspector component establishes a live TLS connection over 443/tcp and retrieves the certificate.
- **Normalized 2 IOC(s) from event ...**
  Two IOCs have been generated from the SSL response: a domain IOC for haydararici.com.tr and an ssl_cert IOC represented by the certificate's SHA-256 hash.

- **Start of the IOC summary section**
  From this point onward, the summary block displays the final state of all normalized IOCs from all sources, together with their scores and additional metrics.
- **Domain IOC: haydararici.com.tr**
  This line shows how the domain IOC has been evaluated:

  ```
  IOC haydararici.com.tr (domain) score={'confidence': 1.0,
  'relevance': 0.0, 'freshness': ..., 'aging_status': 'fresh',
  'data_gravity': 0}
  ```

  The scoring output reflects how the framework evaluates the analytical reliability and contextual value of the domain indicator. Each of the following metrics represents a different dimension of intelligence quality, combining source validation, temporal relevance, and relational context:

  - **confidence: 1.0**
    Because the domain name has been observed from multiple independent sources (live DNS query, WHOIS, SSL), the confidence score approaches the maximum. This is a direct application of the multisource cross-validation logic described in Section 4.4.2.
  - **relevance: 0.0**
    This indicates that the example IR (sector, country, technology stack) doesn't match haydararici.com.tr, meaning that this domain falls outside the operationally targeted profile in this scenario.
  - **freshness ≈ 1.0**
    Because the last_seen timestamp is very close to the analysis time, the freshness value is nearly 1. This indicates that the domain information is current in the time dimension.
  - **aging_status: 'fresh'**
    The AnalyticsEngine labels the domain as fresh in the aging model based on its most recent observation time. This is the practical implementation of the data aging concept described in Section 4.4.1.
  - **data_gravity: 0**
    Because no additional relationships (other IOCs or campaigns) have yet been attached to the entity associated with this domain, its data gravity is currently 0. As

more IOCs are linked to the same entity in the future, this number will increase, making the volume of intelligence concentrated around the domain visible. This directly reflects the data gravity concept explained in Section 4.4.1.

- **SSL certificate IOC**
  This line represents the `ssl_cert` IOC identified by the certificate's SHA-256 hash:

  ```
  IOC 56b44a97b6a315df7a53b9e2c5e32ed6af88b16db83621222cfbf9d283eacdad
  (ssl_cert) score={...}
  ```

  The subsequent attributes explain the identity and context of the certificate:

  - **`*subject: commonName = .haydararici.com.tr`**
    Indicates that the certificate was issued for the wildcard domain `*.haydararici.com.tr`, meaning that all subdomains may share the same TLS identity.
  - **`issuer: countryName = US, organizationName = "Let's Encrypt"`**
    Specifies the certificate authority that signed the certificate. In this case, it's Let's Encrypt, a widely used free CA.
  - **`notAfter: 'Feb 5 11:53:18 2026 GMT'`**
    The certificate's expiration date. This information is critical for tracking how the TLS infrastructure is renewed over time and serves as a core input for the SSL renewal rhythm and operational tempo analysis described in Section 4.2.4.
  - **`subjectAltName: ('DNS', '*.haydararici.com.tr'), ('DNS', 'haydararici.com.tr')`**
    The SAN field lists all domain names for which the certificate is valid. Here, both the root domain (`haydararici.com.tr`) and all subdomains (`*.haydararici.com.tr`) are covered by the same certificate.
  - **`age_days: 0.0, aging_status: 'fresh'`**
    Because the certificate IOC is being observed for the first time in this run, its aging metric is very close to zero days and labeled as `fresh`. This indicates that the SSL IOC is currently a fresh signal.

4.5 Summary

This chapter has shown that OSINT isn't defined by the volume of data collected, but by the discipline with which data is selected, validated, protected, and transformed into intelligence. Effective OSINT begins with a clear IR, proceeds through controlled collection and source verification, and gains value only when information is interpreted within context rather than in isolation. In this framework, OPSEC, ethical and legal boundaries, and evidentiary integrity aren't secondary concerns; they are the conditions that make intelligence reliable and defensible.

The chapter also demonstrated that professional OSINT spans multiple layers, from passive discovery and metadata analysis to infrastructure profiling, social media assessment, active observation, and structured data correlation. Across all of these layers, the analyst's task is to recognize patterns, distinguish signal from noise, and preserve the analytical meaning of what is observed over time. When supported by proper normalization, correlation, and storage practices, OSINT evolves from a collection activity into a durable intelligence system capable of supporting sound operational and strategic decisions.

Next, we'll shift our focus to intelligence-gathering techniques specific to the deep web and dark web.

Chapter 5
Advanced Intelligence Collection from the Deep and Dark Web

The modern threat ecosystem is no longer limited to attack campaigns visible on the surface. The darkest layers of cyberspace have become operational centers for a wide range of actors, from international criminal organizations and state-sponsored threat actors to financially motivated groups and ideological hacker collectives.

The deep web and dark web constitute a *shadow economy* in which strategies are developed, personnel are recruited, technical tools are shared, identities are concealed, and global attack chains are coordinated. This chapter dives into the depths of this shadow economy, comprehensively examining the structural dynamics of dark networks, anonymity mechanisms, data flows and information flows, and the operational signals that must be interpreted within these ecosystems. The deep web and dark web aren't environments that a randomly browsing user might encounter—they present a unique world with its own behavioral language and rhythm, shaped by carefully engineered anonymity layers, operational security (OPSEC) protocols, and integrated crypto-based payment systems.

In this context, the intelligence analyst becomes a specialist who deciphers the behavioral patterns of threat actors who successfully remain anonymous, monitors power relationships between marketplaces, detects data patterns within encrypted communication networks, and produces operational foresight from these insights. The objective is to uncover the connections within the dark ecosystem that aren't visible on the surface, and to interpret the tools, motivations, collaborations, and traces that actors create along their attack chains.

Intelligence collection activities on the deep web and dark web require technical skills, methodological discipline, OPSEC, and knowledge of behavioral analysis. Communication codes developed among threat actors, hierarchies within forums, the economic functioning of marketplaces, and anonymity-execution techniques generate multilayered signals that the analyst must continuously interpret. For this reason, the analyst isn't an external observer, but a researcher with the technical maturity to conduct active reconnaissance in dark networks, maintain composure while assessing operational risks, and manage data flows in a controlled manner.

Accordingly, this chapter provides knowledge across a broad spectrum ranging from threat actors' identity concealment techniques to the economic structures of marketplaces; from botnet management panels to the operational chains of ransomware groups; from statistical analysis methods to natural language processing (NLP)–based content classification. Each topic is constructed as an analysis of real-world threat models and the strategies for producing intelligence value from those models.

5.1 The Invisible Architecture of the Dark Ecosystem

One of the most critical areas of cyber threat intelligence (CTI) is understanding the multilayered structure of the digital world, which is divided into different levels of visibility. The modern internet isn't limited to the surface layer accessible through search engines; beneath it lies a far more complex data ecosystem supported by access restrictions, anonymity protocols, and encrypted communication networks. The deep web and dark web are the most strategic components of this ecosystem in terms of both information density and threat-generation capacity. Therefore, understanding the structural distinction of these layers is essential for grounding intelligence activities conducted in dark networks on a solid foundation.

Within this invisible structure of the digital universe, each layer presents a different world in terms of accessibility, the behavioral rhythm of threat actors, the anonymity tools they use, the nature of data flows, and OPSEC methods. In the following sections, we'll dive into these layers (the surface web, deep web, and dark web) and examine methods used by threat actors in each layer. We'll then explore how to extract intelligence from each layer, walk through some example scenarios, and dive further into the key points to know about dark web ecosystems.

What makes this section significant is that the distinction between the deep web and dark web forms the foundation for understanding the operational models, movement patterns, and digital behavioral intelligence of threat actors. In addition, this section focuses on what these layers are, as well as how actors move within them, through which technical infrastructures they conceal their behaviors, and how they exploit vulnerabilities. The anonymity offered by the dark web isn't an obstacle for the intelligence analyst; on the contrary, when decoded with the right methods, it provides strong clues about threat actors' preferences, modes of movement, and operational weaknesses.

5.1.1 The Operational Differences of the Surface, Deep, and Dark Layers

Understanding the layered structure of the internet requires analyzing which layers threat actors use for which purposes and how they strategically transition between these layers. Each layer contains its own behavioral model, level of visibility, operational opportunities, and risk profile. This structure forms the foundation of a critical analytical capability in threat intelligence known as *layer awareness*.

The *surface web*—namely, the open internet accessible to everyone—is generally where the first traces of threat operations emerge. However, much of the information visible here may not constitute the attack activity itself. Some actors deliberately leave controlled traces on the surface layer to conceal their operations or mislead their targets. For example, raw versions of phishing pages may be tested on private or low-traffic public domains, while the maturation phase of the attack is moved to deeper layers. One of the most critical aspects of the surface web is that it provides fertile ground for social engineering initiatives. Open social profiles, forum posts, blog articles, or fake statements targeting brands function as a psychological preparation space for manipulation processes to be used later.

The deeper layer, the *deep web*, hosts richer content targeted by operators. Data in this layer is unindexed and often requires corporate, personal, or specialized authentication mechanisms for access. For this reason, the deep web is one of the primary areas where attackers collect data. Corporate login systems, customer portals, private application programming interface (API) endpoints, academic archives, and even management panels of Internet of Things (IoT) devices fall within this layer. These structures frequently become targets due to misconfigurations or weak access policies. In fact, a significant portion of credential leaks originates from authentication-pivoting techniques targeting deep web components. Another factor that makes the deep web particularly important from a threat intelligence perspective is unnoticed data circulation. For example, even technical request notes within an internal ticketing system can be highly valuable during the reconnaissance phase prior to a cyberattack.

The *dark web*, on the other hand, represents the most extreme, most isolated, and most speculated-about layer of this structure. However, the true value of the dark web doesn't stem from the illegal content often exaggerated by popular culture, but from the anonymity engineering where actor behaviors are observed most clearly. Tor, I2P, and other privacy networks enable threat actors to hide and to organize. The communication styles, language, terminology, trade models, and community rules used within these networks create a cyber economy that is completely invisible on the surface. The most important distinction of the dark web is that threat actors don't just share tools there; they also form an ecosystem in which operational decisions are made. The target selection criteria of a ransomware group, the pricing strategy of an exploit vendor, the trust score of a card seller, or the moderation rules of a forum administrator are all highly valuable intelligence signals.

The real distinction between these layers lies beyond visibility alone; the rhythm of operations, the tools used, the level of OPSEC applied, and the objectives of the actors are all shaped by the layer in which they operate. The surface web is the noisiest environment. The deep web is an intermediate layer where critical data flows quietly. The dark web, however, is a paradoxical structure built on both isolation and collaboration—it's the most visible form of behavior within the invisibility. Interestingly, most major attacks occur through the convergence of processes across at least two of these layers. For instance, the design phase of a phishing operation may take place on the surface web, credential harvesting may shift to the deep web, and the collected data may then be sold on dark web forums.

The critical point in threat intelligence is the analyst's ability to recognize the transition paths that connect these layers. Actors never remain confined to a single layer; they constantly change direction between layers to avoid leaving traces. Therefore, a superficial analysis is sufficient only to understand what is visible; the true operational picture can be understood through inconsistencies within layers, contextual shifts, communication density, and timing information.

5.1.2 Special Methods Used by Threat Actors

Cyber threat actors use the layered structure of the internet to both hide and reduce the risk of detection by dividing their operations into different phases. This multilayered technical approach requires different tools, different communication methods, and different security levels at each stage. Therefore, the methods used by threat actors combine psychological, behavioral, and operational strategies. Understanding these methods enables an analyst to determine who the actor is and why they behave in a particular way.

We'll now present the key methods used by threat actors:

- **Layered signal distortion**

  One of the most distinctive techniques used by threat actors is the deliberate distortion of traces they leave while moving across layers. This method goes beyond IP or user-agent manipulation; it also includes the intentional alteration of time signatures, linguistic rhythm, writing style, and platform selection.

  The actor ensures that the traces left on the surface web don't align with their behavior on the deep web. For example, an individual who uses English technical jargon on a forum may share a poorly written Turkish text in an open space. The objective is to break the tracking chain.

  One of the most critical aspects of this technique is the use of *time segmentation* microtactics. The actor doesn't maintain consistent time patterns across different layers, thereby neutralizing both machine learning–based behavioral analyses and manual stylistic analyses.

- **Modular personas**

  A threat actor doesn't operate with a single persona; instead, the operational process is often carried out through multiple characters divided into modules. These operational persona modules are each configured to have a different writing style, device footprint, temporal activity pattern, and forum history, for example:

  - Persona A handles the technical preparation of the attack (exploit testing, proof of concept [POC] generation).
  - Persona B publishes fake announcements on the surface web.
  - Persona C sells the obtained data on dark web marketplaces.
  - Persona D conducts propaganda via Telegram or X (formerly Twitter).

Even if all of these identities are managed by a single individual or group, they appear completely independent from the outside. This is a highly advanced privacy method that makes tracking nearly impossible.

- **Contextual anonymity**

  One of the lesser-known methods used by actors is contextual anonymity. Traditional anonymity relies on tools such as Tor, virtual private networks (VPNs), and proxy chains. Contextual anonymity, however, combines technical privacy methods with social-context manipulation.

  With this method, the actor completely changes their behavior depending on the platform they are communicating on in the following ways:

  - Demonstrates high technical competence on technical forums
  - Uses amateur-looking expressions on Telegram channels
  - Prefers entirely neutral and brief communications on marketplaces
  - Avoids cultural references in private messages

  The goal is to invalidate all contextual clues an analyst might use to classify the actor. This technique is particularly common among certain groups originating from Eastern Europe and Southeast Asia.

- **Triple-phase operational isolation method**

  This method is used by advanced persistent threat (APT)–level groups to avoid leaving operational weaknesses. The operation is carried out in three separate isolated phases:

  - Preparation phase (pre-ops layer): Infrastructure development is conducted through hosting environments located in specific countries and protected against legal pressure.
  - Execution phase (active layer): The attack is launched on virtual machines (VMs) connected to Tor, I2P, or dedicated proxy networks. These VMs are completely isolated from one another; even clipboard sharing is disabled.
  - Cleanup phase (burn layer): The used infrastructure, disk images, browser profiles, and wallet addresses are completely destroyed. Some actors even randomize the destruction process using physical entropy devices (entropy dongles).

  One of the hardest-to-detect characteristics of APT groups is their consistent and disciplined application of this three-phase method.

- **Behavioral camouflage**

  This technique aims to disable the analyst's behavioral modeling capability. Using this method, the actor does the following:

  - Imitates their own writing styles
  - Deliberately copies the styles of lower-level threat actors
  - Appears active at times that don't align with global time zones

As a result, the actor's true origin, technical capability, or team structure is misclassified. Many attribution errors seen in threat intelligence reports stem directly from this technique.

- **Silent pivoting**
  This is one of the most powerful yet least-known techniques. The actor doesn't use the accessed data directly; instead, they first perform silent pivoting. That is, without revealing their presence on the target system, the actor does the following:
  - Conceals access traces
  - Reduces data-reading speed (rate limiting)
  - Issues command flows that mimic the system's natural processes

  Thanks to this technique, the intrusion may remain undetected for weeks. It's especially used in the initial phase of the credential theft → deep web sale → dark web collaboration chain.
- **Broken signature method**
  This technique renders malware invisible to signature-based detection systems. The actor does the following:
  - Creates multiple distorted versions of the same malicious code
  - Doesn't functionally alter the code
  - Applies only minor variations designed to deceive machine-learning models

  As a result, threat scanners evaluate the same malware as entirely independent samples. In some exploit kits sold on the dark web, the most valuable components are actually these signature-breaking modules.
- **Compartmented collaborative roles**
  Advanced threat groups operate by dividing each role among separate subteams. No member has access to the complete picture. This method dramatically reduces the following:
  - Risk of operational leakage
  - Likelihood of insider threats
  - Probability of mass compromise

  In some groups, there are subteams dedicated to roles such as these:
  - Infrastructure renters
  - Cryptography specialists
  - Data cleaners
  - Psychological manipulation experts

  This model is common among state-sponsored APTs.
- **Latency-driven activation**
  Advanced actors schedule their actions during periods when systems are least active to

avoid detection. This isn't simply a matter of being active at night. In this approach, the actor analyzes the following factors and select the optimal moment accordingly:

- CPU usage on the target system
- Network traffic patterns
- Log rotation times
- Automated backup schedules

This is why some attacks occur at seemingly meaningless times such as 03:17. It's a deliberate concealment technique.

These specialized methods clearly explain why threat actors are able to operate unseen for extended periods and why certain attack chains are so difficult to unravel. They also provide analysts with valuable clues regarding which behavioral, technical, and time-based signals to monitor when tracking actors moving between the dark web and the deep web.

5.1.3 Techniques for Extracting Intelligence from Each Layer

The value of a CTI analyst is measured by both the data they can access and how they generate meaning from that data. The surface web, deep web, and dark web each contain different types of signals; however, a professional analyst transforms these signals into real intelligence not by consuming them in their raw form, but by interpreting the invisible connections between them. Just as threat actors manipulate their traces while moving across layers, the analyst's task is to detect the inconsistencies created by that manipulation. For this reason, extracting intelligence from each layer is a multilayered analytical discipline.

We'll explore the key intelligence extraction techniques in the following sections.

Intelligence Extraction from the Surface Web: Behavioral Signal Discovery

The surface internet appears to be the simplest and most ordinary information space; however, when analyzed using the correct methods, it contains some of the most critical behavioral traces of threat actors. When evaluating the surface web, the analyst doesn't focus on raw content; instead, they examine the time of content creation, target audience, word choice, metadata, and redistribution curve. Even a phishing page—no matter how amateur it may appear—can reveal the geographic focus of the attack, the language model used by the threat group, and the maturity level of the operation.

One lesser-known technique is *meta-inconsistency analysis*. For example, even if an announcement text is written in Turkish, traces of Hindi or Romanian characters may appear in its meta tags. This serves as a geographic indicator of the actor's development environment and is a critical finding in surface web analysis.

In addition, inconsistencies between profile creation timing and posting rhythms on fake social media accounts help determine the level of automation used in operational persona generation. Although analysts can't derive who conducted the attack from such signals, they can determine which model executed it.

Intelligence Extraction from the Deep Web: Reversing Access Behavior

What determines the value of the deep web isn't just the amount of data it stores: it's the access logic of the systems that generate that data. Systems operating on the deep web often produce the most critical signals prior to an attack—signals that usually go unnoticed.

Professional analysts also apply *access deviation analysis*. This method involves detecting subtle anomalies in the access behavior of threat actors compared to normal user flows within a system. For example, consider a normal user who logs in between 07:00 and 23:00. The following activities may indicate that the account is being prepared for circulation within the threat ecosystem:

- A silent request made at 03:43 using a compromised account
- Low-frequency packet movement
- Abnormal session duration
- Immediate logout patterns

Another advanced technique used on the deep web is *backdoor variance measurement*. Some threat actors quietly introduce minimal test changes into compromised corporate panels. These changes usually don't affect the system's primary functionality, but they create microscopic inconsistencies that generate log entries.

When an analyst detects these inconsistencies, the attack can be prevented while it's still in the deep web phase, which represents the most valuable stage of professional intelligence.

Intelligence Extraction from the Dark Web: Behavioral Signature Decoding

The greatest strength of the dark web is anonymity; however, from an intelligence perspective, this anonymity is also its greatest weakness. The more actors attempt to conceal themselves, the more behavioral signals they produce.

One professional method used here is multiplatform behavioral correlation. If an actor posts a sales listing on a marketplace via Tor at 14:05 and advertises the same offer on a Telegram channel at 14:06, this temporal correlation is an extremely strong indicator for identity matching.

Analysts also conduct *word frequency deviation analysis* on forum posts. Most threat actors, even while maintaining anonymity, unconsciously preserve the same writing patterns:

- Excessive attachment to specific technical terms
- Recurring sentence structures across messages
- Preferred special characters
- Pricing language

Such minor details can allow recognition of the same actor even when multiple aliases are used. A lesser-known technique is the *actor identification via marketplace economics* model. Some threat actors exhibit consistent pricing behaviors:

- Recurring 20% discount patterns
- Bulk dump releases at specific intervals
- Obsessive preference for certain cryptocurrencies

These parameters are often far more powerful identifiers than technical fingerprints. When applied correctly, these methods make it possible to map the behavioral profile of the individual or group behind anonymous dark web accounts.

Cross-Layer Correlation: Mapping the Operational Chain

Intelligence can be extracted separately from each layer; however, true value emerges when the data obtained from these layers is connected. Professional analysts refer to this as *chained intelligence correlation*.

In this model, the following flow is analyzed along a single line: a domain observed on the surface web → a credential movement on the deep web → a sales listing on the dark web → propaganda announced by the actor via Telegram → a sudden movement in a cryptocurrency wallet.

This line provides strong signals regarding the motivation, scope, economic relationships, and the underlying actor network behind the attack. This technique was once known only to government agencies; today, it's increasingly used by advanced CTI teams.

Time-Based Intelligence Production: Operational Rhythm Analysis

Operational rhythm analysis focuses on when, where, and through which communication channels an actor is active. Through this analysis, it becomes possible to accurately determine the following:

- The actor's geographic region
- Team structure
- Whether the operation is manual or automated
- Weekday versus weekend activity differences
- The urgency level of the operation

For example, a group consistently operating between 05:30 and 08:00 strongly indicates an Asia-Pacific origin; however, without behavioral analysis, this alone isn't meaningful. Professional intelligence emerges by combining these temporal indicators with signals from other layers to build a coherent profile.

Extracting Meaning from Data Noise: Negative Space Analysis

Traditional intelligence collection focuses on existing data. Professional techniques, however, analyze missing data. *Negative space analysis* operates through questions such as these:

- Why did a conversation on a forum abruptly stop?
- Why was a particular vendor profile on a marketplace not archived?
- If an actor has been inactive for a long time, is this post-operation silence?
- Why was a specific organization not included in a dump list?

Missing data is often more meaningful than what is present. This technique is especially used to detect post-operation trace-erasure behaviors of dark web actors.

5.1.4 Example Scenarios

As in real threat operations, intelligence chains in which the surface web, deep web, and dark web triad is activated simultaneously reveal a structure so complex that it can easily escape the notice of most analysts. However, analysts who can detect the invisible transitions between layers are able to decipher the technical dimension of a threat, the underlying motivation, the infrastructure used, and the behavioral model of the actors.

Let's see this at work in some practical examples.

Silent Reconnaissance Operation: Phase-Shifted Leakage Chain

Although no data exfiltration from a financial institution's internal network had been detected, the first sign that drew analysts' attention was an extremely ordinary-looking JavaScript file published on the surface web. The file appeared harmless in terms of content (most of the code even belonged to a popular library), but there was a short comment line written using a Polish character set within the meta tags. The unnecessary presence of this line became the starting point for deeper analysis.

One week later, an encrypted announcement appeared on a closed forum on the deep web under the title "high-value corporate internal communication package." The announcement didn't directly list filenames; instead, it used markers such as `ref-127a`. Analysts noticed that these references matched the random string contained in the name of the JavaScript file on the surface web.

Following these two signals, a marketplace on the dark web showed that an unidentified seller was preparing a sale related to the same references. What was striking was that the seller was operating outside their usual behavioral rhythm. Normally active in the evening hours, the seller logged in this time at 04:12 Turkey time. This temporal shift indicated that the seller was conducting an unusual operation and was likely performing a security test via the deep web.

When the tracking chain was validated across all three web layers, the objective of the operation became clear: The attackers didn't steal any files for an entire week; they merely tested the stability of access paths within the systems. The real goal was to prepare a phase-shifted groundwork within corporate systems prior to a major data leak. This is a technique frequently used by professional attack groups; even if the data at hand isn't yet valuable, the operational infrastructure is quietly shaped to guarantee future data flow.

Identity Distortion Operation: Invisible Logistics Shadow

A midsized technology company reported that no data related to them was being shared on the dark web. However, during routine scans, analysts noticed two extremely low-priced accessory files appearing on a marketplace: `auth-key.bin` and `dev-cache.tmp`.

At first glance, the files appeared insignificant. What drew attention, however, was that the seller marked them as "available for bulk purchase." Such low-level files are typically used in identity distortion operations. Before sharing truly high-value data related to a target company, attackers release small files to marketplaces to gauge reactions. If a response comes from an organization or researcher, the actors reduce the visibility of the attack; if no response occurs, they proceed with the full operation.

Analysts recognized this pattern and continued monitoring. Three days later, on a closed developer forum on the deep web, the same seller was observed using specific jargon meaning "access to a hot test environment." This terminology wasn't publicly known and was used only by certain groups.

This indicated that the goal of the attack wasn't direct data sales, but rather gaining long-term covert access to the target organization's development processes. Moreover, while the seller was sharing only test files on the marketplace, they were discussing full access on the deep web forum. In other words, the dark web visibility was deliberately concealing the operational reality occurring on the deep web.

This operational model escapes the attention of many analysts because actors intentionally present themselves as insignificant data sellers while planning the real breach within closed networks.

Chained Economic Manipulation: Signal Distortion Trade

When suspicion of price manipulation arose on a cryptocurrency exchange, the first clues appeared on surface web forums. Randomly created user accounts were repeating the same words with the same mistakes. This indicated bot activity; however, the bots appeared deliberately primitive to conceal the true objective.

At the deep web stage, a far more intriguing signal emerged. Within a closed community, certain users were repeatedly requesting timestamp correction. The purpose behind this request was to manipulate the exchange's order execution times to artificially trigger price fluctuations.

The most critical finding emerged on the dark web. A marketplace was renting out small botnet clusters under the name "disruptive signal packages." These botnets generated low-level transaction traffic to deceive the exchange's algorithmic order systems. When these three layers were combined, the objective of the attack became clear: The actors appeared to be conducting ordinary social media manipulation on the surface, while coordinating on the deep web and renting the technical infrastructure for the operation on the dark web.

Such chained manipulations are among the most difficult operational models to detect in classical CTI processes. Each signal within a single layer appears harmless on its own; but when combined, they reveal a much larger attack chain.

Silent Information Exploitation: Dormant Access Technique

In most attacks, data is stolen or sold. In this scenario, however, the attackers neither used, sold, nor even exfiltrated the data. Instead, a backdoor implanted in a corporate email system collected information for months, and this information was used solely to develop the attackers' operational strategy.

There were no traces on the surface web; even the phishing attempt used at the beginning of the attack was so ordinary that most analysts might classify it as an automated attack.

On the deep web side, the signal was more subtle. Certain error logs belonging to the email server were triggered at times when they weren't normally observed. When these logs were examined, it became apparent that the attacker was performing only read operations, with no write actions or data transfer attempts.

On the dark web, it was observed that the attacker made no disclosures or sales. This indicated that the attack wasn't driven by commercial motivation, but by a long-term intelligence monitoring operation. In such cases, the threat actor doesn't sell the data but instead uses it in operational decision-making processes (for example, to support physical attacks, investment manipulation, or state-sponsored economic analysis), quietly building an information pool over time.

5.1.5 Typology of Dark Web Ecosystems

Many people think of the dark web as an environment where cybercriminal activities take place. However, it's more accurate to think of it as a digital underground economy with its own rules, economic cycles, hierarchies, and behavioral models. Although the structures within this ecosystem can appear similar on the surface, in reality, each component serves a different purpose, follows a different organizational model, and operates with a distinct operational logic. For an analyst to accurately interpret this ecosystem, technical knowledge alone is insufficient; they must also understand actor psychology, risk perception, and economic motivations.

The following sections examine the primary structural components of this ecosystem in greater detail. Each subsection focuses on a distinct operational layer of the dark web economy and explains how these structures function within the broader criminal infrastructure. Understanding these layers allows analysts to interpret both the economic and behavioral dynamics that shape the underground threat environment.

Marketplaces

At first glance, marketplaces appear to be simple venues for illegal trade; however, their true value lies in the fact that they function as sociotechnical hubs that most clearly reflect the dynamics of the dark economy.

Some marketplaces subject vendors to specific tests. In certain cases, invisible *impact coefficients* are assigned to vendors based on transaction volume. These coefficients indicate how seriously a vendor is taken within forums, which communities they are connected to, and which attack tools they have access to.

Some advanced marketplaces go beyond traditional rating mechanisms and employ a *delayed reputation model*. In this model, vendor scores aren't updated immediately; instead, they change after a delay of several days based on calculated risk coefficients. The goal is to prevent manipulation through fake user activity.

One of the most important intelligence functions of marketplaces is their ability to reveal attack trends at an early stage. For example, a sudden increase in the price of certain exploits may serve as an early indicator of an impending attack wave.

Community Forums

While marketplaces represent the economic dimension, forums are the cultural and operational centers of the dark ecosystem. Discussions within these forums are the most powerful means of understanding actors' true intentions, technical proficiency, and long-term objectives.

In forums, an actor's value is measured by the information and how it's presented. In some advanced forums, the following criteria are used to determine an actor's level:

- Line length of text
- Density of technical jargon
- Symbolic signatures embedded within sentences
- Uniform message posting times

These signals are highly valuable for behavior-based profiling in threat intelligence. Subcommunities within forums are also internally differentiated in the following ways:

- Closed collaboration rooms of ransomware groups
- Testing environments for exploit developers
- Operational coordination channels for botnet administrators
- Account cleaning discussions among financial fraud teams

For analysts, what is critical in these areas isn't the content itself, but the micro-interactions that form around it. Often, the signal of an attack is the unusual reaction to the exploit (or the absence of any reaction).

Service-Based Models

One of the most critical transformation points of the dark web is the evolution of cybercrime activities from individual initiatives into an organizational and scalable service industry. This transformation is driven in part by the rental of criminal tools, but more so by the structuring of the entire operational cycle within a professional framework. This structure is now defined through attack as a service (XaaS) models and forms the backbone of the modern cyber threat ecosystem.

The service-based crime economy is designed to optimize the *cost–risk–time triangle*. Even inexperienced attackers can launch complex and professional attacks without requiring high technical competence, simply by paying a certain fee. This doesn't lead to the democratization of the threat ecosystem, but rather to the massification of criminal capability. For example, an attacker can initiate a ransomware campaign without knowing how to write code, purchase an advanced phishing operation externally, or distribute thousands of malware samples for just a few dollars.

One of the most striking aspects of this industry is that each service provider typically specializes in a specific domain:

- Malware as a service (MaaS) operators focus on malware production and updates.
- Ransomware as a service (RaaS) teams manage extortion processes, victim panels, payment infrastructures, and data leak portals.
- Phishing as a service (PhaaS) operators offer sophisticated systems that continuously update fake email templates in real time, test browser compatibility, and provide statistical performance reporting.

The most dangerous aspect of these models is the distribution of operational responsibility. The individual purchasing the service faces a lower direct risk of detection because they don't generate the technical traces of the attack themselves; the service provider, meanwhile, typically operates multilayered infrastructures over Tor, I2P, or dedicated proxy tunnels to evade security researchers. This mutual protection makes the criminal economy as structured and sustainable as a conventional commercial ecosystem.

At a more advanced level, some attack services operate under a corporate model, for example:

- Membership packages
- Performance bonuses
- Affiliate commission schemes
- Technical support
- Service level agreement (SLA)-like service continuity guarantees
- Beacon-based monitoring dashboards
- Customer refund policies

The presence of such elements demonstrates that the dark web crime economy follows a fully corporate organizational logic.

From an intelligence perspective, the most critical aspect of this industry is that attacks generate powerful signals through their technical traces, financial flows, communication patterns, market dynamics, and service provider–customer relationships. While many analysts focus on technical indicators, professional CTI teams can interpret threat actors' true intentions, capacities, and future targets through behavioral signals emerging from this economic structure.

The service-based structure of the dark web crime economy can be understood more clearly by examining its dominant operational models. These models reflect how criminal capabilities are packaged, distributed, and scaled across different actors. Each service category represents a specialized layer of the underground economy, designed to support different stages of cyber operations. As mentioned earlier, here are the most common service models and the intelligence signals they generate:

- **Malware as a service (MaaS)**
  MaaS providers sell more than malware; they also offer packages that include the following:
  - Automated update modules
  - Target-adaptive variant generators
  - Anti-analysis mechanisms
  - Telemetry obfuscation systems
  - Campaign management panels

  In the most advanced models, MaaS operators even provide post-attack data processing and classification services, such as prioritizing stolen credentials based on their value.
- **Ransomware as a service (RaaS)**
  The RaaS ecosystem is one of the most organized structures on the dark web. Some RaaS providers use a performance bonus model; as ransom payments increase, the provider reduces its commission, thereby incentivizing affiliates to carry out more aggressive attacks. Some unique features found in advanced RaaS platforms include the following:
  - Customized victim portal templates
  - Accelerated content delivery network (CDN) infrastructures for data leakage
  - Internal dispute resolution mechanisms
  - Performance analysis reports for affiliate attackers

  These patterns are critically important for intelligence analysts seeking to understand a group's growth rate, technical capacity, and target selection criteria.
- **Phishing as a service (PhaaS)**
  PhaaS enables even low-skill actors to conduct effective phishing attacks. PhaaS services include a hidden mechanism known as the *dynamic template update system*. This system keeps fake templates up to date by automatically tracking the following:

- Logo changes of target companies
- Email signature formats
- Design updates of commonly used software as a service (SaaS) platforms

PhaaS providers also offer hit-rate analysis, generating dark statistics on which types of emails are most successful in which regions. These statistics enable attackers to optimize their campaigns.

Behavioral Models Across Ecosystems

Each ecosystem produces a different behavioral model. In marketplaces, economy-driven relationships dominate, and price fluctuations reveal attack trends. In forums, technical capacity and group dynamics emerge; actors' intentions are most clearly understood here. In service-based models, operational professionalism stands out, indicating that the dark economy has reached an industrial level.

A critical point unknown to many analysts is that these ecosystems form a closed loop that feeds itself. For example, an exploit is first discussed in a forum, then priced in a marketplace, subsequently integrated into MaaS/RaaS infrastructures, and finally manifests as attack campaigns visible on the surface.

An analyst who examines only a single part of this loop only sees the event. An analyst who can interpret the entire loop sees the threat.

Layered Ecosystem Mapping

Professional threat intelligence teams monitor and map dark web ecosystems. To do so, they evaluate the invisible economy as a three-dimensional structure, analyzing the following:

- Actor flow channels
- Relationships between forums, marketplaces, and service providers
- Access hubs
- Geographic time rhythms
- Price stability zones

These maps contain early signals of past attacks and operations that may emerge in the future.

5.2 Accessing Hidden Services and Managing Anonymity

Producing effective intelligence within the complex structure of the dark web ecosystem isn't limited to understanding its layers; how those layers are accessed directly and how securely determines an analyst's operational success. Hidden services—such as Tor, I2P, and similar anonymity networks—are the most critical environments where threat actors congregate, discuss operational plans, and conduct the activities of the illicit economy. Every action taken in these spaces carries the risk of leaving traces and compromising operational integrity. For this reason, access strategies represent far more than technical infrastructure; an analyst's safety, presence, confidentiality, and ability to operate independently depend on the quality of anonymity management.

Access to hidden services represents a decisive threshold for intelligence analysts just as much as it does for threat actors. However, while threat actors have systematically optimized their anonymity infrastructures for years, many institutional intelligence operations still approach these infrastructures superficially. In reality, these networks do more than conceal IP addresses; each network functions as an encrypted micro-internet with its own topology, performance characteristics, obfuscation mechanisms, traffic behavior, and risk model. Rather than simply launching a browser, entry into these micro-networks should be treated as a controlled operational framework supported by multilayered security measures.

The critical objective of this section is to ensure that analysts understand the technical tools required to access hidden services and master the operational logic of these services, their risk profiles, and how threat actors manipulate these networks. Anonymity networks aren't passive tools; every packet traversing them carries behavioral signals that must be analyzed. The timing of Tor circuits, the geographic structure of exit nodes, the directional distribution of I2P tunnels, and the bridge topologies used by actors all serve as early indicators of movement and attack preparation within the dark ecosystem.

Anonymity Management

Anonymity management isn't limited to reducing external visibility; it also requires an inward-facing security mindset. The operating system (OS) used by the analyst, VM isolation, network configuration, Domain Name System (DNS) resolvers, browser components, and even typing rhythm can all become traceable clues monitored by threat actors. For this reason, in modern intelligence operations, OPSEC is a foundational architectural component that is essential for operational sustainability. Once anonymity is compromised in the dark web, it's nearly impossible to restore.

Understanding the access methods of hidden services on the dark web both protects the analyst's identity and enables the analyst to decipher the underlying logic of the methods used by threat actors. For example, if a particular actor consistently accesses services via bridges, this doesn't merely indicate a need to bypass censorship, but rather reflects a deliberate security behavior developed against circuit monitoring. Similarly, the tunnel bandwidth used on I2P or the selection of specific exit nodes on Tor can provide important signals regarding an actor's technical capacity, geographic location, or organizational affiliation.

This section is built precisely on this awareness: Hidden services aren't just addresses being accessed; every access, every connection, and every circuit leaves a trace that shapes the strategy of both the analyst and the threat actor. For this reason, accessing hidden services is a high-level intelligence discipline, a planned risk management practice, and a delicate operational design—not, as many assume, just a purely technical process.

5.2.1 Secure Use of the Tor Infrastructure

Although Tor is commonly defined as an anonymity tool that hides an IP address, in professional threat intelligence and dark web operations, it constitutes a foundational component of a far more comprehensive security architecture. The protection provided by Tor is reliable only when it's configured correctly; otherwise, inconsistencies in circuit behavior, improper system configurations, and browser fingerprinting can turn into serious vulnerabilities that expose the analyst's identity. For this reason, Tor isn't a standalone tool, but a multilayered anonymity ecosystem that operates through the combined interaction of the OS, hardware, network topology, software profile management, and OPSEC practices.

To understand how Tor can be used securely in intelligence operations, the surrounding technical environment must also be considered. The following sections examine the OS layer and the isolation models that determine how effectively Tor protects the analyst's anonymity.

Operating System Isolation

The first step in using Tor securely is reducing the attack surface of the OS. Using Tor in a conventional Windows or macOS environment may appear anonymous in theory, but in practice, it presents a high level of risk due to telemetry mechanisms within the system, time synchronization behavior, DNS cache activity, and application calls that are logged by the OS.

Several specialized OSs have been developed specifically for Tor-based anonymity. Systems such as Tails, Whonix, and Qubes OS integrate Tor into their architecture and isolate the user environment to minimize OS-level data leakage. Each platform implements a different isolation model and therefore offers different advantages for secure dark web operations:

- **Tails**
  Due to its amnesic nature, Tails leaves no traces when shut down; however, it has a critical weakness that many analysts are unaware of. Compared to other anonymity systems, Tails produces more fingerprinting artifacts during clock synchronization errors. For example, when accessing Tor at a national border or within a state-routed network, Tails' Network Time Protocol (NTP) time correction can generate unexpected traffic. In advanced operations, this activity can become a behavioral signature.
- **Whonix**
  This platform is more consistent from an anonymity perspective due to its architecture that chains two machines together. However, Whonix's true strength lies in its log separation behavior. The separation between the gateway and the workstation ensures that

every action performed by the analyst is automatically routed through Tor, thereby almost completely preventing classic risks such as IPv6 leaks, DNS fallback, or WebRTC leaks.

- **Qubes OS**
  This platform adds another layer on top of these: isolation isn't managed virtually, but instead through compartmentalized security domains. A Tor circuit running on Qubes is isolated at both the network level and even at the window manager level. When certain threat actors recognize this architecture, they may even infer the analyst's operational sophistication.

Which of these three systems should be used at which stage is directly related to how closely the analyst is operating to the threat actor.

Tor Circuit Rebuilding (Circuit Isolation)

Randomly renewing Tor circuits is a common mistake made by amateur users. This is because when circuit renewal is performed with a consistent temporal rhythm, it creates a behavioral signature for the analyst.

The truly professional approach is *dynamic circuit diversification*. The following occurs under this method:

- Separate circuits are used for specific pages.
- Fixed circuits are used for specific forums.
- Low-latency circuits are used for marketplaces.
- High-latency circuits are used for profile analysis.

This diversity disrupts the time-series correlations used by threat actors in behavioral tracking. An even more advanced technique is exit node directional analysis. Because exit nodes in certain countries are passively scanned by attackers, connecting through those nodes may cause the analyst to be noticed. Therefore, circuit renewal should be guided by a decision tree based on geographic risk.

Browser Fingerprint Reduction

Although Tor Browser is optimized for fingerprint reduction, browser fingerprinting isn't limited to the user-agent, language, or screen resolution. Threat actors now perform fingerprint measurements at much deeper levels:

- GPU behavior
- Canvas hash variation
- Transport Layer Security (TLS) fingerprint
- HTTP/2 packet ordering
- JavaScript execution delays

- AES-NI hardware acceleration signatures
- Time precision (millisecond deviations)

Most of these techniques are discussed in private forums and closed exploit communities and don't appear in open-source materials.

Rather than attempting to eliminate these fingerprints entirely, analysts aim to blend into the population. Extreme fingerprint values create unique identities even if the IP address is hidden. Tor's strongest protection isn't individual anonymity, but crowd anonymity. For this reason, true experts never modify Tor Browser profile files; default usage is safer than custom configurations that introduce outlier characteristics.

Disabling JavaScript, WebGL, and the Canvas API for OPSEC Purposes

Disabling JavaScript doesn't provide absolute security as many people assume; this belief usually stems from the fact that many browser-based tracking and fingerprinting techniques rely on JavaScript execution. However, in practice, completely disabling JavaScript can itself become a behavioral anomaly, and in some dark web forums, it may result in being flagged as a suspicious user. For this reason, the OPSEC approach isn't about disabling everything but applying layered permission management:

- **Marketplaces**
  JavaScript should be disabled, as fingerprinting attacks are frequently used.
- **Some dark web forums**
  Access may be blocked if JavaScript is disabled. OPSEC in this context requires the use of segmented profiles.
- **Technical communities**
  Canvas and WebGL should be disabled for protection purposes, as GPU-based fingerprinting is widespread.

Analysts also apply time-based expectation disruption. In some forums, JavaScript is expected to be enabled for certain actions. If the analyst needs to perform such actions, they proceed in a controlled manner using different profiles; repeatedly toggling JavaScript on and off within the same profile makes tracking easier.

If the Canvas API isn't blocked, threat actors can use Scalable Vector Graphics (SVG)–based fingerprinting to detect even minor versions of the browser engine used by the analyst. Therefore, OPSEC isn't just about restriction but also about environment control.

Remember!

What matters isn't using Tor, but managing operations over Tor.

The Tor infrastructure is a critical tool in dark web intelligence; however, using it securely isn't a simple privacy practice, but an operational discipline. Leaked circuit timing, improper OS-level configurations, aggressive circuit renewal behavior, extreme fingerprint values, and incorrect OPSEC choices can tear down the analyst's anonymity veil in a single move.

For this reason, professional use is far more than just launching Tor: It requires remaining invisible on Tor, maintaining operational continuity, and staying one step ahead of the analytical methods used by threat actors.

5.2.2 I2P and Alternative Privacy Networks

In the eyes of many security professionals, the dark web has become synonymous with Tor; however, the dark web doesn't operate entirely on Tor. As Tor has become more widespread and detection techniques have advanced, the most professional and cautious threat actors have shifted toward alternative anonymity networks that provide more closed, more isolated, and more directional privacy. At the center of these networks is I2P, which from the outside may appear to offer a hidden service model similar to Tor; however, both its traffic flow model and its operational architecture are products of a completely different security paradigm.

Understanding I2P goes beyond knowing an alternative anonymity tool. The existence of this network reflects the evolution of professional threat actors' OPSEC approaches. State-sponsored groups and high-capability criminal organizations use I2P as an operational universe to minimize traceability risks. This is because I2P doesn't connect users to the internet; instead, it creates a completely closed darknet ecosystem in which users connect only to one another. Traces left within this ecosystem don't become visible through centralized exit nodes as they do in Tor; traffic chains are carried solely through fragmented tunnels between routers. As a result, even if an operation conducted on I2P is detected by external networks, source attribution is practically impossible.

In the following sections, we'll dig deeper into the differences between I2P and Tor. We'll discuss I2P's unique architecture, communities, and strategies.

Differences Between I2P and Tor

For a professional analyst, the most critical distinction of I2P is that it employs an inbound-focused traffic model. Tor anonymizes users while allowing them to access the internet; I2P, by contrast, isolates users from the internet and enables communication only with other services within the I2P network. This provides threat actors with a unique advantage: services hosted on I2P can't be accessed from outside the network. As a result, botnet management panels, command-and-control (C2) infrastructures, ransomware negotiations, and exploit distribution channels can be operated within a closed "privacy island" shielded from external threats.

The most dangerous aspect here is that most CTI teams don't actively monitor I2P for three main reasons:

- **Unpredictable network topology**
  I2P uses constantly changing inbound and outbound tunnels. Traffic doesn't follow a fixed structure.
- **Lack of service indexing**
  There's no centralized directory similar to Tor's onion directories. Each service dynamically constructs its own tunnel map.
- **Inefficiency of automated crawlers**
  Conventional dark web crawlers are nearly useless on I2P. Even if a connection is established, tunnel directions can change within seconds.

For this reason, threat actors using I2P know that the likelihood of detection is significantly lower than on Tor. This difference creates dark zones within attack processes. An analyst who tracks an operation only through Tor may fail to realize that the most critical part of the operation is actually being conducted on I2P.

Another important aspect of I2P's operational use is actor segmentation. While Tor hosts a relatively broad user base, I2P is used by smaller, more technical, and more disciplined communities. The actor profiles typically observed on I2P include the following:

- Exploit developers
- Specialized botnet engineers
- Subcells of state-sponsored APT groups
- Closed marketplaces engaged in high-risk data trade
- Advanced criminal groups building chained tunnel infrastructures

What these actors have in common is that, rather than engaging in the commercial volume–focused activities seen on Tor, they conduct long-term operations with high OPSEC levels. For example, a zero-day exploit shared on an I2P forum is often never seen on Tor marketplaces. Such content is offered only to a limited set of actors who have proven their technical competence.

The primary objective of Tor is to anonymize users while enabling secure interaction with the external internet. In other words, Tor relies on exit nodes. In contrast, I2P employs an entirely inbound-focused architecture, which means the following:

- Tor: Anonymous access to the internet
- I2P: Anonymous service hosting within a closed network

I2P's design prioritizes access not to the external world but to services offered by other I2P users. Therefore, sites with the `.i2p` extension (eepsites) have no connection to real IP addresses; they are protected by tunnels within the network. This distinction is operationally critical. While hidden services on Tor, despite their anonymity, are optimized for internet accessibility, I2P represents a completely closed darknet universe.

Some advanced threat groups conduct operations requiring high anonymity on I2P rather than Tor for two reasons:

- **Highly unpredictable traffic**
  Tunnels in I2P are constantly refreshed, and their directional structure is more dispersed than in Tor.
- **Absence of a centralized index**
  While Tor hidden services can be indexed, visibility on I2P is far more limited.

For this reason, forums, marketplaces, and botnet control panels hosted on I2P are often overlooked by most CTI teams.

Remember!

Viewing I2P as a simple alternative to Tor is a major misconception. The two networks have the following characteristics:

- Serve different types of threat actors
- Exhibit different traffic behaviors
- Offer different OPSEC levels
- Host different economic activities

Decentralized Tunnel Architecture

One of the strongest yet most difficult aspects of I2P to analyze is its decentralized tunnel architecture. While Tor operates on a circuit-based model, I2P uses two separate tunnels: inbound tunnels and outbound tunnels.

Each connection is carried through different tunnels depending on direction. This structure makes it nearly impossible to track actors' real-time traffic behavior.

A lesser-known but critical feature is that I2P tunnels aren't fixed-length like Tor circuits. Tunnel length can change continuously. This significantly reduces traffic predictability and dramatically increases the OPSEC level of attackers.

For this reason, professional analysts don't limit themselves to content scanning when examining I2P; they also monitor more abstract signals such as tunnel behavior over time intervals, packet density, and directional entropy within the network.

I2P Forums, Hidden Marketplaces, and P2P File Areas

A culture distinct from Tor has developed on I2P. Communities here are smaller, more closed, and have a higher technical level. Exploit developers, crypto-theft groups, and APT groups maintaining long-term access particularly prefer I2P forums for the following reasons:

- They provide a low profile.
- Reconnaissance tools are limited.
- Communities enforce aggressive moderation to protect themselves.
- Automated crawlers are almost entirely ineffective.

Some marketplaces prefer I2P for high-value products that don't require data transfer, for example:

- Zero-day exploits
- Code execution bypass tools
- Custom-developed rootkit packages
- Botnet control infrastructures
- Chained proxy clusters

Most of these products aren't sold on general dark web marketplaces; they are distributed only to selected actors via I2P. Peer-to-peer (P2P) file areas are the least known yet most dangerous part of I2P. These areas consist of hidden nodes used by threat actors for payload distribution. Classic hash scanning is insufficient for analysts to detect them; instead, distribution timing, node density, and tunnel direction are far more critical signals.

Bridge and Outproxy Strategies

In I2P, the concepts of bridges and outproxies are more complex than in Tor because the primary goal of the I2P network isn't to access the internet, but to move securely within the network itself. Bridge strategies are used to prevent actors from directly connecting to specific I2P nodes. These strategies include mechanisms such as the following:

- Changing tunnel length
- Skipping specific nodes
- Blocking high-risk routes
- Applying overlay routing

More intriguing is the use of outproxies. In the I2P architecture, an *outproxy* is a gateway node that allows traffic originating from the internal I2P network to exit toward the regular internet, acting as a bridge between the anonymous network and external web services. By design, using an outproxy reduces anonymity within I2P; however, some attack groups do this deliberately. The reason is that when an outproxy is used, traffic can be mixed with Tor or normal internet traffic, making it more difficult for analysts to determine whether the traffic originates from I2P. This method is known as *cross-layer anonymization*.

5.3 Summary

Accessing hidden services within the dark web requires disciplined anonymity management and strong OPSEC. Networks such as Tor and I2P provide the infrastructure where threat actors communicate, coordinate operations, and sustain underground economic activities. For intelligence analysts, operating safely in these environments depends on understanding how anonymity networks function and minimizing technical signals that could expose operational presence.

Secure use of Tor relies on proper system isolation, careful circuit management, and reducing browser fingerprinting risks. At the same time, alternative networks such as I2P represent more closed and less observable environments often preferred by technically advanced actors. Understanding the structural and operational differences between these networks enables analysts to interpret threat actor behavior and extract meaningful intelligence from hidden services within the dark web ecosystem.

In the next chapter, we'll switch our focus from collecting intelligence to mapping threat actor behavior.

Chapter 6

Threat Actor Profiling and Behavioral Mapping

Truly understanding an attack requires the ability to read how the threat actor thinks: their motivations, decision-making processes, and operational methods. This chapter moves beyond classical surface-level definitions of threat actors to examine them within a holistic behavioral, technical, and analytical framework.

Threat actor profiling is a multilayered analysis process formed by combining habits hidden within lines of code, operational timing patterns, technical preferences, tendencies to make mistakes, and the logic behind target selection. An actor's operational rhythm often emerges naturally; recurring connection patterns, consistent ways of using specific command sequences, the signatures of tools they develop, or stability in their communication methods constitute the fundamental building blocks of a behavioral profile. For this reason, profiling answers where the threat came from and also sheds light on what kind of mind is behind it, and what might it do next?

In this analytical process, two globally accepted reference frameworks play a critical role: the MITRE ATT&CK matrix and the Diamond Model. MITRE ATT&CK enables the mapping of behavioral patterns by classifying the tactics and techniques applied by an attacker throughout an operation within a structured framework. The Diamond Model, on the other hand, examines the threat actor through four core components: actor, capability, infrastructure, and victim. This reveals the context of an attack through a holistic approach. In this chapter, we'll examine in detail how both models are used, how they complement each other, and how they gain meaning during an incident analysis.

Threat actors aren't static; they continuously adapt their methods, evolve, and seek new opportunities as defensive mechanisms advance. For this reason, behavioral mapping also includes tracking changes over time. The dynamic nature of profiling is formed by the evolution of techniques used by an actor, consistency across campaigns, and how past mistakes are reflected in future operations. This chapter comprehensively explains how to interpret the tactical transformations of threat groups, what behavioral anomalies signify, and how these signals should be evaluated within a long-term intelligence cycle.

One of the core focuses of this chapter is the role played by profiling methods both before and after an attack. In proactive defense, known habits of threat actors contribute to the development of security controls and the reduction of the attack surface. In retrospective analyses, behavioral indicators are indispensable for identifying potential perpetrator groups, detecting related campaigns, and reconstructing the incident as a whole.

6.1 Introduction to Threat Actor Profiling

Understanding a threat actor often requires far more than examining a single line of code they wrote, a log trace they left behind, or an infrastructure pattern they used. Profiling is the process of analyzing the internal motivation that drives them, the mental model they adopt when planning an attack, and their perception of risk. What an attacker ultimately targets isn't always the compromised system itself; sometimes it's a message, a challenge, or simply the need to make their existence felt. For this reason, threat actor profiling is one of the rare fields where technical discipline intersects with psychological depth.

An actor's objectives are often at least as decisive as the technical tools they employ. For some threat groups, the goal is solely to exert political influence or obtain strategic information. For others, it appears as an equation in which financial gain, rapid reward, and low-risk preferences converge. Target selection is directly related to the operational framework an actor sets for themselves: while a state-sponsored group may prefer critical infrastructures, financially motivated actors may target payment systems or high-volume e-commerce platforms. Hacktivist groups, on the other hand, gravitate toward institutions with symbolic value because, for them, true success lies not in the collapse of systems, but in making the message visible.

Motivation, however, is a far more passionate matter. Attacker motivation often sits at the center of persistent campaigns that can last for years. For example, in advanced persistent threat (APT)–level groups, motivation is intertwined with complex geopolitical interests and may remain unchanged for years in line with a state's strategic objectives. In contrast, the motivation of an independent cybercriminal is based on much more rapidly changing factors such as economic pressure, personal greed, or pure opportunism. Hacktivist motivations are shaped by emotional, social, and ideological components; therefore, they display an unpredictable, sudden, and fluctuating rhythm. This diversity makes the profiling process both complex and highly compelling.

The following sections cover the basics of threat actor profiling: the different types of threat actors, the underlying motivations and dynamics, and how to put it all together into a profiling strategy.

6.1.1 Types of Threat Actors

When actor categories are examined, it becomes evident that each group carries its own distinct operational DNA:

- **APT groups**
  These groups adopt a long-term, low-visibility, and disciplined approach to achieving their objectives. These actors generally possess hierarchical internal structures and conduct their operations with a military or state-level planning culture. Research, development, testing, and deployment phases are clearly separated from one another. Their traceable behavioral patterns often evolve into institutional habits shaped over many years.

- **Cybercriminals**
  Cybercriminals act with a desire for rapid results. For them, a campaign is valuable as long as it delivers a high return on investment. As a result, they frequently change their tools, easily abandon their infrastructure, and quickly attempt new attack vectors when they fail. Their behavioral patterns are more flexible, more commercial, and often more pragmatic. However, this very flexibility makes them difficult to predict.
- **Hacktivists**
  These actors form a unique category at the intersection of technical capability and ideology. Although these groups may appear to lack a consistent behavioral model due to their sometimes fragmented, anonymous, and heterogeneous structures, the nature of their motivation leads them to place far greater emphasis on symbolism and visibility. The techniques used may be relatively simple; however, the timing of their actions and target selection are driven by strong emotional triggers.

In advanced profiling efforts, it's critical to recognize that none of these categories are rigid. Situations in which an APT group outsources work to hired criminals during a campaign, financially motivated groups instrumentalize ideological rhetoric, or hacktivist structures receive technical consultancy are far more common in the field than is often assumed. Therefore, profiling shouldn't be limited to classification alone; it must be a multilayered analytical process that simultaneously evaluates context, actor evolution, and the strategic interests behind an incident.

To understand a threat actor, we often have only fragmented log lines, incomplete command fragments, encrypted connection attempts, and perhaps a few seemingly superficial behavioral traces. Yet when these scattered pieces are brought together using the right methods, they can form a surprisingly clear picture because the people behind attacks leave something that outlives technical tools: the echo of their habits. This echo reveals the enduring character of a group that may remain unchanged for years.

Some actors embed personal preferences even into their command-line usage, for example, a group consistently choosing the same variable name in its PowerShell scripts or another group displaying an obsession with a specific alphabetical ordering when configuring its command-and-control (C2) infrastructure. These are seemingly insignificant details, yet they carry evidentiary value from a profiling perspective. In the advanced stages of threat actor profiling, the significance of these traces goes beyond repeated behavior; they are micro-indicators that reflect the rhythm of the actor's mind.

6.1.2 Operational Psychodynamics

An actor's motivation may appear simple on the surface: money, ideology, competition, influence, sabotage, and so on. Yet in advanced analyses, motivation isn't a standalone concept; it's the psychodynamic driving force behind the entire chain of behavior.

In some APT groups, for example, the historical distribution of operations aligns with the internal work rhythm of the actor's own culture. It's not widely known, but the operational intensity of certain state-sponsored groups shows an inverse correlation with their national holiday calendars. When holidays begin, operations don't stop; on the contrary, the core part of the team continues working, and operations carried out with fewer people often become more creative and less predictable. This phenomenon provides a critical clue in correlating many incidents.

Financially motivated groups, on the other hand, often display impatient and consumer-like structures that are reflected even in technical traces. For instance, when a criminal organization edits the configuration files of the malware it uses in a random add-remove manner rather than following a systematic approach, this indicates a structure that relies more on opportunistic gains than on strategic planning.

Let's dig into these motivations, or operational psychodynamics, in more detail.

Habit Analysis: Unconscious Signatures

Threat actors unconsciously leave behind a unique behavioral signature. This signature is visible in code, work style, timing, error patterns, and even the way files are packaged. The following examples are abstracted from real cases and anonymized:

- **Operational timing: the "silent hours" signature**
  When the attack intensity of an APT group is examined over several years, most operations are found to concentrate within specific time windows. What is notable is that these hours don't correspond to the target countries' timelines, but to the "quiet hours" of the actor's own internal geography. This indicates that the group conducts operations with a small core team rather than a large workforce. Such data doesn't directly reveal geographic location, but it indirectly exposes it through behavioral indicators.
- **Code reuse: a developer's fingerprint**
  In the malware used by some actors, a unique string-formatting style can be observed even when the technical structure changes. For example, two meaningless characters were observed appended to the end of timestamps in log outputs and reappear across different tools written by the same developer. This constitutes the *style trace* of a specific individual within the group, and if that person moves to a different team, the internal profile of the new group can also be identified.
- **The language of the infrastructure ecosystem**
  Some threat actors unconsciously gravitate toward specific linguistic patterns when choosing domain names. For example, financially motivated crime groups more frequently use short, aggressive, imperative words, whereas ideological groups tend to favor symbolic or mythological references. These references sometimes form a consistent thematic pattern across multiple campaigns.

Strategic Decision-Making Patterns of Actors: The Invisible Logic

In advanced profiling, both the technical capability of actors and the logic behind their decision-making processes are analyzed. In some APT groups, operational mistakes are almost nonexistent; for such groups, making an error is a flaw that may carry political consequences.

In contrast, within certain hacktivist collectives, the "culture of error" is far more relaxed. Because visibility is the primary objective, an operation may be considered successful as long as sufficient impact is achieved, even if it's technically flawed.

In financially motivated groups, operational logic is almost always tied to the investment–return balance. The quality of the infrastructure used by a group progresses in direct proportion to the financial gain it expects. For this reason, even infrastructure maintenance behaviors constitute an important part of the profile.

6.1.3 Indicators, Context, and the Right Question

One of the most critical steps in threat actor profiling is asking the right question. Technical analysis often answers the "how" question. However, the fundamental question a threat actor analyst must evaluate is "Why was this method chosen?" This can be framed in a few different ways:

- Why did an actor use the same exploit for months?
- Why did they repeatedly patch and redeploy the same C2 infrastructure?
- Why did they choose a more difficult path when an easier opportunity existed?

Each of these questions points to the actor's mental model. Some APT groups place extreme importance on operational security (OPSEC); others prioritize operational speed. Some financial groups conduct attacks rapidly even at the cost of detection, whereas others refuse to take risks even for minimal profit. These differences form the foundation of behavioral profiling.

In the cybersecurity world, it's a common misconception to view every group as continuously evolving and constantly changing. In reality, many threat groups use nearly the same techniques, the same tools, and the same operational logic for years due to operational culture.

Some groups perceive change as a risk. Even if detection rates increase due to the lack of experimentation with new techniques, internal processes preserve structures with which they are comfortable. Conversely, some groups constantly pursue innovation; these groups are harder to detect, but their behavioral profiles are more distinct because even innovation requires a certain internal rhythm.

6.2 Tactics, Techniques, and Procedures

What a threat actor does is often visible to everyone; however, how it's done tells a story that only those who look carefully can understand. Tactics, techniques, and procedures (TTPs) are used to read this story. A tactic defines the framework of the objective the actor seeks to achieve; a technique reflects the method chosen to realize that objective; and a procedure represents the individual style in which the chosen method is executed. In a sense, TTP is both the actor's fingerprint and their signature logbook.

Tactics may change and techniques may be updated, but procedures—the way the work is actually carried out—carry the most unpredictable and raw signals about the actor's true identity. For this reason, TTP analysis is more than mapping the technical stages of an attack; it's reading the attacker's mental model, focus patterns, impatience, discipline, and even personal habits.

A TTP pattern can't be understood from a single incident. Each incident is just a snapshot of the actor. To truly understand a threat actor, one needs a film composed of those frames arranged sequentially. TTP mapping is the method for revealing that film.

Some threat groups repeatedly use the same tactical sequence over years. For example, the order of target reconnaissance → authentication bypass → persistence → data collection → exfiltration functions almost like a cultural standard for many APT groups. When this sequence changes, analysts often interpret it as a new capability, a team change, or a new strategic direction. Thus, TTP isn't just a technical list; it's the outward expression of an organization's internal dynamics.

Note on Attack Chains

Some financially motivated crime groups dramatically shorten their attack chains due to psychological pressure: the desire to obtain money faster produces shorter chains. In contrast, sophisticated espionage groups build long, meticulous, and patient chains because, for them, the most valuable outcome is remaining undetected.

Now, let's unpack each TTP component and see what analysis techniques can be applied.

6.2.1 What Are TTPs?

A *tactic* is never chosen at random. Every tactic is an outward reflection of the actor's internal strategy, for example:

- If an actor always begins with credential theft, this indicates that the group successfully leverages social engineering.
- Some groups prefer exploit usage because their technical capabilities are high and their OPSEC supports it.
- Others choose low-noise paths, such as abusing only legitimate tools (e.g., Living Off the Land Binaries [LOLBins]). This is a deliberate choice to remain below defensive radar.

Tactics point to the organization's way of thinking at an institutional or collective level. Therefore, when a group's tactical repertoire changes over time, this shift is often not a technical necessity but a sign of a strategic rupture.

Techniques are the methods an actor selects to carry out an attack. The same tactic can be implemented with entirely different techniques by different actors. Technique is the first true mirror reflecting an actor's capability level, for example:

- APT-level groups typically use zero-day vulnerabilities, custom-written exploit sets, or multistage complex chains.
- Cybercrime organizations are cost-oriented; they prefer rented exploit kits, modified remote access trojans (RATs), or customized versions of open-source tools.
- Hacktivists often use off-the-shelf tools, focusing on visibility and noise rather than technical sophistication.

In TTP analysis, examining techniques can reveal an actor's internal structure. A specific loop structure within a piece of code may point to a particular developer within a group. For this reason, technical analysis can be used to both resolve incidents and perform actor identity matching.

A *procedure* is the way an actor performs a technical action. When viewed in isolation, it may seem insignificant; however, in profiling, it's the most decisive element. Each of these constitutes the actor's behavioral DNA:

- The compression ratio a group uses when packaging payloads
- Variable-naming habits in PowerShell
- The ordering of specific parameters in C2 configuration files
- The level of rigor applied in log-cleaning methods
- The timing intervals used when executing commands in the target environment

There's a fact that isn't widely known but is used by advanced threat analysts: procedures change less than techniques. This is because procedures are learned operational reflexes. An actor may learn new techniques and update tactics; however, the way work is carried out often remains constant. For this reason, procedure analysis is a powerful correlation tool.

6.2.2 TTP Analysis

TTP analysis examines how threat actors transform TTPs into repeatable operational behavior patterns. Rather than focusing solely on isolated technical artifacts, it seeks to understand the structured logic behind adversary actions and the consistency of their operational methods. In this way, TTP analysis provides the analytical bridge between observable technical indicators and the broader behavioral profile of the actor. We'll discuss different methods of TTP analysis in the following sections.

Advanced TTP Correlation: Profiling Through Timelines

TTP mapping is used to determine which technique was used as well as when, how often, and under which conditions. In practice, this mapping is performed by correlating observed behaviors with standardized frameworks such as MITRE ATT&CK, documenting each technique across incidents, and aligning the associated timestamps and contextual indicators within a chronological dataset.

These rhythmic data discreetly reveal the following details about the actor:

- Internal team structure
- Operational tempo
- Workload distribution
- Weekly/monthly rhythm
- Campaign model

Example: Operational History

Following is a threat group's operational history spanning six years:

- Attacks are concentrated between the 23rd and 28th of the month.
- No attacks occur on Mondays.
- Low volume is noted on Friday evenings.
- Minimal activity occurs between 02:00–05:00.
- New techniques are typically introduced in the first quarter of the year.

When analyzed, this data can reveal the operational team's internal working rhythm, holiday cycles, and even staff changes. Such rhythmic analyses are applied by very few analysts worldwide because only timestamps are visible on the surface. Yet behind those timestamps lies the actor's work culture.

TTP Analysis from a Defensive Perspective

TTP analysis isn't used solely for post-incident correlation. When read correctly, it enables defense to shape the future, for example:

- If an actor's lateral movement techniques are limited, adjusting segmentation strategies can blunt the actor's ability to advance.
- If a group is known to rely consistently on the same credential-harvesting method, multifactor authentication (MFA) policies can directly target that weak point.
- Some groups depend on specific exploits; when those exploits are pushed out of the market, the group's impact decreases dramatically.

The real value of TTP analysis lies in forcing defense to compel the attacker to change structure. A threat actor's TTP set is more than an inventory of attack techniques. When analyzed correctly, it conceals the actor's operational maturity, internal communication style, access to resources, and even technical debt. Examining TTP patterns can sometimes provide stronger attribution than pages of indicators of compromise (IOCs). IOCs are easy to change, but behavior is the shadow of the human—it's harder to change and nearly impossible to hide.

TTPs as a Gateway to a Group's Internal Organizational Structure

From the outside, many threat groups appear as a single entity; however, internally there are roles, areas of expertise, decision-makers, testing teams, operational executors, and individuals responsible for infrastructure. This internal structure doesn't directly manifest outwardly, but TTP analysis can reveal this invisible organizational model.

Example: Deconstructing Internal Structure Through Tactical Flow

If a group's attack chain proceeds as follows:

- The target scanning phase is fast and aggressive.
- The credential collection phase is highly systematic.
- The lateral movement phase is slow and controlled.
- The data exfiltration phase uses high bandwidth.

This structure clearly indicates the following:

- Reconnaissance team operates aggressively: Typically inexperienced or automation heavy
- Credential collection is highly orderly: Presence of a highly skilled red teamer internally
- Lateral movement is slow: High OPSEC awareness
- Exfiltration is high-volume: The person responsible for infrastructure has a high-risk appetite

In other words, the rhythm of an attack is the pulse of the internal organization that carried it out. This analysis method is used by very few threat intelligence analysts because the connection between TTPs and organizational structure isn't intuitive at first glance. Yet this connection lies at the core of advanced profiling.

Micro-Errors in Procedures

The most critical point in advanced TTP analysis is identifying the traces of micro-errors left by the actor. These errors are much more than technical weaknesses—they are psychological windows.

Example: Unnecessary Spaces

In most of a group's C2 configurations, the parameter order follows a consistent logic, yet there's an unnecessary trailing space at the end of the `fallback_server` parameter. This space appears across three separate campaigns and in two different malware families. This implies the following:

- The same person is editing this section.
- That person most likely uses copy-and-paste.
- The operational step may be a last-minute edited task.

In other words, that unnecessary space is an anonymous yet unique signature belonging to a specific individual within the group. Such procedural anomalies are like unintentional handwriting left behind by actors.

TTP Evolution

In many analyses, changes in TTPs are commonly interpreted as they learned new techniques, but in reality, the situation is far more internal. A group's TTP evolution is often the result of invisible variables such as the following:

- Internal staff changes
- Increases or decreases in funding
- Political context
- Development capacity
- External pressure
- Operational fatigue

Real-World Example: Identifying Internal Team Rotation Through TTP Changes

When the operations of an APT group between 2020–2024 are examined, the following is revealed:

- 2020–2021: Heavy obfuscation + custom-developed exploit chains
- 2022: Obfuscation quality drops, exploits weaker
- 2023: Return to high quality
- 2024: Mixed quality, high variation

On the surface, this may look like technical inconsistency, but it actually points to something much deeper:

- There may have been a developer change within the team in 2022.
- A previous or more capable team may have returned in 2023.
- The team may have expanded in 2024, leading to stylistic diversification.

Such details are never written explicitly in reports; yet through TTP analysis, it's possible to detect even internal human-resource fluctuations within a group.

Reverse Reading of TTPs Through Target Countries

In TTP analysis, the usual approach is actor to target. Expert-level intelligence analysts also do the reverse by deriving meaning from target to TTP. To uncover this relationship, analysts examine a set of guiding questions that connect technique selection with geographic and operational context:

- Why did an actor choose this country?
- Why was this target attacked using these techniques?
- Why does this tactic repeat in specific geographies?

Example: Geography-Driven TTP Variation

If a threat group uses credential harvesting + LOL techniques against Latin American countries, but prefers exploit chains + stealth exfiltration in Europe, this reveals the following:

- Defenses in Latin America are weaker. The actor prioritizes rapid credential harvesting and quick system access before defensive monitoring and response mechanisms can react.
- Defenses in Europe are stronger. The actor prioritizes stealthier techniques and slower, low-visibility data exfiltration to remain undetected for longer periods.
- The same group has two different behavioral profiles for two continents.

This indicates that the group's operational planning uses a geography-dependent risk model. Such analyses play a critical role in predicting an actor's strategic objectives.

Temporal Flow in TTP Correlation

Reading an attack chain from top to bottom is easy. The difficult part is understanding its breathing rhythm. In some attacks, the following are true:

- Reconnaissance is very fast.
- Initial compromise is slow.
- Persistence is established hastily.
- Data exfiltration is extremely systematic.

This rhythm can reveal human parameters such as whether the actor was in a hurry, being cautious, experiencing panic at a certain moment, using automation, or performing manual actions.

This type of analysis is applied only in a limited number of highly advanced threat intelligence centers because it sits at a complex intersection where nontechnical human behavior meets technical action.

Predicting the Future Through TTP Analysis

How a threat actor will attack in the future can be inferred from past techniques. However, advanced prediction doesn't rely solely on technical traces; it focuses on the actor's behavioral model, for example:

- If an actor rapidly adopts new techniques, the likelihood of using aggressive new methods under defensive pressure is high.
- If an actor persistently uses old techniques, technical debt is high and the group may quickly become ineffective against modern defenses.
- If a group's procedures haven't changed for two years, operational stability is very high and the group may be state-sponsored.

Such inferences provide high accuracy in threat assessment models.

Just as a person's body language reveals what words don't say, a threat actor's TTPs convey character, rhythm, and a mental world beyond technical details. For this reason, TTP analysis isn't merely a component of cyber threat intelligence (CTI); it's the digital behavioral anthropology of the actor.

6.3 Applying the MITRE ATT&CK Framework

CTI is a vast universe in which events that appear disconnected on the surface are in fact quietly linked to one another. Transforming this universe into a meaningful and traceable structure isn't possible with technical tools alone; it requires a strong, systematic, and behavior-focused perspective. MITRE ATT&CK emerged precisely to meet this need. For many people, MITRE ATT&CK looks like a large table listing tactics and techniques, but this table is just the visible layer of a much deeper framework.

The true value of MITRE ATT&CK lies in turning every step taken by an attacker—from the initial point of contact to the final data exfiltration—into a mental journey. Being able to see why an actor chose a specific technique at a specific stage offers a unique opportunity to understand how that actor thinks. In this sense, MITRE ATT&CK is far more than a way to trace an attack's footsteps; it's a guide for reading the behavioral anatomy of a threat actor.

One of the framework's most important contributions isn't breaking an attack into pieces, but revealing the relationships between those pieces. Attacks are never composed of a single technique; each technique prepares the next and is the consequence of the previous one. Analysts with field experience see more than tactic names when they look at the MITRE ATT&CK matrix; they also read what the actor prioritizes, where the actor behaves more cautiously, where risks are taken, and the tempo of the operation itself. This transforms the framework from a static list into a living model of thought.

Using MITRE ATT&CK effectively doesn't mean memorizing the techniques listed in the table. The real skill lies in understanding where those techniques sit within an actor's operational habits. Which techniques a group has relied on for years, which it has never preferred, and which it has recently adopted all reveal the group's experience, team knowledge, and even its internal dynamics. When applied correctly, MITRE ATT&CK improves defensive strategies and provides a comprehensive behavioral map that follows a threat actor step by step.

In the following sections, we'll discuss the MITRE ATT&CK techniques and subtechniques, as well as apply the framework in various scenarios.

6.3.1 Techniques and Subtechniques

The techniques used by threat actors may appear, at first glance, as independent steps—a command line execution, a Windows Management Instrumentation (WMI) call, a credential extraction operation. In advanced analysis, however, each of these techniques becomes a node within the actor's behavioral structure. When these nodes are brought together on the same map, the attacker's operational logic becomes almost tangible. Seeing which clusters of techniques an actor feels comfortable with opens a window into how that actor thinks.

For some actors, revolving around certain techniques is almost reflexive. For example, credential dumping techniques (T1003.x), command interpreters (T1059.x), remote execution mechanisms (T1047), or Server Message Block (SMB)–based lateral movement techniques (T1021.1) may form a group's technical comfort zone. To an analyst, this zone isn't a collection of randomly scattered moves. On the contrary, it's the outward expression of years of accumulated knowledge, habits, team experience, and operational culture.

For this reason, along with being an attack step, an actor's choice of techniques is also an indicator of style. The choice of subtechniques, in particular, reveal a great deal because it reflects the preferences of the technical lead within a group: how they write code, how they test, which tools they favor, and which methods make them feel more secure. In short, subtechniques are subtle behavioral signals that reflect the folds of the actor's mind.

For an analyst trying to understand a threat actor, these signals are invaluable. If techniques are words, subtechniques are the actor's accent. Intonation, preference, and rhythm remain consistent; moreover, they form a silent language that conveys a group's internal dynamics to the outside world.

The following examples illustrate a small subset of techniques commonly observed in real-world intrusion campaigns. They are presented not as an exhaustive list, but as representative cases that demonstrate how specific techniques and their subtechniques can reveal an actor's technical habits, operational preferences, and behavioral signatures.

T1003.x: Credential Dumping

Credential dumping is in the first-choice category for many threat actors due to the following characteristics:

- Sits at the center of privilege escalation chains
- One of the fastest ways to gain value in a target environment
- Becomes a critical building block for establishing persistence on the target system

Advanced actors test Local Security Authority Subsystem Service (LSASS) access not with a single tool, but through multiple variants, for example:

- Common tools such as Mimikatz are considered low OPSEC because they are widely known, heavily documented, and frequently detected by modern security solutions, making their use easier to identify during incident response.
- Custom memory scrapers typically provide higher OPSEC because they are developed specifically for the operation, allowing actors to modify code, evade known detection signatures, and reduce the likelihood of tool-based attribution.
- Minidump combined with scripted extraction techniques represents a midlevel stealth approach, as it relies on legitimate system functionality but still leaves recognizable forensic traces during memory dumping.
- Cobalt Strike Beacon built-in modules can provide a relatively quiet but effective method because they are designed to integrate credential access capabilities directly within the post-exploitation framework, reducing the need to introduce additional external tools into the environment.

For some actors, the credential dumping procedure is so fixed that even the order in which tools are executed becomes predictable. This is a powerful indicator for linking related campaigns.

T1059.x: Command and Scripting Interpreter

This technique family is one of the areas where the character of threat groups is most clearly revealed. The choice of subtechnique can expose the following:

- The actor's scripting expertise
- The importance they place on operational speed
- Whether they develop their own tools
- How sensitive they are to defensive detection mechanisms

Example behavioral indicators are as follows:

- Preference for single versus double quotes in PowerShell commands
- Block lengths used in Base64 encodings
- Bash variants preferred over Python
- Eval-heavy JavaScript code indicating aggressive and fast campaigns
- More structured, modular script design indicating the presence of a skilled developer

These small microtraces within such techniques are like a threat actor's digital handwriting.

Example: Actor Differentiation Through Different Implementations of the Same Technique

Command and Scripting Interpreter (T1059) is a very commonly used parent technique. However, the way its subtechniques are implemented carries strong signals about an actor's identity. Consider the following two scenarios:

- **Actor A – T1059.001 (PowerShell)**
  - Always prefers single quotes in command chains
  - Base64 encoding always has fixed block length
  - Obsessively leaves a semicolon at the end of the final line
- **Actor B – Same subtechnique**
  - Commands are predominantly double quoted
  - Encoding length varies
  - Fills logs with random characters for evasion

MITRE ATT&CK simply states "PowerShell was used." But the difference between these two implementations forms the actors' respective "handwriting." This information is invaluable for real-world incident analysts.

T1047: Windows Management Instrumentation (WMI) Execution

WMI is one of the favorite techniques of advanced threat groups because of the following characteristics:

- Can execute remote commands without leaving logs
- Low-noise method for lateral movement
- Legitimate system component means detection becomes more difficult
- Integrates easily with persistence mechanisms

For advanced analysis, what matters most is how WMI is used:

- The use of WMI event subscriptions often indicates persistence, as attackers can register permanent event consumers that automatically execute malicious code when specific system events occur.
- Standalone WMI command execution is commonly associated with fast operational tasks because it allows attackers to run remote commands quickly without deploying additional tools to the target system.
- Chained operations executed through the WMI command-line (WMIC) often suggest a level of automation, as actors may link multiple WMI commands together within scripts to perform repetitive administrative or lateral movement tasks.
- Consistently using the same parameter ordering in WMI queries can serve as a behavioral signature of a specific actor because such patterns often reflect habitual scripting practices or reused code across multiple campaigns.

Some APT groups even build WMI namespace structures in exactly the same way. This is an excellent indicator for identification.

T1021.1: SMB/Windows Remote Services

SMB-based remote execution works with built-in tools, requires no additional software, enables fast lateral movement, and is extremely effective in domain environments. What distinguishes actors who use this technique is usually their credential usage style:

- Some groups follow a *command → control → return* model.
- Some repeatedly query the same SID structure.
- Some prefer only administrator accounts.
- Some always test access in the same order (`IPC$`, `ADMIN$`, `C$`).

Techniques executed over SMB directly reveal an actor's level of knowledge within the domain and their risk appetite.

> **Remember!**
>
> Every technique observed in an attack (the chosen subtechnique, the variable names used, a small parameter order in a WMI call) reflects a fragment of the actor's inner world and provides a behavioral signal.

6.3.2 Applications of MITRE ATT&CK

MITRE ATT&CK is widely recognized as one of the most comprehensive frameworks for understanding adversary behavior. It provides analysts with a structured language to describe how threat actors operate across different stages of an intrusion. When used systematically, as we'll discuss in the following sections, MITRE ATT&CK enables analysts to reveal patterns, behavioral consistencies, and operational strategies in observed activities that would otherwise remain hidden.

Combining MITRE ATT&CK with the Time Dimension

MITRE ATT&CK isn't a static list; when a time dimension is added, it becomes a far more powerful tool. When a threat actor's campaigns span multiple years, their internal evolution can be traced through technique selection.

Let's examine an insight abstracted from real incidents:

- 2018: T1133 (External Remote Services) dominant
- 2019: T1190 (Exploit Public-Facing Application)
- 2020: Increase in exploit chain complexity
- 2021: Significant increase in LOL techniques

- 2022: Growth in timestomping and log manipulation techniques
- 2023: Command execution almost entirely WMI-based

This sequence isn't random. It reflects efforts to increase OPSEC, adaptation to defensive pressure, growth of internal team capacity, and the onboarding of new developers. MITRE ATT&CK mapping makes these internal transformations visible.

Threat Actor Attribution with MITRE ATT&CK

Many advanced security teams believe MITRE ATT&CK is used only in analytical presentations. In reality, MITRE ATT&CK is a unique tool for forensic attribution. When an actor's campaigns are mapped, the following converge into a single table:

- Technique clusters used
- Order of techniques
- Subtechnique preferences
- Density of variations
- Operational time windows

This table is often a stronger linkage mechanism than hashes, IP addresses, or domain chains because MITRE ATT&CK maps behavior, not IOCs. As noted earlier, behavior is the hardest thing to change.

Capability Gap Analysis Using MITRE ATT&CK

Deriving capability gaps from a threat actor's MITRE ATT&CK map is a rarely used method worldwide. What does this mean? The set of techniques an actor *does not* use reveals weak points in their capability map. Some groups are strong in lateral movement but weak in privilege escalation. Others are highly creative in data exfiltration but mediocre in persistence techniques. Identifying these gaps must not be overlooked because defensive strategy can be built directly on top of them. This method is used especially by national cyber defense teams, but it rarely becomes public.

MITRE ATT&CK from a Defensive Perspective

Many organizations use MITRE ATT&CK to understand the attacker; however, its true power lies in redesigning defense. When the MITRE ATT&CK table is inverted and used as a defense-gap analysis tool, the following questions are used:

- Which techniques can't be detected?
- Which subtechniques produce weak logging?
- Under which tactics are there no controls at all?
- In which behavior chains are no alerts triggered?

The answers appear plainly. This is the most objective representation of defensive weakness. What is lethal for a threat actor is a defense that memorizes the actor's MITRE ATT&CK rhythm. Once you read the player's moves, the game is over.

6.4 Using the Diamond Model in Threat Profiling

Although there are many methodologies in cyber threat analysis, most of them make only a specific facet of an attack visible. The Diamond Model breaks this limited perspective and enables the analyst to perform a four-dimensional reading: adversary, capability, infrastructure, and victim. When these four elements are brought together, not only how an attack occurred but also why it occurred in that way, by whom it was conducted, and against whom it was executed becomes clear within a holistic framework.

Applying the Diamond Model at a truly advanced level isn't about examining these four elements separately and moving on; it requires understanding the rhythm, direction, intensity, and continuity of the relationships between them. The real power of the model lies in reading how these relationships move and interact within the attack ecosystem.

We'll unpack the four dimensions of the Diamond Model in the following sections.

6.4.1 Adversary: The Entity Conducting the Attack

The most dynamic and complex component of the Diamond Model is the adversary. An adversary could be the name of an APT group, a cybercrime organization, or a hacktivist collective. In advanced threat intelligence, however, the adversary is a dynamic center that makes decisions, defines strategy, selects techniques, and determines the operational rhythm. In the hands of an analyst, the adversary isn't a static label, but a living construct; one whose behavior constantly changes, has its own internal culture, and sometimes is predictable, yet often requires intuitive interpretation.

An adversary's behavior is never static. When campaigns conducted by the same group across different years are examined, changes in operational speed, deviations in technical preferences, transformations in infrastructure usage, or increases in error rates all demonstrate that the adversary evolves over time. Some adversaries seek rapid results; their attacks are short-lived, they don't shy away from risk, and they see detection as a natural part of the operation. Others proceed cautiously; their campaigns may last months or even years, and remaining undetected becomes a core success criterion. This difference is one of the clearest ways to understand an adversary's internal culture.

The true character of an adversary can be read along five main axes:

- **Motivation**
  An adversary's motivation gives meaning to the entire operation chain. Motivation may be shaped by political objectives, financial expectations, ideological orientations, the desire to gain competitive advantage, or state-directed strategic interests.

Motivation explains why the adversary attacks and reveals the invisible logic behind target selection.

- **Risk tolerance**
  Each adversary approaches risk differently. Some prefer rapid progress using high-noise techniques; for these actors, detection is an acceptable cost. Others have low risk tolerance; they conduct complex, silent, multistep operations and treat even the smallest mistake as a reason to halt the operation. Risk tolerance directly determines the adversary's operational style.
- **Operational culture**
  Operational culture is the outward reflection of an adversary's internal discipline. This culture invisibly determines the following:
  - Which tools are preferred
  - How techniques are applied
  - How infrastructures are managed
  - How roles are distributed within the team
  - What types of errors occur during operations

  Some groups have a culture that is visibly structured and orderly; others display a scattered, hasty, or amateur rhythm.
- **TTP continuity**
  An adversary maintaining the same TTPs over years—or repeatedly using them with limited variations—is one of the strongest indicators of a group's technical identity. This continuity provides analysts with a critical advantage in linking campaigns. Breaks in TTP continuity, on the other hand, may indicate internal team changes, the addition of new developers, operational pressure, or accumulated technical debt.
- **Technical capacity level**
  An adversary's technical capacity can't be inferred solely from the sophistication of the exploits it uses. Code quality, preference for modular design, level of obfuscation, persistence mechanisms, error rates, and the technical responses given to obstacles encountered during operations all provide deep signals about capacity. High-capacity adversaries rapidly adopt new techniques, whereas low-capacity ones remain dependent on the same tools and procedures.
- **Internal team structure**
  An adversary's internal structure is often invisible, but behavioral analysis can reveal how the team is organized. Some groups have multilayered structures that include the following:
  - Separate reconnaissance teams
  - Dedicated exploit development units
  - Operational executors
  - Infrastructure managers

In more amateur groups, all tasks may be handled by a single individual, resulting in a much more variable operational rhythm. Team structure directly affects TTP continuity and the likelihood of errors.

In the Diamond Model, the adversary is the determining force that permeates all three of the other elements for the following reasons:

- The adversary defines the capability set.
- Infrastructure usage style reflects the adversary's culture.
- The victim profile reveals the adversary's motivation.
- The TTP rhythm expresses the adversary's operational behavior.

For this reason, understanding the adversary isn't only the starting point of the Diamond Model; it's the key to understanding the entire model.

When an analyst resolves the adversary, it becomes far easier to predict the group's future actions, see where attack chains are heading, and comprehend the relationships among the other elements.

6.4.2 Capability: The Tools, Techniques, and Knowledge in the Adversary's Hands

Capability defines the real limits of an adversary. Some threat groups can use zero-day vulnerabilities; others rely on basic credential phishing. Capability determines what is possible, regardless of what the group desires.

Reading this element at an advanced level isn't done by just looking at the tools used. To understand a group's capability, the following questions are asked:

- Do they develop their own tools?
- Do they customize open-source tools?
- Do they use modular architectures or monolithic attack chains?
- Is code quality consistent?
- How quickly can they adopt new techniques?
- Is their OPSEC level high?

The true strength of capability reflects the technical maturity of the adversary's mindset. For example, if a group has been using the same exploit chain for years, this means one of two things: either their technical debt is high, or the method they possess is still effective in the geography they target.

6.4.3 Infrastructure: The Invisible Backbone Carrying the Attack

For many analysts, infrastructure consists only of IP addresses, domains, and C2 servers. In advanced profiling, however, infrastructure is one of the most critical components defining the adversary's operational style. Some groups rely on infrastructures such as the following:

- Multilayer proxy chains
- Geographically distributed virtual private server (VPS) networks
- Rented malware-hosting providers
- Tor combinations
- Custom-configured C2 protocols

Others use more scattered, disorganized, and inexpensive infrastructure. This alone provides major clues for understanding the adversary's motivation and operational quality.

One reality is that there are actors who use the same Autonomous System Number (ASNs), the same domain registration patterns, and the same certificate behaviors for many years. Such habits show that even when victims change, infrastructure usage style remains consistent. For analysts, this is a powerful correlation mechanism.

6.4.4 Victim: The Target of the Attack and the Reflection of the Profile

The victim element is often perceived as a passive component. In reality, victim selection is one of the strongest indicators reflecting the adversary's strategic focus. The victim profile can reveal the following:

- Adversary's target geography
- Sector dependency
- Economic interest model
- Political orientation
- Operational scope
- Professional capacity

If an adversary focuses exclusively on the financial sector, even with high technical capability, this indicates that their strategic objective is money. Some groups attack only government institutions, which points to a political or intelligence-driven structure.

For analysts, the most critical detail is that victim selection is never random. In the Diamond Model, the victim element exposes the adversary's most naked intent.

6.4.5 Diamond Model Relationships

The true power of the model emerges from the dynamic relationship network among the four elements of adversary, capability, infrastructure, and victim:

- **Adversary → Capability**
  This is the complete toolset the adversary develops or acquires to achieve its objectives.

- **Adversary → Infrastructure**
  The adversary's OPSEC, risk perception, and discipline are directly reflected in the infrastructure.
- **Adversary → Victim**
  Target selection is the technicalized form of motivation.
- **Capability → Infrastructure**
  As capability increases, infrastructure becomes more complex; low-capability groups usually use simple infrastructure.
- **Infrastructure → Victim**
  Some adversaries consistently use the same infrastructure fractions in specific geographies. This relationship isn't limited to the origin of the attack; it indicates the area where they gain advantage.
- **Victim → Capability**
  The target's defense level influences the adversary's capability selection. When a target environment has relatively weak defensive monitoring, adversaries often rely on simpler and quieter techniques that allow them to gain access quickly without needing complex attack chains. When a target environment has stronger security controls and detection capabilities, adversaries tend to employ more sophisticated technique chains, combining multiple methods to bypass defenses and maintain access.

Each relationship reveals a portion of the adversary's mental map.

High-level threat intelligence analysts use the Diamond Model not just to draw relationship maps but to extract the adversary's rhythm. This rhythm consists of the following:

- Which victims are attacked and when
- Which techniques are used at what intensity
- How frequently infrastructure is refreshed
- During which periods operational silence occurs
- Which tactics yield high success rates

Once this rhythm is extracted, the adversary's future attack tendencies can be predicted with high accuracy. This method can be applied by very few people in practice because it requires combining technical data with behavioral patterns. However, when applied correctly, the Diamond Model begins to show both who the adversary is and what they will do next.

This four-element framework transforms an attack from a simple incident into a complex structure explainable through the triad of human behavior, technical capability, and strategic objective.

6.5 Behavioral Indicators and Fingerprints

The technical traces left behind by threat actors can fade over time; IPs are shut down, domains are abandoned, and malware is repackaged. However, there are certain traces that are repeated unconsciously, no matter how careful the actor is. These are *behavioral indicators*; that is to say, silent traces that reveal the unseen aspects of an attacker. They expose a group's technical capability, decision-making style, operational discipline, and internal team habits through deep signals.

As we'll see in the following sections, these indicators manifest themselves in the structure of code, word choices, development rhythm, tool usage preferences, timing patterns, and the form of OPSEC mistakes. For this reason, behavioral traces are far more valuable than technical IOCs. IOCs can be changed, but behavior is difficult to change.

6.5.1 Code Reuse

One of the most valuable signals encountered when examining a threat actor's codebase is code reuse. Although it may appear at first glance to be a simple habit, these repetitions are often silent signatures that reveal the actor's identity. Every developer, whether consciously or not, produces code within certain cognitive patterns. A person's way of thinking, problem-solving approach, and technical reflexes are inevitably reflected in the lines they write.

Analyses conducted over many years have shown that even the most sophisticated threat actors repeatedly exhibit their own distinctive stylistic traces. These traces go beyond determining which programming language is being used to reveal the mental rhythm of the person who produced the code.

For example, some groups, often unknowingly, behave in the following ways for years:

- Prefer the same loop patterns
- Never abandon a particular variable-naming scheme
- Copy function skeletons into other projects
- Follow their own distinctive structure in string formatting
- Place unnecessary whitespaces according to a specific logic
- Use the same exception blocks for error handling

These patterns are like a developer's handwriting, or a digital fingerprint. Therefore, such traces play a critical role both in correlating threat actors and in identifying individuals within a group.

According to anonymized findings obtained from the real world, even across different malware developed by some APT groups over the years, the same microscopic preferences recur. Examples include the following:

- Chronically recurring two lines of unnecessary whitespace in a JSON parsing function
- The use of a specific prefix in string processing functions
- The persistent continuation of the same faulty formula in certain mathematical operations
- A particular try/catch model used in command execution functions
- Rhythmic gaps between code blocks that the developer unconsciously leaves behind

These small details are often not even noticed by the threat actors themselves. However, for an analyst—especially an intelligence professional performing behavioral profiling—these traces are strong signals proving that the same individual reappears across different campaigns.

Code reuse also provides clues about a group's internal structure. If the same patterns are found across multiple malware samples, it can indicate the following:

- There's a single core developer within the team.
- Developers share the same style guide.
- The codebases are derived from a common source repository.
- The group's tool-development culture hasn't changed much over the years.

Code reuse isn't a matter of a threat actor's laziness, haste, or habit. These repetitions show how they think while producing software, what reflexes they possess, and even within which cognitive boundaries they operate. Therefore, these traces are one of the most valuable behavioral components of threat actor profiling.

Example

When five different pieces of malware used by an actor in different years are analyzed, you notice that the following has never changed within a JSON parser function contained inside them:

- Two unnecessary blank lines
- A single space left before every } character
- The same two-character prefix used in parameter naming

Such microtraces are almost biometric in nature for distinguishing the actor from others.

6.5.2 Linguistic Patterns

Code is a linguistic work, along with being a technical product. Certain patterns can also be found in a threat actor's use of language. These patterns manifest within software and in the following:

- Phishing emails
- Infrastructure configuration files
- C2 command syntax
- Log-tampering tools
- Debugging texts hidden inside payloads

Linguistic indicators examined by advanced analysts include the following:

- Recurring word choices
- Punctuation habits
- Consistency of incorrect grammatical structures
- The tone of comments used within code
- Linguistic motifs embedded in infrastructure domain names
- The rhythmic structure of C2 commands
- Behavioral choices in prefix/suffix usage

Some threat groups have been observed maintaining a specific phonetic order in their domain names. This points to the rhythm of the person behind the domain generation algorithm. This is evidence that language ceases to be a technical element and becomes an inner voice the actor is unaware of.

6.5.3 OPSEC Errors

No matter how advanced, disciplined, or well-equipped a threat actor may be, at some point, errors inevitably emerge as a result of being human, which is where OPSEC errors gain meaning. The attacker's entire effort is to conceal their identity, obscure their infrastructure, mask their behavior, and leave minimal traces behind. Yet a flawless operation is never truly possible. Even the most professional actors unknowingly create a weak link. When properly analyzed, this weak link turns into a powerful profiling component that reveals the actor's true nature.

On the surface, most OPSEC errors appear to be very small and insignificant. Yet in the hands of an analyst, they become one of the most direct ways to recognize an actor. For example, a forgotten debug string inside a payload may look like a very simple oversight. However, the word structure of this string, the language used, habitual character sequences, or the formatting of the line may contain important clues about the style of the person who wrote the code.

Another common mistake is the use of an incorrect timestamp in a C2 configuration file. Such a timestamp may indicate a disconnect between the planning phase and the execution phase of the operation. In some cases, inconsistencies in timestamps can even reveal the actor's geographic location. Time zone differences, local time settings, and even the language pack of the installed system can unintentionally leak diary-level information.

Similarly, recurring patterns in domain registrations shouldn't be underestimated. Even if a threat actor registers domains under different names in each new campaign, when they repeatedly use the following, it reveals their operational habits:

- The same configuration template
- A similar email format
- A distinctive registrar preference

This pattern serves as a silent clue regarding who manages the infrastructure and what types of automation are being used.

A virtual private network (VPN) connection dropping for a few seconds may also be part of a deeper story. This brief interruption may lead to the leakage of the real source IP or draw attention to technical issues within the VPN infrastructure being used. Professional actors usually employ layered anonymity to prevent such errors; however, even a single connection fluctuation can provide the analyst with a valuable signal about the geography from which the operation is conducted or the service provider being used.

One of the most prominent examples of OPSEC errors is remaining tied to the same ASN for years. Even if an actor appears to change infrastructure, if they continue to operate within the ASN blocks of the same hosting provider, this indicates a certain habit dependency. Such dependencies reveal the actor's budget structure, access capabilities, and search for operational comfort.

Sometimes, even a simple `sleep()` command can be a critical OPSEC trace. Sleep durations reflect the following:

- Actor's quality control approach
- Automation tools they use
- Rhythm of the operation

Some groups use fixed sleep values; others prefer functions that appear random but are mathematically repetitive. These choices are a reflection of the group's internal coding culture.

Finally, the reuse of the same user agent across different campaigns is also an important signal. A group that has the capacity to generate a new user agent for each operation, yet still uses old user-agent patterns, demonstrates an internal habit or a weakness in task distribution. Such recurring traces are a significant part of the actor's digital footprint.

All of these examples show that OPSEC errors are powerful indicators that reveal the actor's true identity, working style, and operational maturity. For experienced analysts, these errors are more valuable than hash values in identifying an actor because technical traces can be erased, infrastructures can be shut down, and protected servers can be changed, but behavioral flaws are doomed to repeat.

Real-World Example

An infrastructure chain that an APT group had concealed very well for years was unraveled solely due to an unusual organizational unit label used in the Transport Layer Security (TLS) certificate on a single server. The label reflected the personal habit of a specific individual within the group.

6.5.4 Correlation of Behavioral Traces

A single indicator is never sufficient to understand a threat actor. However, when code traces, linguistic patterns, and OPSEC errors are examined together, a clear behavioral profile emerges:

- Code style: Technical background + cognitive reflexes
- Linguistic patterns: Cultural background + personal habits
- OPSEC errors: Internal team discipline + operational pressure
- Timing patterns: Operating geography + team size
- Technical choices: Skill level + comfort zone

This combined model is used to identify the actor *and* anticipate their future actions.

A threat actor's behavioral profile doesn't remain completely unchanged. Microlevel changes emerge when the team grows, developers change, funding increases, or operational pressure rises. However, an individual's cognitive rhythm, level of technical competence, and software production style don't change entirely. For this reason, behavioral traces can't be fully broken; they can only be concealed. Threat analysts follow these semi-fixed traces because they reveal the actor's ultimate identity.

Shut-down servers, changed domains, and repackaged malware can't erase the behavioral signature hidden in the way a human thinks. The flow of code, the choice of words, or a small OPSEC mistake reveal the true face behind an actor's mask. Therefore, behavioral profiling is the psychology of digital behavior.

Behavioral indicators offer an extremely powerful analytical layer for understanding threat actors; however, when used without control, this power can push the analyst away from the truth rather than closer to it. The main reason is that, unlike technical indicators, behavioral traces aren't absolute or directly verifiable; they are highly open to context, interpretation, and the analyst's mental framework. A hash value or an IP address is objective data; but indicators such as code style, linguistic habits, or operational preferences always pass through a human interpretive filter. At this point, the analyst's biases, assumptions, and prior experiences can have a far more destructive impact than technical shortcomings.

The greatest risk in behavioral profiling is the analyst unconsciously projecting their own mental patterns onto the subject of analysis. Labeling a behavior as a habit, inexperience, cultural trace, or deliberate choice often stems not directly from the data itself, but from the analyst's past observations and expectations. This can lead to false correlations and incorrect actor profiles, especially in campaigns analyzed with limited datasets. This is where the danger of behavioral traces lies: they appear sufficiently convincing, yet when misread, they can lead the analyst to conclusions that seem highly consistent but are fundamentally unfounded.

For this reason, behavioral profiling isn't a technical matching activity; it's an interpretive process that requires constant skepticism, self-control, and contextual awareness. The analyst must approach every observed pattern by asking "Under what conditions could this have emerged?" instead of "What does this resemble?" Otherwise, temporary operational pressure, tool constraints, or intentionally planted misdirection may be mistaken for the actor's characteristic behavior. Behavioral indicators don't produce certainty, but tendencies, which gain meaning only through careful, multilayered, and time-spanning analyses.

Behavioral profiling tests the analyst's cognitive discipline more than their technical competence. Errors made here usually stem from hasty inferences and unexamined assumptions, not from lack of information. Therefore, the fundamental principle when working with behavioral indicators is not to force the data to speak and also to accept what the data can't say. Only when this balance is achieved does behavioral analysis become a truly valuable and reliable intelligence tool.

The following principles highlight several common analytical pitfalls that you must consider when interpreting behavioral indicators during threat actor profiling:

- **Don't assume that behavior = identity.**
  One of the most common mistakes is assuming that a specific behavioral pattern directly belongs to a single actor. In reality, many behavioral patterns may be derived from the following:
  - Open-source code
  - Leaked malware kits
  - Shared underground forum cultures
  - The same training pools

  Therefore, a single code style, linguistic motif, or OPSEC error isn't proof of identity. These should be evaluated only as signals that narrow the field of possibilities. The analyst must position behavioral traces not as evidence, but as hypothesis enhancers.
- **Don't isolate the time factor.**
  In behavioral analysis, time data detached from its context can completely distort a profile. The same actor may behave hastily during crisis periods, disciplined in long-running operations, and carelessly under financial pressure. Therefore, OPSEC errors observed in short campaigns must not be weighed equally with orderly behavior seen in long-term infrastructures.

  The analyst must read behavior together with the operational phase. Otherwise, temporary pressure-induced flaws may be mistaken for permanent character traits.

- **Don't confuse technical debt with cognitive habit.**
  Code reuse or the use of old functions doesn't always indicate a developer's habit. In some cases, this may arise due to the following:
  - The deliberate carrying of technical debt
  - Pressure for rapid delivery
  - The conscious decision not to change the toolset

  Here's the critical question: "Is this repetition a choice, or a necessity?" Any behavioral conclusion drawn without making this distinction renders the analysis superficial.
- **Cleanse your own cultural bias.**
  When analyzing linguistic patterns, the most dangerous trap is the analyst assuming their own language and cultural perception to be universal. A grammatical error may be the effect of a native language, residue from a translation tool, or deliberate obfuscation.

  Therefore, linguistic indicators must not be used alone to infer nationality, direct geography, or ethnic origin. Language gains meaning only when it overlaps with other behavioral layers.
- **Avoid the OPSEC error = inexperience fallacy.**
  An OPSEC error doesn't always indicate a low-level actor. On the contrary, some advanced groups do the following:
  - Deliberately leave certain traces
  - Exhibit distracting behaviors
  - Produce controlled leaks designed to mislead analysts into false attribution

  Therefore, the analyst must ask the following question: "Is this truly an error, or a guided trace?" Any OPSEC assessment made without asking this question remains incomplete.
- **Confirm behavioral traces across multiple campaigns.**
  Deriving a behavioral profile from a single campaign isn't intelligence: It's speculation. Behavioral traces gain meaning only if they recur across different years, target sets, or infrastructures. Any behavior that doesn't repeat must not be made part of the profile.
- **Don't accept automation outputs blindly.**
  Behavioral profiling is well-suited for automation; however, automation is blind to context. Systems that produce code similarity, linguistic analysis, or timing correlations can't establish cause-and-effect relationships, can't understand operational pressure, and can't interpret the human factor. Therefore, automation outputs *support* the analyst's decision; they don't *replace* it.
- **Read trends instead of conclusions.**
  The goal of behavioral profiling isn't to say "This actor is definitely X." The goal is to anticipate the following:
  - The direction in which they may evolve
  - Which mistakes they are prone to repeat
  - Under which conditions they may exhibit OPSEC fragility

When this approach is adopted, behavioral analysis becomes a proactive rather than reactive intelligence tool.

Behavioral indicators, when used correctly, enable access to the deepest layer of a threat actor. However, this field also forces analysts to confront themselves because every mistake made here isn't a technical error, but a cognitive one. For this reason, behavioral profiling isn't a test of tool knowledge, but a test of analyst maturity.

6.6 Summary

Cyberattacks can't be fully understood by examining technical artifacts alone. Threat actor profiling focuses on the behavioral, operational, and strategic patterns behind an attack, revealing how adversaries think, make decisions, and conduct operations. By analyzing habits embedded in TTPs, analysts can identify recurring patterns that expose an actor's operational style and internal structure.

Frameworks such as MITRE ATT&CK and the Diamond Model provide structured methods for interpreting these patterns. MITRE ATT&CK enables the systematic mapping of attacker techniques and their evolution over time, while the Diamond Model reveals the relationships between adversary, capability, infrastructure, and victim. Together, these frameworks transform isolated technical observations into a coherent analytical picture.

Behavioral indicators (code reuse, linguistic patterns, timing habits, and OPSEC errors) often act as persistent fingerprints that remain even when technical indicators change. However, such signals must be interpreted carefully and validated across multiple campaigns. When used responsibly, behavioral profiling enables analysts to understand past attacks and anticipate the future behavior of threat actors.

We'll move on to another key aspect of CTI in the next chapter: threat intelligence feeds.

Chapter 7

Integrity, Poisoning, and Enrichment in Threat Intelligence Feeds

In the world of threat intelligence, trusting data in constant motion is often a reflex and sometimes a dangerous habit. Feeds appear as simple lines flowing across security operation center screens; yet behind every line lies a chain of sources, a claim of accuracy, and (unfortunately at times) an unseen intent to manipulate. So, how much can we truly trust this stream?

A poisoned feed does more than generate an incorrect indicator of compromise (IOC); it creates a vibration at the core of the decision-making process, a small but critical shift. An advanced persistent threat (APT) group deliberately spreading fake infrastructure or an old, reheated indicator being served as "new" by different sources are sufficient to distort the threat picture constructed in analysts' minds. Sometimes, this distortion goes unnoticed; sometimes it creates a feeling within the team that "something isn't right, but what?" That tiny discomfort—that intuitive crack—is the analyst's greatest ally.

Understanding feed integrity is both a technical validation process and a behavioral examination: Is the source consistent over time? Is there an abnormal surge in observation volume? Are multiple feeds coincidentally highlighting the same IOC? (In real life, coincidence is rare, especially in threat intelligence.)

This chapter seeks to untangle this complex equation. At times, we step back to read the bigger picture; at other times, we must focus on the lifecycle of a single IP address. Because assessing the reliability of a feed sometimes requires examining millisecond-level data flow and at other times months of infrastructure evolution, this rhythm is a struggle in and of itself.

In operational reality, threat actors attack, mislead, distract, and generate noise. For example, an artificially inflated reputation score can be enough to collapse the prioritization logic of a newly established security operations center (SOC) team. Or a misattributed domain can steer an analyst toward an entirely different tactical profile, which is the confusion the attacker desires. These manipulations are sometimes skillfully concealed; other times, they are overlooked because they are too obvious.

In the later sections of this chapter, we'll unpack the validation methodologies applied by analysts—from source-weighting models to pattern variance analysis. For now, it's enough to keep one question in mind: A feed may be telling us something-but what is it actually trying to say?

The answer is often found not in the data itself, but in the journey the data has taken. The purpose of modern cyber threat intelligence (CTI) pipelines—especially those designed to integrate with security orchestration, automation, and response (SOAR) platforms, which automate security workflows and coordinate actions across multiple security tools—is to make this journey visible, filter out the noise, and leave behind only what's truly actionable.

In the end, this chapter will instill an unsettling yet necessary awareness: The most fragile point of intelligence isn't the information itself, but uncontrolled trust in that information. That awareness is the strongest line of defense for a professional analyst.

7.1 The Anatomy of a Threat Intelligence Feed

A threat intelligence feed when viewed from the outside, appears to be a calm stream of data, a sterile JSON output, and sometimes even just a single line of information. Yet beneath those lines lies a complex biology. Like an organism, it has cells, vessels, weak points, and sometimes unnoticed mutations. That's why we speak of anatomy because a feed isn't merely data, it's a structure with its own breathing rhythm. Recognizing this is the distinguishing capability of the experienced analyst.

So why is it so difficult to understand the true structure of a feed? The answer is because data is rarely honest. Or rather, when detached from context, data says nothing on its own. For example, a single IP address appearing at midnight within a Structured Threat Information eXpression (STIX) bundle seems insignificant at first glance. But what if that same IP was marked as "low-confidence" three days earlier in a different Trusted Automated eXchange of Intelligence Information (TAXII) collection or a different provider feed? And what if yesterday, it produced a sudden pivot in passive Domain Name System (pDNS) records? At that point, a small data point suddenly sits at the center of a large cloud of suspicion. This transition is silent and fast. It creates a faint feeling in the analyst's core to wait—there's something here.

Understanding the anatomy of a feed is, in fact, the intersection of three distinct domains: structural correctness, protocol reality, and distribution behavior. Some feeds flow stably; others vibrate, as if generating noise within themselves. That vibration says a lot. Sometimes, the behavior of a feed tells more than the rhythm of the attacker itself.

Threat actors manipulate more than systems—they manipulate data flows as well. Therefore, feed anatomy examines both how a feed is produced and how it lives. Some generate high-entropy outputs that are highly diverse, inconsistent, and scattered. Others are excessively orderly, or almost too clean. Excessive order is a dangerous sign for a feed that it may be designed to deceive machines, not humans.

Real-World Example

Once, while comparing IOC packages from two different sources, I noticed only a 14-second difference in the "first-seen" timestamp of a domain. It seems tiny, yes. But considering the geographical infrastructures of the two sources, this was an impossible simultaneity. That small inconsistency ultimately revealed the first step of a large deception campaign. Sometimes, to understand anatomy, it's enough to look at a deviation of a single second.

Now ask yourself this: To understand what a feed is really telling you, which layer of the data are you looking at? Raw content? Distribution protocol? Metadata? Or the invisible behavioral model formed by all of them together?

In this section, we'll try to uncover that invisible model by exploring the data structures, protocols, and distribution models. The anatomy of a feed isn't a purely technical category trapped within the rigid structure of STIX/TAXII, but a multilayered structure read through its origin, context, transmission method, consistency, and intent.

Keep in mind that a feed isn't what it says. It's what it implies.

7.1.1 Data Structures and Content Models

Most analysts attempting to understand the inner workings of a threat intelligence feed instinctively begin by looking at IOC lists. But this is only the visible layer, which is just a thin shell of the data ecosystem. The real depth lies in how an IOC is represented, generated, contextualized, and invisibly related to other data points. Data structures are the skeletal system of these invisible layers; a single mis-modeled field can distort the meaning of the entire intelligence chain.

IOC types are the core bones of this structure: IP addresses, domain names, URLs, hash values, email indicators, and so on. However, the meaning they carry on their own is almost always incomplete because an IOC isn't a static entity; it's a dynamic signal that takes shape within operational context and gains meaning according to the attacker's behavioral rhythm. In the wrong context, even the most accurate IOC can lead the analyst astray.

Consider this example: An SOC analyst working the night shift detects two consecutive IP addresses. The first appears clean. The second is marked as malicious in open-source intelligence (OSINT) sources. Under normal circumstances, the matter would end there. But both IPs are located within the same Autonomous System Number (ASN) block, generate similar packet density within the same time window, and show an unusual drop in time-to-live (TTL) values, as if traffic were being forcibly pushed out of an encrypted tunnel. At this point, the question changes from "Is the second IP malicious?" to "Are these two IPs actually orchestrated by the same attacker?"

This shift demonstrates that data structures don't just carry data; they build webs of meaning among data points. This is where observables, patterns, and relationships come into play.

- **Observables**
  Observables are the rawest, most tangible signals of the real world: the first-seen time of

an IP, the IP resolved by a domain, the file a URL redirects to, and so on—small, but critical signals. There's a tension here: Observables often appear overly simple, yet the true intent is frequently hidden beneath that simplicity.

- **Patterns**
  Patterns reveal the attacker's rhythm. A malware family that repeats the same hashing behavior, command-and-control (C2) beacons constrained to specific time intervals, rhythmic anomalies in DNS query flows, and so on are invisible but consistent traces. Sometimes, the analyst doesn't consciously read this rhythm, yet intuitively feels that something isn't right.
- **Relationships**
  Relationships are the connective tissue that constructs the entire story. IOCs are meaningless in isolation; however, when placed side by side, they form a graph that reveals the attacker's strategic intent. For example, when you combine the following, they reveal the skeleton of an invisible campaign:
  - Domain
  - Hash
  - Command sequence used within the same tactic
  - Simultaneous change observed in pDNS

  This relational structure can even provide clues about the attacker's *intent level*, which is the extent to which observed technical signals reveal a purposeful operational objective, such as reconnaissance, persistence building, or preparation for disruption, rather than random or opportunistic activity. This makes it one of the most difficult building blocks to obtain in CTI.

In the modern intelligence landscape, the concept of a single IOC has low operational value because attackers no longer leave a single IOC. Under the *IOC diffuse strategy*, they scatter hundreds of low-value indicators randomly to drown analysts in noise. Therefore, the contextual IOC approach—understanding who, when, where, and with which attack technique or infrastructure component an indicator is associated—has become a critical necessity.

To make this concrete, let's consider an advanced example: A single SHA-256 hash in your possession represents only a file. But let's say that hash does the following:

- Is packed using a known packer used by an Iran-based APT
- Shows low uniqueness on VirusTotal
- Repeatedly matches a specific JA3 signature in TLS fingerprinting during C2 communication
- Indicates a particular family through mutex creation patterns in behavioral analysis
- Targets the financial sector

In this case, the hash ceases to be merely the identity of a file; it becomes an operational signal resonating within the attacker's ecosystem.

The difference is immense. A single IOC is fragile, but a contextual IOC is resilient: It lives, produces meaning, and feeds the decision-making mechanism.

At the professional level of CTI, one truth remains unchanged: the only way to understand an attacker's intent is to evaluate correlated, contextual, model-based IOCs. Let's walk through some strategies for analysis:

- **The time dimension of an IOC**

  The value of an IOC isn't static; it changes over time. For this reason, professional CTI systems treat an IOC as an entity that exhibits behavior over time not just a data point. The IOC time model generally consists of four parts:

  - First seen: The moment the IOC is first observed
  - Last seen: The most recent moment the IOC is confirmed
  - TTL: The operational validity period of the IOC
  - Decay function: The gradual decrease of the IOC's confidence score over time

  These decay functions may behave logarithmically, linearly, or, even in some attacker groups, in a burst pattern (burst decay). For example, in APT28 infrastructure, domains sometimes behave extremely actively within the first 24 hours and then fall completely silent. If the analyst is unaware of this, they may mistake the IOC for a passive domain. Therefore, the temporal model of an IOC is one of the most critical components of context.

- **Structural entropy and data anomaly detection**

  The string-based content of IOCs (domain, URL, file path, hash variations, etc.) contains specific entropy levels, that is, measurable degrees of randomness or structural unpredictability within the character sequence. Entropy analysis reveals whether the attacker is using automated generation, for example:

  - A human-created domain like `paylasim-merkezi.com` has low entropy.
  - A domain generation algorithm (DGA) domain has high entropy. (A DGA domain is generated by malware using an automated algorithm that continuously produces large numbers of pseudo-random domain names for C2 communication, like `jks9a0dk2nsd.com`.)

  Professional CTI analysts use IOC entropy for three purposes:

  - Detecting DGAs
  - Identifying automated IOC generation (feed poisoning)
  - Grouping similar campaigns to create correlation clusters

  For this reason, data structures carry mathematical traces.

- **Source consistency of an IOC**

  The same IOC may originate from multiple sources. However, each source assigns different attributes to the IOC. In CTI, the following five consistency metrics are used:

  - Source frequency: How often does the source report this IOC?
  - Source diversity: Is the IOC repeated across different feed types?

  - Confidence symmetry: Do different sources assign similar confidence levels?
  - Temporal overlap: Do sources report the IOC within similar time windows?
  - Semantic alignment: Does the threat category of the IOC align across sources?

  If inconsistencies appear in even three of these metrics, the IOC is marked as suspicious in professional CTI platforms.

- **The deep model of the observables layer**
  Observables aren't just surface-level signals, as each observable contains micrometric data beneath it. For example, an IP observable actually has 12 distinct attributes:
  - ASN
  - Border Gateway Protocol (BGP) prefix
  - Delegation date
  - Reverse DNS (rDNS)
  - Pointer (PTR) record
  - Geolocation
  - Classless Inter-Domain Routing (CIDR) range
  - Risk history
  - Port activity
  - Connection density
  - Edge node behavior
  - pDNS correlations

  The combination of these small attributes reveals the true ecosystem shadow of an IOC. Therefore, observables aren't superficial data; they comprise a layer where the attacker's tactical movement traces are stored at a microlevel.

- **The directionality of relationship models**
  In CTI graph structures, IOC relationships are evaluated as directed because the link an IOC forms with another IOC isn't always symmetrical, for example:
  - A file hash may point to a C2 domain when malware associated with that hash communicates with that infrastructure, creating an active relationship in which the file directly interacts with the domain.
  - However, the reverse relationship isn't always valid; the presence of a C2 domain doesn't necessarily imply that it's linked to a specific file hash because the same infrastructure may be used by multiple samples or campaigns.

  Analysts who don't understand this difference may build the threat chain incorrectly. Directed relationships are the only way to understand the chained structure of an attack. Some CTI platforms use special labels for directed relationships:
  - `uses`
  - `communicates-with`
  - `downloads-from`

  - `spawned-by`
  - `controlled-by`

  These labels indicate which stage of the attack the IOCs belong to (and facilitate MITRE ATT&CK mapping; see Chapter 6, Section 6.3).

- **The behavioral layer of contextual IOCs**
  Traditional IOCs are typically treated as static indicators that represent a single observable artifact, such as an IP address, domain, or file hash. In contrast, contextual IOCs are interpreted within a behavioral framework that reflects how the indicator functions within an attack. Behavioral IOCs therefore include attributes that describe the operational role and activity patterns associated with the indicator:
  - Initial purpose of use
  - Role in the attack chain
  - Applied tactic
  - Applied technique
  - Operational intensity
  - Associated infrastructure
  - Attacker intent

  Some advanced CTI systems model IOC behavior using a four-dimensional approach, called 4D IOC mapping:
  - Temporal: Time
  - Spatial: Geographic distribution
  - Functional: Attack role
  - Operational: Frequency of use

  Correlating these dimensions makes it possible to determine whether an IOC represents inactive or obsolete infrastructure (dead), infrastructure that is currently participating in ongoing activity (active), or a bridge to discover additional infrastructure or artifacts within the same campaign (pivot).

- **Lifecycle anomalies in IOC data structures**
  An IOC passes through the following lifecycle: creation → distribution → use → observation → expiration. However, advanced CTI analysis teaches one key lesson: Attackers manipulate this lifecycle. There are three types of manipulation:
  - Lifecycle shortening manipulation: The IOC is kept active for a very short time, making detection difficult.
  - Lifecycle extension manipulation: Attackers "reheat" old IOCs to create confusion.
  - Lifecycle distortion manipulation: The IOC is distributed in a way that violates chronological order, causing analysts to build an incorrect timeline.

  These manipulation techniques can only be detected through a contextual IOC model.

- **Metacontext usage in data structures**
  Every IOC carries invisible metadata. This metadata reveals the unseen profile of the IOC. The following are some example metacontexts:
  - Geographic origin of the feed provider
  - Political/operational bias of the feed
  - Intelligence collection method (honeypot, customer telemetry, OSINT, collaboration network, etc.)
  - Data integrity history
  - Manipulation history

  This metacontext represents the highest interpretive layer that determines the true value of an IOC because it reveals the conditions under which the data was collected, the reliability of the source that produced it, and the potential biases or limitations embedded in the feed. In other words, two identical IOCs may carry very different intelligence values depending on who observed them, how they were collected, and whether the underlying data source has a consistent history of integrity.

Data structures and content models are the intelligence infrastructure of a security operation. If this foundation is flawed, even the best intelligence can lead to wrong decisions.

7.1.2 Standardized Protocols and Formats

Consuming threat intelligence data in its raw form is easy, but representing it in a format that is transferable across teams, automation-friendly, and nonconflicting across infrastructures is where the real challenge begins. If the representation is incorrect, the data appears flawed and, in some cases, even works in favor of the attacker. For this reason, standards such as STIX and TAXII form silent architectures that most analysts don't notice, yet constitute the foundation of every product and every integration. We'll discuss them in detail in the following sections.

Understanding STIX and TAXII

Structured Threat Information eXpression (STIX) 2.x carries the semantic model of threat data. Cyber observable objects (COOs) represent the abstracted form of everything observable in the digital environment. An IP address isn't just a value; within a COO, it's carried together with dozens of micro-elements such as timestamps, associated protocols, the type of data transmitted, and the source of observation. This demonstrates that an IOC isn't just an identity, but an entity that carries traces of operational behavior.

Relationship objects (ROs) are even more interesting, as they reveal the true power of STIX 2.x. They model how entities are connected, for example, linking a hash to a file, a domain to a C2 server, or a threat actor to a particular malware family it uses.

These relationships are directional; they aren't symmetric. This is a critical detail in the operational analysis of threat intelligence because the state of one indicator "using" another isn't the same as being "affected by" it. STIX doesn't conflate these. It provides structural correctness rather than intuitive interpretation to analysts.

However, the real strength of STIX lies in its constraint system. STIX objects can be validated by the schema and are based on rules; as a result, incomplete or incorrect modeling can often be caught at the production stage, and many platforms can automatically reject such data. When a proper validation and policy layer is established, this enables feed poisoning attempts to be detected earlier.

Trusted Automated eXchange of Intelligence Information (TAXII) 2.x serves as the transport mechanism for STIX data, meaning that STIX defines how threat intelligence is structured while TAXII provides the standardized protocol used to exchange that structured intelligence between systems.

In practice, TAXII 2.x is understood around three carrier concepts:

- **Collections**
  The logical container of a specific dataset.
- **Channels/endpoints**
  The interfaces that define the direction of feed flow.
- **Filtering mechanisms**
  The selection layer that ensures only the necessary data is delivered to the consumer.

Thanks to the collection-based model, an organization can isolate and distribute highly sensitive data to specific subgroups. In other words, the feed consumed by the SOC can be entirely different from the feed consumed by the red team. This isolation reduces the risk of intelligence generating false alarms in the wrong hands.

An advanced feature of TAXII is its incremental sync capability. With this feature, the consumer retrieves only new changes; thus, network bandwidth isn't wasted while processing massive CTI packages. In real operations, this difference is dramatic: in large-scale intelligence traffic, the incremental retrieval approach—by pulling only changes—can drastically reduce data transfer and significantly shorten synchronization time.

The bundle, indicator, and sighting models represent three critical stages of the CTI data lifecycle:

- **Bundle**
  Bundle allows all STIX objects to be carried within a single container. It appears simple, but it's important to remember that bundles preserve chronological integrity for attack chain analysis.
- **Indicator**
  Indicator contains actionable signals. What many don't realize is that STIX indicators aren't limited to IOCs; they can also carry pattern-based detection signatures. In other words, an indicator can do more than say "This IP is malicious"; it can say "Generate an alert when this behavior pattern occurs." This is the foundation of the CTI → detection engineering bridge.

- **Sighting**
  Sighting is confirmation from the real world. Something isn't just claimed to exist; its actual observation has been observed recorded. When used correctly, sighting objects reveal the true operational capacity of a threat actor. The difference between an IOC appearing in hundreds of feeds and the existence of a real sighting can sometimes change an organization's entire defensive approach.

STIX's relationship and metadata fields (e.g., confidence, markings, and provenance) allow uncertainty to be modeled; probabilistic interpretation and decisions about how much to trust which claim are often handled by the analytical layer. This technical detail is known only to high-level intelligence architects. STIX relationships aren't limited to a single claim of truth; through confidence scores, they support multivariate probability models. This reflects the true nature of CTI more realistically because intelligence doesn't deal in absolute truth. It deals in probability, weighting, and context.

Through this capability, CTI analytics takes on a structure that increasingly converges toward machine learning–based adversary behavior modeling. This is a critical evolution for the future of CTI engineering.

When the semantic correctness of the STIX data model, the distribution behavior of TAXII, and the operational rhythm of the bundle/indicator/sighting triad come together, CTI pipelines add meaning to data. Without these standards, threat intelligence would be no different from scattered notebooks. Thanks to these standards, CTI becomes a mathematical language for security operations.

The Hidden Power of STIX

STIX objects have an independent capability that most analysts overlook, yet which constitutes one of the most critical defensive lines of the CTI ecosystem: *semantic consistency enforcement*. Thanks to this capability, STIX not only requires data to be defined in the correct format but also mandates that it be modeled in the correct context, with correct relationships, and in a manner consistent with real-world logic.

In other words, STIX is both a container for transporting data and a security layer that preserves the internal logic of the data. This semantic layer is one of the most important components of STIX's schema validation mechanism. For example, if within the same bundle an indicator object's pattern contradicts the claims of other objects, the validator/policy layer can generate a *silent inconsistency*. Often without raising a dramatic alarm, it signals that there's a contextual fracture in the data. In other words, without breaking the system, without shouting, and without presenting the user with a dramatic alert—yet in a very critical way—it indicates that the data contains contextual inconsistency.

This is important because one of the techniques attackers frequently use in feed poisoning attempts is injecting semantic contradictions. Rather than generating false IOCs, attackers aim to confuse the CTI analyst by pairing correct IOCs with incorrect contexts. Let's consider a real example:

- A domain indicator is marked in STIX as "a C2 domain used by APT29."
- In the same STIX bundle, that same domain is associated with a tool object.

This relationship is technically impossible; a domain can't simultaneously be defined as an APT-specific C2 infrastructure and as a generic tool component. These two roles are mutually exclusive behavioral categories within the STIX ontology. This is where STIX intervenes.

Such a contradiction can be detected by the validator/policy layer that validates STIX and can generate an inconsistency signal for the analyst. This signal is recorded as a silent validation error and often functions as a hidden warning. However, when CTI analysts recognize the intent behind this signal, they realize that this isn't a modeling error; rather, it may be a manipulation injected into the feed.

Some advanced CTI platforms mark STIX objects containing semantic errors as suspicious and, in certain cases, may even apply automatic quarantine. This layer also prevents attackers from filling the STIX structure with "noisy but seemingly valid" data because attackers can imitate the format, but imitating semantic relationship consistency is far more difficult.

Selective Trust Propagation

TAXII 2.1 transports STIX objects in a standardized manner; the preservation of trust and the propagation of trust are typically achieved through additional mechanisms such as signing, provenance, access control, and integrity logging. In other words, a feed provider can mark the trust level of the STIX objects it sends in a way that allows this trust to be propagated. This trust level is automatically carried downstream to all consumers.

For example, if a provider assigns a `confidence: 0.22` label to an indicator, this value can be carried and preserved downstream; however, whether it truly remains unchanged in practice depends on the application of security measures such as signing and integrity checks. This is a critical mechanism for data integrity.

A TAXII endpoint can also automatically filter out indicators that don't comply with the recipient's trust policy. For example, consider a SOC team that includes only IOCs with `confidence` > `0.5` in its processes. TAXII can automatically apply this filter, preventing the team from dealing with unnecessary noise.

So, does your organization still process all IOCs with equal importance today? If so, you may possess the attacker's favorite habit: trusting noise.

Advanced Use of Bundles

Bundle objects are packages that carry the chronological story of an attack. Advanced CTI platforms reprocess objects within a bundle for the following purposes:

- Timeline validation
- Event-chain reconstruction
- Detection of actor behavioral continuity

Whether an attack occurred within a single day or over months can be inferred from the metadata within a bundle. For example, consider the following within the same bundle:

- An indicator's `first_seen` value: `2024-05-01`
- A malware object's `last_seen` value: `2023-12-11`

The inconsistency between these two dates indicates that the attacker is reusing old infrastructure (infrastructure recycling). An analyst who notices this can measure the actor's evolution speed.

Indicator Pattern Engine

In many organizations, indicator patterns are assumed to carry only IOCs. However, the STIX pattern language is an incredibly powerful tool for detection engineering and isn't used to its full capacity by most experts. For example, the following pattern can be defined in STIX:

```
[file:hashes.'SHA-256' = 'abc...123'] FOLLOWEDBY
[network-traffic:dst_port = 443] FOLLOWEDBY
[process:name = 'powershell.exe']
```

This represents a three-stage behavioral signature of an attack chain. In practical terms, this pattern describes a behavioral sequence in which a specific file hash is observed first, followed by network communication over port 443, and then the execution of a PowerShell process. Instead of relying on a single indicator, the pattern captures the ordered relationship between multiple events that together represent a suspicious activity chain. Security platforms such as security information and event management (SIEM) or endpoint detection and response (EDR) systems can use this type of pattern to detect correlated behaviors across telemetry streams, triggering alerts only when the defined event sequence occurs.

So, how many of your organization's detection rules still rely solely on a single IOC? Meanwhile, attackers have already moved to behavioral chains.

The Hidden Strategic Value of Sighting Objects: Threat Prevalence Modeling

A sighting isn't merely a "seen" record. In advanced CTI architectures, sighting objects are used to calculate an actor's spread radius. For example, if a single indicator generates sightings across three different continents, this can indicate one of two things:

- The attacker is running a global campaign.
- The indicator is so widespread that its operational value has decreased.

This distinction is determined solely by sighting density. Sighting objects also detect a phenomenon known as false prevalence illusion. Some threat actors generate fake sightings to make an IOC appear more widespread than it truly is.

In advanced CTI systems, such fake sightings are detected through entropy modeling. In practice, entropy modeling measures the statistical variability of sighting attributes such as time distribution, geographic origin, and reporting frequency. When sightings appear artificially uniform, excessively repetitive, or clustered in patterns that deviate from normal operational telemetry, the entropy score drops, indicating a potential manipulation attempt. Conversely, authentic sightings typically display a more natural distribution across time and infrastructure sources. If the following observations are inconsistent with one another, the system raises an alert:

- Frequency of sightings
- Geography of the sighting source
- Sighting time window

The Cognitive Integrity Model of STIX and TAXII

More than just technical data standards, STIX and TAXII are a cognitive alignment framework shaped according to how threat intelligence analysts think, how they model threats, and how they make decisions. This concept is important because threat intelligence isn't just data—it's data interpreted by humans.

If the data model is distant from the way the human brain forms relationships, the analyst is forced to constantly add extra explanations, manual links, and mental modeling. This dramatically increases cognitive load. This is where STIX comes into play. The following characteristics of STIX values are all designed according to the human tendency to seek cause-and-effect relationships:

- How they are named
- Which fields are mandatory
- The direction of relationships
- The verbs preferred in relationship types (uses, controls, communicates with)
- The alignment of tactic and technique fields with MITRE ATT&CK mental models

This is especially visible in relationship chains such as threat-actor → uses → malware → delivers → payload → impacts → sector. For example, a threat actor may use a specific malware family, that malware may deliver a malicious payload, and the resulting activity may impact a particular industry sector. This structure allows analysts to follow the operational logic of an attack as a connected sequence rather than as isolated indicators.

This chain is an optimized flow that enables analysts to construct the threat story in their mind without effort. The human brain assigns meaning not to raw data but to stories. STIX presents data in a form that can be turned into a story.

This is perfectly aligned with the directional pattern of causality in the human brain. An actor uses a tool; a tool doesn't use an actor. This simple yet critical directionality significantly reduces the analyst's cognitive load by aligning data structures with natural causal reasoning patterns—a principle widely discussed in cognitive load theory and human-centered information modeling research.

TAXII creates a similar cognitive balance. The collection logic of feeds makes it easier to mentally categorize data. While trying to understand the data, the analyst doesn't think "What does this IOC mean?" but rather "Which collection's story does this IOC belong to?" This subtle difference accelerates the decision-making rhythm.

Detecting Stealth Access Patterns in TAXII

An attacker may quietly subscribe to TAXII endpoints in an attempt to steal the intelligence produced by an organization. A gateway/log/user and entity behavior analysis (UEBA) layer positioned around the TAXII service can reveal suspicious subscription patterns by monitoring the following behaviors:

- Timing patterns of the subscribing client
- Types of collections being pulled
- Request volume anomalies
- Authentication attempts
- Cross-collection correlation attempts

An attacker's intelligence-gathering operation can be exposed through these signals. This capability isn't widely known in the industry, yet it's critical for strategic intelligence protection.

For example, an organization may place an application programming interface (API) gateway or logging proxy in front of its TAXII server to record all subscription and collection access requests. These logs can then be analyzed by a UEBA or SIEM system to identify unusual client behavior. If a newly registered client begins systematically querying multiple high-sensitivity collections or repeatedly requests updates at abnormal intervals, the monitoring layer can flag this activity as a potential intelligence-harvesting attempt.

7.1.3 Feed Distribution Models

Many teams view a threat intelligence feed simply as the channel through which IOCs arrive; yet this is only a small part. No matter how accurate an IOC's content may be, if it's delivered through a flawed distribution model, the analyst may misinterpret it, automation systems may process it incorrectly, or critical time windows may be missed. In other words, how threat intelligence is delivered is often just as decisive as what it delivers.

Many organizations believe feed models consist of only four main categories (streaming, polling, event-driven, federated). This is a superficial classification as beneath these categories lies a much deeper architecture that defines the access behavior of threat intelligence. This architecture silently controls the timing, intensity, contextual prioritization, and internal consistency of data flows.

Therefore, the distribution methods examined in the following sections are more than technical transport mechanisms. Each one shapes an organization's threat intelligence metabolism: how data is digested, interpreted, and acted upon. When this metabolism is strong, the organization can respond quickly and accurately to new threats. When it's weak, even the most critical IOC may go unnoticed in time, incident response (IR) is delayed, and the attacker's operational freedom expands.

The following sections examine the most common feed distribution models and the operational logic behind each approach.

Continuous Streaming Feeds

The continuous streaming model is a constantly flowing data pipeline, similar to price streams in financial markets. Real-time alerts, IOC updates, and sighting notifications flow through this channel.

Contrary to popular belief, however, not all streaming feeds are the same. Advanced CTI providers apply the following three parameters in streaming feeds:

- **Tick frequency**
  This is the update frequency.
- **Delta content**
  This refers to transmitting only changes in the data.
- **Noise shaping**
  This refers to using signal-processing algorithms to reduce data noise.

If the signal-to-noise ratio (SNR) drops in streaming feeds, SIEM and SOAR systems begin generating excessive alerts. For this reason, some vendors balance streaming data with operational silence periods.

> **Real-World Example**
>
> An APT group's C2 infrastructure exhibits an active waveform for 45 minutes and then suddenly stops. A streaming feed transmits this waveform as a time series. The curve of this waveform can reveal the attacker's operational rhythm to the analyst. Such behavioral patterns are visible only in streaming architectures.

Scheduled Feed Polling

This time-based pull model is based on pulling feeds at regular intervals. However, many organizations mistakenly believe or assume that polling means simply fetching data once per minute. Real CTI platforms apply deterministic data consistency rules during polling:

- **Consistency windows**
  Tolerance for timing misalignment in feed updates.
- **Metadata drift detection**
  Halting the polling cycle if contradictions appear in object metadata.
- **Adaptive polling rate**
  Dynamic timing adjustment based on threat intensity.

Some advanced platforms automatically accelerate or slow polling rates according to threat density. This is especially effective for optimizing data traffic in large-scale infrastructures.

Real-World Example

In a financial institution, when intense attack activity is detected at night, polling switches to burst mode. In this mode, the system pulls feeds every two seconds; when activity subsides, it returns to its previous stable level. This behavior is rarely implemented, but it should be.

Event-Driven Feed Distribution

The event-driven feed model functions as the reflex system of threat intelligence infrastructures. In this model, data doesn't flow continuously; the flow begins only when an event is triggered and stops automatically once the event's impact ends. In this respect, event-driven distribution both reduces noise and sharpens operational awareness by delivering critical information at the moment it occurs.

Most people associate the "event" concept solely with attack signals. However, in STIX-based systems, the mechanisms that can trigger an event are far broader and often extend well beyond direct attack indicators. In reality, an event can be any of the following:

- **Creation of a new RO**
  For example, establishing a new directed relationship between a malware sample and a C2 domain can trigger intelligence flow. This means one more step of the attack chain has become clear.
- **A change in an indicator's confidence score**
  Confidence is both a quality metric and a behavioral indicator signaling attacker activity intensity or campaign transition. (See our discussion in Section 7.4.3.)
- **Association of a threat actor with a new tactic, technique, and procedure (TTP)**
  This event may indicate that the actor's capability set has expanded, the attack surface has shifted, or operational tactics have been updated.
- **Sighting density exceeding a threshold**
  Sightings are real-world confirmed observations. When a certain threshold is crossed, it indicates that the attack's spread momentum is increasing.

These triggers demonstrate that the event-driven model does more than just generate alerts. It's an intelligence delivery method aligned with the rhythm of the attack chain. For SOC teams, the strategic value of this approach is immense because what the SOC needs most is the right data, at the right time, within real-time context.

The true strength of event-driven distribution, however, lies deeper: This model makes the topological changes of attacker behavior visible. Topology here refers more to the attacker's decision-making patterns, relationship density among threat objects, campaign transitions, and microshifts in operational rhythm.

Real-World Example

If an indicator's confidence score rises from 0.12 to 0.52 within a short time, this doesn't just mean the indicator has become more reliable. It may also indicate the following:

- The attacker is reactivating the same infrastructure.
- A new campaign is being launched.
- New sightings are emerging in different regions.
- The threat actor's operational intensity is increasing.

Such dynamic changes can be detected in real time only in event-driven architectures. In systems with latency, this signal is lost; by the time the analyst notices the behavioral spike, the attacker may already have moved to the second phase of the operation.

The event-driven model isn't just a distribution method; it's the only intelligence flow mechanism capable of resolving attack dynamics in real time. Failure to see such changes at the moment they occur can close critical decision windows and cause irreversible delays, especially in rapidly spreading campaigns.

Internal and External Feed Federation Architectures

Federation architecture is the silent champion of the threat intelligence world. Without federation, no intelligence ecosystem can scale. Internal federation allows different units (SOC, threat hunting, red team, forensics) to consume feeds with varying sensitivity levels according to their own context. The true value of federation, however, emerges in interorganizational collaboration. Advanced federation architectures do the following:

- Define trust tiers based on feed access levels
- Determine which data goes into which collections
- Add obfuscation layers during transfer to mask sensitive information
- Apply attribution sanitization to prevent internal IOCs from leaking externally

Real-World Example

For example, federated threat-sharing platforms such as MISP Threat Sharing (MISP) enable cross-organization pattern merging during intercompany intelligence exchange. When multiple organizations contribute STIX objects that exhibit similar behavioral or infrastructural patterns, the platform can correlate these indicators and generate a shared threat model that reflects activity observed across different participants. This means that if similar patterns are found across STIX objects from different organizations, the system automatically creates a collective threat model. This model provides a level of threat visibility that no single organization could ever achieve on its own.

7.2 Feed Poisoning and Manipulation Techniques

In the world of threat intelligence, the most dangerous attacks are usually the ones that occur silently. System vulnerabilities can be patched, malicious files can be detected, and fake traffic can be filtered, but if the intelligence feed itself is poisoned, the heart of the defense is being tampered with. A poisoned feed does more than just produce incorrect data; it manipulates the analyst's reality, infiltrates decision-making processes, and subtly distorts the defense strategy without being noticed.

The concept of feed poisoning is often associated with fake data produced in IOC generation. But in real-world attacks, manipulation techniques are far more complex, layered, and target focused. These techniques focus not only on adding or changing data but on attacking the structural integrity of intelligence, relationship models, contextual weights, and trust scores. In other words, the target is no longer a single IOC. It's the entire CTI ecosystem.

For modern attackers, feed poisoning is no longer limited to injecting false indicators; it's a tactic aimed at influencing the analytical and decision-making processes that threat intelligence feeds support. In the following sections, we'll familiarize ourselves with what motivates this type of feed poisoning and manipulation, learn about the techniques to generate manipulated IOCs, and zoom out to understand the complete attack infrastructure.

7.2.1 Objectives of Advanced Manipulation

Let's begin with a simple example: If a threat actor distributes the same domain across different STIX feeds with different confidence scores, the shared perception of that domain among analysts is distorted. If Company A's feed says "high risk" and Company B's feed says "uncertain risk," two independent analysis processes will handle that domain differently.

As a result, defensive processes fragment into segments, and decisions become inconsistent. This is the attacker's favorite ground: the defense breaking apart internally.

But feed poisoning isn't limited to this. Advanced manipulation techniques generally pursue three main objectives, which we'll discuss in the following sections.

Structural Manipulation

Attackers can mislead systems without breaking STIX or TAXII structure by making semantic relationships contradictory. STIX schema validation strongly enforces formal consistency (format, required fields, type compatibility); however, semantic consistency (the contextual meaning of relationships, actor–infrastructure–TTP alignment) often requires an additional rule/ontology layer on most platforms, so it remains a shared responsibility of analysts and analytical controls in practice.

For example, a domain could be associated both with a threat actor and with a benign-tool object. The model may appear correct, but the context is contradictory. This is a sign of poisoning.

Contextual Manipulation

Contextual manipulation (*context poisoning*) is one of the most insidious and dangerous forms of feed poisoning. Because the attacker doesn't touch the IOC itself; the visible surface is perfectly clean, properly formatted, and often technically valid. But the context around the IOC (i.e., the metadata that gives it meaning) has been deliberately distorted. If fields are filled in the correct format, this manipulation can pass schema validation. Because the error is semantic rather than formal, it may not be automatically detected without additional context verification and consistency controls, yet the content tells the wrong story. For this reason, contextual manipulation is an attack on perception and directly targets the analyst's mental model.

The context attackers target consists of the critical elements that determine an IOC's operational value, and it can be distorted through the following manipulations:

- **Incorrect `first_seen` time for an indicator**
  The `first_seen` field indicates when an IOC became active. By manipulating this time, the attacker can cause the analyst to misjudge the attack phase. Example manipulation details include the following:
  - The IOC may be taken from an old campaign, but the `first_seen` value is shown as current, so the analyst assumes the attack has just begun.
  - The IOC may be newly created, but is backdated, so it's believed the attack has been ongoing for much longer.
  - Temporal drift techniques are used to artificially extend or shorten the IOC's lifecycle.

  These changes are highly effective for concealing the threat actor's operational rhythm.
- **Artificially increasing sighting counts**
  As we've discussed, a sighting is a real-world observation of an IOC and is one of the most critical operational signals. By manipulating sighting volume, the attacker can do the following:
  - Create a false perception of prevalence for an IOC
  - Mislead automation systems into believing a high-priority threat exists
  - Collapse SOC incident triage processes and steer teams toward false alarms

  Advanced attackers use the following for sighting manipulation:
  - Building honeypot-like fake observation sources

  - Generating fake traffic via botnets
  - Sending simultaneous observation simulations from different locations

  This technique targets the prioritization mechanism of defense by corrupting intelligence metrics.

- **Linking unrelated TTPs to a threat actor**
  Associating an IOC with incorrect TTPs is one of the most severe distortions of intelligence context. Because TTPs are an actor's behavioral signature, incorrect mapping leads to the following:
  - The analyst classifying the attacker incorrectly
  - Defense teams preparing against the wrong MITRE ATT&CK techniques
  - Incorrect attribution
  - Misunderstanding the motivation behind the attack

Real-World Example

Imagine a finance-focused APT group that is normally associated with banking malware and credential harvesting. If manipulated intelligence data falsely adds supply-chain compromise techniques to this group's TTP profile, a SOC analyst may interpret the threat as a potential compromise of third-party software vendors rather than a targeted financial intrusion. This misinterpretation can lead the organization to escalate the alert level and initiate unnecessary IR actions. Such manipulations are especially common in professional campaigns such as state-sponsored threat actors, intelligence operations, and competitive cyber espionage.

- **Artificially raising or lowering the confidence score**
  The confidence score represents the accuracy and reliability level of an IOC. Manipulating this score directly affects defensive decisions. In the case of artificial elevation, the following happens:
  - A low-importance IOC is perceived as critical.
  - SIEM/SOAR systems generate unnecessary alerts.
  - Defensive resources are directed toward the wrong targets.

  In the case of artificial reduction, the following occurs:
  - An actually critical IOC appears low priority.
  - The attack isn't noticed in the early stage.
  - The threat actor's attack chain progresses comfortably.

  Confidence-score manipulation is often applied through microdeviations within a STIX bundle that aren't easily noticed. For example, even a small change such as from 0.68 to 0.72 can completely change the automation's IOC prioritization.

Behavioral Manipulation

Advanced attackers may target both the content and the feed's consumption rhythm (event flow, trigger density, update waveform) to increase the SOC's workload and reduce the visibility of critical signals. Although this vector is covered more limitedly in the literature, its practical impact is high.

Real-World Example

In an event-driven threat intelligence feed, an attacker may inject fabricated ROs that falsely link a benign domain to a known malware family. When the feed is ingested by a SIEM or SOAR platform, the system interprets this relationship as an active threat indicator and automatically triggers detection rules or correlation alerts. As a result, the SOC may generate multiple high-priority alerts for an infrastructure that isn't actually malicious. While analysts spend time investigating these false correlations, genuinely critical indicators may remain unnoticed within the alert queue.

An even more sophisticated technique is known as temporal poisoning: IOC timestamps are manipulated to distort the true timeline of the attack. While the analyst believes the attack is active, it has actually already ended or vice versa.

The most dangerous aspect of feed poisoning attacks is that they are extremely difficult to detect because the attacker doesn't destroy the data; they inject an alternative meaning into it. For this reason, preserving feed integrity requires epistemic accuracy. The analyst must evaluate not how correct the information is, but how much potential it has to be manipulated.

7.2.2 Generation of Fake or Manipulated IOCs

Fake IOC generation is often evaluated in the cybersecurity literature as a simple act of adding incorrect data. Although accurate, this definition is far too superficial. In real operations, the attacker's goal isn't to produce a single fake indicator; rather, it's to construct a carefully designed data architecture that disrupts the defense's threat perception, distorts relationship maps, and targets the analyst's intuition.

We'll walk through the different strategies attackers use to generate fake IOCs in the following sections.

Noise IOC Injection

Noise IOC injection is the most well-known manipulation type; however, most analysts fail to recognize how calculated this noise actually is. The attacker's goal isn't usually to create an abundance of IOCs, but to blind the defense through a specific density and distribution.

Noise IOCs are generated for the following purposes:

- Reduce the priority of critical IOCs
- Weaken the analyst's pattern-recognition ability
- Divert correlation engines into unnecessary branches
- Overwhelm the triage mechanisms of SOAR systems

A lesser-known tactic advanced actors use is to generate noise IOCs in a way that passes entropy analysis, creating semi-human domain/URL strings. Let's consider a few examples:

- **`myfiles-update-check.com`**
  This domain resembles common software-update or cloud-service infrastructure. Terms such as `files`, `update`, and `check` frequently appear in legitimate service domains used for synchronization or update validation. As a result, the domain may appear benign at first glance and may not immediately trigger suspicion during initial triage.
- **`dataverse-secure-auth.net`**
  This structure mirrors naming patterns commonly used in authentication or identity-management services. Words such as `secure` and `auth` are often associated with login portals or API authentication endpoints. Because such terminology aligns with legitimate service naming conventions frequently exploited in phishing campaigns, the IOC may appear plausible within normal internet infrastructure.

In a more advanced version of noise generation, the attacker injects IOCs in a waveform spread over time. This waveform can be sufficient to deceive SIEM correlation engines because those engines generate alerts based on specific density thresholds. This is where manipulation begins.

Reputation False Inflation

IOC reputation scores, especially in commercial feeds, are one of the primary determinants of defensive decisions. Attackers know this very well. Therefore, instead of directly generating new IOCs, they manipulate the reputation of existing ones. This manipulation is carried out through three techniques:

- **Sending fake sightings**
  The IOC is reported as having been observed across many different geographies. The system generates a perception of widespread threat.
- **Impersonating high-trust sources**
  The attacker produces fake STIX bundles that resemble the style of popular OSINT sources. The platform may assume the source is legitimate and raise the confidence score.
- **Injecting a benign IOC into the threat ecosystem**
  A harmless IP is deliberately associated with malicious infrastructures. When this association is repeated sufficiently, its reputation degrades.

Reputation manipulation is particularly effective at derailing threat-hunting processes. An analyst may spend hours investigating an IOC that poses no real risk. This gives the attacker an invaluable advantage: time consumption.

Deliberate Miscontextualization of Threat Clusters

Which threat cluster an IOC belongs to plays a critical role in understanding the threat actor's intent, capability, and TTP set. For this reason, attackers deliberately distort the context of threat clusters. This manipulation occurs in the following ways:

- Disinformation is added to the known TTP set of a specific actor group.
- Inappropriate relationships are introduced.

Real-World Example

Consider a financial services organization monitoring an APT group historically known for deploying fileless PowerShell-based malware during credential-harvesting campaigns. Suppose manipulated threat intelligence data introduces a fabricated relationship linking this actor to a newly registered domain through the relationship type `communicates-with`. In reality, the domain is unrelated to the actor and was inserted into the feed as part of a manipulation attempt.

When this relationship is ingested into a CTI platform or SIEM correlation engine, the system may interpret the domain as an active C2 infrastructure. Analysts may then initiate investigations into network traffic associated with that domain, deploy new detection rules, or escalate the IR posture.

In practice, however, the attacker's real infrastructure may lie elsewhere. By forcing defenders to focus on a fabricated infrastructure node, the manipulated relationship diverts analytical attention away from the actual operational assets of the campaign.

The purpose of this manipulation is to do the following:

- Create a misleading perception of the threat actor's infrastructure
- Cause defensive controls to be deployed against irrelevant indicators
- Conceal the real direction of the campaign
- Distort attribution analysis

Contrary to popular belief, this type of contextual manipulation isn't limited to highly sophisticated actors. Competitive organizations, hacktivist groups, or even opportunistic attackers can introduce low-level context poisoning into threat intelligence ecosystems.

7.2.3 Mirrored, Masked, or Deception-Oriented Infrastructures

One of the most dangerous forms of manipulation created by threat actors is concealing what the attack infrastructure truly is and reconstructing it to resemble something entirely different. At this point, the attacker doesn't target a single data point—they transform the entire observation surface, break the defense's perception mechanism, and fundamentally shake what the analyst refers to as threat visibility.

Such infrastructures can achieve a high level of evasion against purely static IOC correlation and superficial signature-based controls; therefore, detection approaches that aren't reinforced with behavioral, time-series, and topological analysis layers are far more likely to miss them. Their goal isn't to hide, but to be visible as something other than what they truly are. Sometimes, they imitate a completely legitimate service; sometimes they precisely replicate real sinkhole behaviors; at other times, they generate deceptive signals so sophisticated that they can weaponize honeypot data against the defender.

The analyst's problem is that these infrastructures don't feel wrong. On the contrary, they are so natural, so statistically normal, so behaviorally consistent that they raise no suspicion at first glance. A C2 server appearing as content delivery network (CDN) traffic, a malicious domain behaving like a sinkhole, or honeypot data circulating as an innocent threat signal all create not only detection difficulty but perceptual distortion.

This is why mirrored, masked, and deception-based infrastructures represent one of the most sophisticated, multilayered, and dangerous battlefields in the modern CTI landscape. Threat actors are no longer concerned solely with bypassing technical controls, so they are attacking the analyst's mental model. These infrastructures share three critical characteristics:

- **Extraordinary proximity to reality**
  Behavioral, topological, and temporal-metric properties are made nearly indistinguishable from legitimate systems.
- **Purpose-driven adaptation**
  Network behavior can dynamically change in response to defensive reactions through finely tuned mechanisms. For example, when scanning activity is detected, traffic may instantly shift into a passive mode.
- **Operations on the perceptual architecture**
  They introduce subtle microdeviations into analysts' validation flows, relationship models, and correlation logic that are extremely difficult to notice.

For this reason, an analyst shouldn't relax upon seeing a sinkhole domain nor become complacent when encountering a legitimate-looking telemetry model. Modern deception architectures are no longer parodies of reality; they have become its twin. Here's the danger: When an IOC is wrong, suspicion arises; when a relationship graph is broken, attention is drawn; but when everything looks exactly correct, that's when the real problem begins.

Now, let's dig into the different ways attackers can manipulate your infrastructure, methods to detect these manipulations, and relevant countermeasures.

Types of Infrastructure Manipulations

Modern CTI must address the fight against mirrored, masked, and deception-oriented infrastructures as an epistemological problem; that is, the problem of distinguishing what is real from what is fake. The following characteristics illustrate how such infrastructures are designed to closely resemble legitimate network behavior while concealing their malicious function:

- **C2 infrastructures that imitate legitimate traffic**

  C2 infrastructures imitating legitimate traffic are no longer a basic technique; advanced threat actors have elevated this imitation to a multilayered identity level. Such C2 infrastructures can be characterized as follows:

  - Appear like CDN traffic
  - Mimic the TLS fingerprints of popular software as a service (SaaS) services
  - Exactly match HTTP header ordering used by legitimate browsers
  - Deepen their visibility by proxy cascading through real, legitimate endpoints

  In other words, the traffic both looks and behaves legitimately. When the data surface encountered by the CTI analyst appears superficially normal, IOC-level detections become completely ineffective.

  An even more dangerous technique is known as *reflection-based mirroring*. In this method, the C2 temporarily records the response patterns of a legitimate service and generates its own responses as unaltered copies of those patterns. In such cases, network intrusion detection system (IDS) devices can't distinguish between the real and the fake.

  Some APT groups go as far as browser telemetry, applying session entropy mimicry: by replicating the randomness levels of browser sessions, they make variations in C2 requests extremely difficult to interpret.

- **Malicious domain sets disguised as sinkholes**

  Sinkholes are normally infrastructures that give defenders an advantage: a malicious domain is taken over and redirected to a safe destination. However, modern threat actors have inverted this concept. Some malicious domain sets are now deliberately made to resemble sinkholes:

  - TTL values are tuned to match typical sinkhole behavior.
  - pDNS records are crafted to resemble a seized infrastructure.
  - IP blocks similar to those used in SOAR sinkhole whitelists are employed.
  - Fake passive records are created to make the domain appear first active and later dropped.

  The objective is clear: to trigger the defender's assumption that "this domain is under our control now; there's no threat." A deeper version of this deception technique is known as dual-state domain deception. Here, the domain simultaneously exhibits two different behaviors:

  - A sinkhole appearance to defensive telemetry
  - A real C2 appearance to the attacker's client

  This dual behavior is typically implemented through regional DNS manipulation and is extremely difficult to detect.

- **Deliberate mislabeling of honeypot-derived data**

  Honeypots are normally data-collection mechanisms for defense. However, threat actors can exploit the influence honeypots have on automated intelligence ecosystems. The manipulation works as follows:

  - Controlled but harmless attacks are sent to honeypots.
  - The honeypot reports this traffic as genuine attack behavior.
  - The report is distributed across the wider ecosystem in STIX/TAXII format.
  - Threat clusters on the defender side place these IOCs into incorrect categories.
  - Over time, honeypot-derived data forces analysts to focus on an incorrect threat profile.

  This model is known as honeypot poisoning. A more sophisticated version operates as follows: The attacker applies only the initial stage of a real TTP to the honeypot, for example, reconnaissance, low-rate scanning, or environment profiling, but deliberately doesn't complete the attack. This incomplete attack behavior distorts the defense's TTP evaluation and causes the threat actor to be placed in the wrong category.

  Here's another example: An actor attacks a honeypot using TTPs that actually belong to a different group. This creates false attribution. While defense teams pursue the wrong actor, the real attacker remains hidden.

Detection Methods for These Infrastructures

Mirrored, masked, and deception-based infrastructures are architectural constructs that manipulate the detection process itself and target the defense's observation capacity. Therefore, detection methods require layers that go far deeper than classical IOC correlation—layers that are behavioral, temporal, and often highly intuitive. The following methods represent the most advanced approaches used to make such infrastructures visible:

- **Traffic identity modeling**

  Every network communication carries an identity. Mirrored or masked infrastructures may perfectly imitate this identity at the surface level; however, the deep structure of that identity is inevitably distorted. To extract this deep structure, the following metrics are analyzed:

  - **JA3/JARM fingerprint variance analysis**

    C2 infrastructures that imitate legitimate services can replicate certain fingerprint components with high accuracy; however, microdynamics such as session renewal/renegotiation patterns, timing jitter, post-handshake flow rhythm, and environmental latency distributions are much harder to reproduce consistently and therefore provide discriminative signals.

- **Session cadence (session rhythm)**
  Imitating human behavior–based rhythm is extremely difficult. For example, browser traffic typically exhibits a burst + pause model. Infrastructures imitating C2 traffic can't reproduce this rhythm in a stable way.
- **Payload silence analysis**
  Some mirrored C2 infrastructures silence payloads entirely to reflect legitimate traffic. When this silence becomes "too perfect," it's a sign of suspicion.

Without traffic identity modeling, mirrored/masked infrastructures can remain invisible for long periods in many environments because they generate low anomaly levels in IOC-based controls.

- **Time-based deviation detection**
  Deception infrastructures can't fully simulate the flow of time while imitating network behavior. For this reason, microlatency anomalies are among the strongest detection signals.

  Latency variance depends on factors such as geography, anycast routing, cache state, and peering; therefore, instead of fixed thresholds, the shape of the distribution (extremely low variance = overly regular, extremely high variance = overly chaotic) and its consistency over time should be analyzed comparatively.

  This entropy difference emerges regardless of how well the attacker hides. Time-based microdeviations (jitter, burst + pause patterns, latency distribution, and rate of change) are frequent weaknesses in deception architectures because real systems naturally vary due to environmental factors, and maintaining such variability consistently over time is costly for the attacker.
- **Topological behavior analysis**
  The most powerful way to understand the true nature of an infrastructure isn't to look at what IOCs say, but at how IOCs are connected to one another. Mirrored and masked infrastructures introduce three critical distortions in relationship graphs:
  - Excessive symmetry: Legitimate systems don't operate symmetrically.
  - Missing boundary nodes: Certain relationships are deliberately severed.
  - Low-variance degree distribution: Node-degree distribution is unnatural.

  This technique is commonly used in academic graph analysis but hasn't yet become widespread in CTI. Advanced platforms can use graph morphology to identify masked nodes within the infrastructure. This is one of the most powerful methods for microlevel deception detection.
- **Comparative analysis via sinkhole simulation**
  One of the most advanced techniques used to detect malicious domains behaving like sinkholes is simulating legitimate sinkhole behavior and comparing it. Sinkhole domains typically exhibit the following behaviors:
  - Low TTL
  - Fast resolution stability

  - Specific geographic IP blocks
  - Stable routing patterns

  Masked domains imitate this behavior but can't fully replicate the following:

  - DNS resolution ordering
  - Authoritative DNS response times
  - Start of Authority (SOA) record variances

  These differences exist at the millisecond level, yet they are of enormous importance in CTI analytics.

- **Honeypot interaction reverse engineering**
  To detect manipulated data originating from honeypots, a technique known as *interaction reversal* is used. This technique does the following:

  - Takes the behavior recorded by the honeypot
  - Back-calculates where this behavior would fit within a real attack chain
  - Models possible variants of this chain, and performs consistency checks

  If honeypot data exhibits patterns that are excessively perfect, behaviorally incomplete, too short to align with a TTP flow, or unsynchronized with metadata, it's flagged as manipulated honeypot output.

  A reality many analysts overlook is that honeypot data manipulation is often detected not from the honeypot itself, but from the systems consuming the feed.

- **Cognitive inconsistency analysis**
  This technique isn't well known in the CTI world because it's cognitive rather than purely technical. Sometimes, whether an infrastructure is real or not is revealed not by technical signals, but by the consistency of the story the data tells, for example:

  - C2 behavior is real, but the domain lifecycle is fake.
  - pDNS records are natural, but authoritative records are contradictory.
  - TTP mappings are realistic, but sighting density is inconsistent.
  - Indicator timing is natural, but actor relationships feel forced.

  When these microcontradictions accumulate, a cognitive inconsistency forms in the analyst's mental map. This inconsistency exposes deception layers that technical tools may miss. CTI analysts often apply this method intuitively, without realizing it.

- **Event-driven anomaly correlation**
  Mirrored and masked infrastructures generate unexpected gaps or excessive conformity in real-time distribution layers (especially in event-driven feeds). If the following observations are made, this indicates that the infrastructure was designed for deception:

  - Event triggers that should normally be variable are overly stable.
  - Relationship changes don't follow a natural waveform.
  - Indicator confidence changes are abnormally linear.

Because the real threat ecosystem is chaotic. Deception infrastructures create controlled chaos—and the two aren't the same.

When behavioral models, topological distortions, time-based anomalies, cognitive inconsistencies, and honeypot counter-analysis are combined, reality becomes visible. Defense systems may use only some of these methods; professional CTI analysts must use all of them.

Countermeasures and Defense Strategies

Mirrored, masked, and deception-oriented infrastructures mislead the defense like a military unit moving through fog, distorting vision, disrupting orientation, and warping operational reality. Therefore, countermeasures shouldn't focus solely on detection; they must preserve reality itself, protect the perceptual integrity of the defense, and build cognitive, technical, and behavioral defense layers that collapse the attacker's deceptive architecture.

For modern CTI, threat analysis is the art of constructing a counternarrative; that is, dismantling the false story created by the attacker. This is achieved using the following strategies:

- **Multilayer context validation**

  The most effective defense against context-targeting attacks is cross-validating context across multiple independent sources:

  - The indicator timeline is compared against at least three independent telemetry sources.
  - Sighting density is verified across different geographies.
  - Actor relationships are tested via STIX and also through human intelligence (HUMINT), open-source intelligence (OSINT), and signals intelligence (SIGINT) contributions.
  - Reputation data is normalized and removed from single-vendor dependency.

  The critical principle is that no single context is accepted as true on its own. Every context is treated as a hypothesis under debate in light of other contexts. This neutralizes the strongest attack vector of deception infrastructures: single-context manipulation.

- **Signal-to-noise ratio (SNR) management**

  Against the attacker's noise IOC injection strategy, the defense must apply SNR engineering. The goal isn't to suppress noise, but rather to make the signal itself more visible. This requires advanced analytics such as the following:

  - Entropy maps of IOC behavior
  - Domain lifecycle coefficients
  - C2 traffic density slope analysis
  - Anomaly filtering via time-series decomposition

  As a result, the defense disentangles the chaos created by the attacker and elevates real signals.

- **Topological consistency auditing**
  In graph-based threat models, deception infrastructures deliberately distort relationship maps. The defense must therefore retest the graph's structural consistency daily and sometimes hourly. This involves metrics such as these:
  - Centrality values
  - Cluster cohesiveness
  - Betweenness anomaly index
  - Relationship directionality maps

  Through these controls, the defense can eliminate fake nodes and fake edges silently injected by the attacker. Unnatural graph behavior is often difficult to detect visually; however, metrics such as centrality, cohesiveness, and directionality make structural anomalies more systematic and repeatable.
- **Behavioral identity parameterization**
  C2 infrastructures that imitate legitimate traffic may appear benign at the surface, but behavioral identity can't be fully replicated. The defense can counter this by constructing a fingerprint of a C2 session. This fingerprint consists of dozens of micrometrics, including the following:
  - Session density
  - Jitter variance
  - Post-handshake packet rhythm
  - TLS parameter jitter profiles
  - Burst + pause distributions

  The combination of these metrics dismantles the attacker's traffic mask. This technique is notably new and has been integrated into the operational pipeline by very few organizations.
- **Reversing deception behavior**
  Some infrastructures apply masked/mirrored behavior so skillfully that detection alone is insufficient; the behavior must be reverse-engineered to reconstruct the real infrastructure beneath. In this method, the following occurs:
  - Masked traffic produced by the infrastructure is recorded.
  - This traffic is passed through decomposition algorithms to separate masked behavioral layers.
  - The true traffic model beneath the mask is mathematically reconstructed.

  This approach is particularly effective against mirror C2 techniques used by APT-level actors.
- **Telemetry segmentation against honeypot poisoning**
  Telemetry produced by honeypots is one of the most valuable data sources in threat intelligence because it reflects attacker behavior directly and often without filtering.

However, this advantage is also a critical weakness because when honeypot data is manipulated, the CTI pipeline is directly misled.

Attackers know that honeypot systems automatically feed data into the STIX/TAXII ecosystem. Therefore, by conducting deliberate, low-intensity attacks against honeypots, they can make the data flow appear as real attack behavior. This manipulation leads to IOC-level errors and to severe inaccuracies in deeper intelligence layers such as TTP mapping and actor profiling.

For this reason, defense mechanisms apply two critical techniques (collectively referred to as *telemetry disambiguation*) to preserve the integrity of honeypot telemetry:

- **Honeypot-origin flagging**

  This technique separates honeypot data from normal telemetry and automatically applies context de-weighting to honeypot-derived data. How does it work?

  Each data source (sensor, honeypot, endpoint, gateway) is ingested into the system with an origin tag. Honeypot-originated data is evaluated in reduced trust mode within the CTI pipeline. If an IOC is reported only via honeypot sources, it receives low contextual weight. If the same IOC is also supported by real endpoint telemetry, its weight is increased again.

  Through this method, the defense eliminates two major risks:

  - Fake TTPs generated as a result of honeypot poisoning being accepted as real
  - Honeypot-specific behaviors being mistakenly attributed to a threat actor

  Honeypot-origin flagging is especially effective at preventing the error of artificial threat actor creation that can occur during attribution processes.

- **Pattern interruption analysis**

  Attackers who manipulate honeypot behavior typically simulate only a portion of the attack chain. As a result, honeypot data may appear incomplete, overly regular, overly flawed, or overly superficial compared to real-world attack chains. Pattern interruption analysis is designed precisely to detect these anomalies. This technique does the following:

  - Models the expected stages of an attack chain (reconnaissance → delivery → execution → persistence → C2 → impact)
  - Compares honeypot data against these stages
  - Identifies points where the chain is broken or artificially truncated
  - Flags these breaks as deception signatures

  The following are a few examples:

  - Reconnaissance present, but execution absent
  - Delivery present, but persistence absent
  - C2 handshake present, but no follow-up activity

  A real attacker chain doesn't operate this incompletely; such behavior is an indicator of honeypot manipulation. This analysis detects both missing behavior and excessively perfect behavior. Advanced deception infrastructures sometimes simulate the attack chain in an overly ideal manner: flawless stages, regular timing, low entropy, and so on. This contradicts the nature of real attack behavior. Therefore, pattern interruption analysis detects both deficiency and excess.

When these two layers operate together, the defense can effectively prevent attackers from using honeypot data to do the following:

- Generate fake TTPs
- Distort threat actor profiles
- Misrepresent the attack chain
- Create artificial sighting density
- Misdirect new threat clusters

Honeypot poisoning is a complex threat type that goes unnoticed in many organizations but produces critical consequences. Telemetry disambiguation techniques have therefore become indispensable in modern CTI pipelines to make this threat visible.

- **Advanced threat model cross-validation**

  Deception infrastructures often corrupt only a portion of the threat model. As a result, the defense model doesn't appear to collapse entirely; however, small inconsistencies begin to irritate the analyst's intuition. To resolve this, advanced models cross-validate one another, including the following:

  - Actor capability trees: Structured models that map the technical and operational capabilities historically associated with a threat actor, allowing analysts to verify whether newly observed behaviors align with the actor's known skill set and tooling patterns.
  - Kill-chain alignment: The comparison of observed activity with the expected stages of an attack lifecycle (reconnaissance, delivery, execution, persistence, etc.) to determine whether the sequence of events follows a coherent operational progression.
  - Tactic drift analysis: The examination of gradual shifts in tactics or techniques across related incidents to determine whether changes represent natural evolution of an actor's methods or potentially manipulated signals.
  - Campaign continuity heuristics: Analytical rules used to assess whether infrastructure, tooling, and behavioral patterns remain consistent across different phases of a long-running campaign.

  If any model contradicts the others, this inconsistency signals the possibility of masked infrastructure. This method is particularly effective for uncovering the true profile of long-running APT operations.

- **Trust-weighted dynamic filtering**

  Defense systems should no longer operate on static trust scores. Deception attacks have proven to be especially effective against such approaches. The new model is dynamic, behavior-driven trust weighting, as follows:

- An IOC gains weight not based on source history, but on observed behavior.
- An actor relationship is evaluated not by reputation, but by relational consistency.
- A domain is analyzed not by DNS records, but by lifecycle continuity.

This filtering model neutralizes the trust manipulation hidden by deception infrastructures. Countermeasures aren't designed solely to find malicious infrastructure; they are meant to preserve the defense's perception of reality and dismantle the false narrative written by the attacker.

Defense has now become a competition of perception. Whoever reconstructs reality more accurately wins the battle. Table 7.1 shows the primary layers targeted by feed poisoning attacks, along with typical manipulation vectors, observable anomalies, operational impacts, and recommended defensive controls.

Target Layer	Attacker's Objective	Typical Manipulation Vectors	Symptoms/ Anomalies	Impact	Recommended Controls
Structural manipulation	Distort the relationship model to break the threat narrative	Inconsistent links in STIX relationship semantics; inconsistencies in actor → infrastructure → tool chains; the same IOC associated with conflicting objects	Model appears schema-valid, but relationship graphs show forced links, weak clustering, and abnormal bridge nodes	Incorrect correlation, wrong attribution, flawed hunting hypotheses	Semantic validation rules; ontology-based relationship constraints; graph consistency checks (centrality/ cohesiveness); source-weighted relationship scoring
Context manipulation	Corrupt the meaning of the IOC rather than the IOC itself (perception operation)	Shifting `first_seen`/`last_seen` timestamps; sighting inflation; micro-adjustment of confidence scores; incorrect TTP mapping; narrative shifting via tag/sector/geo metadata	Timeline drift; abnormal sighting distribution; linear/artificial patterns in confidence changes; inconsistencies in MITRE ATT&CK mappings	Prioritization breaks down; SOC triage flows to wrong targets; true campaign rhythm becomes invisible	Multisource validation; time-series consistency checks; sighting source trust and normalization; confidence stability analysis; TTP–actor alignment testing

Table 7.1: Feed Poisoning: Target Layers, Manipulation Vectors, and Defensive Controls

Target Layer	Attacker's Objective	Typical Manipulation Vectors	Symptoms/ Anomalies	Impact	Recommended Controls
Behavioral (consumption/ tempo) manipulation	Disrupt consumption rhythm to create cognitive overload; drown critical signals in noise	Fake relationship waves in event-driven flows; alarm storms via trigger density; temporal poisoning to invert active/finished perception	Waveform patterns in alarm generation; trigger anomalies; abnormal threshold breaches in event volume; stable/calculated repetition in feed update cadence	SOC distraction; missed critical IOCs; automation making incorrect decisions	SNR management; rate/ volume anomaly detection; feed cadence baselining; event waveform analysis; automation kill-switches and de-weighting policies
Trust layer (scoring) manipulation	Hijack the platform's trust computation	Source impersonation; reputation inflation via fake sightings; degrading reputation by associating benign IOCs with malicious ecosystems	Inconsistent trust profiles for the same IOC across sources; sharp score divergence by source; rapid reputation jumps	Distorted whitelist/blacklist decisions; unnecessary blocking or missed threats	Source authentication; provenance/ traceability; source-dependent score separation; anomaly-sensitive dynamic trust weighting

Table 7.1: Feed Poisoning: Target Layers, Manipulation Vectors, and Defensive Controls (Cont.)

7.3 Detecting Low-Quality or Malicious Threat Intelligence Feeds

The value of threat intelligence feeds is far more fragile than most organizations assume. An IOC may be incorrect, which can be tolerated. A domain name may be outdated, and this is also normal. However, when the overall structural quality of a feed deteriorates, the defense begins to operate within a distorted threat reality.

Most analysts evaluate feed quality as if it were a one-dimensional metric: "Is there a correct IOC or not?" In reality, quality rests on a much deeper structure. Determining whether a feed is trustworthy requires more than validating indicator accuracy; it requires examining the rhythm of the data, the behavior of the source, the consistency of context, the logic of relationships, and the feed's evolution over time.

Low-quality or malicious feeds rarely reveal themselves overtly. They usually manifest as small anomalies interspersed among legitimate data:

- Illogical confidence score spikes
- Semantically conflicting actor–IOC relationships
- Unnatural distributions of timestamps
- Unexplained increases in the SNR
- Metadata behavior inconsistent with the source's historical profile
- Synchronization breaks between the feed and other sources

None of these deviations constitutes a critical alert on its own. Yet when they converge, they form the signature of a weak or malicious feed. The most dangerous aspect of malicious feeds is that they appear technically valid. The STIX schema is flawless, TAXII endpoints function correctly, and metadata flows in the proper format, but the content tells a false story, much like a professionally crafted yet manipulative propaganda piece.

For example, an artificially inflated sighting density designed to manufacture reputation may lead an analyst to believe that an IOC is operationally active. Similarly, a deliberately miscategorized TTP may cause an overlooked tactic of a threat actor to appear far more widespread than it truly is.

Even worse, some low-quality feeds are simply misconfigured, rather than malicious. Yet the outcome is the same: the CTI pipeline degrades, SOAR systems are misdirected, and the defense team expends its time in the wrong areas. This, in turn, grants attackers exactly what they desire most: time.

In threat intelligence, the most dangerous thing isn't false information—it's information whose falsehood goes unnoticed. The following sections present key analysis techniques to avoid this worst-case outcome.

7.3.1 Structural and Statistical Quality Analysis

Understanding the quality of a threat intelligence feed isn't limited to verifying whether the data is correct. Quality is measured by how the data behaves, how it evolves, and how consistently it aligns with other data points. For this reason, structural and statistical quality analysis represents one of the most critical validation layers in the CTI domain, typically conducted through a combination of high intuition and high mathematics.

Low-quality or manipulated feeds often appear technically flawless in form; however, they are behaviorally unnatural. Data carries its own rhythm. Consequently, quality detection begins by identifying distortions in that rhythm, as we'll discuss in the following sections.

Data Density Anomalies

The normal data density of a feed follows a recognizable distribution over time: daily, weekly, or campaign-based fluctuations. In malicious or low-quality feeds, however, this density exhibits unnatural spikes. These anomalies typically manifest in three forms:

- **Volume spikes**
  A source suddenly begins producing 5–10 times its normal volume of IOCs within a short period. This doesn't merely indicate an aggressive campaign; more often, it signals attempts at noise injection, reputation manipulation, or feed poisoning.
- **Volume silence**
  The source produces no data for an extended period. Even real threat actors rarely go completely silent; therefore, prolonged silence often indicates pipeline disruption, deliberate manipulation, or data filtering errors.
- **Entropy imbalance**
  IOC diversity normally remains within a stable entropy range. Under attacker manipulation, entropy either rises excessively (noise generation) or drops sharply (pattern hiding).

The common denominator across all three anomalies is that the feed's rhythm is disrupted, which is one of the strongest indicators of invisible manipulation.

Pattern Distortions in Repeating IOC Clusters from the Same Source

A feed provider typically exhibits a distinct IOC production style: its own OSINT methodology, sensor infrastructure, regional telemetry sources, and collection rhythm. For this reason, IOC clusters originating from the same source normally display similar patterns. When manipulation occurs, this pattern breaks. For example, the following situations indicate pattern distortion:

- The TTL values of domain IOCs from the same source have changed dramatically compared to previous months.
- The signature variations of hash IOCs don't align with the source's historical telemetry.
- The pDNS distribution appears overly flat.
- Actor relationships that the source has never reported before suddenly emerge.
- The distribution of confidence scores doesn't match the source's typical curve.

Pattern distortion is often the first signal of feed poisoning due to the following:

- The attacker can't perfectly imitate the source's style.
- The relationship density of manipulated data is unnatural.
- Some fields appear excessively prominent while others appear excessively sparse.
- A style break emerges within the correlation graph.

For advanced CTI analysts, this discrepancy stands out immediately, much like a stylistic inconsistency in a written text, but it remains invisible to most automation systems.

Inconsistencies in Reputation Change Curves

An IOC's reputation score changes regularly over time. These changes occur through natural processes such as an increase in sightings, the strengthening of actor relationships, and the broader adoption of the campaign in which it's used. However, in manipulated feeds, reputation change curves exhibit unnatural delta spikes. These inconsistencies fall into three categories:

- **Nonlinear spike**
  An IOC's reputation may rise from 0.1 to 0.85 within a very short period. Such an abnormal increase, especially when not accompanied by corresponding sightings, is a clear indicator of manipulation.
- **Reverse drift**
  A decrease in an IOC's reputation despite being supported by increasing contextual evidence is most often the result of a poisoned feed. This is a tactic used by attackers to slow down defensive response behavior.
- **Stable delta masking**
  Instead of disrupting the reputation curve, some actors generate a stability that appears natural. This creates a false sense of trust. The analyst assumes the IOC is irrelevant, while in reality it's actively in use.

The consistency of the reputation curve is one of the strongest indicators of feed quality and is highly sensitive to manipulation techniques. Low-quality or malicious feeds don't reveal themselves directly; their schemas are correct, their formats are valid, and they appear technically legitimate. However, their rhythms are distorted, distributions are inconsistent, delta values are unnatural, and patterns don't align with the source.

For this reason, structural and statistical analysis is one of the most strategic skills in the CTI domain. It teaches analysts to focus not on the indicator itself but on the behavior of the indicator. When that behavior is disrupted, the traces of manipulation inevitably surface.

7.3.2 Heuristic Validation Techniques

Detecting low-quality or manipulated feeds isn't always a deterministic process. Some deviations exist that no schema validation, no static control, and no valid format check can reveal. This is because manipulation is often hidden not in the structure of the data, but in the story the data tells, its behavior over time, and the relationships it forms with other sources. For this reason, heuristic validation techniques constitute the strongest line of defense for feed integrity.

The essence of the heuristic approach is that the data may be correct, but it doesn't look correct. Alternatively, the data may be incorrect, but it doesn't appear incorrect. Resolving this zone of ambiguity requires methods that go beyond rule-based validation and are more abstract, more behavioral, and more contextual. Next, we'll present three of the most critical of these methods.

Risk-Weighted Quality Metric (Weighted Confidence Score)

Evaluating the reliability of an IOC is often done simply by looking at its confidence label. However, this is one of the most common examples of a flawed approach because confidence is determined on the producer side, whereas weighted confidence is recalculated on the consumer side.

The weighted confidence score is derived through the dynamic weighting of the following factors:

- Historical consistency of the source
- Changes in the IOC's behavior over time
- Whether the sighting density is natural
- IOC's role within the attack chain
- Semantic consistency of relationships (relationship coherence)
- Degree of overlap with other sources
- Real-world stability of the IOC value (infrastructure volatility score)

The critical point of this metric is that confidence is no longer a fixed number; it's a function that evolves within context. Even a confidence value of 1.0 may drop to 0.41 in a weighted model because the IOC's lifecycle, the reliability of the source, or its contextual alignment demands it.

Low-quality feeds typically exhibit extreme weight collapse under this model. Attacker manipulation, on the other hand, usually produces artificial inflation. These two extremes constitute the strongest detection signals of weighted scoring.

Multisource Cross-Validation

In the modern threat intelligence ecosystem, no IOC can be accepted as true in isolation. Attackers no longer target only systems; they target intelligence itself—injecting false relationships, manipulating context, distorting timelines, and even deliberately sabotaging reputation metrics. For this reason, the most powerful way to validate an IOC is to compare it against the narratives told by other sources; that is, to apply multisource cross-validation.

Cross-validation measures how closely different sources' perceptions of the threat landscape align. The same IOC may carry entirely different meanings across three different ecosystems. What matters is whether these differences are natural or manipulative.

Cross-validation is also a measurement of epistemic consistency across sources. When this consistency breaks down, it indicates either a degradation in feed quality or the onset of deliberate manipulation.

This approach consists of five primary validation layers, and each layer has its own distinct signal-to-noise separation characteristics:

- **Temporal cross-check**
  The `first_seen` and `last_seen` values of an IOC should be similar across different sources; however, they don't need to be identical. A natural telemetry source may observe an IOC earlier or later than another. But CTI analysis recognizes one fundamental principle: If a timeline contains major discrepancies, the cause isn't faulty telemetry—it's manipulation.

Let's examine two example anomalies:

- While one source marks an IOC as `first_seen: 03:12`, another source states `first_seen:` five months earlier, which means a different narrative is being told.
- A `last_seen` date that doesn't align with the trend observed in another source produces a signal that the IOC lifecycle has been distorted.

Temporal cross-checking appears to be the simplest layer, yet it has one of the highest manipulation-detection capabilities because attackers often manage to imitate form but fail to imitate time.

- **Topological matching**
  The relationship graph an IOC forms with other objects isn't the language of a single source, but the shared language of the entire ecosystem. For this reason, the strongest validation method is to compare how the same IOC is positioned across different feeds.

  The following questions are critical:

  - Is the IOC associated with the same actors across all sources?
  - Is it linked to the same malware family, or is the relationship entirely missing in some sources?
  - Does the graph structure appear natural, or is the relationship depth in one source suspiciously shallow compared to others?

  An advanced manipulation technique known as *relationship poisoning* becomes immediately visible through topological matching. An attacker may incorrectly associate an IOC in a single location, but can't do so consistently across all sources.

  A natural relationship graph is irregular, organic, and grows over time. A fabricated relationship graph is symmetric, context-deficient, and topologically too clean.

- **Behavioral similarity scoring**
  More than a static value, an IOC carries behavior: traffic profile, sighting cadence, TTL fluctuations, DNS resolution intensity, role within a campaign, and more. This behavior should appear similar across different sources. If an IOC exhibits high-risk behavior in one source, appears completely passive in another, and isn't visible at all in a third source, this behavioral inconsistency is a classic indicator of manipulation.

  One of the most commonly used metrics in advanced analysis is behavioral *cosine similarity*. Behavioral vectors are compared across different sources, and the behavioral identity of the IOC is derived. In practice, this comparison is performed by representing observable attributes of the IOC (such as request frequency, DNS query patterns, TTL variability, geographic distribution of sightings, and protocol usage) as numerical features within a behavioral vector. These vectors are then compared across different intelligence sources or telemetry datasets using cosine similarity to measure how closely their behavioral patterns align.

Cosine similarity values between behavioral vectors must be calibrated according to the organization's own telemetry distribution. In practice, in many environments, high similarity (e.g., ≈ 0.8 and above) indicates natural behavior, while noticeable drops (e.g., below ≈ 0.5) suggest contextual inconsistency and a potential for manipulation. Very low similarity values, on the other hand, raise the risk of feed poisoning.

- **Source deviation index (SDI)**

  Each feed source develops its own identity over time: the types of IOCs it produces, actor relationships, geographic telemetry density, behavioral variance, and so on. For this reason, a source suddenly diverging significantly from others isn't normal.

  The SDI metric measures how closely a feed's behavior over time aligns with the aggregated behavioral baseline of other sources. In practice, this metric is derived by comparing key feed attributes, such as IOC type distribution, sighting frequency, geographic telemetry patterns, and relationship density, against the multisource average, and calculating the statistical deviation between the source and this baseline, for example:

  - SDI < 0.25 = source is behaving normally
  - SDI 0.25–0.45 = low inconsistency
  - SDI 0.45–0.65 = abnormal fluctuation
  - SDI > 0.65 = high likelihood of manipulation

  This value reflects the quality of the feed and the possibility that the source itself has been compromised. This is because attackers often target feed producers directly, injecting their own contradictory IOCs into the system.

- **Confidence consensus**

  The confidence score is often underestimated, but in the world of cross-validation, it represents the pulse of ecosystem-wide consistency. For example, an IOC that presents as follows poses a problem:

  - High in Source A
  - Medium in Source B
  - Low in Source C
  - Unknown in Source D

  The narratives told by the data aren't aligned. Such discrepancies are particularly common in reputation poisoning attacks. An attacker artificially raises or lowers the confidence score of an IOC in one source, while other sources fail to converge accordingly.

  Confidence consensus algorithms evaluate the following:

  - Is there a logical correlation between the scores?
  - In which source do sudden jumps occur?
  - Do score changes align with the normal behavior patterns of the sources?

  This technique can even reveal the social engineering variant of feed manipulation because some sources may be more easily deceived by attacker-crafted false context.

Entropy-Based Anomaly Scoring

Every IOC has an inherent level of structural randomness, for example:

- DGA domains exhibit high entropy.
- Human-generated domains have low entropy.
- Malicious URL parameters carry medium entropy.
- C2 communication patterns exhibit temporal entropy.

Entropy-based anomaly scoring is used to detect distortions in these natural randomness levels. This technique is applied across three dimensions:

- **String entropy analysis**
  Entropy levels of IOC components such as domains, URLs, and file paths are calculated. Automated generators used for fake IOC production typically generate outputs that are either overly regular or excessively chaotic.
- **Temporal entropy analysis**
  The observation density of an IOC over time is examined. An IOC belonging to a real campaign follows a recognizable wave pattern with a certain rhythm. Manipulated IOCs typically appear overly linear, overly symmetric, or excessively random. These patterns are unnatural.
- **Relational entropy analysis**
  Variance within an IOC's relationship graph is measured. Real relationships are irregular; fake relationships appear too orderly, meaning low entropy.

A critical insight here is that feed manipulation techniques such as relationship poisoning can be detected with high precision through entropy imbalances and topological irregularities in relationship graphs. For this reason, relational entropy analysis stands out as one of the most powerful methods for detecting feed poisoning.

Numerical models are unforgiving; it's nearly impossible for an attacker to fabricate a relationship topology that appears naturally organic.

7.3.3 Source Reliability Modeling

Threat intelligence isn't gold that can simply be mined and refined into a single ingot; it's a composite of inputs, not a product that can depend on a single source because in the CTI universe, every source is fallible. Some degrade over time, some are manipulated, some are limited by partial operational visibility, and others deliver high accuracy only within specific regions. For this reason, modern CTI architecture no longer focuses on the IOC itself. Instead, it evaluates the character of the source that produces the IOC.

Source reliability modeling isn't a superficial good source versus bad source distinction. It's a multidimensional evaluation architecture that measures a source's behavior over time, consistency, data rhythm, in-context performance, entropy balance, and long-term persistence. Most importantly, source reliability evolves over time like a living entity.

Source reliability is composed of three foundational elements, which we'll discuss in the following sections.

Source Trust Tiers

In the threat intelligence ecosystem, the question of how accurate an IOC is can no longer be answered solely by examining the IOC's intrinsic properties because no data source is absolute, no feed is always correct, and no telemetry surface provides full visibility. Above all, no source is completely immune to manipulation.

So, when measuring IOC quality, modern CTI pipelines and analytical engines evaluate the identity, behavior, history, and consistency of the source that produces the data, not the data itself. *Source trust modeling* is a multidimensional analysis that defines a source's epistemic weight within the threat intelligence ecosystem.

In contemporary CTI infrastructures, in addition to reliability, sources are differentiated by dozens of metrics such as the following:

- Visibility capacity
- Contextual integrity
- Topological relationship accuracy
- Temporal consistency
- Reputation history
- Tendency to generate noise
- Level of manipulation risk
- Manner of contribution to the intelligence lifecycle

To manage this complexity, CTI architectures classify sources into the following four fundamental trust tiers. This classification does more than assign a label to a source; it also produces a multiplier effect that directly influences the final risk value of an IOC.

- **T0 - Epistemic core sources**

  The T0 tier represents the closest layer to ground truth within the intelligence universe. These sources include the following:

  - Direct internal organizational telemetry
  - Endpoint sensors
  - Network gateway, DNS, proxy, and firewall logs
  - Internal sighting data
  - Raw event records from EDR and extended detection and response (XDR) infrastructures
  - Organization-owned host-based behavioral telemetry

  These sources generate firsthand observational data. The most critical characteristic of T0 sources is that the data doesn't pass through intermediaries; it directly touches the event itself. For this reason, T0 data has the highest contextual accuracy, the lowest risk of manipulation, and the highest correlation quality.

However, this doesn't mean that T0 sources are perfectly reliable. Advanced attack scenarios such as insider threats, log manipulation, and endpoint impersonation can also mislead T0 sources. Therefore, even T0 data isn't considered absolute truth, but rather high-probability truth.

- **T1 - High-trust external sources**

  The T1 tier includes external feeds that are the most consistent, have the lowest false-positive rates, and produce highly accurate data from a relationship-graph perspective. The common characteristics of these sources include the following:

  - Long-term behavioral stability
  - High accuracy within similar IOC types
  - Campaign-based telemetry consistency
  - Minimal fluctuation in confidence scores
  - High accuracy in topological relationship models

  T1 sources are typically provided by the following:

  - Enterprise threat intelligence services
  - Global sensor networks
  - Corporate reputation data providers
  - Premium TAXII/STIX feed producers

  What makes T1 data valuable isn't the volume of data it produces, but the fact that the data it produces is statistically and topologically consistent.

- **T2 - Medium-trust sources**

  T2 sources represent the gray zone of CTI analytical processes. These sources are heterogeneous in nature and include the following:

  - OSINT-based feeds
  - Community-generated IOC clusters
  - Regional telemetry providers
  - Semiautomated scanning infrastructures
  - Narrow-scope sources focused on specific threat categories

  The most characteristic feature of T2 is that it produces both highly valuable insights and significant noise. For this reason, T2 data isn't used alone for decision-making, but provides very high contextual value when combined with T0/T1 data.

  T2 sources are difficult to analyze because, while they produce high-quality data during certain periods, at other times, they carry a high risk of manipulation. Analysts typically spend the most time validating T2 data. At this tier, cross-validation is mandatory.

- **T3 - Low-trust/high manipulation-risk sources**
  The T3 tier represents the most fragile trust layer of the intelligence ecosystem. Sources in this category include the following:
  - Honeypot-only feeds
  - Random OSINT websites
  - IOC lists shared on community forums
  - Academic or experimental datasets
  - Open-source automated scanners prone to manipulation
  - Sources exhibiting persistent stylistic inconsistency

  The problems associated with T3 data include the following:
  - High false-positive rates
  - Relationship inconsistencies
  - Sharp fluctuations in reputation scores
  - Abnormal entropy levels
  - Traces of artificial telemetry
  - Susceptibility to timestamp manipulation

  IOCs originating from these sources typically cap at a weighted confidence score of approximately 0.3–0.4 and rarely exceed this threshold. Therefore, T3 data is never used on its own for decision-making. However, when placed in the correct context, especially during early detection phases, it can carry high strategic value.

Source reliability is a multiplier mechanism that determines an IOC's final confidence score, for example:

- An IOC's base confidence value = 0.9
- Source T3 effective confidence = 0.9 × 0.4 = 0.36

On the other hand:

- Base confidence = 0.6
- Source T1 effective confidence = 0.6 × 0.85 = 0.51

In other words, a weak source can devalue a strong IOC, while a strong source can provide context to a weak IOC. Most CTI analysts know this intuitively but don't apply it mathematically. Modern CTI systems now apply this model automatically.

Time-Decay Model

Every IOC loses value over time, much like a biological organism. For this reason, modern CTI doesn't measure the IOC's risk value itself; it measures the rate at which that risk value erodes over time. The time-decay model operates on the following principles:

- **IOC's effective lifetime window**
  Different types of IOCs have different operational lifetimes. The effectiveness of an indicator depends not only on its accuracy but also on how long it remains relevant within the threat environment. Some indicators become obsolete quickly as attackers rotate infrastructure, while others remain valid for extended periods because they are tied to persistent artifacts such as malware binaries. For this reason, analysts must consider the expected lifetime window of each IOC type when prioritizing detection and response actions:
  - Short for IPs (hours–days)
  - Medium for domains (weeks)
  - Long for hashes (years)
- **Risk decay function**
  Some IOCs lose value linearly over time; others exhibit an exponential decline, for example:
  - C2 domains – fast decay: C2 domains tend to lose their intelligence value quickly because attackers frequently rotate or abandon infrastructure once it's exposed or blocked. As a result, the operational lifetime of such domains is often short.
  - Phishing URL – medium decay: Phishing URLs usually remain active for a limited campaign window. They may persist for several days or weeks while the phishing operation is ongoing, but are typically removed or abandoned once the campaign is detected.
  - Malware hash – low decay: Malware hashes represent static artifacts tied to a specific binary. Even if the attacker changes infrastructure or campaign details, the hash of a previously observed sample remains a reliable identifier for that exact file over long periods.
- **Decay + sighting synergy**
  If an IOC appears to be aging but sightings are still active, the decay model flags this contradiction.
- **Decay manipulation detection**
  Some malicious feeds manipulate an IOC's `first_seen` value to create a fresh appearance. The time-decay model catches this manipulation because the math is definitive: Artificial freshness doesn't produce natural behavior.

Time is the greatest enemy of malicious feeds because manipulation can't be sustained for long. The time-decay model exposes this weakness.

Feed Stability Index

Every feed exhibits characteristic behavior, much like a human being. A source's long-term stability is its most critical trust indicator. This behavior is measured across three layers:

- **Structural stability**
  Over time, does the source remain consistent in the following areas?
  - Types of IOCs it produces
  - Shapes of its relationship graphs
  - Reputation distribution
  - Confidence curve
  - Metadata completeness rate

  Or, is it good in one period and poor in another? Manipulated sources typically show abrupt style breaks. This break doesn't match the feed's normal behavioral signature.
- **Behavioral stability**
  How consistent are the feed's patterns over time in the following:
  - Periodic rhythm
  - Distribution of sighting data
  - Correlation trends
  - Anomaly patterns

  Reliable sources exhibit a natural rhythm; manipulated sources exhibit a rhythm that is too regular or too noisy.
- **Source integrity drift**
  This concept is almost never discussed in CTI literature, yet it's one of the most important detection tools. If a source does the following over time, that indicates integrity drift:
  - Begins producing fewer relationships
  - Starts showing inconsistencies in relationship directions
  - Shows weakening actor matches
  - Exhibits reputation curves that behave contradictorily compared to earlier periods

  Integrity drift is one of the strongest indicators that a source has been compromised or has operationally degraded.

Key Considerations

Source trust modeling, no matter how mathematical it may appear, carries one of the highest risks of producing false confidence. When a model is poorly constructed, it measures an organization's epistemic habits, not reality; that is, which sources you instinctively trust, which noise you have long accepted as normal, and which gaps you fill with assumptions but not data. The following considerations don't just comprise a quality checklist; they form a defensive framework designed to protect the CTI system against deceiving itself:

- **The model itself is an attack surface (model-poisoning perspective).**
  In CTI, feed poisoning most often targets the scoring layer, not the IOC layer. For an attacker, the most efficient objective isn't to make the IOC wrong, but to make its weight calculated incorrectly.

For this reason, the trio of source trust tiers + weighted confidence + decay isn't an algorithm set but a decision engine. Targeting decision engine occurs through the following:

- Threshold manipulation: Producing a large number of IOCs near the organization's automation threshold so the model repeatedly fails around the critical boundary (e.g., dense production in the 0.49–0.51 band).
- Distribution-shape imitation: Distorting variance without altering the mean. This bypasses simple alarm mechanisms, but a well-designed model can read the language of variance.
- Targeting the control surface: Attacking the variable with the greatest influence in the model. If the source tier multiplier accounts for 60% of the score, the attacker targets the source; if time-decay dominates, `first_seen`/`last_seen` values are manipulated.

Your scores aren't results; they are measurement instruments. If the instrument is broken, even correct data will be misread.

- **T0 data isn't absolute truth.**
  T0 (internal telemetry) is treated in many architectures as the core reality. Although this is correct, it's incomplete. Internal data can also produce the following:
  - Visibility drift (sensor version changes, log format drift)
  - Coverage drift (new segments, new proxies, new EDR policies)
  - Collection bias (overrepresentation of a specific environment)
  - Operational noise (tests, penetration tests, red team activity, automation errors)

  If you treat internal data as the gold standard when evaluating external feeds, internal drift may unjustly label external data as bad—or do the opposite. Therefore, T0 data also requires self-validation.

  A good practical check is to track the incident-to-sighting ratio (incident request [IR]/service request [SR]) trend within T0. If sightings increase but incidents do not, two possibilities exist: Either defenses have significantly improved, or sightings have lost meaning.

- **Confidence isn't a label, but a production-process signature.**
  Treating confidence as a simple numeric value is a classic mistake. In most sources, confidence is a combination of the producer's visibility capacity, methodology, analyst culture, and commercial motivations. Thus, confidence describes the personality of the source before it describes the correctness of an IOC. When different sources assign different confidence values to the same IOC, the question isn't "Which is correct?" but "Which source's confidence-production behavior is consistent in this context?"

  A critical technical detail to monitor is the tail behavior of the confidence distribution. Manipulation usually affects the tails, not the mean:
  - An inflated high tail = reputation manipulation
  - An inflated low tail = reputation suppression
  - An overpopulated middle band = threshold-adjacent tactics

- **Cross-validation must be an alignment test, not a match test.**
  The most dangerous fallacy in multisource validation is the reflexive belief that if it appears in many sources, it must be true. This is the attacker's objective: to create a consensus illusion.

  Cross-validation must therefore be performed across three layers:

  - Value matching: Is the IOC the same?
  - Narrative alignment: Does the IOC play the same role (phishing, C2, scanning, etc.)?
  - Topological alignment: Does the IOC occupy a similar position in the graph? (Does it fall into the same cluster? Connect to similar bridge nodes?)

  An IOC may appear in three sources, but if it plays a different role in each, this usually reflects semantic divergence rather than quality.

- **Topological cleanliness is a reason for suspicion.**
  Real relationship graphs are messy and irregular: missing edges, asymmetric growth, and unexpected bridges. In fake or manipulated graphs, two patterns commonly appear:

  - Excessive symmetry (overly neat clusters)
  - Excessive minimalism (context stripped away, schema intact)

  Therefore, when assessing relationship quality, *more relationships = better* is a false assumption. The real metric is whether relationship growth aligns with data rhythm. If a source suddenly begins producing relationships, this may indicate genuine maturation, but more often, it's the first signal of relationship poisoning attempts.

- **Attackers can't imitate the time layer, but you may misread it.**
  As correctly emphasized earlier, attackers can imitate form but struggle to imitate time. The critical risk here is that time is often misread not by the attacker, but by the consuming system. Here are a few examples:

  - `first_seen` is sometimes not the true first observation, but the moment the producer recorded it.
  - `last_seen` may not represent the actual last observation, but the last update time.
  - Some feeds apply time zone normalization, rounding, or aggregation.

  Therefore, when performing temporal cross-checks, the exact meaning assigned to time-related fields must be clearly defined across the intelligence pipeline. In many feeds, timestamps such as `first_seen`, `last_seen`, or `observed_at` may refer to different events depending on how the data was collected and published. In some cases, the timestamp represents the moment a sensor detected activity; in others, it may indicate when an analyst recorded the observation or when the feed provider published the data. Without clearly understanding this temporal context, analysts may incorrectly interpret the chronology of events and draw misleading conclusions from otherwise valid intelligence.

Caution

If this contract is unclear, the time-decay model may misinterpret normal behavior as manipulation.

- **Beware of false confidence in entropy analysis.**

  String entropy is powerful for use cases such as DGA detection; however, for CTI quality, it's not evidence but a suspicion generator. Attackers can tune entropy, and many legitimate infrastructures (CDN subdomains, tracking parameters) naturally exhibit high entropy.

  Entropy analysis must therefore satisfy three conditions:

  - Class condition: What entropy band is normal for this IOC class?
  - Source condition: What entropy distribution has this source historically produced for this class?
  - Context condition: Is the IOC's graph role and temporal behavior consistent with the observed entropy?

  If these three conditions aren't met together, entropy is highly prone to false positives.

- **SDI can be misled.**

  SDI is conceptually strong; however, if the reference set is polluted, SDI becomes polluted as well. An attacker may operate in two stages:

  - First, inject widespread noise into the ecosystem (via OSINT or community sources).
  - Then, create a baseline that makes the target source's deviation appear normal.

  Therefore, the SDI reference set should neither be static nor randomly changed. It should consist of T1-heavy core validation sources combined with carefully controlled T2 sources.

- **Over-amplifying the tier multiplier blinds the model.**

  If the source multiplier is too aggressive, the following occurs:

  - Genuine early signals from T3 sources are suppressed.
  - Noise from T1 sources appears overly valuable.
  - The system loses its discovery capability and becomes merely a validation machine.

  This is exactly what the attacker wants: a reactive defense. Therefore, the tier multiplier shouldn't be an absolute reduction but a context-sensitive softening mechanism. For example, T3 data should never trigger automatic blocking, but it should remain valuable for hypothesis generation.

- **The model produces a decision load, not a finding.**

  The most critical point is that while CTI modeling improves quality, it simultaneously generates a decision burden on SOC/SOAR operations. If the model produces too many ambiguous outputs, analysts burn out. If the model produces too many high outputs, automation becomes dangerous. Finally, if the model produces too many low outputs, blind spots emerge.

Therefore, the success of quality analysis must be measured not only by accuracy but by decision ergonomics:

- Validation time per alert
- Operational time lost per false positive
- Cost of missed true positives
- Automation rollback requirements

CTI quality must optimize human time as much as data accuracy. Otherwise, even the most accurate feed exhausts the organization and weakens its defense.

In threat intelligence, aiming for 100% accuracy is misleading in practice because the threat landscape is dynamic, observation surfaces are partial, and adversaries are adaptive. But this is achievable: making errors visible. That is where the true strength of this section lies: in systematizing the detection of error.

If your CTI pipeline makes rhythm disruption, relationship topology forgery, temporal contract deviations, consensus illusions, and source integrity drift visible, then you're doing more than consuming feeds; you're building an intelligence system that audits feeds themselves.

7.4 Data Enrichment Techniques

In threat intelligence, no data is ever complete on its own. An IP address, a domain name, a hash value, or an email indicator that appears harmless all carry only an initial signal in their raw form. When that signal is detached from context, it's weak, its lifespan is short, and it often leads the analyst in the wrong direction. Raw data is like an object standing alone in a dark room: its existence is evident, but what it represents is unclear.

This is where data enrichment comes into play. Enrichment is the process that dispels this uncertainty, giving the signal a lifeline, a story, and a direction. Through enrichment, a single IOC ceases to be an isolated indicator and becomes a gateway into the unseen layers of the threat ecosystem.

Enrichment is more than collecting additional information. Its true function is to reveal the following:

- Which attack chain the IOC plays a role in
- Which actors it's associated with
- Which campaigns it's connected to
- How it behaves over time
- Which infrastructures it's linked to
- Its true position within the threat ecosystem

Without this context, an IOC is nothing more than a data point. Once context is added, it transforms into a decision-making instrument, a generator of foresight, and often an early signal of an attack that hasn't yet occurred.

Data enrichment isn't a passive task in threat intelligence; on the contrary, it's one of the most critical cognitive mechanisms through which an analyst learns to understand what they see. An IP address on its own conveys very little meaning, but when that same IP is combined with the following enrichment layers, the analyst begins to understand what it is and why it matters:

- Links to past campaigns
- An ASN history that aligns with adversarial behavior
- Anomalous geographic mobility
- Sudden spikes in its reputation curve
- Resemblance to C2 topologies previously used by a known APT

Sometimes, a single enrichment layer—such as the timing of a domain's registration coinciding with the start of a campaign—can force the analyst to reconstruct the entire picture. At other times, a small pDNS record reveals that an IOC isn't just a minor component of an attack chain, but rather the primary node of the C2 infrastructure at the center of that chain.

The following sections explore techniques to enrich IOC context and to correlate that enriched content with threat profiles.

7.4.1 IOC Contextual Enrichment

The raw form of IOCs never provides the depth an analyst requires. IP addresses, domains, and hashes gain meaning only through the context woven around them. When context enrichment is performed incorrectly, a threat actor may appear to represent an entirely different profile; when done correctly, the IOC clearly reveals the role it plays within the attack chain.

Let's walk through three fundamental enrichment branches that are the most critical context generators in modern CTI architectures.

Context Deepening with ASN, WHOIS, GeoIP, and RDAP

An IP address or domain is considered incompletely analyzed unless the network autonomy under which it operates is understood. The ASN provides a high-confidence indicator of which infrastructure operates a given IOC.

ASN blocks frequently preferred by attackers become recognizable over time; some are inexpensive, some are rapidly disposable, and others are known for short-lived abuse windows. For this reason, ASN analysis strengthens the true identity of an IOC by revealing the following:

- The historical abuse rate of the hosting provider in use
- Whether the IP address is associated with adversarial communities
- The routing behavior and stability of the network
- The ownership change history of the network

WHOIS data provides critical metrics such as domain age, registrar, name server history, registration format, and registration timing. For example, a domain registered at 03:11 UTC may reflect a routine registration pattern observed among specific botnet operators. These timestamps are often overlooked, but research shows that certain actors consistently favor specific time windows.

GeoIP reveals the physical geography of an IOC, but on its own, it's insufficient. The true power of enrichment emerges when GeoIP is combined with temporal movement analysis. An IP address rapidly jumping between regions often indicates the use of CDNs, proxies, or attacker-controlled infrastructure rather than natural network distribution.

Registration Data Access Protocol (RDAP), the modern and more verifiable successor to WHOIS, offers a critical advantage by exposing organizational relationship chains. Historical ASN transfers, IP block resale tendencies, and suboperator acquisition records of the organization associated with an IP are embedded within RDAP data. In many investigations, these records reveal which sub-ISP an attacker is attempting to use to remain unnoticed.

In operational CTI analysis, infrastructure context is typically enriched through a combination of ASN ownership data, domain registration records, geographic hosting indicators, and allocation history. Analysts commonly begin by identifying which autonomous system operates an IP address. For example, a query such as `whois -h whois.cymru.com " -v 185.220.101.1"` can reveal the ASN number, the organization responsible for the network, and the associated network prefix. If the same ASN repeatedly appears across multiple malicious indicators, it may suggest an attacker-preferred hosting infrastructure or a provider that historically experiences higher abuse levels.

Domain registration records provide another important layer of context. A simple query such as `whois suspicious-domain-example.com` can reveal key attributes including domain creation date, registrar information, name server configuration, and the presence of privacy-protected registrant fields. Analysts frequently examine these elements to identify patterns across malicious campaigns. For instance, many phishing operations rely on recently registered domains that share similar naming structures or reuse the same name server infrastructure.

Geographic hosting data also contributes to contextual analysis. Using tools such as `geoiplookup 185.220.101.1`, analysts can determine the approximate geographic location of an IP address. However, a single location snapshot is rarely sufficient. When infrastructure associated with the same campaign appears to move rapidly between geographically distant regions, this often indicates the use of proxy relays, VPN infrastructure, CDN masking, or attacker-controlled routing layers rather than a stable hosting environment.

Additional ownership and allocation insight can be obtained through RDAP queries. For example, `curl https://rdap.arin.net/registry/ip/185.220.101.1` retrieves structured records describing the organization responsible for the IP block, along with allocation hierarchy and registration metadata. These records often reveal whether infrastructure is directly owned by a hosting provider or suballocated through reseller networks. In many investigations, such information helps analysts determine whether an attacker is leveraging leased infrastructure, secondary providers, or transient network allocations to remain less visible.

By correlating ASN ownership data, WHOIS registration patterns, GeoIP distribution indicators, and RDAP allocation records, analysts can move beyond simple IOC identification and develop a clearer understanding of the infrastructure ecosystem in which an indicator operates. This enrichment process frequently reveals infrastructure reuse patterns, hosting environments commonly favored by threat actors, and organizational relationships that help explain how adversaries structure and conceal their operational infrastructure.

Deriving Lifecycle Intelligence Through Passive DNS Correlation

Discussed at length in Chapter 4, pDNS is one of the most powerful contextual tools for IOC enrichment. pDNS does more than show which IP addresses a domain resolves to; it reveals temporal changes, resolution density, connectivity continuity, and infrastructure transition patterns.

Observations about a domain allow the IOC to be viewed not as a static domain, but as a behavioral entity, for example:

- Moving across 14 different IP addresses within two days
- Resolving exclusively to specific ASN blocks at the same time
- Intersecting with IPs previously used in known APT campaigns
- Exhibiting abnormal increases or decreases in TTL values
- Becoming active only during specific attack windows and remaining completely silent otherwise,

pDNS data provides unique value particularly in areas such as these:

- C2 infrastructure rotation
- Fast-flux networks
- DGA detection
- Hijacking attempts
- Attackers' backend IP rotation rhythms

A lesser-known fact is that a domain's TTL behavior often reflects malicious intent more accurately than the IP address itself. Sudden drops in TTL values frequently coincide with periods when attackers rapidly rotate infrastructure to evade detection.

Mapping the Adversary Ecosystem Through Reverse Infrastructure Mapping

Reverse infrastructure mapping involves tracing an IOC backward to extract the components it's connected to, such as domain clusters, IP blocks, Secure Sockets Layer (SSL) certificate fingerprints, hosting operators, CDN bridges, routing architectures, and subnet relationships.

This technique enables the analyst to understand the attacker's infrastructural topology. An IOC by itself may appear small; however, the network it's connected to may represent a much larger operation. For example, the following are all critical signals that become visible through reverse mapping:

- The same SSL certificate being used across 19 different domains
- The same ASN block repeatedly appearing in C2, phishing, and spam infrastructures
- The same IP address being shared by different APT groups at different times (*infrastructure cohabitation*)
- The same certificate chain being deployed across both malicious and benign servers

The greatest strength of reverse infrastructure mapping is that an attacker can change a single IOC, but hiding the entire relational ecosystem of the infrastructure is nearly impossible. For this reason, infrastructure traces are among the most resilient sources of context against feed poisoning.

Putting It All Together

Contextual enrichment transforms an IOC from a solitary value into a component of attacker behavior, infrastructure topology, and campaign evolution. When the following five layers are combined, an IOC becomes a living component of an attack narrative:

- ASN analysis provides identity.
- WHOIS/RDAP delivers historical grounding.
- GeoIP reveals geographic movement.
- pDNS illuminates the lifecycle.
- Reverse mapping exposes the ecosystem.

7.4.2 Correlating with the Threat Profile

No matter how completely an IOC's context is enriched, it only gains true value when correlated with the correct threat profile. This is because the operational meaning of an indicator isn't defined solely by its technical properties, but by which actor used it, for what purpose, and at which phase of which campaign. In this sense, the process of correlating with a threat profile is a critical cognitive stage that transforms scattered signals in the analyst's hands into a coherent attack narrative. In this context, the term *threat profile* refers not to a specific adversary identity, but to the behavioral and operational characteristics associated with an attacker or group of attackers. While a threat actor represents the entity conducting the attack, a threat profile captures the patterns that describe how that actor typically operates, including preferred infrastructure, TTP usage, operational timing, and campaign style. Mapping IOCs to MITRE ATT&CK techniques, TTP patterns, and adversary cluster models goes beyond the surface form of the indicator and makes the attacker's intent, methodology, and behavioral evolution visible.

We'll discover the complete correlation process and walk through a detailed use case in the following sections.

Components of the Correlation Process

The most powerful component of this correlation process is the *MITRE ATT&CK framework.* As we discussed in Chapter 6, Section 6.3, MITRE ATT&CK provides a common vocabulary by categorizing adversary behaviors into specific techniques and tactical objectives used by actors worldwide. Seeing which MITRE ATT&CK techniques intersect with a given IOC enables the analyst to determine both what is happening and why it's happening this way. For example, a domain being active at specific hours may be associated with credential harvesting campaigns that previously exhibited similar behavior. Or, IP addresses routed through particular ASN blocks may align with a known lateral movement technique. When an IOC shows semantic alignment with MITRE ATT&CK techniques, its role within a campaign becomes clearer: Is it an initial access vector, a persistence mechanism, C2 communication, or a data exfiltration stage? Recognizing this relationship transforms the IOC from an isolated signal into a functional component of the attack chain.

TTP pattern enrichment represents a more refined and advanced layer of this process. TTPs reveal both what threat actors do and how they do it (see Chapter 6, Section 6.2, for an in-depth discussion). Behavioral patterns are often stronger identifiers than IOCs because attackers can change technical indicators, whereas behavioral patterns tend to be repetitive and characteristic. For instance, some actor groups prefer specific packet sizes in C2 traffic; others register domains during particular time windows; and others repeatedly apply routine parameter configurations to specific versions of open-source C2 frameworks. When these subtle behavioral details are matched to an IOC and mapped to a TTP template, the analyst moves one step closer to understanding the attacker's methodological identity.

TTP enrichment goes beyond behavioral matching to read the relationship between time and context. Whether an IOC matches a historically known TTP, whether its behavioral flow reflects the attacker's past habits, or whether it signals a new variant of an existing TTP are all evaluated by the enrichment engine. This process surfaces known behaviors and deviations within TTPs. Deviations matter because advanced threat actors often reuse the same TTP with small variations. These variations may indicate a transition to a new campaign, a tactical optimization, or an attempt to evade specific defensive mechanisms.

Adversary cluster mapping is the highest layer of threat profile correlation. At this stage, the objective isn't simply to determine who may have used an IOC, but to understand which actor cluster, infrastructure community, operational style, and behavioral profile the indicator most closely aligns with. Actor clusters often yield more accurate results than formally named threat groups because attackers can change group names much faster than they can change behavioral traits such as infrastructure usage, tool preferences, development routines, operational hours, coding styles, recurring mistakes, and automation signatures.

In this context, cluster mapping analyzes the operational characteristics behind the IOC, not its surface form. For example, a specific SSL certificate may initially appear to resemble infrastructure previously associated with both APT33 and APT34 because certain observable attributes (e.g., certificate issuer naming conventions, reuse patterns across related domains, or similarities in TLS configuration) can overlap across multiple campaigns. However, deeper cluster analysis based on factors such as usage rhythm, certificate rotation behavior, pDNS history, and the density of relationships with known infrastructure may reveal that the certificate is actually linked to only one actor cluster. This level of analysis focuses on the attacker's cognitive and operational traces, which is why cluster mapping is often considered the most reliable attribution method.

Threat profile correlation is particularly effective in complex cases where IOCs are meaningless in isolation, yet carry the attacker's signature when viewed within behavioral and contextual relationships. For instance, some actor groups generate TLS certificates with specific parameter structures: the issuer Common Name (CN) may reference a variant of a fictitious organization, or the validity period may be set to an unusual value such as 397 days. On its own, such a certificate isn't a threat indicator; however, when cluster mapping reveals that the same parameter configuration was previously used in an Iran-linked operation, the trace immediately gains attributional power. These contextual traces are technically observable, but their true value emerges only through behavioral continuity.

Another advanced example is the use of DNS patterns in *actor correlation*. Some actors prefer specific consonant–vowel sequences in domain names; others leave no pDNS footprint during the first 48 hours of a domain's lifecycle; and others route DNS resolution exclusively through three specific ASN blocks. These small but recurring behaviors produce a threat profile far stronger than the IOC's surface form. When analysts observe such patterns, they gain insight which operational habits governed a domain's creation.

Another advanced technique used in matching IOCs to TTP variants is the extraction of parametric behavioral subclusters. In this context, parametric behavioral subclusters refer to identifiable micropatterns within the execution of a broader TTP (i.e., consistent command syntax, tool invocation parameters, process creation chains, or timing characteristics) that reveal how a particular actor implements a technique. While each TTP represents a technical definition, actors don't implement these techniques in identical ways.

For example, one group may perform Local Security Authority Subsystem Service (LSASS) memory dumping (`T1003`) using standard tools, while another executes the same technique exclusively via PowerShell using a distinctive command chain. When an IOC matches such variants, it reveals both the technique used and how the actor customized it. Customizations are often more persistent than the actor's name.

The most striking aspect of adversary cluster mapping is its ability to go beyond technical signals and reveal the cognitive traces of an operation. Some groups repeatedly use specific linguistic structures in tool error messages; others consistently apply the same timing jitter range in C2 communications. Some operators deliberately randomize certain metrics, while others turn even that randomness into a consistent fingerprint. For example, it's known that a particular APT group has added a fixed two-second delay to every new version of its C2 framework since 2019. This minor detail alone can link an IOC to the correct cluster, even in the absence of other indicators.

Cluster mapping is also used to detect *infrastructure cohabitation* between actors. If two different APT groups are observed using the same IP blocks at different times, this often indicates that they rely on the same underground infrastructure provider. If the same IP has been used by both a financially motivated group and a state-sponsored actor, analysts can reconstruct which actor was active during which period by examining operational timing, pDNS patterns, and TLS fingerprints. Such analysis is critical for IOC correlation, operational chronology, and intent inference.

The most advanced level of threat profile correlation occurs when an IOC doesn't directly match any known TTP or cluster, yet its behavioral rhythm closely resembles that of a specific threat actor. This phenomenon—often described as behavioral resonance—typically signals a new campaign, a new variant, or a previously undocumented toolset. When analysts recognize this resonance, they are no longer identifying the attacker's tools, but the attacker's way of thinking. This level of correlation is considered one of the highest-accuracy attribution methods in CTI because, while technical indicators change over time, an attacker's operational mindset tends to remain remarkably stable.

Example: DNS Plus Infrastructure-Based CTI Assessment

To illustrate how these analytical principles are applied in practice, the following example demonstrates a CTI-oriented assessment of a real-world domain infrastructure. The case highlights how DNS records, ASN attribution, and hosting-provider relationships can be examined together to reveal the operational structure behind a domain.

In our example output shown in Figure 7.1, the `dns_cti_analyzer.py` tool analyzes both the DNS records of the domain `xxxx.com.tr` and the underlying IP infrastructure behind those records, and then interprets the findings from a CTI perspective.

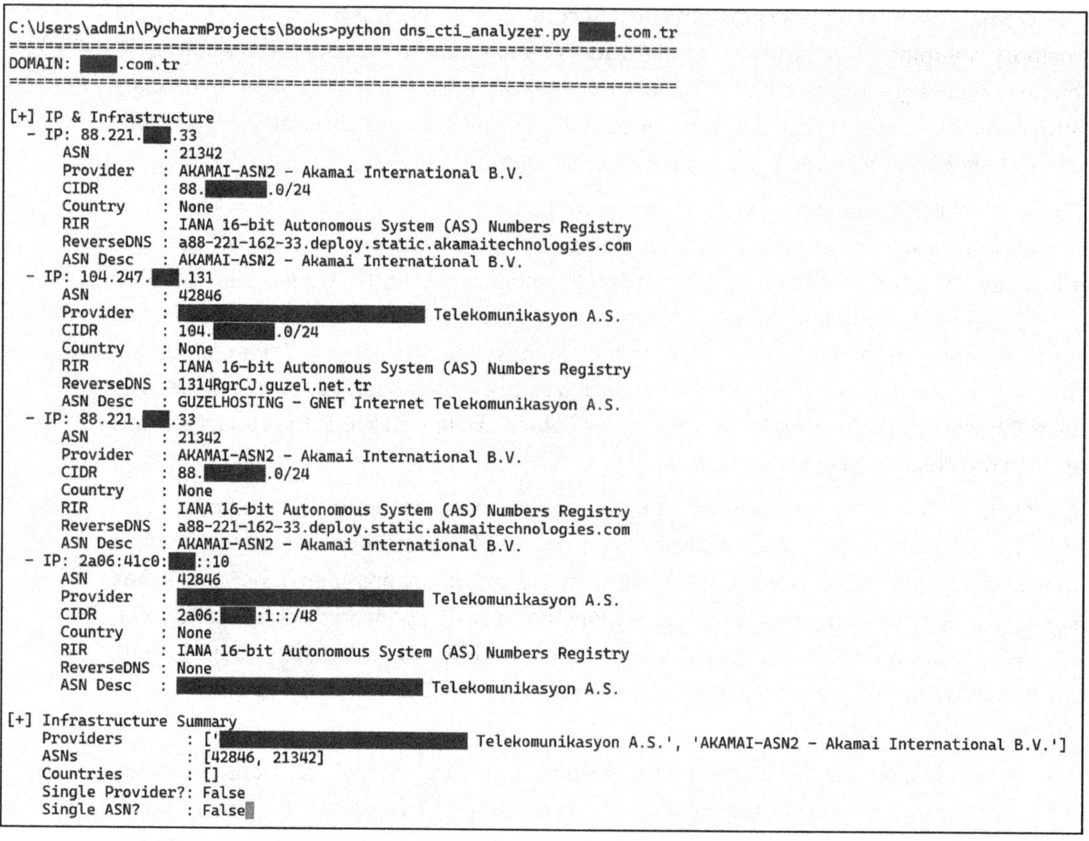

```
C:\Users\admin\PycharmProjects\Books>python dns_cti_analyzer.py ███.com.tr
==========================================================================
DOMAIN: ███.com.tr
==========================================================================

[+] IP & Infrastructure
  - IP: 88.221.███.33
      ASN        : 21342
      Provider   : AKAMAI-ASN2 - Akamai International B.V.
      CIDR       : 88.███████.0/24
      Country    : None
      RIR        : IANA 16-bit Autonomous System (AS) Numbers Registry
      ReverseDNS : a88-221-162-33.deploy.static.akamaitechnologies.com
      ASN Desc   : AKAMAI-ASN2 - Akamai International B.V.
  - IP: 104.247.███.131
      ASN        : 42846
      Provider   : ██████████████████ Telekomunikasyon A.S.
      CIDR       : 104.███████.0/24
      Country    : None
      RIR        : IANA 16-bit Autonomous System (AS) Numbers Registry
      ReverseDNS : 1314RgrCJ.guzel.net.tr
      ASN Desc   : GUZELHOSTING - GNET Internet Telekomunikasyon A.S.
  - IP: 88.221.███.33
      ASN        : 21342
      Provider   : AKAMAI-ASN2 - Akamai International B.V.
      CIDR       : 88.███████.0/24
      Country    : None
      RIR        : IANA 16-bit Autonomous System (AS) Numbers Registry
      ReverseDNS : a88-221-162-33.deploy.static.akamaitechnologies.com
      ASN Desc   : AKAMAI-ASN2 - Akamai International B.V.
  - IP: 2a06:41c0:███::10
      ASN        : 42846
      Provider   : ██████████████████ Telekomunikasyon A.S.
      CIDR       : 2a06:████:1::/48
      Country    : None
      RIR        : IANA 16-bit Autonomous System (AS) Numbers Registry
      ReverseDNS : None
      ASN Desc   : ██████████████████ Telekomunikasyon A.S.

[+] Infrastructure Summary
    Providers        : ['██████████████████ Telekomunikasyon A.S.', 'AKAMAI-ASN2 - Akamai International B.V.']
    ASNs             : [42846, 21342]
    Countries        : []
    Single Provider?: False
    Single ASN?      : False
```

Figure 7.1: DNS, ASN, and Provider Relationship Mapping for Domain xxxx.com.tr

In the first section of the output, the IP addresses associated with the domain and their infrastructure details enriched via RIPEstat are shown:

- **`88.221.xxx.33`**

  This IP address contains the following information:

  - `ASN: 21342`
  - `Provider: AKAMAI-ASN2 - Akamai International B.V.`
  - `CIDR: 88.221.xxx.0/24`
  - `ReverseDNS: a88-221-xxx-33.deploy.static.akamaitechnologies.com`

  These details indicate that a portion of the domain's traffic is served through the Akamai CDN. In other words, user traffic is routed not directly to the origin server, but to Akamai's edge infrastructure.

- **`104.247.xxx.131`**

  This IP address contains the following information:

  - `ASN: 42846`
  - `Provider: XxxTelekomunikasyon A.Ş.`

  - CIDR: `104.247.xxx.0/24`
  - ReverseDNS: `131aRgC3.guzel.net.tr`

  This IP belongs to a Turkey-based hosting provider. It most likely represents the network block where the actual application/web servers are hosted.

- **`2a06:41c0:1:4::10 (IPv6)`**
  This IP address contains the following information:
  - ASN: `42846`
  - Provider: again `XxxTelekomunikasyon A.Ş.`

  This points to the same provider's IPv6 infrastructure. The fact that the domain is reachable over both IPv4 and IPv6 indicates the use of modern network support.

The IP `88.221.xxx.33` appears twice in the output; this is due to references to the same IP via different DNS records (e.g., A record and Mail Exchange [MX] record, which we'll touch on later in this example).

This table tells us the following: `xxxx.com.tr` uses a hybrid infrastructure. Akamai CDN is used on the frontend, while hosting-owned IP blocks are used on the backend. This model is a common enterprise pattern preferred for performance, scalability, and distributed denial of service (DDoS) resilience.

In the `Infrastructure Summary` section, an aggregated view of the infrastructure is presented:

- **`Providers`**
  At least two different autonomous systems/providers are in use:
  - `Telekomunikasyon A.S.`
  - `AKAMAI-ASN2 - Akamai International B.V.`

  This indicates that the setup isn't dependent on a single hosting company and instead combines local hosting with a global CDN.

- **`ASNs: [42846, 21342]`**
  The IP addresses are confirmed to be distributed across two separate ASNs:
  - `21342`: Akamai
  - `42846`: Hosting

- **`Countries`**
  `None` means the tool didn't directly retrieve country information for the IPs via RIPEstat. Unknown here doesn't mean there's no country; it means the country field wasn't populated in this query set.

- **`Single Provider?/Single ASN?`**
  `False` is used for both. The infrastructure isn't concentrated under a single provider or ASN. From a CTI perspective, this implies a distributed infrastructure design with a lower single-point-of-failure risk.

To provide a more integrated view of these findings, the next stage of the analysis combines DNS records, infrastructure relationships, and risk indicators into a unified assessment model. As illustrated in Figure 7.2, this consolidated view allows analysts to evaluate how DNS topology, provider distribution, and infrastructure concentration interact within the domain's overall operational structure.

```
[+] Infrastructure Summary
    Providers        : ['                         , '                         .', 'AKAMAI-ASN2 - Akamai International B.V.']
    ASNs             : [42846, 21342]
    Countries        : []
    Single Provider?: False
    Single ASN?      : False

[+] DNS Records
    A/AAAA: ['88.221.    .33', '104.247.    .131', '88.221.    .33', '2a06:41d1:        ::10']
    NS:
      - ns1.                 com: ['88.221.    .33', '104.247.    .69', '88.221.    .33']
      - ns2.                 com: ['88.221.    .33', '165.232.    .61', '88.221.    .33']
      - ns12.                .com: ['88.221.    .33', '128.199.    .206', '88.221.    .33']
      - ns11.                .com: ['88.221.    .33', '134.199.    .27', '88.221.    .33']
    MX:
      - 10, mail exchanger = mail.               com.tr: ['88.221.    .33', '88.221.    .33']

[+] Email Security
    SPF   : None
    DMARC: None
    DKIM : False

[+] DNSSEC
    Enabled: False

[+] Pattern Analysis (C/V/D Behavioral Profile)
    Raw Pattern      : VCCVCVC  (C=Consonant, V=Vowel, D=Digit)
    Length           : 7
    Digit Ratio      : 0.000  → no digits; typical of corporate / readable domain names
    Entropy          : 0.985  → medium (mixed but not fully random)
    Max Run Length   : 2  (maximum consecutive repetition of the same symbol)

    Interpretation:
      - Domain structure is readable with low randomness; closer to a corporate/brand-style naming pattern.

[+] FINAL CTI RISK SCORE (Component-wise)
    Infrastructure Risk      : 0.000  (ASN/topology, single provider, single ASN, etc.)
    DNS / Email Surface Risk: 0.350  (NS/MX concentration, missing SPF/DMARC, etc.)
    Pattern Risk             : 0.100  (C/V/D distribution, entropy, digit ratio)
    ------------------------------
    Weighted Total Score     : 0.145  → Risk Level: Low
    Note: Overall risk is low; current signals do not indicate a structure deliberately optimized for malicious activity.
gs.
================================================================================
```

Figure 7.2: Integrated DNS, Infrastructure, and Risk Assessment for Domain xxxx.com.tr

In the `DNS Records` section, the DNS role distribution of the domain is shown:

- **A/AAAA**

  Both CDN IPs (Akamai) and origin hosting IPs (Hosting) are used together:

  `['88.221.xxx.33', '104.247.xxx.131', '88.221.xxx.33', '2a06:41c0:1:4::10']`

- **NS**

  All name server (NS) records point to hosting name servers:

  - `ns1.hosting.com` (multiple IPs [hosting + Akamai block])
  - `ns2.hosting.com` (similarly, multiple Ips)
  - `ns3.hosting.com`
  - `ns4.hosting.com`

This means that DNS hosting for the domain is consolidated under a single service provider. Although the IPs appear mixed, authoritative DNS control is centrally organized.

- **MX**
  The mail exchanger (MX) record indicates that email traffic is routed through `mail.xxxx.com.tr` to an IP in the Akamai block:
  `Priority 10, mail exchanger = mail.xxxx.com.tr → ['88.221.xxx.33', '88.221.xxx.33']`
  A notable point here is that email traffic is also associated with a CDN IP; this either indicates a reverse proxy/TCP relay design or suggests that the records haven't yet been fully optimized.

From a CTI perspective, the partial concentration of NS and MX IPs within the same block increases the tool's NS/MX concentration component in the DNS risk score because attackers frequently use this type of consolidation in malicious infrastructures. However, because the providers involved are well-known enterprise vendors (Akamai and Hosting), this alone isn't an indicator of malicious intent; rather, it produces a signal more in the nature of a hardening/improvement recommendation.

Let's move on to the `Email Security` section, where you can see the following:

- `SPF: None`
- `DMARC: None`
- `DKIM: False`

This means no Sender Policy Framework (SPF) policy is published in the domain's DNS records, no Domain-based Message Authentication, Reporting, and Conformance (DMARC) policy is defined, and no TXT record indicating DomainKeys Identified Mail (DKIM) presence is detected.

This situation indicates weak protection against email spoofing during outbound email delivery. From a CTI standpoint, if the `xxxx.com.tr` domain were to be impersonated in phishing campaigns, detection or blocking based on standard email security controls would be more difficult. For this reason, the code produces a relatively high score of 0.350 in the `DNS / Email Surface Risk` component, most of which is driven by the absence of SPF and DMARC.

In the `DNSSEC` section, you can see `Enabled: False`, meaning there's no Domain Name System Security Extensions (DNSSEC) signature for the domain. DNS responses aren't cryptographically signed, so they aren't protected at the DNS layer against spoofing attacks (e.g., cache poisoning, man-in-the-middle [MITM]). This omission doesn't make the domain malicious by itself, but it's considered a negative factor in terms of security maturity.

Next, the `Pattern Analysis (C/V/D Behavioral Profile)` section contains some key results:

- **`Raw Pattern`**
  The tool first extracts the consonant–vowel structure of the domain name to identify linguistic patterns that often distinguish human-created domains from automatically generated ones:

  `VCCVCCVC (C = Consonant, V = Vowel, D = Digit)`

  For `xxxx.com.tr`, this pattern corresponds to `a(V)`, `n(C)`, `f(C)`, `a(V)`, with other segments, yielding a total pattern length of seven characters.
- **`Length: 7`**
  This is a short and readable name; typically seen in brand/corporate domains.
- **`Digit Ratio`**
  The digit ratio metric measures the proportion of numeric characters within the domain name, a common signal used to distinguish human-readable domains from algorithmically generated ones:

  `0.000 → "no digits; typical of corporate / readable domain names"`

  The preceding indicates that the domain contains no digits. DGA-generated domains often contain many digits, whereas this name is fully alphabetic and clearly human-selected.
- **`Entropy: 0.985`**
  Entropy analysis evaluates the degree of randomness in the domain string, helping analysts determine whether the name resembles natural language structure or automated generation:

  `"medium (mixed but not fully random)"`

  On a theoretical scale of roughly 0 to ~1.6, 0.985 indicates a structure that isn't fully random but also not overly repetitive. In other words, this is a readable and meaningful name without signs of random generation.
- **`Max Run Length: 2`**
  This shows that the same type of symbol (`C/V/D`) repeats at most twice consecutively. This supports a natural naming rhythm; there are no long repetitions such as "CCCCCC" or "DDDD".

The `Interpretation` line summarizes this as follows: `Domain structure is readable with low randomness; closer to a corporate/brand-style naming pattern.`

Thus, based solely on the name pattern, the model concludes the following about `xxxx.com.tr`:

- Behaves like a classic corporate/brand domain
- Doesn't exhibit the random, digit-heavy patterns typical of DGA or mass phishing domains

The final section, `FINAL CTI RISK SCORE`, provides an overall risk score composed of three main components:

- **`Infrastructure Risk: 0.000`**
  The infrastructure assessment indicates that the domain operates across a distributed hosting structure, as reflected in the following characteristics:
  - Multiple providers: Akamai and Hosting
  - Multiple ASNs: There's no concentration on a single "dark ASN"; therefore, the infrastructure risk component is scored as zero.
- **`DNS / Email Surface Risk: 0.350`**
  The DNS and email surface risk score is primarily driven by the following observations:
  - Partial concentration of NS/MX IPs within the same block
  - Absence of SPF and DMARC records

  From a CTI perspective, this reflects insufficient email security hardening rather than malicious actor behavior.
- **`Pattern Risk: 0.100`**
  The pattern-based assessment highlights the following characteristics of the domain name structure:
  - The short, readable, digit-free, medium-entropy name is low risk.
  - The 0.1 score can be interpreted as a conservative margin from the automated system; there's no indication of DGA or aggressive randomness.
- **`Weighted Total Score: 0.145`**
  This combined score corresponds to a low risk level and is calculated using the following weighting model:
  - 30% infrastructure
  - 30% DNS/email
  - 40% pattern

As a result, a low total risk score of 0.145 is produced. The summary line states the following:

```
Overall risk is low; current signals do not indicate a structure
deliberately optimized for malicious activity.
```

In other words, based solely on DNS, infrastructure, and naming pattern analysis, the tool concludes that the `xxxx.com.tr` domain doesn't exhibit characteristics intentionally optimized for malicious activity.

7.4.3 Risk and Threat Scoring

In threat intelligence, the first step beyond the raw value of an IOC is to understand whether the indicator exists and, most importantly, what level of risk it carries. Not all IOCs have the same weight: some remain at a low threat level, while others can single-handedly open the door to a critical attack chain. For this reason, risk and threat scoring is the core mechanism that models an IOC's operational impact in both numerical and behavioral terms. When scoring is done correctly, an organization's defensive prioritization becomes sharp; when done incorrectly, real threats disappear into noise.

There are three main types of scoring:

- **Static scoring**
  Traditional scoring methods are often based on static attributes. Such static scoring models look only at the IOC's intrinsic properties: whether an IP has been used maliciously in the past, the age of a domain, a hash's association with a known malware family, or an ASN's abuse history. These metrics are static and change infrequently. However, attackers now know how easy it is to manipulate static signals, so scoring based solely on these methods is almost always incomplete. Domain age can be changed, ASNs can be migrated, and hash variants can be continuously generated. Static signals have become the weakest links and the attackers' primary targets.
- **Behavioral scoring**
  For this reason, modern CTI systems combine static scoring with behavioral scoring. Behavioral scoring focuses both on the values and how those values change—how quickly they evolve, the context in which they appear, and even their anomalies over time. For example, a domain remaining on the same IP for an entire day may be ordinary; however, switching ASNs every 45 minutes is a strong indicator that the attacker is running a fast flux–like operation. Similarly, a sudden spike in connection requests to an IP from specific geographies may indicate campaign activation timing. The real strength of behavioral scoring lies in making this behavioral topology behind static values visible.
- **Confidence scoring**
  Confidence scoring models form an epistemological layer that determines how trustworthy an IOC is. An IOC's threat level and its correctness are independent dimensions: a highly dangerous indicator can be false, while a seemingly low-threat indicator may be highly accurate. Confidence scoring dynamically calculates trustworthiness based on the source, contextual coherence, timestamps, relationship density, and enrichment results. In advanced models, this score isn't static. It's continuously recalculated as the IOC's behavior changes, the source's consistency degrades, or new relationships emerge. Because one of the most common attacker manipulations targets the perception of an IOC's reliability, a strong confidence model is a cornerstone of high-quality threat intelligence.

One of the most advanced components of risk and threat scoring is the *threat impact coefficient (TIC)*. TIC acts as an operational multiplier that models both how dangerous an IOC is and how much damage it could cause in specific parts of an organization if it materializes. The same IP address can have completely different TIC values in two organizations. If it's associated with a critical payment gateway in a large financial institution, it receives a high TIC; however, the same IOC affecting only a test environment doesn't carry the same weight. TIC accounts for the IOC's relationship to organizational assets, its potential impact on business workflows, the sensitivity of the data it could target, and how resilient the security infrastructure is against that threat—thus producing true operational risk.

Advanced CTI teams combine these three models (static scoring, behavioral scoring, and confidence/impact scoring) into a single hybrid threat-scoring matrix. This matrix simultaneously evaluates not only what an IOC did in the past, but how it behaved, what it indicates now, and what it may evolve into in the future. The result isn't static risk, but evolutionary risk. Threats aren't static; an IOC that appears benign one day may become the entry point of a major campaign the next. Systems that detect such evolutionary risks free defenders from noise and focus analyst attention on the right points.

In the modern threat landscape, risk and threat scoring is no longer limited to measuring static or behavioral properties of an IOC; it has become a multilayered structure that also models the IOC's evolutionary behavior within the threat ecosystem and its relationships with other indicators. An IOC's risk profile never exists in isolation; it gains meaning through its relationships with other indicators, actor behaviors, and campaign dynamics. Therefore, advanced scoring systems treat an IOC not as an isolated data point but as an interactive component within a dynamic threat network. Let's explore some more advanced scoring techniques:

- **Risk Propagation Gradient (RPG)**
  One of the most important concepts emerging here is RPG, which measures how quickly the risk level of a single IOC propagates to other related IOCs. For example, if a domain is known to carry high risk, IP addresses linked to it via pDNS may have their risk scores increased by the RPG multiplier, even if they would normally be considered low risk. This approach assumes that attacker behavior chains are mutually reinforcing, and in practice, it delivers very high accuracy. While attackers may hide individual indicators, concealing the topological trace of risk propagation is far more difficult.
- **Temporal risk drift analysis**
  Temporal risk drift analysis examines how an IOC's threat level changes over time. This analysis helps determine whether changes in risk score are natural or manipulated. Some IOCs may maintain a low-risk profile for days and then suddenly spike. In attacker manipulations, however, such spikes often occur without contextual support—no increase in sightings, no behavioral activity—yet the risk is artificially inflated. Temporal drift analysis detects this inconsistency and prevents scoring systems from being misled by attacker-driven perception manipulation.
- **Inter-Indicator Volatility Index (IIVI)**
  Going deeper into risk modeling reveals the Inter-Indicator Volatility Index (IIVI), a model that measures volatility relationships between IOCs. IIVI analyzes how synchronized behavioral changes in one IOC are with changes in related IOCs. This approach is particularly effective against APT campaigns, where infrastructures tend to move together. For instance, if activity increases on one C2 IP, other C2 nodes belonging to the same actor often show similar movement in the same time window. If an IOC behaves out of sync

with this pattern, it usually indicates either a manipulated indicator or a degraded component of the attacker's infrastructure.

- **Contextual suppression factor**
  Another dimension of advanced risk modeling is the contextual suppression factor, which dynamically calculates the influence of contextual factors that raise or lower an IOC's risk level. For example, a hash may belong to a known ransomware family, but behavioral enrichment may show that it has only been observed in a test laboratory; the contextual suppression factor then lowers the risk score accordingly. Similarly, some benign domains are repeatedly misclassified as malicious across campaigns; for such IOCs, the contextual suppression factor acts as a balancing mechanism that prevents artificial risk inflation.
- **Organizational Sensitivity Mapping (OSM)**
  One of the most critical yet often overlooked aspects of threat scoring systems is OSM, which mathematically explains why the same IOC can represent completely different threat levels across organizations. An Operational Technology (OT) network controlling production lines and a demilitarized zone (DMZ) consisting only of web servers can't be affected in the same way by the same IOC. An IP address may pose low risk to a generic enterprise network, while targeting assets in the financial sector dramatically increases its risk. True threat scoring therefore never produces a generic risk; it always yields a contextualized risk calculated against the organization's own asset sensitivity map.
- **Predictive Threat Trajectory (PTT)**
  When all of these advanced models are combined, a new concept emerges that goes beyond traditional IOC scoring: the Predictive Threat Trajectory (PTT). PTT focuses not on what an IOC is today, but on what it has the potential to become. Early anomalies in an IOC's behavioral curve may indicate that an attacker is preparing to scale a campaign, strengthen infrastructure, or shift tactics. The PTT model captures these signals and calculates the approaching risk of a threat that isn't yet fully active. This approach transforms defense from reactive to proactive.

All of these advanced techniques serve a single goal: deriving risk from the IOC's behavioral, contextual, and operational reality, not form its superficial form. In modern CTI, the threat doesn't reside in the IOC itself—it's hidden in the story the IOC tells.

7.5 Summary

Threat intelligence feeds are channels that deliver indicators, but more than that, they are analytical structures whose reliability depends on context, integrity, and trust. An IOC gains real operational value only when it's interpreted together with its behavioral patterns, temporal characteristics, source reliability, and relationships with other indicators.

This chapter emphasized that feed poisoning is rarely limited to inserting false indicators. More often, attackers attempt to manipulate the perception layer by distorting relationships, context, timing, or trust scores within the intelligence ecosystem. As a result, technically valid data may still lead analysts to incorrect conclusions.

For this reason, effective CTI requires multilayer validation and contextual enrichment. Techniques such as structural consistency checks, behavioral correlation, infrastructure analysis, and source verification help analysts transform raw indicators into meaningful intelligence while reducing the risk of manipulation.

Ultimately, the strength of a threat intelligence system depends not only on how much data it collects but also on how accurately it can preserve the meaning and integrity of that data during analysis.

Next, we'll move on to a complex but critical topic in CTI: network-centric forensic intelligence.

Chapter 8
Network-Centric Forensic Intelligence

Network-centric forensic intelligence focuses on a very honest layer of the digital world: traffic flow. No matter how skillfully an attacker attempts to erase their traces, they can never completely conceal the subtle vibrations they leave behind while passing through the network. This chapter tells the story of analysts who are able to hear these silent vibrations.

Network-centric forensic intelligence focuses on one of the most reliable layers of the digital environment: network traffic. Even when attackers attempt to erase traces of their activity, the interactions they create while communicating across a network inevitably leave behind subtle signals. Every packet, flow, and query becomes a fragment of evidence that can help analysts reconstruct the path of an intrusion. When interpreted correctly, network traffic forms a continuous chain that reveals how an incident unfolded.

In modern threat environments, attacks rarely appear as a single visible action. Instead, they unfold through sequences of small and often inconspicuous interactions across systems and networks. Following these traces can resemble tracking footsteps through a crowded city—difficult, but not impossible. The network rarely lies; the challenge lies in understanding how to read the patterns hidden within it. Network-centric forensic intelligence therefore seeks to transform raw traffic observations into structured insight about attacker behavior.

Tools such as Wireshark, Zeek, and NetFlow exporters provide complementary perspectives that enable this transformation. Packet-level inspection reveals the microscopic structure of communication, behavioral analysis interprets events within context, and flow records provide a large-scale view of traffic relationships across time. When these perspectives are combined, isolated signals begin to form a coherent narrative that describes how an attacker interacted with the network.

At first glance, network forensics may appear to be a purely technical and mechanical discipline. In practice, however, it also requires analytical intuition. Analysts must learn to distinguish between normal operational patterns and subtle deviations that indicate preparation, experimentation, or covert communication. A seemingly routine connection in a packet capture may in fact represent the early stages of a command-and-control (C2) channel or the quiet probing phase preceding lateral movement. Recognizing such signals requires both technical expertise and the ability to interpret behavioral context.

Network-centric analysis therefore extends beyond identifying individual indicators. It involves asking deeper questions about traffic behavior: why a connection began, why it repeated at specific intervals, or why a communication suddenly stopped. Timing patterns, packet size regularities, DNS query bursts, or unusual metadata relationships may each appear insignificant in isolation, yet together they can reveal the operational rhythm of an attacker.

These rhythms often expose important characteristics of adversary behavior. Some threat actors operate rapidly and aggressively, producing visible spikes in scans and connections. Others move slowly and deliberately, attempting to blend into normal network activity for extended periods. In both cases, the structure of traffic flow provides clues about strategy, patience, and operational intent.

For analysts, the goal of network-centric forensic intelligence is to both detect malicious activity and interpret the relationships hidden within network data. Packet structures, flow metadata, and log correlations together form a map that reveals how actions are connected across time and infrastructure. Through this process, network evidence can explain what occurred during an incident and why it unfolded in a particular way.

As this chapter progresses, the focus will shift from conceptual foundations to practical analytical techniques. We'll explore how network traffic, metadata relationships, and behavioral patterns can be used to reconstruct attacker activity and uncover hidden stages of an intrusion. By learning to interpret these signals, analysts gain the ability to detect emerging threats and understand adversary behavior long before the full impact of an attack becomes visible.

8.1 Introduction to Network-Centric Digital Forensics

At first glance, network traffic appears mechanical, even emotionless: packets flow, protocols communicate, sessions close. Yet behind this rhythm, the shadows of human behavior move: Every attack ultimately carries the trace of a human decision. Network-centric digital forensics is the discipline that uncovers this trace, a quiet art of observation that seeks to decipher the reason behind stillness, the meaning of a delay, and the intent behind a deviation.

The primary purpose of this introductory section is to create a new focal point in the analyst's mind: Network data is no longer just a record; it's the invisible conversation itself between the attacker and the defender. To hear this conversation, the analyst's perspective must change. The question is no longer "Is there traffic?" but rather "Why is the traffic behaving this way?" This transformation occurs slowly for many people because seeing the behaviors behind a behavior requires both technical knowledge and an intuitive attentiveness developed over years of experience.

With this in mind, the following sections will dive into the objective of network-centric forensics, how to define the scope of your investigation, and how the analysis is valuable in cyber threat intelligence (CTI).

8.1.1 The Fundamental Objective of Network Digital Forensics

Network digital forensics seeks to understand the outcomes of an attack and the logic of its formation. Most of the time, the attacker's first move contains the weakest signal:

- A single anomalous query blended into ordinary Domain Name System (DNS) traffic
- A Transport Layer Security (TLS) handshake that appears normal yet has a distorted rhythm
- A very small increase in data volume that no one is expected to notice

These signals are like the opening sentence of a story. To understand the story, the beginning must first be found. The analyst's goal is to make this beginning visible and, by integrating this coherence into CTI processes, to illuminate the intent of the threat actor.

For this reason, network digital forensics both determines what happened and addresses why it happened this way, when it began, and what the next step might be. This multilayered approach adds value to every level of the CTI ecosystem, from strategic analysis to operational decision making.

8.1.2 Defining the Scope

One of the most critical stages of forensic analysis processes is defining the scope of the investigation correctly. However, in network digital forensics, scope goes far beyond a classical boundary-definition exercise. Here, scope refers to tracing the clues that guide the attacker's movements because within a network structure, every segment, every session, and every log line carries potential context even though only a very small portion of them point to the essence of the incident.

In real-world attack scenarios, an analyst may be confronted with hundreds of gigabytes of traffic. Within this mass of data, the attacker's footprints often appear as only a few kilobytes of anomalous concentration. For example, a single outbound connection deviation observed in just 1 minute of a 48-hour traffic window may reveal the entire operational chain. Conversely, a routine fluctuation in a large wide area network (WAN) segment may give the impression of an attack while, in reality, carrying no meaning at all. For this reason, for an expert analyst, scope is defined not by data volume, but by what the data is saying.

The following three questions constitute the frameworks at the very center of the analytical strategy:

1. **Which network segment may have been subjected to a behavioral change?**
   This question aims to map the attack behaviorally, not geographically. Advanced persistent threat (APT) actors typically gravitate toward low-visibility segments: servers

with low patch levels, background service networks, forgotten virtual local area network (VLANs), or legacy test environments.

During lateral movement, attackers seek silent paths between segments to expand their access privileges. Server Message Block (SMB), Windows Remote Management (WinRM), Remote Desktop Protocol (RDP), remote procedure call (RPC), and even Internet Control Message Protocol (ICMP) tunnels can indicate these segment transitions.

Expert analysts evaluate segment behavior not only by traffic volume but by traffic texture, which can include the following:

- Packet size statistics
- Connection duration variance
- Protocol usage density
- Source–destination asymmetry

Small deviations in these metrics point to the segment in which the attacker is operating.

2. **Which log types complete the context of the incident?**

   Each log type illuminates a different layer of the attack. Therefore, scope definition requires the ability to align log types with their respective technical depths.

   Firewall logs reveal connection start/end rhythms, port/protocol anomalies, and lateral movement attempts.

   DNS logs carry the earliest signals of attacker intent:

   - Domain generation algorithm (DGA) behaviors (randomness detectable through language models)
   - `NXDOMAIN` bursts
   - Short time-to-live (TTL) values
   - Rapidly changing A records (indicators of fast-flux infrastructures)

   NetFlow/IP Flow Information Export (IPFIX) records function as the attacker's movement map, revealing the following signals with high accuracy through flow analysis:

   - Low-volume periodic beaconing
   - Sudden outbound volume increases during exfiltration
   - Pivot points used by the attacker within the network

   When combined with endpoint detection and response (EDR)/host logs, the chain of connection attempt → process creation → credential access is completed. (This contextual integration maps directly to MITRE ATT&CK techniques; see Chapter 6, Section 6.3.)

   Therefore, the scope considers which logs to look at and how the attacker's story appears when these logs are combined.

3. **How should the time window be constrained?**

   Time is one of the most reliable indicators of attacker behavior, for example:

   - Fast attacks: Credential stuffing, scanning, and brute-force attempts become visible within seconds or minutes.

- Stealth-focused APT operations: These may generate low-volume but rhythmic traffic over hours, days, or even months. In such cases, the time window must be broad and pattern analysis must be applied.

Time-based correlation techniques such as the following play a critical role in determining how long the attacker has been present:

- Session stitching: Reconstructing fragmented network sessions by linking flows that share temporal proximity, infrastructure reuse, or protocol continuity so that analysts can rebuild the attacker's activity timeline even when sessions are intentionally broken or distributed across multiple connections.
- Time-weighted anomaly scoring: Assigning different analytical weights to anomalies depending on when they occur and emphasizing persistent or recurring deviations over time rather than isolated spikes, which helps reveal slow-moving or stealth-oriented operations.
- Sequence drift detection: Identifying gradual deviations in the normal ordering or timing of network events (e.g., request–response patterns, beacon intervals, or authentication sequences), which may indicate the progressive presence or adaptation of an attacker within the environment.

For an expert analyst, the time window is a parameter that reveals the breathing rhythm of the attack. If chosen incorrectly, the entire attack chain may remain invisible.

When scope is defined correctly, the complex traffic of the network begins to take shape like geometry: connection clusters, movement paths, anomaly density zones, and attacker pivot points emerge like a complete puzzle. This enables CTI professionals to produce tactical (tactic, technique, and procedure [TTP]), operational (attack chain), and strategic (threat actor models) outputs. Incorrect scope, however, produces a single critical outcome: The attacker remains present where attention has lapsed, not where visibility exists.

8.1.3 The Value of Forensic Analysis in CTI Scenarios

In the CTI world, network forensics is often perceived as an auxiliary component of the process; however, in real operational environments, its role is far more decisive. Every finding produced by CTI analysts—the validity of an indicator of compromise (IOC), the certainty of a TTP, the behavioral profile of a threat actor—only attains its true meaning when it aligns with network data in the final stage because the network is the one place an attacker can't fully conceal. The layer that reflects behavior, intent, and operational discipline in their most unfiltered form is the silent flow of packets.

Some attacks are concealed with such sophistication that they leave no obvious traces in logs, EDR outputs, or host artifacts. Yet the vibration of attacker behavior reveals itself in the microrhythms of network data. This vibration is what creates the difference between an uncertain hypothesis and a provable reality in CTI analysis.

The following contextual examples illustrate why this is critical:

- **Behavioral similarity analysis: Mathematical traces of collective memory**
  Every sophisticated threat group has its own distinctive operational rhythm. This rhythm becomes visible in beacon intervals, session initiation delays, data exfiltration patterns, and the pulse line of C2 communications, for example:
  - An APT group's beaconing interval may repeat with a jitter of approximately ±50 ms.
  - In some China-origin threat groups, connection intervals include slight shifts aligned with local business hours.
  - In certain state-sponsored operations, login patterns follow a weekly cycle.

  If traffic analyzed in a new incident matches this characteristic rhythm at the millisecond level, attribution ceases to be a purely theoretical assumption and becomes a quantitative certainty.

  This method is one of the most powerful tools on the behavioral analytics side of CTI. Because code can change, infrastructure can change, and domain rotation can occur, but completely changing an attacker's behavioral rhythm is nearly impossible.
- **Covert channel deconstruction: Detecting the whisper within silence**
  Exfiltration techniques are becoming increasingly sophisticated. Attackers attempt to mask low-volume data leaks embedded within legitimate application traffic, most commonly using HTTPS, DNS, Quick UDP Internet Connections (QUIC), or custom tunneling methods.

  The task of a network forensic analyst is to hear the whisper within this silent flow, for example:
  - Low-entropy payloads, especially in DNS-based tunnels, deviate from the statistical model of normal query sequences.
  - In HTTPS tunnels, unusually homogeneous packet sizes increase the likelihood of data concealment.
  - Some actors perform exfiltration using micropacket sequences that function like a time-based Morse code. Normal traffic doesn't behave with such precision.

  These unnoticed micro-anomalies constitute a critical alert for CTI teams that this attacker may be testing a new exfiltration technique. When detected early, this signal enables advanced inference not only about the incident itself but also about the threat actor's innovation capacity.
- **Infrastructure correlation: Mapping the attacker's operational window**
  C2 infrastructures are rarely random. Time pressure, operational discipline, and the level of automation used by the attacker leave distinct traces in infrastructure behavior. The following parameters can reveal a threat group's operational DNA:
  - Autonomous System Number (ASN) history: Some groups frequently rotate ASNs; others deliberately maintain a low profile within a single ASN for operational stealth.

- Density curves: Repeating rise/fall cycles in traffic volume can indicate the attacker's working hours, operational frequency, and even internal shift patterns.
- Connection periodicity: Repeating connection intervals (such as beaconing every few minutes or at fixed time gaps) can reveal how the implant communicates with its C2 infrastructure. By analyzing the timing, regularity, and jitter of these connections, analysts can infer whether the malware uses scheduled beaconing, adaptive check-ins, or operator-triggered communication patterns.

Real-World Examples

Let's consider a few examples to demonstrate infrastructure correlation:

- **27-minute beacon cycles**
  In several investigations associated with the Lazarus Group (a North Korea–linked threat actor known for financially motivated and espionage-oriented cyber operations), analysts observed C2 communication patterns that repeated roughly every 27 minutes. This type of periodic beaconing typically originates from a malware implant that is programmed to contact its C2 server at fixed intervals to check for new instructions or to transmit status data. While the exact timing may vary slightly due to network jitter, the persistence of a near-constant interval can reveal the internal scheduling logic of the malware. When such patterns are consistently observed across multiple victims or infrastructure nodes, they may provide useful clues about tooling reuse or shared operational practices within the threat group.
- **10-second microbeaconing**
  In other cases, analysts encounter extremely short communication intervals—for example, beaconing events that occur approximately every 10 seconds. This pattern is often associated with lightweight implants, meaning small, resource-efficient malware components designed to maintain persistent connectivity while generating minimal network overhead. These implants typically communicate with low-latency C2 infrastructures, allowing the attacker to maintain near real-time awareness of the compromised system. Because the communication packets are usually small and frequent, this microbeaconing behavior can be difficult to detect through traditional volume-based monitoring alone. However, once the periodic pattern is identified through timing analysis, it can become a strong indicator of automated C2 communication.

Such infrastructure correlations provide CTI teams with both insight into who did it as well as the predictive capability regarding how and when they will act.

Network forensics provides unique insights into a threat actor's decision-making style, likelihood of making mistakes, operational comfort zone, and risk tolerance. Ultimately, this discipline reveals the following:

- Continuity of attacker behavior
- Level of operational flexibility
- Weak links that define operational fragility
- Likely timing of future attacks
- Group's logistical capacity

For this reason, forensic analysis is a powerful mechanism for exposing the threat actor's mental model. It makes the invisible visible, gives voice to silence, and at the very moment when the attacker attempts to hide intent within behavior, it turns data into evidence. Isn't this precisely our job?

The Subtle Balance That Makes Network Forensics Unique

Behind every piece of forensic insight lies a small emotion: suspicion. No matter how professional an analyst may be, it's that small inner spark of "What if?" that elevates the process beyond the ordinary. Network forensics operates on this spark because when an attacker's pursuit of perfection collides with the analyst's suspicion, truth emerges.

Sometimes a single packet disrupts all calm. Sometimes silence itself begins to shout. Sometimes a seemingly flawless network flow carries, beneath the surface, the vibration of a hidden machine at work. This balance is what distinguishes network forensics from other disciplines. Here, data is sensed, weighed, and questioned—much like a criminal investigator noticing a silent but telling detail at a crime scene. Network forensics is the unseen backbone of CTI. The trace of an attacker's first step is found here; the proof of the final step is hidden here as well. Most importantly, the mind of the threat actor becomes readable through this data.

8.2 Traffic Capture and Protocol Analysis

At the heart of network-centric forensic intelligence lies the process of traffic capture because every action an attacker attempts to mask ultimately has to flow across the network. Even in environments where logs can be silenced, EDR sensors can be disabled, and system artifacts can be erased, network traffic acts like a briefly opened curtain that reveals the attacker's true intent. Even if this curtain is lifted for only a few packets, the raw data captured in those moments can be powerful enough to illuminate the entire anatomy of the attack. For this reason, packet capture (PCAP) files are the attacker's behavior reduced to its purest form.

In real attack operations, obtaining PCAP data often requires meticulous planning. The point at which traffic is collected, the layer at which it's captured, the quality of the Test Access Point (TAP) or Switched Port Analyzer (SPAN) infrastructure used, packet loss rates, timestamp accuracy are all factors that directly determine the analytical value of the captured data. A delay of just a few milliseconds can lead to misinterpreting an attacker's beaconing rhythm. The loss of a few packets can render traces of TLS handshake manipulation invisible. For an advanced analyst, therefore, the PCAP collection process isn't simply a matter of issuing a `start capture` command; it's an effort to record the attacker's breathing from the correct vantage point.

The value of PCAP lies less in revealing content than in exposing behavior. Most modern attacks operate through encrypted channels such as TLS, QUIC, or VPN tunnels. Although payloads may remain hidden, attacker activity can still appear through behavioral signals, including irregular connection patterns, anomalies in packet length distributions, jitter deviations, or timing artifacts such as TCP timestamp variations created by implants behind Network Address Translation (NAT).

Even brief anomalies within otherwise normal traffic may signal critical events, such as the first beacon of an implant, reconnaissance activity, or an exfiltration attempt. These deviations often provide the earliest indication that malicious activity is present. For this reason, PCAP collection is a cornerstone of network forensic analysis. While logs and artifacts provide context, packet-level data preserves observable attacker behavior and enables investigators to reconstruct how an intrusion unfolded.

The following sections will explain the importance of traffic capture in detail. We'll walk through important techniques with the popular open-source network packet analyzer, Wireshark, and discuss analysis approaches for DNS and TLS protocols. We'll round out the section with an in-depth example of protocol analysis.

8.2.1 The Power of Raw Traffic

The true power of raw traffic doesn't come, as many assume, from seeing everything, but from correctly choosing where and how you see it. The moment you decide to collect raw traffic, you're no longer dealing with abstract analysis; you're sharing the same line as the attacker. For this reason, the first critical question is always the same: "Where in the network should I listen?"

Your position within the network topology doesn't just present you with different packets; it tells an entirely different story. Capturing traffic at the gateway level is like a watchtower observing the boundary between the outside world and the internal network. What you typically see here are inbound exploit attempts, scanning traffic, brute-force attempts, initial outbound connections to malicious infrastructure, and the external leg of C2 communication rhythms. This is where you most clearly observe how the attacker probes the door, which IP ranges are scanned, and which TLS fingerprints are used. However, this level often misses the quiet movements taking place inside.

Capturing traffic within internal segments reveals a different reality. Here, rather than external attacks, you witness the silent progression of a threat actor already positioned inside the environment: subtle changes in Kerberos traffic toward domain controllers, unusual authentication attempts over SMB, unexpected concentration of RDP connections, command traffic pushed through WinRM or Windows Management Instrumentation (WMI), and so on. Lateral movement rarely manifests as a loud explosion, but instead unfolds as a quiet crawl. The only place you can reliably capture this crawl is deep inside internal segments. No external intrusion detection system (IDS) can meaningfully observe these internal vibrations from the gateway level.

Capturing traffic close to servers, on the other hand, exposes the psychology of the application layer. A TAP or SPAN placed near application servers reveals how the attacker interacts with workloads, performs parameter manipulation, issues anomalous HTTP request patterns, maintains long-lived idle-but-open TCP sessions, and sometimes embeds data exfiltration attempts into application logic. Requests originating from the same IPs yet following a distinctive Uniform Resource Identifier (URI) pattern, login attempts differing from normal user behavior by microscopic margins, and variations in API parameter lengths are all details that expose attacker intent at the application level.

Beyond all this, true expertise lies in focusing but on *timing*, not on traffic volume. Collecting gigabytes of PCAP doesn't always equate to good analysis. What matters is capturing the exact moment the attacker "touches" the system; that is, the microrhythmic disruption left by that touch. Most advanced attacks begin with a critical maneuver that occurs within a very small fraction of a second: a single anomalous DNS query, a Secure Sockets Layer (SSL) handshake arriving just slightly earlier than expected, or millisecond-level jitter manipulation during TCP session establishment. The attacker tries to keep this maneuver as brief and invisible as possible, and PCAP is the only evidence capable of recording that instant.

At this point, timing quality becomes critical. Timestamp accuracy in PCAP files is far more important than most analysts realize when reconstructing the anatomy of attacker behavior. If the capture device isn't properly synchronized via Network Time Protocol (NTP) or Precision Time Protocol (PTP), attempts to analyze microlevel attacker rhythms can lead to false conclusions. Especially with advanced actors who manipulate jitter, attacker intent is read not from packet contents, but from the temporal gaps between packets. Connection intervals shifted by milliseconds, beacon intervals that appear random but actually follow a specific pseudo-random number generator (PSRNG) model, and hidden signals blended into normal traffic can all be analyzed with confidence only if timestamps are sufficiently precise and consistent.

Some advanced implants leave traces in subtle fields at the TCP/IP layer. Few people realize that many specialized implants manipulate TCP timestamp fields or partially emulate operating system (OS) TCP stacks. After each C2 communication, they may introduce a fixed offset or a slightly chaotic yet repeatable pattern in timestamp values. This behavior doesn't appear in standard Linux or Windows TCP implementations and becomes a signature specific to the attacker's framework. If you examine traffic only at the application layer, you'll never notice this signature. But once you analyze PCAP at the raw header level, you can conclude that the packets may be trying to look like a real browser, but their TCP behavior betrays the identity they are masking.

Jitter manipulation is equally critical. Some threat actors attempt to evade network detection and response (NDR)/IDS systems that detect beaconing by making connection intervals appear random. However, the randomness used is rarely truly random; it often adheres to a specific distribution, seed, or minimum/maximum range. By capturing long-term PCAP data and statistically analyzing timing intervals, you can uncover a repeating structure beneath this pseudo-randomness. A properly applied time-series analysis strips away the attacker's randomness mask and provides direct insight into how the implant's behavioral engine operates.

Raw traffic also enables you to analyze source–destination relationships without stripping them of context. While summary flow records (e.g., NetFlow) map movement (see Section 8.3.1), full PCAP produces evidentiary artifacts at both content and header levels. For example, traffic consistently flowing from a specific server to a narrow IP block within the same ASN, always with roughly similar packet sizes, can be strongly reinforced as potential exfiltration—even when encrypted with TLS—through PCAP details. Server Name Indication (SNI) fields, JA3/JA3S fingerprints, the ordering of Client Hello extensions, and fragmentation behavior are all forensic traces you can use to document the identity of the attacker's toolset.

8.2.2 Wireshark Techniques

Wireshark is often perceived as merely a tool that displays packets; however, for a true CTI analyst, Wireshark functions as a microscope that transforms the abstract traces of attacker behavior into concrete evidence. In the right hands, Wireshark reveals how traffic flows *and* how the attacker thinks. No matter how professional an attacker may be, they inevitably leave small, involuntary rhythm distortions in the manipulations applied to traffic. This is where Wireshark's real power lies: its ability to capture these micro-anomalies.

The techniques we'll discuss in this section go beyond traditional analysis methods and represent advanced approaches used to uncover threat actors' C2 mechanisms, implant behaviors, and operational routines.

Time-Based Behavioral Analysis: The Cryptography of Connection Rhythms

Attackers may change tools, infrastructure, or tactics across different operations; however, perfectly mimicking timing patterns is nearly impossible. Wireshark provides several graphical analysis views for examining packet timing and flow behavior, including the Time-Sequence Graph (Stevens), the TCP Time-Sequence Graph (`tcptrace`), and the round-trip time (RTT) graph. In addition, analysts often review TCP timestamp behavior within packet details to assess timing consistency across a session. Together, these views help reveal the temporal structure of C2 communication by making retransmissions, latency changes, burst patterns, and periodic check-ins easier to detect.

When a CTI expert examines time graphs, they may observe the following:

- **Millisecond-level jitter manipulation**
  Advanced APT implants configure jitter to appear random; however, due to the nature of the PRNG algorithm used, certain intervals repeat. For example, a beacon interval varying between 120 and 350 ms that exhibits a consistent 20 ms deviation every eight cycles indicates a specific PRNG modulus flaw.
- **Micro-anomalies during the TCP slow-start phase**
  In normal TCP slow-start behavior, the congestion window grows exponentially. C2 implants, however, often use custom TCP stacks, and this growth curve contains direct imitation. This imitation isn't genuine, and the curve appears distorted in graphs.
- **Timestamp drift in C2 channels**
  Some implants manipulate OS time. This drift appears as a shift in the Time-Sequence Graph, which Wireshark can clearly capture.

None of these behaviors are visible at the log level; they are revealed only through raw PCAP combined with Wireshark analysis.

Deep Analysis of Reconstructed TCP Sessions

In encrypted protocols such as TLS or QUIC, payloads are invisible, but behavior is not. Wireshark's Follow TCP Stream feature, when interpreted for packet sequencing, enables surprisingly powerful inferences about the tools used by the attacker. Figure 8.1 shows a reconstructed TCP session view in Wireshark using the Follow TCP Stream feature, showing the reassembled byte stream and payload structure of a network communication.

Key aspects that can be analyzed at an advanced level include the following:

- **Packet size signatures**
  For example, specific Cobalt Strike Beacon configurations generate fixed-size patterns ranging between 273 and 295 bytes. The Empire framework, by contrast, produces wider-variance burst patterns. This measurement alone can indicate which implant family is in use.
- **Burst peaks**
  Some malicious tools periodically generate short bursts of high packet density. This often appears as mini data bursts during the exfiltration phase.
- **Jitter behavior (inter-packet delay analysis)**
  While normal user traffic exhibits high variance in jitter values, malware traffic often contains unintended regularity. Using these techniques, behavioral fingerprints with very low error margins can be extracted.

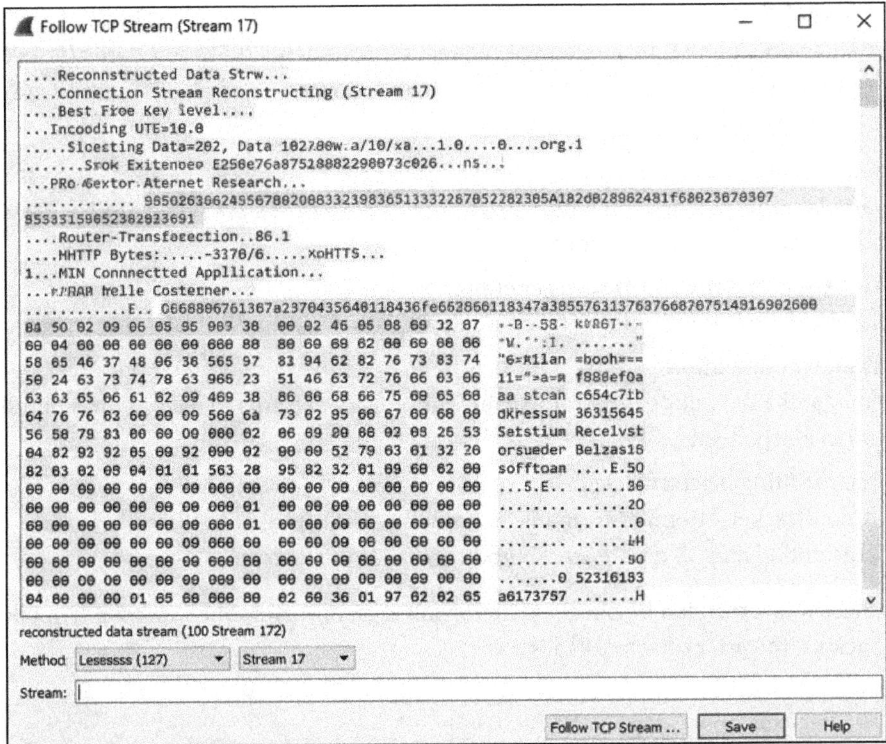

Figure 8.1: Wireshark: Follow TCP Stream Feature

Entropy Analysis

Many analysts measure entropy only at the payload level; however, for advanced threat analysis, the truly critical factor is time-based entropy. An attacker may perfectly encrypt payloads, but masking communication timing with the same level of perfection is far more difficult.

Wireshark doesn't allow direct entropy measurement on packet timing, but with PCAP export, the following analyses can be performed:

- **Interarrival time (IAT) entropy**
  By calculating the entropy of packet arrival intervals, C2 behavior can be identified. High entropy indicates normal user behavior, while low entropy can indicate script-based or programmatic communication.
- **Sequence entropy analysis**
  Variance in command transmission cycles can reveal which implant architecture is being used.
- **Periodic noise extraction**
  Even if beaconing is concealed, recurring troughs become visible in entropy graphs.

More importantly, sudden drops in entropy often appear as signals of exfiltration preparation or the initiation of lateral movement.

Header Anomalies

Most C2 frameworks don't fully implement request for comments (RFC) standards for two primary reasons: speed and stealth. As a result, the following anomalies are frequently observed in attacker tools:

- **Unusual TCP window size values**
  For example, some custom implants use deterministic window sizes such as 512 or 1,024 to evade network detection systems.
- **Incorrect yet functional checksum generation**
  This is a characteristic indicator of fake traffic generators.
- **Timestamp modification**
  Some implants manipulate timestamp values in specific patterns to disrupt time-based correlation mechanisms.
- **TTL fingerprinting discrepancies**
  Even if an attacker attempts to mimic a browser, the TTL distribution often doesn't match the underlying OS or TCP stack, producing an artificial trace.

None of these indicators can be observed in firewall logs; however, once Wireshark opens the raw packets, the attacker's mask falls away.

> **The Wireshark Philosophy**
>
> For an expert CTI analyst, Wireshark isn't just a tool; it's an interrogation platform. Every packet asks the same question: "Am I behaving the way I am supposed to be here, or is someone shaping me for a hidden purpose?"
>
> Answering this question correctly requires both technical knowledge and an intuition capable of reading adversary behavior. An attacker may change their code, switch from ICMP to HTTPS, or update tunneling techniques hundreds of times. But they can never fully conceal the rhythm of their behavior, their timing, or their micro-errors.

TLS Fingerprint Analysis (JA3/JA3S)

With the rise of encrypted traffic, attackers may successfully hide packet contents; however, the fingerprint revealed during the TLS handshake is one of the most critical identity markers they can't conceal. The JA3 and JA3S fingerprinting methods for TLS traffic don't identify an encrypted session; they characterize how it behaves. This behavior often carries traces of the implant's underlying libraries, the compilation environment, and even the developer's preferences.

JA3 is the fingerprint of the `ClientHello` message, while JA3S corresponds to the `ServerHello`. When analyzed together, these two fingerprints reveal behavioral characteristics on both the client side and the opposing infrastructure used by the attacker.

Many assume JA3 is just a user agent–like identifier, but, in reality, JA3 touches one of the most fragile points of the TLS ecosystem: the inability of applications to perfectly replicate TLS stacks. Attackers may hide many things, but reproducing TLS behavior with absolute fidelity is nearly impossible.

In some cases, an attacker may attempt to spoof a JA3 value to create a mask. However, the forged JA3 value inevitably collides with other inconsistencies: the SNI structure doesn't resemble a real browser, cipher suite ordering is incorrect, or the elliptic curve list is suspiciously short. In such cases, the JA3 value may appear legitimate, but the behavioral fingerprint unmistakably exposes its falseness.

An even more advanced observation is that some implant frameworks, even after altering their JA3 values, still leak micro-errors such as TCP timestamp jitter, modified maximum segment size (MSS) values, or missing Application Layer Protocol Negotiation (ALPN) extensions. These flaws represent momentary leakage points where JA3 spoofing fails.

JA3 is mathematically a hash, but using it as a superficial identity check is a serious mistake. When the components behind a JA3 value are examined, the attacker profile becomes far clearer, for example:

- **Cipher suite ordering**
  This reveals the timidity of the TLS library in use. In this context, *timidity* refers to how conservatively a TLS implementation proposes cipher suites during the handshake. Some libraries prioritize only a small set of well-established and widely compatible cipher suites at the top of the list, avoiding experimental or rarely used options. This cautious ordering behavior reflects the design philosophy of the library and can therefore act as a subtle fingerprint of the underlying TLS implementation. OpenSSL and BoringSSL follow different ordering schemes. Go-based implants exhibit distinct patterns, while some Rust-based tools leave careless errors in elliptic curve lists.
- **Elliptic curve preferences**
  These preferences often expose themselves through unnecessary option usage in the elliptic curve cryptography (ECC) range. Many malware developers copy curve ordering from real browsers but forget to optimize key share parameters. The result is a JA3 profile that resembles Chrome, but feels distorted.
- **Extension order**
  This is one of the strongest indicators in JA3 analysis. TLS libraries don't randomize extension ordering; they generate a natural sequence tied to specific build configurations. Attackers often attempt to imitate this order but fail to implement Generate Random Extensions And Sustain Extensibility (GREASE) extensions correctly. Incorrect GREASE usage is an immediate red flag. GREASE is a mechanism defined in the TLS ecosystem to prevent protocol ossification by periodically inserting reserved, nonfunctional values into TLS handshake fields. Modern TLS implementations, such as those used in contemporary browsers, include these placeholder values to ensure that servers remain

tolerant of unknown extensions. Because legitimate libraries handle GREASE in a consistent and standards-compliant manner, incorrect placement, formatting, or absence of these values can reveal imperfect TLS impersonation attempts.

For this reason, JA3 analysis doesn't just determine if the fingerprint is malicious but whether the fingerprint was produced by a human user or by a hastily patched attack tool, which is the much more critical data for a CTI analyst.

JA3S is often neglected, yet it represents a powerful second half of the equation by revealing the TLS behavior of the opposing side, namely the C2 infrastructure. Fake C2 servers commonly exhibit the following traits:

- Forced TLS 1.2 usage
- Broken `ServerHello` extension ordering
- Uniform curve selection typical of sloppy reverse-proxy deployments
- Unnecessary padding in the ALPN field

As a result, JA3S analysis often exposes deception that JA3 attempts to conceal—like a stain bleeding through the background.

When JA3 and JA3S are evaluated together, the resulting picture resembles a true biometric trace of attacker behavior: forensic DNA embedded within encrypted communication.

A further advanced observation is that even highly capable APT groups, despite manipulating JA3 fingerprints, can't alter TLS negotiation time. Handshake duration depends on numerous parameters—from CPU clock speed to library optimization levels—and is among the most difficult behavioral signatures to fake. Consequently, advanced analysts don't examine JA3 in isolation, but rather analyze the JA3 + RTT + handshake-duration triad.

In recent years, a technique gaining popularity within CTI communities is JA3 collision profiling. Different tools sharing the same JA3 value are behaviorally differentiated by jointly evaluating the following:

- Connection periodicity
- Extension negotiation ordering
- ALPN preferences
- Session resumption behavior

This approach is exceptionally effective, especially in distinguishing between Cobalt Strike, Sliver, and Mythic. These platforms are widely used to deploy and manage implants on compromised systems and to maintain communication with attacker-controlled infrastructure. Although they may produce similar JA3 fingerprints (either by design or through deliberate fingerprint imitation), their underlying implementations still generate subtle differences in connection timing, extension negotiation patterns, and session management behavior.

HTTP/2 and QUIC C2 Communication Analysis Techniques

The fact that attackers have increasingly shifted toward HTTP/2 and QUIC protocols in recent years is no coincidence. These protocols provide performance advantages and, more importantly, incorporate a multilayered architecture that significantly complicates analysis. During the HTTP/1.x era, an attacker's actions were clearly visible through static header structures, connection lifecycles, and classic TCP behavior. However, HTTP/2's frame-based flow and QUIC's encrypted, User Datagram Protocol (UDP)–based structure, which appears connectionless yet behaves in a stateful manner, elevate C2 communication analysis to an entirely different level.

For expert analysts to detect C2 activity over these protocols, inspecting packet contents alone is no longer sufficient. They must instead observe microbehavioral distortions left in frame sequencing, abnormal use of flow-control mechanisms, unnatural jumps within QUIC's packet number space, and rhythm anomalies concealed beneath the TLS-in-UDP handshake.

This section explains the advanced techniques required to decode these invisible rhythms.

The Hidden Depths of HTTP/2 C2 Analysis

HTTP/2 is a protocol particularly favored by attackers because its frame-based structure blinds most traffic analysis tools at the content level. However, when it comes to analyzing attacker behavior, HTTP/2's advantages turn into liabilities.

Although HTTP/2 encrypts application content, its frame structure remains observable. The order, size, and frequency of frames often reflect the behavior of the client implementation. While legitimate browsers follow consistent patterns, many C2 implants produce simplified or irregular frame sequences, making behavioral detection possible even when payload data is encrypted:

- **Frame pattern fingerprinting**

  In HTTP/2, each request and response is split across multiple frames. Normal browser behavior follows a predictable sequence:

  ```
  SETTINGS → WINDOW_UPDATE → HEADERS → DATA → DATA → END_STREAM
  ```

  Variations in this sequence remain within defined bounds. A C2 implant, however, typically does the following:

  - Keeps `HEADERS` frames minimal
  - Produces `DATA` frames of uniform size
  - Sends `WINDOW_UPDATE` frames either too infrequently or excessively

  For example, in some Sliver implants' HTTP/2 beacons, single-byte `DATA` frames appear at regular intervals. No legitimate browser generates `DATA` frames that small. Even with encrypted content, this directly represents a C2 heartbeat indicator.

A little-known nuance is that some amateur C2 frameworks fail to properly update the `HPACK` header compression dictionary, causing `HEADERS` frames to become unnecessarily large. This exposes HTTP/2 C2 traffic within the first 30 seconds.

- **Flow control violations**

  HTTP/2 uses a flow-control system where the client and server explicitly tell each other how much data may be sent. This mechanism is so sensitive that even minor flaws in attacker code can become visible. Some C2 implementations do the following:

  - Fail to send sufficient `WINDOW_UPDATE` frames to the server so the connection repeatedly stalls and resumes in an anomalous manner.
  - Periodically send unnecessary `WINDOW_UPDATE` frames, producing an unnatural mini-burst behavior.

  This behavior can even reveal the quality of the HTTP/2 library used by the attacker.

Real-World Example

An HTTP/2-based C2 channel used by a Korean threat group continued sending a `WINDOW_UPDATE` frame every 45 seconds, even while the connection was idle. The frame carried no actual traffic. This behavior stemmed from a fixed timer task inside the implant. A single anomaly was sufficient to fingerprint the entire C2 architecture.

- **`PRIORITY` frame anomalies**

  Browsers manage HTTP/2 streams using defined priority hierarchies; attackers, however, often neglect or misapply `PRIORITY` frame usage. One detection technique focuses on the weight value in `PRIORITY` frames. Browsers exhibit characteristic distributions:

  - Chrome typically uses values between 16 and 256.
  - Firefox generally falls within 32 to 220.

  C2 implants, by contrast, often use a single static weight value, for example, 16 or 255. This repeatedly used weight value exposes the attacker's identity like a birthmark.

The Hidden Depths of QUIC C2 Analysis

QUIC is a protocol designed for high speed, operating over UDP while integrated with TLS 1.3. Its advantages for C2 are clear: rapid connection establishment, a handshake that is difficult to detect, the seemingly connectionless nature of UDP, and mechanisms that obscure retransmissions.

Although QUIC encrypts much of its control information and operates over UDP, its traffic still follows structural and timing patterns determined by the client implementation. Legitimate applications tend to produce consistent packet pacing and flow behavior, whereas many C2 implants generate simplified or irregular transmission patterns. By examining these timing and size relationships, analysts can uncover behavioral indicators even when the protocol's internal fields remain encrypted:

- **Packet number space analysis**
  Even though packet numbers in QUIC are encrypted, timing and size relationships can still expose the attacker. For example, normal QUIC implementations send ack-eliciting packets in a specific rhythm. C2 implants, however, generate a pattern that varies with CPU behavior—traffic that appears irregular at first glance but in fact follows a repeating computational rhythm. Because many implants schedule packet generation using internal timers or event loops, small fluctuations in processing time create uneven packet spacing. Over longer observation windows, however, these variations repeat in consistent cycles, revealing the underlying algorithmic pattern behind the traffic.

  A hidden indicator observable in a C2 channel is that every 10 packets, the packet number range returns to its average value. This is a trace of a *wrapback correction* mechanism in the implant's packet sequencing algorithm.

  This technique represents an observation that hasn't yet even entered the broader literature.

- **QUIC handshake profiling**
  In the QUIC handshake, the `ClientHello` is concealed within TLS 1.3, but QUIC's own frames still carry the rhythm of the handshake. C2 implants tend to do the following:

  - Misuse 0-RTT data. This QUIC feature allows a client to send application data during the first round-trip without waiting for the full handshake to complete. Legitimate implementations use this mechanism cautiously and only when session resumption conditions are met, whereas poorly implemented implants may attempt to send 0-RTT data in contexts where it shouldn't appear.
  - Send incomplete salt values in Retry packets. In QUIC, a Retry packet is a server message that asks the client to restart the handshake using a server-provided token, typically as part of address validation. Improvised C2 frameworks sometimes construct these packets incorrectly, leaving fields such as the retry integrity or salt values inconsistent with standard implementations.
  - Generate initial packet sizes that differ from real QUIC implementations. The initial packet is the first client packet that begins the QUIC handshake and normally follows strict size conventions defined by the protocol to support path validation and amplification limits.

  For example, a Chrome-based QUIC client most often sends an initial packet of around 1,200 bytes. A C2 implant, by contrast, typically generates packets in the range of 600 to 800 bytes. This difference alone reveals that the attacker is imitating QUIC but failing to fully implement it.

- **Flow ID/stream ID patterning**
  QUIC stream ID values follow a specific logic. In normal applications, stream IDs increase sequentially at defined intervals.

  A common mistake observed in C2 implants is for stream IDs to follow a fixed pattern such as 0, 4, 8, 12, and so on. This results from a misinterpretation of the QUIC protocol and becomes an implant-specific fingerprint. In an APT campaign observed several years ago, QUIC-based C2 traffic was detected specifically because of this quadruple increment error in stream ID usage. The mistake indicated that the developer had referenced an outdated RFC version tied to an early QUIC draft. This type of detail is a signal only a CTI analyst deeply familiar with the protocol can recognize.

HTTP/2 and QUIC Behavioral Combination

Advanced threat actors sometimes use HTTP/2 and QUIC together to evade detection mechanisms. The critical observation here is that attackers may imitate protocols, but they can never flawlessly replicate behavioral synchronization between two protocols.

Consider an example hybrid analysis finding, as follows:

- HTTP/2 beacon interval: 17 seconds
- QUIC stream creation interval: 18 seconds

This 1-second desynchronization indicates that the attacker is running two separate modules in a loosely coupled, clock-based manner. This discrepancy reveals a microservice-like design and a modular C2 architecture. No log will reveal this detail. However, PCAP + Wireshark + behavioral analysis can expose it within minutes.

HTTP/2 and QUIC have become the most powerful masks of modern C2 communication. Yet when examined with the correct techniques, these masks can be torn away through the following means:

- Micro-errors in frame sequencing
- Abnormal flow control behavior
- Rhythm distortions within QUIC's packet number space
- Unconcealable patterns during the handshake phase

Detection of DNS over HTTPS C2

DNS over HTTPS (DoH) has become a favorite concealment channel among attackers because DoH both encrypts DNS queries and embeds them within what appears to be ordinary HTTPS traffic. As a result, to security devices, the traffic looks like an innocuous web request, not like a DNS request.

However, an experienced CTI analyst knows that even the best-hidden activity leaves behind behavioral traces, especially when a historically lightweight, predictable, and rhythmic protocol such as DNS is forcibly wrapped inside HTTPS.

When an attacker implant uses DoH, it attempts to behave like a normal browser, but fully reproducing the DoH structure correctly is difficult. Over time, microdistortions emerge within the HTTPS flow that don't align with DNS's natural rhythm. This is where the essence of DoH C2 detection begins: the ability to hear what looks like HTTPS but behaves like DNS.

Several observable characteristics can help analysts distinguish legitimate DoH activity from covert C2 communication:

- **Query-length morphology**
  Many people assume that DoH traffic is completely encrypted; however, the size of DoH queries isn't encrypted. Moreover, the size distribution of DNS queries is an extraordinarily stable statistical characteristic. Normal DoH traffic has the following characteristics:
  - Query sizes are concentrated roughly between ~50 and 120 bytes.
  - AAAA queries are slightly larger than A queries.
  - TXT records appear only for specific services.
  - There's minimal variance between queries made for the same host.

  In a C2 implant, however, it's common to observe the following:
  - Query sizes that follow a repeating pattern (e.g., 87 → 87 → 87 → 120 → 87)
  - Extremely low variance in query sizes (because they are bot-generated)
  - In some implants, fully fixed-size, Base64url-encoded embedded command payloads

  This statistical pattern is golden for DoH C2 fingerprinting. Very few analysts measure this directly using Wireshark or Zeek scripting; yet once measured, the attacker's fingerprint becomes visible even through encryption.
- **Query frequency and timing analysis**
  In the normal DNS world, timing jitter reflects human-like randomness. In DoH C2 traffic, however, the rhythm exhibits excessive regularity that imitates machine behavior. Microcharacteristics that even advanced implants can't hide include the following:
  - Fixed beacon intervals (e.g., one DoH request every 22 seconds)
  - Jitter patterns that don't occur in human behavior
  - The inability to realistically simulate DNS caching behavior

  For example, in Chrome-based legitimate DoH traffic, the time difference between DoH queries to the same site varies significantly due to user behavior and caching mechanisms. In a C2 implant, by contrast, here's the variance:
  - 22.031 s
  - 22.100 s
  - 22.002 s
  - 22.045 s

  This rhythm isn't human behavior; it's a task loop. Such regular deviations can expose C2 traffic hidden even in the darkest corners of HTTPS.
- **SNI and host header behavioral analysis**
  Many malicious DoH implants attempt to imitate popular DoH provider API endpoints:
  - Cloudflare (`mozilla.cloudflare-dns.com`)
  - Google (`dns.google`)

- Quad9
- NextDNS

Attackers, however, often make the same mistake: The SNI looks realistic, but the contents of the HTTP/2 `HEADERS` frame deviate from real client behavior, for example:

- Browsers aren't required to send a user agent, but most do; bots often leave it empty.
- Chrome DoH queries usually include `Accept-Encoding: gzip, deflate, br`; in malicious implants, this list is often in the wrong order.
- Some threat groups fail to correctly order the ALPN extension (`h2`, `http/1.1`).

These micro-errors are like contour lines beneath a disguise. Easy to miss, but instantly recognizable to a careful CTI analyst.

- **HTTP/2 frame identification**

  Because DoH is most commonly transported over HTTP/2, HTTP/2 frame usage is a highly reliable source for revealing the attacker's handwriting. A genuine Chrome or Firefox DoH query typically follows this sequence:

  - `SETTINGS`
  - `WINDOW_UPDATE`
  - `HEADERS`
  - `DATA`
  - `DATA`

  Distortions observed in C2 implants include the following:

  - `HEADERS` frames that are overly uniform and minimal
  - `DATA` frames with unnecessarily equal sizes
  - `STREAM` ID correlations that don't match browser behavior
  - `PRIORITY` frames that are usually not used at all

  Some DoH implants don't update the `HPACK` compression dictionary within `HEADERS` frames, resulting in frame signature drift. This is an extremely subtle trace that reveals the attacker's coding habits.

- **DNS layer extraction**

  Even though DoH is encrypted, the overall structure of the DNS message remains unchanged. Indirect indicators can be derived from this structure:

  DNS message total length = HTTPS payload length – HTTP/2 framing overhead – TLS overhead

  Using this calculation, it's possible to determine the following with meaningful accuracy:

  - Which DNS record type is being used
  - Which domain chain the attacker is operating on
  - Whether exfiltration is occurring based on payload size

Here's an example:

- Normal A record: ~60–75 bytes
- Long TXT records: 150–500 bytes
- DoH used for exfiltration: 900+ bytes (a clear upper-bound violation)

■ **Cache miss pattern**
This is one of the most sophisticated DoH C2 detection techniques and is applied by almost no one. Real DNS clients exhibit predominantly cache-hit-heavy behavior. C2 implants, however, must generate a new query for each command, resulting in continuous cache misses. Therefore, we have the following:

- `DoH_cache_hit_ratio` < 5% → ~95% probability of C2 behavior
- Normal user traffic → 40%–70% cache-hit ratio

This mathematical difference constitutes a forensic signature that pierces all layers of encryption.

■ **Beaconing waveform analysis over DoH**
Even in encrypted DoH traffic, the beaconing rhythm can be transformed into a waveform: time → packet length. This produces repeating ripples in the graph. Signatures observed in DoH beacons of certain APT groups include the following:

- Low-amplitude but perfectly periodic waveforms
- Amplitude increases when commands are received
- An elongated waveform backbone during exfiltration

This resembles the heartbeat graph of a C2 channel. When visualized, the hidden channel inside HTTPS becomes plainly visible.

DoH is a tunnel for the attacker, but for the CTI analyst, when examined properly, it's one of the protocols that produces the most behavioral errors.

8.2.3 Protocol Dissection

Protocols are mechanical; however, the way an attacker uses them isn't. For this reason, protocol dissection isn't about reading the clauses of a standard, but about examining behaviors that deviate from that standard. The following sections discuss areas that are especially critical in advanced CTI analyses.

DNS Protocol

Long TXT records, multisegment domain chains, faster-than-normal query response cycles, and low-TTL manipulations are all indicators that the attacker is performing DNS-based tunneling. Expert analysts can determine which queries are generated by automation by measuring the variance of DNS query lengths.

Let's examine a case in which an APT group uses a DNS-based tunneling mechanism (DNS exfiltration) to leak data from an internal network to the outside. All behavioral signatures used in the example are based on real attack techniques.

Within a corporate network, security analysts notice an unusual DNS query pattern on the security information and event management (SIEM). At first glance, the DNS traffic appears normal, but the statistical behavior of some queries doesn't align with the natural DNS rhythm. As the analysis deepens, the following anomalies are identified:

- **Excessive use of long TXT records**
  The analyst encounters a well-known attack technique frequently mentioned in SANS Institute research: abuse of TXT records.

  The following example TXT query draws attention:

  ```
  lsdkfj3409sdlfkj2309.sdlkfjsd09sdfj.sldkjf0293sdfj.domain-attacker-controlled.com   TXT?
  ```

  This query is unusual for the DNS protocol because it has the following characteristics:

  - A total length of 220 bytes
  - A six-segment fully qualified domain name (FQDN)
  - High entropy values (7.9 bits) across its segments

  This structure is a typical example of a data exfiltration mechanism encoded with Base32/Base64. The attacker embeds sensitive data from the server into DNS record names by splitting it into small chunks.

- **Anomaly in segment count**
  Normal DNS queries typically contain 2–3 segments (e.g., www.google.com), whereas the domain chains used in this attack vary between 6 and 10 segments (what we call an *over-chained domain*).

  The following command is used for analysis:

  ```
  tshark -r suspect.pcap -Y "dns.qry.name" -T fields -e dns.qry.name |
  awk -F'.' '{print NF}' | sort | uniq -c
  ```

  The output is as follows:

  - 3 3
  - 2 4
  - 198 7
  - 205 8

  This result shows that DNS chains generated by automation systematically use 7 and 8 segments. This is a signature of the attacker's data-chunking algorithm.

- **Faster-than-normal query-response cycles**
  While attempting to bypass DNS over TLS (DoT)/DNS over HTTPS (DoH) tunneling controls, the attacker's response cycle time (RCT) behavior becomes artificial.

Consider the following observations:

- Average DNS query response time: 5–12 ms
- Normal corporate DNS latency average: 22–35 ms

This speed difference indicates that the attacker is hosting the DNS proxy on a local UDP-based forwarder, not in the external world. In other words, the victim machine is sending DNS requests directly to the attacker's server. This is a well-known anomaly caused by the speed optimization common in DNS tunneling operations.

- **TTL manipulation**

  During tunneling, the analyst observes the following TTL behavior:

  - TTL = 0
  - TTL = 1
  - TTL = 2

  Normal DNS record TTL values, however, range from 60 to 86,400, and even in CDNs, TTL values rarely drop below 30.

  Keeping TTL values intentionally low indicates an attempt to bypass caching because the attacker wants every DNS request to reach their server directly.

- **Identifying machine-generated query length variance**

  The expert analyst measures the variance of query lengths using the following command:

  ```
  tshark -r suspect.pcap -Y "dns" -T fields -e dns.qry.name | awk
  '{print length}' | stats
  ```

  The output is as follows:

  - Count: 403
  - Mean: 182.4
  - Standard deviation: 1.76

  This result confirms that the DNS traffic contains fixed-size data blocks generated by a machine. Such a low standard deviation is impossible in human- or hardware-driven DNS traffic. It's observed only in automatically encoded exfiltration data packets.

Remember!

When the following five behaviors appear together, tunneling is confirmed:

- Abnormally long TXT records
- Consistent domain chains of seven to eight segments
- Unusually fast DNS response times
- Manipulated TTL values ranging from 0 to 2
- Nearly constant query length variance

These findings demonstrate that the attacker is using the DNS protocol as both a C2 channel and an exfiltration pipeline.

TLS Protocol

The cryptographic confidentiality provided by TLS has, for many years, effectively become a curtain behind which attackers operate to their advantage. Content is encrypted, data flows become unreadable, and packet inspection layers are forced into silence. However, what most attackers fail to realize is that TLS can hide content, but it can't hide behavior. That behavior is a trace ledger where every implant, every framework, and even every developer's coding habits leak outward through subtle vibrations.

For a true analyst, these vibrations are the shadow of the identity the attacker is trying to conceal:

- **TLS `ClientHello`**

  `ClientHello` is the very first moment a client says, "This is who I am." It's also usually the weakest point of attacker-generated traffic. Over the years, browsers have developed highly natural and consistent extension ordering patterns. Each browser, such as Chrome, Firefox, Edge, and Safari, carries a signature born from its own TLS stack.

  Attacker tools, on the other hand, do one of the following:

  - Fail to perfectly replicate this ordering
  - Apply an incomplete subset of the real browser's extension set
  - Rely on a framework library that generates extension order in a fixed, standardized pattern

  For a CTI analyst, extension ordering is more than a list; it's an identity sentence and often a behavioral DNA sequence. Experienced analysts know very well that extension ordering can never be imitated perfectly.

  The difference between the ordering of a real Chrome browser's 22 extensions and a Sliver implant attempting to mimic the same list is like grammatical errors in a language. Small distortions appear everywhere.

- **JA3 fingerprint variations**

  JA3 values are hashes; however, behind that hash lies a much broader behavioral landscape that exposes the attacker. The implant framework, its libraries, and even the developer's OS preference are reflected in JA3 variations, for example:

  - Python-based implants tend to leave the same elliptic curve ordering.
  - Go tools often contain inconsistencies in extension ordering that look like unnecessary human interference.
  - Some APT groups replicate Chrome's JA3 value but fail to correctly imitate the ALPN sequence.

For this reason, JA3 is more than a fingerprint. When you remove the mask, it becomes the visible surface of the underlying behavioral engine. When attackers attempt to spoof JA3, they most often make mistakes not on the JA3 side, but on JA3S: the `ServerHello` extension ordering in many frameworks remains static and doesn't align with real browser behavior. This single small flaw can expose the entire operation.

- **Session resumption behaviors**
  Browsers and legitimate clients use session resumption to improve performance. Attackers, however, most often misuse this mechanism, disable it entirely, or directly bind resumption ticket timing to the implant's task loop.

  A real browser makes resumption decisions based on network conditions, cache lifetime, and user behavior. An implant, by contrast, operates according to the machine's internal rhythm. That rhythm typically looks like this:

  - C2 Poll #1 → No resume
  - C2 Poll #2 → Resume
  - C2 Poll #3 → Resume
  - C2 Poll #4 → No resume

  Browser behavior, however, follows a more consistent trend: No → Resume → Resume → Resume. This discrepancy indicates that the TLS stack used by the attacker's implant is primitive.

- **Low-entropy padding**
  Although TLS 1.3 supports padding mechanisms; that is, the ability to insert additional, nonfunctional bytes into a TLS record to mask the true length of the encrypted payload and reduce the effectiveness of size-based traffic analysis—most attackers still expose themselves. In legitimate implementations, padding is occasionally used to make encrypted traffic less predictable by introducing variability into record sizes and preventing observers from inferring application-layer behavior from packet lengths alone. However, correct use of padding requires closely mimicking the natural variability produced by real client software. This is because correct use of padding helps conceal traffic, whereas incorrect use reveals that the attacker has overengineered the flow. Low-entropy padding typically arises from the following conditions:

  - The developer adding padding based on a fixed-size packet logic
  - Padding being applied at a constant size rather than reflecting real browser behavior
  - Certain C2 frameworks being unable to simulate TLS record length variance

  Real browser traffic carries organic variability. Attacker traffic, by contrast, appears sterile—almost as if it were produced in a lab environment. This sterility is one of the strongest pieces of evidence for a forensic analyst.

- **Header motifs**
  An attacker may imitate a browser, but they can't perfectly reproduce the complex header motifs that evolve over time within real browsers, for example:

- Chrome slightly updates its ALPN ordering with each release.
- Firefox introduces new GREASE extensions during certain epoch changes.
- Safari's TLS behavior contains micropatterns that vary by macOS version.

When an attacker attempts to follow these motifs, they either miss the current browser signature or add unnecessary extensions, internally cracking their own disguise.

This is where the expert analyst's role truly begins—hearing the unnatural disharmony within these small motifs. TLS hides content, but it can't hide behavior. From extension ordering to padding entropy, from session resumption rhythm to JA3 variation, every point can forensically expose the identity of the implant framework being used.

For a true CTI analyst, the real question is "Does this traffic truly belong to a browser, or is someone merely trying to behave like one?"

SMB, RDP, and Internal Network Protocols

Within enterprise environments, protocols such as SMB and RDP form the operational backbone of internal communication. These protocols are designed for legitimate administrative tasks—file sharing, remote system management, and user session access. However, the same features that make them indispensable for daily operations also make them attractive tools for attackers seeking to move laterally within a compromised network.

In SMB traffic, subtle variations within the negotiate protocol and session setup exchanges can reveal the presence of attacker tooling. Many offensive frameworks implement only partial or simplified versions of the SMB protocol stack, which results in distinctive negotiation patterns. Differences in dialect selection, capability flags, or security negotiation parameters may therefore act as behavioral fingerprints of the tools used during lateral movement.

RDP traffic can provide similar indicators. Changes in encryption levels, unusual transitions in security negotiation, or irregularities in channel establishment may suggest that a session has been manipulated or proxied through attacker-controlled infrastructure. In some cases, automated tools generate RDP sessions that lack the nuanced interaction patterns of genuine user activity, producing connection behaviors that appear technically valid but operationally artificial.

For a CTI analyst, the critical skill isn't just understanding protocol specifications but recognizing how attackers bend those protocols to serve their objectives. The protocol itself is rarely malicious; rather, the misuse of its negotiation sequences, authentication flows, and session behaviors exposes the presence of lateral movement. When examined through this behavioral lens, internal network protocols become silent witnesses to the attacker's operational methods.

Bringing It All Together

When these elements are analyzed together, they provide the analyst with a unique advantage:

- The temporal axis of the attack becomes visible.
- The operational rhythm emerges clearly.
- Intuitive yet powerful inferences about the implant architecture can be made.
- The internal structure of TTPs is behaviorally validated.
- The technical evidence required for attribution becomes unambiguous.

8.2.4 Example: IATI-Based DNS Tunneling and Multistage C2 Rhythm Analysis

Our example in Figure 8.2 shows the output of the `traffic.pcap` file analyzed with the `network_forensics_advanced.py` tool. The JSON we see on the screen contains four main blocks produced by the network-centric forensic intelligence engine: `network_fingerprint`, `dns_tunneling`, `beaconing`, and `attack_reconstruction`. In this section, we'll interpret this output step-by-step, in both a technical and intelligence context.

```
C:\Users\admin\PycharmProjects\Books>python network_forensics_advanced.py traffic.pcap --json
[+] Running Network Fingerprint Engine...
[+] Running DNS Tunnel Analysis (IATI)...
[+] Running Advanced Beaconing Detection...
[+] Reconstructing Multi-Stage Attack Timeliline...

===== FINAL ANALYSIS RESULT =====

{
    "network_fingerprint": {
        "TLEV": 2.7246202328169525,
        "FlowAsymmetryIndex": 18.364367672206956,
        "TTLVariance": 0.0
    },
    "dns_tunneling": {
        "iati_score": 4.517379977489550805,
        "domains_analyzed": 90,
        "findings":
           "High IATI score -> Likely DNS tunneling (encoded exfiltration)."
    }, ]
    "beaconing" {
        "beacon_detected": true,
        "variance": 0.0036848001673057 5725,
        "autocorrelation": 0.999999898956211,
        "fft_periodicity": 1.2915714036745192,
        "severity": "MEDIUM"
    },
    "attack_reconstruction"; [
        "Stage 1 - Reconnaisssance / DNS-based capability check"
        "Stage 2 - Irregular but present beaconing attempt"
    ]
[+] Saved: analysis_output.json

C:\Users\admin\PycharmProjects\Books>
```

Figure 8.2: Network Fingerprint, DNS Tunneling, Beaconing, and Attack Reconstruction JSON Output

We start with the network fingerprint, which consists of the following code:

```
"network_fingerprint": {
  "TLEV": 2.7246...,
  "FlowAsymmetryIndex": 18.3643...,
  "TTLVariance": 0.0
}
```

Let's break down the rows in this section:

- **TLEV ≈ 2.72**
  This value summarizes the entropy of packet sizes. A very low transport-layer entropy vector (TLEV) means nearly uniform traffic (e.g., fixed-size beacon packets); a very high TLEV means a completely scattered flow resembling human behavior. A value around 2.7 indicates that the traffic is neither fully random nor sterilely constant; it suggests a structure that carries traces of automation but isn't entirely machine-like. This fits hybrid scenarios (C2 activity embedded within legitimate traffic) quite well.
- **`FlowAsymmetryIndex` ≈ 18.36**
  This metric is based on the standard deviation of the number of flows between source-destination pairs. A high value indicates that some relationship pairs are used far more heavily than others. Here, a value around ≈18 indicates a clear concentration between a few IP pairs; that is, the traffic is piling up around a small core. From an attacker perspective, this can typically be a sign of behaviors such as these:
  - Regular access to a single/small number of C2 servers
  - Continuous loading against a specific resolver/DNS infrastructure
- **`TTLVariance` = 0.0**
  A TTL variance of zero indicates that the TTL value is completely constant across the inspected packets. This can mean three things:
  - The traffic is originating from the same OS/proxy/hardware chain (a homogeneous source).
  - The PCAP duration is short and capture is one-directional.
  - It was produced in a controlled lab scenario.

  In all three cases, the interpretation is the same: From a network topology perspective, the noise is low, and there's a clean sample suitable for behavioral analysis

Next is the DNS tunneling section, which identifies suspicion of a covert channel:

```
"dns_tunneling": {
  "iati_score": 4.5137...,
  "domains_analyzed": 90,
    "findings":
      "High IATI score → Likely DNS tunneling (encoded exfiltration)."
}
```

Instead of classic single-metric approaches for DNS tunneling, the tool uses a lightweight but multisignal model we call incremental anomalous token inference (IATI). This score is the combination of three components:

- Entropy of query names: Are they random/encoded?
- Distribution of domain segment counts: Are there long chains with seven to eight parts?
- Character-class transitions: Anomalies appear in letter–digit–symbol sequences.

With `iati_score` ≈ 4.51 and a threshold of > 4.5, the engine reaches the following conclusion: "`High IATI score → Likely DNS tunneling (encoded exfiltration).`"

There are a few additional details to unpack:

- A `domains_analyzed` of `90` means that in a short but meaningful window, 90 queries were examined; this is sufficient volume for the statistical model to gain reliability.
- The fact that the score is above threshold but not extremely high places this as not definite tunneling, but rather within a strong high suspicion layer.

Next is the advanced beaconing section, which contains C2 rhythm analysis:

```
"beaconing": {
  "beacon_detected": true,
  "variance": 0.0036848...,
  "autocorrelation": 0.9999998...,
  "fft_periodicity": 1.29157...,
  "severity": "MEDIUM"
}
```

This block evaluates the C2 beacon rhythm derived from TCP timing using a multistage model:

- **`variance`≈ 0.0037**
  The IAT variance is quite low, but not a perfectly flat line. This is overly regular compared to human behavior, but suggests automation that has been masked with some jitter.
- **`autocorrelation` ≈ 0.999999**
  Nearly perfect autocorrelation means the IAT sequence remains highly similar when shifted over itself. This implies a cyclical, repeating rhythm, which is very close to a classic beacon signature.
- **`fft periodicity`≈ 1.29**
  This is the ratio between the dominant frequency and the average power in the Fourier spectrum. A value of 1.29 indicates periodicity is present, but not yet as rigid as a single-frequency, military-discipline channel.
- **`severity` `MEDIUM`**
  Therefore, the engine selects "`MEDIUM`" severity: A clear beacon exists, but both duration and sample density aren't strong enough to push it into a critical/definite band.

Overall, `beacon_detected: true` and `severity: "MEDIUM"`. The tool judges that an active C2 heartbeat exists in this PCAP because it sees a low-jitter, highly correlated, repeating TCP rhythm; however, it classifies it as a beacon in a preparation/stabilization phase rather than a high-risk, high-volume channel.

This brings us to the final stage of attack reconstruction, which presents an automated attack timeline:

```
"attack_reconstruction": [
  "Stage 1 - Reconnaissance / DNS-based capability check",
  "Stage 2 - Irregular but present beaconing attempt"
]
```

This section combines all previous metrics and converts the flow into a phase-based narrative:

- **`Stage 1 - Reconnaissance / DNS-based capability check`**
  The high IATI score and anomalies in DNS query morphology suggest that, in the first stage, capability discovery/infrastructure testing was performed over DNS.

  The attacker's goal here is to test whether data can be carried reliably via DNS, determine which resolver path is more stable, and to probe TTL/cache behavior.
- **`Stage 2 - Irregular but present beaconing attempt`**
  The beacon engine output (`beacon_detected: true` with `MEDIUM` severity) indicates a second stage: a C2 attempt that isn't fully settled but has become detectable.

  The phrase `Irregular but present` implies the attacker may still be experimenting with jitter settings, interval tuning, and so on; the channel is operational but not yet fully optimized.

> **Remember!**
> Rather than looking at individual metrics one by one, it's far more valuable to see the operational story the network engine derives from those metrics. The same JSON output can become a correlation rule on the SIEM side; on the CTI side, it can turn into a forensic summary that captures the attacker's tactical behavior.

8.3 Flow-Level Analysis

At first glance, network flows are just sequences of numbers: byte counts, packet quantities, session durations, ingress-egress directions, and so on. Yet in reality, flow-level analysis is one of the most sophisticated and most intuitive domains of network-centric forensic intelligence because it examines the behavior of traffic instead of the traffic itself. Packet contents are no longer visible; numbers seem as though they say nothing. But the rhythms of flows reveal the microclues that attackers leave behind, either deliberately or unintentionally.

Every system, after a while, produces its own traffic heartbeat. Through the combination of user activity, application load, and system processes, a natural rhythm forms within the network. This rhythm maintains a largely predictable structure, even at different times of the day. This is where attackers struggle to remain hidden: the disruption of this natural order at the flow level.

A thin outbound flow from an otherwise ordinary server, a long-lived connection that doesn't fully align with the machine's normal behavior, a light data transfer that repeats regularly at the exact top of every hour, and more may appear random on the surface. Yet when flow behavior is examined in context, they resemble the operational breathing patterns of an attacker.

To observe and analyze these subtle flow-level rhythms, analysts rely on telemetry systems specifically designed to summarize network behavior over time. Among the most widely used of these are NetFlow and IP Flow Information Export (IPFIX), which transform raw traffic activity into structured flow records suitable for behavioral analysis. We'll discuss these systems in the following sections, along with key flow-level analysis techniques.

8.3.1 Introducing NetFlow and IPFIX

NetFlow and IPFIX are two powerful observation tools developed expressly to capture these rhythms. In modern networks, where encryption layers have expanded and content visibility has diminished, these tools have become even more important. Protocols such as HTTPS, HTTP/2, and QUIC hide data packets entirely behind a veil of secrecy; DoH tunnels render classic DNS invisible; and TLS 1.3 completely darkens payload content. But attackers forget one thing: Encryption hides content, not behavior.

Flow-level behavior contains traffic volumes, packet sequencing over time, directional asymmetries, microfluctuations in connection duration, and interflow silences. This information can't be encrypted because it's inherent to the protocol itself.

An attacker can hide commands, data content, tunnel structure, and even the client's identity. But one thing can't be hidden: the language of the flow. This language carries the attacker's operational reflexes, left behind without their awareness.

This flow-level language is often far more powerful than packet-level analysis. Modern techniques used in C2 communications have made content inspection nearly impossible. Yet NetFlow/IPFIX records the attacker's behavior on the network in the form of a numerical trace:

- Low-volume but regular outbound flows indicating exfiltration preparation
- Millisecond-level silence intervals that expose beaconing rhythms
- Recurring flow-duration values reflecting the implant's execution cycle
- Continuous but low-volume connections toward specific ASNs
- Extremely low-variance packet size distributions seen in bot-based traffic
- Minimal yet perfectly timed connections observed during command polling phases

Attackers know how to conceal content; however, they often fail to decorate these mathematical traces at the flow level with human-made randomness. For this reason, flow analysis remains one of the strongest foundations of forensic detection, even within encrypted C2 channels. Visualizing flows further simplifies the understanding of this complex behavior. The purpose of a flow-graph design is to transform the attacker's rhythm into a visual form. When density clusters, directional heat maps, time-series lines, and entropy curves come together, attacker behavior becomes readable "by eye."

For example, the following silent signatures indicate the presence of a C2 channel:

- Mini outbound spikes recurring every minute in a 24-hour flow graph
- Short-lived but high-frequency connection attempts within the internal network
- Directional asymmetries passing through a proxy that don't match natural proxy usage
- Long-established connections that carry minimal data volumes

Such anomalies may not mean much on their own, but when evaluated in context, the attacker's entire operational chain emerges from among these quiet traces.

The greatest strength of NetFlow and IPFIX is their ability to capture behavior regardless of the method used by the attacker. Whether it's DoH, a TLS tunnel, HTTP/2, or QUIC, it doesn't matter because even if the content isn't visible, the shape of the flow is.

If an attacker sends pulses to a C2 at fixed intervals, the flow graph produces a heartbeat like an EKG trace. If an implant performs exfiltration, waveforms rise in outbound flows. If there is pivot movement, sudden bursts appear in internal segments. If a threat actor moves slowly and covertly, flows exhibit a cycle of dead silence + sudden activity.

Reading these rhythms is almost a form of intuition. An analyst who listens to the behavior of the network can sense the presence of an attacker even without packet content. And in many cases, what stops an attack isn't the visibility of the content, but that slight disturbance in the rhythm of the flow.

To make these behavioral rhythms visible, analysts rely on the following techniques that transform raw flow records into visual patterns that can be examined over time:

- **Flow visualization**

  When you visualize NetFlow data, the first thing you notice is that some flows are "lonely", that is, small veins that don't resemble normal traffic, as if they don't belong to the natural fabric of the network. Attackers often conduct their C2 communications directly over these lonely flows.

  For an advanced CTI analyst, flow visualization is like a microscope for threat detection. For example, the following behavioral signatures indicate an attacker is breathing toward the C2:

  - A thin flow fluctuating between 0.0 and 2.5 kbps throughout the day
  - Long-lived TCP flows that aren't seen in normal users
  - Momentary outbound traffic that rises and falls like a pulse at regular intervals

A specific observation here is that some QUIC-based C2 channels leave very subtle traffic density increases of 6–8 ms in IPFIX records, which could be described as microburst anomalies. Packet content is invisible; however, these microfractures can be clearly observed in flow graphs.

- **Anomaly detection**

  Anomaly detection at the flow level makes it possible to decipher the attacker's behavioral model without performing content analysis. The critical parameters that CTI experts frequently examine are as follows:

  - Flow duration: Flows that remain open for extended periods while carrying very small volumes of data often indicate beaconing or heartbeat behavior typical of C2 communication.
  - Flow directionality: Asymmetric connections originating from the internal network toward external destinations may signal early stages of data exfiltration or preparation for outbound data transfer.
  - Bytes-per-packet variance: Packet size distributions that remain unusually uniform and lack the natural variability of human-driven traffic can suggest automated communication generated by malware or scripted tools.
  - Interflow silence interval: Repeating periods of silence between flows that occur at consistent intervals may indicate an implant that is periodically waiting for commands from a remote controller.

  Each parameter alone may not say much; but when combined, the attacker's mind, code, timing, and mode of operation begin to emerge. For an analyst, the real magic is that an anomaly isn't data—it's intent. NetFlow and IPFIX are among the rare mechanisms capable of revealing this intent without content.

- **Threat attribution**

  The greatest power at the flow level is the ability to determine who is talking even when there's no content.At first glance, this seems impossible; yet flow behavior carries threat group signatures, for example:

  - The Lazarus group frequently uses an ultrashort burst + long silence form visible in flow graphs.
  - MuddyWater generates medium-volume beaconing flows with jitter manipulation in the 8–13 second range.
  - Gamaredon can be recognized by very distinct fixed 1–2 kb outbound heartbeats.
  - Cobalt Strike Beacons typically produce tight patterns clustered around 150–250 bytes per packet.

  None of these signatures require packet content. Simply looking at the rhythm of the flow is sufficient. An advanced flow analyst, upon seeing this rhythm, can tell who the attacker is—much like a cardiologist recognizing a patient from an EKG trace.

Example

A security operations center (SOC) analyst notices something abnormal during peak hours of the day:

- Flow duration: 45 minutes
- Total outbound volume: Only 31 KB
- Flow count: 1 connection per second, stable
- Bytes-per-flow: Nearly constant
- Interflow gap: 22.101 ± 0.020 s

When these values are combined, the picture that emerges is very clear, even though the content isn't visible: This is a DoH-based beaconing C2. This conclusion is derived using only NetFlow/IPFIX data: no content or payload is required, only behavior.

The Real Power Provided by NetFlow/IPFIX

Flow-level analysis enables analysts to detect suspicious activity and to infer key characteristics of the attacker's operation, including operational duration, degree of stealth, level of automation, and the maturity of the C2 protocol. In modern CTI practice, this capability is critical: While attackers may conceal content, their behavioral patterns within network flows remain observable.

8.3.2 Flow Morphology

The morphology of network flows is a discipline that is still underused in forensic intelligence, yet powerful enough to strip away all the masks of attackers. Because a flow isn't composed solely of bits and bytes, like the DNA of a living organism, it carries a behavioral structure, an evolutionary pattern, and a unique biological rhythm. Even when packet contents are concealed, this rhythm leaves subtle whispers about how professional the attacker is.

So, what happens if we treat flows truly as living entities? Which behavior, which deviation, which silence reveals the attacker's signature? And can the attacker really never succeed in hiding this? This section examines the anatomy of these questions in depth.

Normal Versus Malicious Flows

Normal user traffic isn't a straight line; on the contrary, it resembles a chaotic biological process. The following factors together form an organic flow DNA:

- Inconsistencies in human behavior
- Natural variance in application delays
- Microjitters caused by browsers' memory management
- Cyclical breathing patterns of background services

There are no straight lines in this flow DNA. Everything is wavy, fragmented, and filled with soft irregularities, which is why genuine user traffic is completely distinct from low-entropy robotic behavior.

Attackers, without realizing it, contaminate their flows with a sterile order because of the following:

- C2 implants don't imitate human behavior; they merely execute a task cycle.
- Botnets minimize variance to optimize traffic.
- Data exfiltration mechanisms operate in fixed-size blocks.
- QUIC/HTTP2/DoH tunnels produce programmed deterministic behavior.

This creates a measurable, distinctive sterilization at the flow level. No matter how professional the attacker is, they can't make the morphology of the flow fully human-like.

Flow Genetics: Duration, Packet Size, and Density

Can time expose the attacker? Absolutely, yes. Real user sessions display completely irregular duration distributions due to sudden interruptions, UI wait times, background updates, and random browser behaviors.

C2 implants, on the other hand, produce a burst → silence → burst → silence cycle. This cycle takes a recurring form resembling a biological heartbeat. The variance of silence intervals remains within the 1%–2% range. For an expert analyst, flow-duration genetics means observing the attacker's task cycle and measuring the "implant pulse.

The next key consideration is packet-size sequence motifs. Browser traffic generates high variance in packet sizes. This is because the following factors all create natural variation, typically in the 15%–27% range:

- Rendering processes
- TCP segmentation
- Script execution
- Background threads

In attacker frameworks, however, we have these factors:

- Packet-size variance is usually below 1%.
- Some C2 tools use completely fixed packet sizes.
- During exfiltration, data is always sent in blocks of the same length.

This is like the attacker's genetic signature: monotony in size patterns. So, what is a lesser-known advanced finding?

Some modern APTs believe they increase variance by randomizing packet sizes. Yet even in random sequences, a Gaussian distribution appears. Real human traffic, however, isn't Gaussian; it behaves in a multimodal manner. This difference is forensic gold for the advanced analyst.

The final main factor in the flow genetics is flow density. *Flow density* shows how a flow consumes oxygen over time. This density is a silent graph that reveals attacker behavior.

Normal user traffic has homogeneous density, sometimes increases and sometimes decreases, and displays a distribution that's naturally correlated with application behavior.

C2 communication leaves the following impacts on flow density:

- Microburst density peaks are produced.
- In QUIC-based C2, recurring sudden density islands appear in the 4–8 ms range.
- These signatures are etched into IPFIX records even when content is encrypted.

A critical detail that very few people know is that these microbursts are often related to the processing order of the QUIC library being used. For example, in Go-based QUIC implementations, an unnatural drift known as burst cadence drift occurs. This drift is noticeable only in flow-density graphs.

Recommended Approach

If you're looking for a method that teaches you to see the attacker without looking at content, flow morphology is the starting point of that path. Most analysts focus on the data inside packets; yet the attacker's true face is hidden in behavior.

Here's the recommended approach:

- See every flow as a living being.
- Read every deviation as a trace of trauma.
- Interpret every silence as the implant "holding its breath."
- Consider every burst as the attacker's "reflex movement."

Because flow behavior is both statistical and emotional, the attacker unknowingly imprints their own rhythm onto the traffic.

8.3.3 Rhythmic Deviation Analysis

At the flow level, time is the most powerful forensic resource. In modern attacks, content may disappear completely, but distortions in time reveal everything. This section focuses on timing-based irregularities that frequently emerge in attacker-controlled communications at the flow level. By examining how flows pause, resume, and cluster over time, analysts can identify patterns that differ from normal user-driven activity. These temporal deviations often reveal the operational logic of implants, including command polling intervals, synchronization errors, and subtle latency anomalies introduced by imperfect protocol implementations.

Let's walk through them:

- **Interflow silence fingerprinting**
  Command-waiting implants often produce periods of machine-stable silence, typically

in the 18–30 second range, between successive outbound flows. These silence intervals arise because many implants rely on fixed or semifixed polling mechanisms to check for new commands from the C2 server. As a result, network flows appear in a repeating pattern: a short burst of communication followed by a consistent pause before the next connection attempt.

Human-driven activity rarely exhibits such stable timing. In normal user behavior, pauses between flows are influenced by multiple unpredictable factors, including user interaction, application processing delays, background system tasks, and network conditions. These factors create highly irregular silence distributions rather than evenly spaced intervals.

When analysts examine flow timelines, this difference becomes significant. Repeated silence intervals that remain tightly clustered within a narrow range often indicate automated command polling rather than organic user activity. Over longer observation windows, these patterns can form a recognizable behavioral signature, allowing analysts to identify potential C2 beaconing even when the content of the traffic remains fully encrypted.

- **Epoch-synchronization errors**
  Some implant frameworks fail to correctly synchronize their internal timing logic with the system clock or with standard time references used by the OS. As a result, scheduled communication events, such as periodic beaconing or command polling, may exhibit small but measurable deviations around predictable time boundaries, particularly near the top of the hour.

  In legitimate applications, time-based operations are typically coordinated through well-maintained system timers or synchronized network time sources such as NTP. This produces relatively stable timing behavior across sessions and hosts. Improvised or poorly implemented implants, however, often rely on simplified timing routines, internal counters, or loosely synchronized clocks. Over time, these mechanisms introduce microdrifts, causing flows to appear slightly early or late relative to expected time intervals.

  When observed across multiple flow records, these subtle shifts can reveal patterns of imperfect time management within the implant's scheduling logic. For an experienced analyst, recurring microdrifts near predictable time boundaries may therefore serve as an indicator of automated command polling or task scheduling embedded within attacker-controlled tooling.

- **Latency microburst signatures**
  In some QUIC-based C2 channels, short latency spikes (typically in the 6–12 ms range) can appear within the flow timeline. These microbursts occur when packets are transmitted in small clusters rather than at a smooth pace. Legitimate QUIC implementations usually maintain stable packet timing, whereas poorly optimized malware libraries may generate these clustered bursts due to simplified processing routines. Over time, such patterns can appear as repeated spikes in flow timing graphs and may indicate automated C2 communication.

8.3.4 Directional Asymmetry Analysis

Every connection truly tells a story, but the most accurate sentence of that story often emerges from which direction speaks more. Sometimes inbound data flow is as small as a whisper, while outbound traffic continuously flows as an invisible thin line. Sometimes the connection is like an incomplete breath-in/breath-out cycle—it receives but doesn't give, or it gives but there's no return. This is exactly where directional asymmetry analysis comes into play by making it possible to read the attacker's intent solely by observing flow direction and without ever looking at traffic content.

This field is underestimated by many analysts because it appears simple. Yet in advanced attacks, microdistortions in directional changes can reveal the implant's architecture, communication model, command retrieval method, and data exfiltration technique. Directional asymmetry says far more than what lies inside the packet. What is the connection not trying to say, but ends up saying unintentionally?

Let's explore the key techniques that fall under directional asymmetry analysis.

Exfiltration Signature Mapping

Data exfiltration traffic exhibits unique behavior even when content is fully encrypted in the following ways:

- Outbound traffic is low-volume but regular.
- Inbound traffic is almost nonexistent and often limited only to response ACKs.
- Byte sequences aren't symmetrical; outbound packets are mostly small and homogeneous.
- Exfiltration is usually performed via short microflows rather than session reuse.

An even more advanced observation is that some APT groups use single-packet-per-second outbound pulses during exfiltration. These packets aren't high-entropy because the data is encoded piece by piece. On a flow graph, this forms a distinct rhythm in the shape of 1–2 kb outbound pulses.

We must ask why a legitimate user would continuously send fixed-size packets between 120 and 160 bytes outbound for 45 minutes? There's no natural user scenario that can explain this behavior. Therefore, directional asymmetry allows exfiltration to expose itself within the very first seconds.

C2 Handshake Asymmetry

In pull-based C2 architectures, the implant receives data but rarely sends data. This leaves a very clear trace in directional behavior:

- Outbound functions as a command request.
- Inbound contains only an empty response or a small command payload.
- Data flow isn't symmetrical; it resembles a completely one-sided conversation.

An advanced C2 protocol may implement strong encryption, mask packet sizes, and even diversify timing jitter. However, it can't change the command-retrieval direction. A pull-based implant always says, "I'm here, is there a command?" This behavior appears in most frameworks as regular calls every 20–30 seconds.

Here's an observation: Some APTs attempt to evade detection by masking inbound command packets with equal-sized padding. Yet even this can't break the mathematics of directional asymmetry:

- Outbound traffic → more frequent
- Inbound traffic → less frequent
- Outbound flow count → 5–20 times higher than inbound

This ratio enables a forensic analyst to resolve the attacker's C2 method even when content is encrypted.

Direction-Weighted Ratio Anomalies

Direction-weighted ratio (DWR) is a forensic metric that determines the inbound/outbound ratio within a flow. In normal user traffic, this ratio is balanced:

- HTTP traffic: 40% inbound/60% outbound
- TLS traffic: 50% inbound/50% outbound
- Streaming traffic: 80% inbound/20% outbound

In other words, natural traffic seeks balance. C2 traffic, however, completely loses this balance:

- 90% outbound/10% inbound
- 95% outbound/5% inbound

In some frameworks, even 100% outbound/0% inbound can be observed.

This imbalance isn't just the result of attacker behavior. It stems from the architecture of most implant frameworks:

- Cobalt Strike Beacon is predominantly outbound.
- Mythic is inbound at session start, but then one-way outbound.
- Sliver is heavy outbound and minimal inbound.
- Gamaredon is almost entirely outbound polling.

Thus, DWR effectively reveals the attacker's "language."

Note

My personal assessment is that sometimes DWR alone is clear enough to allow the identity of a threat actor to be inferred—no content, no payload, no fingerprint, only traffic direction. Yet this direction reflects the attacker's character.

The Hidden Layer of Directional Asymmetry

The rarely discussed part of this topic is that even when attackers attempt to correct directional asymmetry, they still make mistakes, for example:

- They try to equalize command packets with outbound traffic, but the timing sequence breaks down.
- They generate fake inbound traffic, but entropy inconsistencies give them away.
- They vary outbound volume, but flow frequency doesn't appear natural.
- They apply symmetric padding, but response timing isn't realistic.

These microconfessions guide the forensic analyst onto the right path even without content. Attackers can hide the command, data, content, protocol, domain name, and JA3 fingerprint, but there's one thing they can't fully hide: which direction the connection is speaking in.

Directional asymmetry is perhaps the purest, least corrupted, and most honest forensic source in C2 detection because the data flow direction isn't a product of the attacker's will, but a mandatory outcome of their architecture.

8.3.5 Flow Entropy and Variance Analysis

Flow-level analysis is more complicated than just reading shapes, graphs, or visual curves. In reality, flows in their purest form are a numerical poem consisting of a rhythm, a disharmony, and a small fracture. The entropy within this rhythm makes all the behavioral patterns visible that the attacker wants to hide.

Here's a secret: No matter how much encryption attackers use, how much tunneling they apply, or how completely they darken packet content, they can't deceive mathematics. Entropy reveals the identity of attacker behavior without any effort. Always avoid listening to what a flow is saying, and instead listen to its tone of voice. Content can be hidden, but tone can't.

Entropy can be categorized in a few different ways, as we'll discuss in the following sections.

Bytes-per-Flow Entropy

Each flow produces its own volume pattern. Natural user behavior has high entropy for the following reasons:

- Instantaneous data bursts occur while pages load.
- Background scripts constantly change sizes.
- User keyboard–mouse interactions increase data volume in small but irregular ways.
- Browser cache behavior creates unpredictable variance.

In short, humans are chaotically beautiful and flawed. The traces of this chaos permeate flow volume. In contrast, attacker implants produce low-entropy flows:

- Packets move in fixed sizes.
- Application cycles are deterministic.
- Data volume changes very little during C2 beaconing.
- Exfiltration blocks are always sent in the same size.

A lesser-known fact is that some advanced APTs attempt to artificially raise entropy. However, most randomization algorithms produce a uniform distribution. Real human traffic distributions aren't uniform, but multimodal. This difference can only be understood at an expert level, yet it definitively exposes the attacker.

Interarrival Time Entropy

The time gap between packets is what attackers struggle most to conceal. In normal user traffic, we get the following:

- Intervals between packets resemble a natural rhythm of life.
- Factors such as CPU load, disk latency, and human reaction constantly alter them.
- Jitter distribution has high variance.
- The flow of time is never smooth.

In malware traffic, however, this rhythm collapses because implants are scheduled by a task cycle; they aren't irregular like human behavior, for example:

- 22.000 ms
- 22.015 ms
- 21.998 ms
- 22.030 ms

This regularity says "This isn't a user," even if the content is fully encrypted. This is machine communication trying hard to stay hidden.

If you observe the time intervals within a flow long enough, you can feel the attacker's impatience, patience, and even operational maturity. Some implants become impatient when no command arrives and the intervals rapidly shorten. Others value stealth more, so the jitter increases excessively, but mathematical consistency is lost. Time reflects the attacker's state of mind.

Packet-Length Entropy

Packet-length entropy is one of the strongest signatures that reveals the identity of an attacker framework, even when content is invisible. This is because of the following:

- Cobalt Strike, in certain modes, produces an overly consistent window of 150–250 bytes.
- Sliver traffic contains small but fixed command gaps.

- Mythic flows are modular, and each module produces different length variance; this modular variance informs the analyst about the implant's configuration.
- Low-quality APT tools such as Gamaredon behave amateurishly with completely fixed packet sizes.

But what is far less known is that even when some implants operate inside TLS tunnels, they can't mask record-length entropy because TLS record framing can't be fully controlled by attacker tools. As a result, a perfectly smooth line never forms, even at the TLS level. In real user traffic, the record-length graph is scattered. In attacker traffic, it appears homogeneous, rhythmic, and with minimal variance. This graph is the trace left behind by mindless automation, even when content is encrypted.

The Deep Layer of Flow Entropy

Flow entropy isn't limited to just three metrics. When analyzed deeply, low-level patterns emerge that expose the attacker's thread structures, code loop cycles, command scheduler, and even network stack implementation. Some of these advanced techniques include the following:

- **Temporal instability index (TII)**
  Measures how natural timing appears over continuity. Low TII indicates a bot, while a high TII indicates a human.
- **Frequency drift factor (FDF)**
  Measures how much timing intervals drift over long periods. Implants keep drift close to zero; humans are masters of drift.
- **Multivariate flow entropy (MFE)**
  Calculates the entropy of the correlation between volume, time, direction, and packet length within a flow. This method is used only by advanced CTI analysts and is capable of distinguishing attacker frameworks with 90%+ accuracy.

These techniques are rarely mentioned even in academic literature, yet their value in operational threat analysis is invaluable.

Yes, it may sound metaphorical, but it's true. When looking at the entropy of a flow, we get the following:

- Poorly written code means the entropy collapses meaninglessly.
- A rushed operation means the time variance narrows.
- A patient and experienced attacker means the values attempt to imitate natural variance, but still fail.

Mathematics is sometimes a mirror; it reflects back at the attacker the very side they try to hide. Flow entropy is a more reliable indicator in modern threat intelligence than packet content. Attackers can hide content, tunnel traffic, and encrypt everything, but they can't convincingly imitate entropy in its natural form.

For this reason, bytes-per-flow behavior, IAT variance, and packet-length entropy—when read together—clearly reveal the following about the attacker:

- Human or machine
- Advanced or amateur
- Patient or impatient
- Stealthy or careless

8.3.6 Example: Deriving Attacker Behavior from Mathematical Flow Traces and Advanced Flow Entropy Analysis

To make the discussion concrete, the following example shows how behavioral conclusions can be derived from a single flow by combining entropy-based measurements with spectral and timing-oriented flow features. Figure 8.3 is the console output of a Python-based analysis script executed against a PCAP file (`traffic.pcap`).

```
C:\Users\admin\PycharmProjects\Books>python flow_entropy_analyzer.py traffic.pcap
{
    "('10.0.0.5', '8.8.8.8', 6)": {
        "byte_entropy": 3.906890595608519,
        "iat_entropy": 3.807354922057604,
        "iat_variance": 0.0010301644672926606,
        "packet_length_entropy": 3.906890595608519,
        "TID": 0.00026724206052987777,
        "FDF": 0.0,
        "MFE": -10.424497979787088,
        "MJSD": 0.07692307692307693,
        "BED": 0.0,
        "PLDF": 1185.0,
        "VSA": -0.000989059185866033,
        "BIV": 0.028047163877809752,
        "TFAD": 0.0,
        "TPS": 1.502286321844014,
        "CHRI": 0.5176285010566304,
        "FMS": 16.79146592271955,
        "autocorrelation": [
            0.014422302542097247,
            0.0019272730736370437,
            -0.00439361160111122,
            0.0013367337772295714,
            0.0032718156128734035,
            -0.0010102136993853084,
            -0.002975712576558784,
            -0.0011979424196104965,
            -0.00018091825274045827,
            -0.0028359104521977854,
            -0.002400991676794621,
            0.0013839571372245146,
            8.325118085858983e-05,
            -0.00021888137447307113
        ],
        "frequency_spectrum": [
            0.5720710754394531,
            0.16071412557085946,
            0.0858998488149513,
            0.13233708684281567,
            0.17845718774288546,
            0.11954728439999732,
            0.05354166662613079,
            0.049597740173339844,
            0.05354166662613079,
            0.11954728439999732,
            0.17845718774288546,
            0.13233708684281567,
            0.0858998488149513,
            0.16071412557085946
        ],
        "flow_psychology": "Exfiltration\u2013Chunked"
    },
}
```

Figure 8.3: Advanced Flow Entropy and Spectral Feature Extraction Output for Behavioral Flow Profiling

In practical terms, the workflow is straightforward: The PCAP is first parsed into individual flows using the standard five-tuple logic (source IP, destination IP, protocol, source port, and destination port, or an equivalent reduced tuple where appropriate), and each flow is then processed to extract statistical, temporal, and frequency-domain characteristics.

These features include entropy values derived from packet sizes and IATs, variance-sensitive descriptors, autocorrelation values, and frequency-spectrum outputs used to identify rhythmic or burst-driven communication patterns.

The purpose of this example is to show how the raw numeric output can be translated into behavioral interpretation. In operational use, an analyst would typically run the feature-extraction script against a PCAP, review the resulting metrics on a per-flow basis, and then compare those values against expected patterns for ordinary user traffic, automated polling, beaconing, or staged exfiltration. The figure therefore serves as a bridge between theory and practice: It demonstrates how mathematically derived flow traces can be used to characterize a connection and to assess whether its behavior is more consistent with benign activity or with attacker-controlled communication.

The result shown in the image belongs to the following: `('10.0.0.5', '8.8.8.8', 6)`. In other words, a flow has been detected from source IP `10.0.0.5` to destination `8.8.8.8` over TCP (`6`). On this flow, both classical and advanced flow entropy, jitter symmetry, variance behavior, spectrogram analysis, TLS-framing response, phase shift, and behavioral profile have been evaluated.

Now, let's examine the meaning of each metric, what the value shown in the figure indicates, and how closely this flow mimics malicious or nonmalicious traffic:

- **`byte_entropy` (3.90)**
  This value measures the randomness of packet sizes.
  - A value of 3.90 is close to user traffic.
  - There are no repetitive or fixed-size packets.
  - Exfiltration chunks haven't formed an excessively regular pattern.

  The flow produces a naturally appearing volume distribution.
- **`iat_entropy` (3.80) + `iat_variance` (0.00103)**
  IAT between packets is as follows:
  - High entropy (3.80) means time gaps show diversity.
  - Low variance is found, but the diversity is at a very microlevel, with no large fluctuations.

  This combination is seen in only two cases: human traffic and well-concealed exfiltration. This example is closer to the second case (supported by the following metrics).
- **`TII`**
  The TII value measures the naturalness over time of the flow. In this case, 0.00026 is extremely low. Such a low TII indicates that the flow doesn't leave equilibrium and doesn't change behavior over time. In human behavior, TII increases because user behavior evolves over time. Here, it does not, which is a trace of automation.
- **`MJSD`**
  Microjitter symmetry deviation (`MJSD`) represents the degree of symmetry in jitter fluctuations. An extremely low value such as 0.07 indicates that microjitter isn't symmetric, meaning packet timing has unnatural micro-adjustments.

Typical ranges include the following:

- Human traffic: 0.20–0.35
- Stealth C2 implants: 0.01–0.09

The value 0.076 here indicates stealth automation.

- **BED**

  Behavioral entropy drift (BED) measures how much the entropy of the flow drifts from start to finish. Drift is low in this example. High drift indicates user behavior, and low drift indicates machine/automation. This shows that the flow remains in a single behavioral mode over time.

- **PLDF**

  The packet length differentiation factor (PLDF) is high, indicating the following:

  - Normal for flows involving large volume differences such as exfiltration
  - Not seen in fixed-size beaconing

  This flow produces variable-volume, exfiltration-like traffic, not C2 beaconing.

- **BIV**

  The burst irregularity value (BIV) is low, which means bursts are very regular. In human traffic, 0.10–0.30 is common. Here, 0.02 indicates robotic behavior.

- **TFAD**

  TLS framing anomaly detection (TFAD) shows the TLS record-length variance is almost zero. This indicates the following:

  - The flow isn't operating inside a TLS tunnel.
  - TLS frames haven't been manipulated.
  - Or, there is insufficient data for TLS frame analysis.

  This can also be observed in transparent exfiltration scenarios

- **TPS**

  The temporal phase shift (TPS) shows phase deviation is high, meaning the flow's timing signal contains some degree of shift. This may be natural on the user side, but when combined with other metrics, small artificial shifts may have been added on top of a very regular rhythm. This is common in stealth exfiltration (to make behavior appear human).

- **CHRI**

  The harmonic resonance index (CHRI) represents the sum of Fast Fourier Transform (FFT) harmonic powers. This metric helps analysts distinguish different communication behaviors by examining how strongly periodic signals appear in the flow:

  - Extremely high in C2 beaconing
  - Unstable in human traffic
  - Moderate in exfiltration

  Here, 0.5 is a sign favoring exfiltration.

- `FMS`

  The flow morphology score (`FMS`) aggregates multiple behavioral metrics into a single composite value that reflects the overall structural characteristics of the flow. In this example, the calculated score is approximately 16.7, which falls within the following general behavioral ranges:

  - C2 beaconing: 5–8
  - Human traffic: 10–20
  - Exfiltration: Typically, 15–25

  Thus, this score is consistent with exfiltration behavior.

Flow Outcome → Flow Psychology

The tool detected this automatically. Strong foundations for this classification include the following:

- Low TII: Stable automation
- Low MJSD: Unnatural microjitter
- Medium–high IAT entropy: Rhythm attempting to remain hidden
- Very high PLDF: Block-based exfiltration patterns
- Low drift: No behavioral change
- Low BIV: Robotic burst structure
- Medium CHRI: Not beaconing, but still rhythmic
- FMS range: Within the exfiltration zone

Although this example flow appears externally like an ordinary client-DNS or HTTPS-like communication, deep flow entropy analysis shows that the traffic behavior actually resembles chunk-based data exfiltration.

Despite the high-looking time-gap entropy, the variance remaining almost constant indicates that the attacker added microlevel jitter to mimic user behavior, but failed to capture the natural degradation tendency of real time series.

The extreme difference in packet sizes (PLDF = 1185) indicates that the flow transmits data in blocks. The extraordinarily low TII confirms that the traffic doesn't change behavior over time, which is a mechanical rhythm that wouldn't be seen in human behavior.

By combining all of these data points, the flow psychology classifier categorized the flow as "Exfiltration Chunked." This behavior closely resembles modern data exfiltration techniques used by advanced threat actors to evade detection.

8.4 Correlation of Logs and Network Metadata

A log line that goes unnoticed sometimes whispers a secret that flow analysis can't solve. Analyzing network flows is a powerful method for perceiving the rhythm of attacker behavior; however, no attack consists of a flow alone. When an attacker infiltrates a network, their presence leaves microscopic traces scattered across the logs of different systems. These traces are meaningless on their own, but when combined within the proper context, the path the attacker walked becomes clear.

For this reason, advanced forensic intelligence doesn't focus solely on packets, jitter, and flow entropy; it correlates firewall logs, proxy records, DNS-level metadata, and the native telemetry of devices with one another. This correlation constitutes a unique multilayered verification mechanism that reveals both the technical and behavioral integrity of the attacker. We'll dive into each of these topics in the following sections.

8.4.1 Firewall Logs

Firewall logs are considered, in most organizations, a log class in which noise is intense. They are heavily simplified on SIEM platforms through aggressive filtering; most analysts focus only on blocked traffic, while the rest fades into the background.

Yet for an advanced CTI or forensics expert, firewall logs represent a critical stage where the attacker first probes the network boundary, exhales before putting on their mask, and leaves early clues about operational rhythm. The rhythm observed at the flow level often leaves a more raw and less manipulated behavioral footprint in firewall logs. Even if the attacker succeeds in concealing packets within flows, manipulating firewall behavior is significantly more difficult.

Even in their most professional operation, the attacker touches the firewall first. The trace they leave at the network gate is cleaner and more natural than the one they leave inside. That trace is the most valuable clue.

C2 traffic may appear as a regular heartbeat in flow analysis; however, firewall logs reveal far more clearly what type of connection behavior that heartbeat actually produces. In a typical C2 scenario, the following patterns emerge in firewall logs:

- Repeated connection attempts to the same external IP address
- Persistent use of the same port (commonly 443, chosen for masking purposes)
- Narrow-band behavior in packet sizes that mirrors the flow-level profile
- Each connection attempt occurring within nearly identical time intervals

This behavior may manifest as 30-second polling, 5-second jitter, or as brief 400-millisecond sudden connection clicks. Interestingly, these 400-millisecond microscopic connection attempts are usually initial warm-up probes performed during the early establishment phase of the C2 channel. Whenever this behavior is observed, it seems as though the attacker is performing one final check before crossing the bridge. The attacker's objective is to determine whether the path to the outside world remains safe. For this reason, small SYN-ACK loops appearing every few seconds in firewall logs are never coincidental.

An attacker may deploy masked C2 profiles internally, employ JARM/JA3 manipulation, encrypted tunnels, or proxy-tampering techniques. Yet what the firewall observes is the raw connection itself—a reality independent of encryption, headers, or payloads.

Firewall logs answer these fundamental questions:

- Where is the connection going?
- How frequently does it occur?
- Why is the same port always used?
- Why is the direction consistently the same?
- Why does the traffic fail to resemble natural user behavior?
- Why is there such low variance in the connection lifetime?
- Why is inbound return traffic negligible to the point of near nonexistence?

Each of these questions carries direct behavioral indicators regarding the attacker's C2 design.

TTL Manipulation to Avoid Firewalls

Some APT groups employ TTL manipulation to evade firewall-based detection mechanisms. TTL manipulation implies that the attacker can do the following:

- Directly alter the packet's effective lifetime
- Control how it appears across the network
- Influence which device it appears to have terminated on

Therefore, if a Linux system that should normally originate traffic with a TTL value of 64 appears in firewall logs with a TTL of 48, then that packet shouldn't be considered natural.

Flow analysis examines only the temporal sequencing of packets; on its own, it can't reliably detect TTL deviation. Firewall logs, however, record end-to-end TTL behavior in a directly observable manner. This indicates that the attacker is deliberately attempting to weave around and bypass a network segment. Observations show that certain actor groups (particularly variants associated with Lazarus and APT29) correlate TTL manipulation with operational time windows: reduced TTL values during specific hours and normal TTL values during others. Even this pattern alone constitutes a highly effective forensic correlation signature.

When encountering such anomalies, it's good to think that the packet is behaving too deliberately to be normal. In addition to indicating whether a connection occurred, firewall logs narrate the story of that connection, for example:

- A sudden surge in outbound traffic from an unexpected user
- Activation of a port that's normally never used

- Dozens of connection attempts within a short time frame
- An external connection occurring in the absence of any other concurrent traffic
- Connection density that doesn't align with the associated ASN or geographic region
- An outbound rhythm observed in one network segment repeating across others

When these signals converge, firewall logs function as a forensic compass. Flow rhythm indicates direction; firewall logs reveal where deviation occurs. Attackers typically attempt to remain silent once inside the network, but when external communication becomes necessary, they are forced to interact with the firewall. This interaction represents the following:

- The first fracture in a carefully concealed C2 channel
- The initial vibration of exfiltration
- The first externally visible trace of a pivot operation
- The residue of a failed reconnaissance attempt

Example: Advanced TTL Manipulation Detection Report: Anomaly-Oriented

The outputs shown in Figure 8.4 were generated by the advanced TTL manipulation analysis engine after processing the PCAP file `ttl_traffic.pcap` with the command `python ttl_detector.py ttl_traffic.pcap`. During this analysis, the engine parses all packets within the capture, extracts their TTL values, and evaluates them against known OS TTL profiles, statistical distribution models, and behavioral consistency checks to identify anomalies such as spoofed sources, hidden proxy usage, or attacker-driven route manipulation.

```
C:\Users\admin\PycharmProjects\Books>python ttl_detector.py ttl_traffic.pcap

================ ADVANCED TTL MANIPULATION ANALYSIS =================

[SOURCE] 192.168.1.50
→ Likely OS Family: Unix-like (Linux/macOS/Android)
→ OS Default TTL Class: 64
→ Example OS in this class: ['Linux', 'Unix', 'Android', 'macOS']
→ Avg TTL: 52.00
→ TTL Variance: 10.47
→ TTL Jitter: 16.00
→ Avg Anomaly Score (vs default): 12.00
→ Estimated Hops (from mode TTL): 0
→ TTL Range: 40-64 (Δ=24)
→ TTL Mode: 64 (count=1)
→ Outliers: 3 (42.9% of packets)
→ Multiple TTL Clusters: False
→ Raw TTL Distribution: {64: 1, 58: 1, 42: 1, 59: 1, 41: 1, 60: 1, 40: 1}
→ APT-like Hourly TTL Pattern: False
→ Hourly Stats: {9: {'avg': 52, 'std': 10.472185381603339, 'count': 7}}
→ Risk Score (0-10): 7
→ Manipulation Likely: True
→ Reasons:
    - Average TTL anomaly score is elevated (12.0).
    - TTL variance is notable (10.5).
    - TTL jitter indicates fluctuation (16.0).
    - TTL jitter is critically high (>15).
    - Significant portion of TTL values are outliers.
→ Sample TTLs: [64, 58, 42, 59, 41, 60, 40]

====================================================================
```

Figure 8.4: Advanced TTL Manipulation Analysis Output Highlighting OS-Based Profiling and Anomalous TTL Behavior

Let's break down this example output, starting with the source traffic information:

```
Source IP: 192.168.1.50
```

The TTL examination performed on the flow associated with this IP has been evaluated in terms of traffic naturalness and OS consistency.

We can then see the probable OS family in the next rows:

```
Likely OS Family: Unix-like (Linux/macOS/Android)
Default TTL Class: 64
Example operating systems in this class: ['Linux', 'Unix', 'Android',
'macOS']
```

This determination means the following:

- The maximum TTL value of the packets is 64.
- Therefore, the traffic is consistent with the expected TTL class for Unix-based systems.
- No indicators associated with Windows (128 TTL) or network device/router (255 TTL) classes have been observed.

This brings us to the baseline TTL behavior, which includes several metrics that we can interpret, as shown in Table 8.1.

Metric	Value	Interpretation
`Avg TTL`	52.00	This is 12 hops lower than the default TTL; may be natural or manipulated.
`TTL Variance`	10.47	TTL distribution is inconsistent; such volatility can't be considered natural for traffic originating from the same device.
`TTL Jitter`	16.00	Jitter ≥ 15 is critical and increases the likelihood of simulated path diversity or deliberate manipulation.
`Avg Anomaly Score`	12.00	Produces significant deviation relative to 64; nonnatural jumps are present.
`Estimated Hops`	0	Because the mode TTL equals 64, a packet appears to be at 0 hops distance; this may indicate manipulation.

Table 8.1: Baseline TTL Behavior

Next, we arrive at the detailed examination of TTL distribution:

- **`TTL Range: 40-64 (Δ = 24)`**
  Such a wide range isn't typical for traffic originating from a single device.
- **`TTL Mode: 64`**
  This indicates that the highest TTL value is also the packet attempting to appear the most natural.
- **`Outliers: 42.9%`**
  Nearly half of all packets deviate by more than 10 from the mode TTL. This outlier ratio is a known indicator of the following attack behaviors:
  - Making packets appear as if they originate from different hop distances
  - Artificial distribution of traffic
  - IDS/firewall evasion through TTL spoofing
  - Obfuscating the detection of C2 traffic
- **`Multiple TTL Clusters: False`**
  No multiple strong clusters were detected; however, this doesn't imply the absence of manipulation. On the contrary, high jitter combined with a high outlier ratio alone already constitutes a strong signal.

Let's move on to the behavior and time-based analyses. Table 8.2 shows the two analyses that were performed and their results. The low TTL at specific hours pattern observed in actor types such as APT29/Lazarus wasn't detected here.

Analysis	Result
APT-like hourly TTL pattern	False: No hourly TTL fluctuations were observed; the attacker may not have used an operational time window.
Hourly stats	`{9: {'avg': 52, 'std': 10.47, 'count': 7}}`: All traffic was observed within the same time window; therefore, time-based variation isn't expected.

Table 8.2: Behavioral and Time-Based Analysis

Next, we see the following risk assessment results:

```
Risk Score (0-10): 7
Manipulation Likely: True
```

A risk score of `7` indicates that the traffic carries a high likelihood of manipulation due to the following:

- Inconsistent TTL values
- Aggressive jitter
- Wide TTL distribution range
- High outlier ratio

The `Reasons` rows show the detection rationale. Factors that led the engine to conclude manipulation include the following:

- Average TTL anomaly score is elevated (12.0).
- TTL variance is notable (10.5).
- TTL jitter indicates fluctuation (16.0).
- TTL jitter is critically high (>15).
- A significant portion of TTL values are outliers.

When these five findings are evaluated together, it becomes evident that the traffic didn't traverse a single natural network path or was deliberately distorted by an attacker.

Finally, we see the `Sample TTLs`:

```
[64, 58, 42, 59, 41, 60, 40]
```

In this sample set, higher TTL values (e.g., `64`, `59`, and `60`) create the impression of a near hop. Lower TTL values (e.g., `40`, `41`, and `42`) simulate a distant hop.

This mixed TTL design is a known characteristic of C2 traffic attempting to evade behavior analysis–based detections employed by modern firewalls. This analysis clearly demonstrates the following about the TTL manipulations:

- Not a single random deviation
- Systematic and deliberate variations
- Potentially intended to conceal the true location and routing path of the traffic source

8.4.2 Proxy Logs

Proxy logs are among the richest sources of behavioral metadata in modern organizations. If an attacker attempts to conceal C2 communication within HTTP/HTTPS, the following elements are inevitably recorded by the proxy:

- User-Agent
- URI patterns
- Header ordering

- Session behavior
- Keep-alive durations

The underlying reality revealed here is that flow-level analysis shows the rhythm, whereas proxy logs expose the seams of the disguise. Analysts should always ask the following question: Is this User-Agent truly consistent with this behavior? For example, when the following three indicators converge, you come very close to the true profile hidden beneath the attacker's mask:

- Traffic sends a Chrome user-agent.
- Deterministic 20-second beaconing is shown at the flow level.
- The `Host` header in proxy logs points to an unusual ASN.

At times, C2 frameworks fail to accurately replicate header ordering. The proxy directly records small anomalies in this ordering, for instance, an incorrect placement of the `Accept-Language` header. These minor deviations are of critical value in forensic analysis for understanding an attacker group's level of protocol impersonation capability.

8.4.3 DNS Metadata

DNS is the only protocol an attacker is inevitably forced to communicate with. Whether they use DoH, tunneling, or domain fluxing, at some point, DNS behavior leaks. DNS metadata completes the gaps that flow analysis alone can't resolve:

- Anomalous TTL values may indicate that the attacker infrastructure has been newly established.
- Variance in DNS query lengths reveals the level of automation.
- Repeated queries within the same second indicate C2 lookup behavior.
- Domain chaining (DGA/dynamic rotation) provides a direct clue to the attacker's operational style.

A lesser-known but critical fact is that some DoH-based C2 infrastructures still trigger a normal DNS fallback mechanism in the background. This fallback may occur within only a millisecond-scale interval; however, it leaves a quiet trace in DNS server logs. For this reason, DNS metadata serves as a behavioral atlas that enables understanding of the attacker's infrastructure quality, operational speed, level of anonymity, and even operational hours.

8.4.4 The Power of Correlation

A truly advanced CTI analyst knows that no single data source exposes the attacker; instead, it's the harmony created by data sources together. Therefore, correlation isn't just a technical process, but a demanding intuitive journey. Consider the following:

- Flow analysis reveals the rhythm.
- Firewall logs reveal boundary interactions.
- Proxy records reveal the traces of the mask.
- DNS metadata reveals the attacker's infrastructure breath.

The moment these sources converge, the following questions become answerable:

- Is it actually possible that this connection was initiated by a real user?
- Even if the attacker has concealed the content, are the direction, timing, and domain behavior consistent?
- Where does the gap between masked traffic and the underlying real C2 begin?
- Which log corroborates or refutes which flow?
- Is there structural consistency between the attacker's rhythm and DNS behavior?

Scenario: The Breaking Point Where Three Sources Converge

Let's see this in practice with an example. An SOC analyst examines an incoming alert. At first glance, there appears to be ordinary outbound HTTPS traffic. However, the convergence of three sources presents the following picture:

- **NetFlow**
  Regular 22-second beaconing.
- **Proxy**
  Chrome User-Agent imitation, but incorrect header ordering.
- **DNS**
  A domain with TTL 60 that is renewed every two minutes.

These three indicators may appear benign on their own, but together, they expose both the attacker's rhythm, their disguise, and their infrastructure. This is correlation at its finest: The true shape of the attacker emerging from within a silent cloud.

Flow analysis tells what an attacker is doing. Firewall and proxy logs explain how it's being done. DNS metadata, on the other hand, shows where the attacker comes from and how they hide. Together, the three bring the attacker's intent, behavioral quality, weak points in disguise, protocol compliance errors, C2 infrastructure maturity, and operational timing onto a single plane of truth.

Example: Microlevel Behavioral Profiling of Beaconing Flows Using Python

The screen output in Figure 8.5 shows three distinct beaconing flows identified by the correlation engine running on `test.pcap`, based solely on analysis at the flow layer. Because there are no additional anomalies on the DNS, proxy, or firewall sides, the sections labeled `=== DNS/Proxy/Firewall Anomalies ===` are empty; in other words, at this stage, only flow rhythms are speaking.

```
C:\Users\admin\PyCharmProjects\Books>python advanced_cti_correlation.py test.pcap

=== DNS Anomalies ===

=== Flow Anomalies ===

 [BEACONING] score=1.00 | 10.0.0.13 -> 8.8.8.8:53 | events=10, avg_interval=60.0s, cv=0.00,
      duration=540.0s, bytes=1042, bytes=1042
    [Advanced] severity=high, profile=DNS_LIKE_BEACON, human_likelihood=0.10
    [Hypothesis] long-running, low-jitter, automated beacon pattern. Likely C2 or bagond
        health-check/heartbeat traffic.
 [BEACONING] score=1.00 | 10.0.0.20 -> 203.0.113.1:443 | events=100, avg_interval=22.0s, cv=0.00,
      duration=2178.0s, bytes=413600, bytes=413600
    [Advanced] severity=high, profile=WEB_LIKE_BEACON, human_likelihood=0.10
    [Hypothesis] long-running, low-jitter, automated beacon pattern. Likely C2 or bagond
        health-check/heartbeat traffic.
 [BEACONING] score=0.21 | 10.0.0.30 -> 93.184.216.34:80 | events=200, avg_interval=6.0s, cv=0.24,
      duration=1195.0s, bytes=1635800, bytes=1635800
    [Advanced] severity=low, profile=WEB_LIKE_BEACON, , human_likelihood =0.70
    [Hypothesis] Although there is some regularity in the rhythm, the jitter and as duration
        indicate it may be mixed with normal work.

=== Proxy Anomalies ===

=== Firewall Anomalies ===

=== Correlation Results ===

C:\Users\admin\PyCharmProjects\Books>
```

Figure 8.5: Flow-Level Correlation Output Showing Microbehavioral Profiling of Beaconing Activity

Let's break down the different elements of this output:

- `10.0.0.13→8.8.8.8:53`

  This entry describes a flow identified by the correlation engine as a potential beaconing pattern between the internal host 10.0.0.13 and the external DNS service 8.8.8.8 over port 53. The following metrics summarize the temporal behavior and statistical characteristics of this communication:

  - Score: 1.00 (maximum confidence)
  - Number of events: 10
  - Average interval: 60.0 seconds
  - Jitter (`cv`): 0.00, virtually zero fluctuation
  - Total duration: 540.0 seconds (approximately 9 minutes)
  - Total data: 1,042 bytes
  - Profile: `DNS_LIKE_BEACON`
  - Human likelihood: 0.10 (i.e., 10%)

  This flow behaves like a perfectly calibrated one-minute metronome: Every 60 seconds, it sends a pulse to 8.8.8.8 over DNS. The jitter being very close to zero creates the impression that you're observing not a human, but a process breathing with millisecond precision. The total data volume is quite small, which suggests that the flow may be a heartbeat traffic sending an "I am alive" signal rather than carrying actual data. For this reason, the analysis engine's hypothesis leans toward `Likely C2 or background health-check/heartbeat traffic`. For the analyst looking at the screen, this line quietly evokes the feeling that something is too regular; such a perfect rhythm isn't human.

- **10.0.0.20→203.0.113.10:443**
  This entry represents a long-lived outbound flow from the internal host 10.0.0.20 to the external server 203.0.113.10 over HTTPS (port 443). The correlation engine classifies this communication as a high-confidence beaconing pattern based on its temporal regularity, persistence, and statistical characteristics summarized here:
  - Score: 1.00
  - Number of events: 100
  - Average interval: 22.0 seconds
  - Jitter (`cv`): 0.00
  - Total duration: 2,178.0 seconds (approximately 36 minutes)
  - Total data: 413,600 bytes
  - Profile: `WEB_LIKE_BEACON` (beacon over HTTPS)
  - Human likelihood: 0.10

  The second flow stands out not only because of its rhythm but also due to its duration and data volume. For approximately 36 minutes, it establishes outbound connections to the same destination (`203.0.113.10:443`) at flawless 22-second intervals and transfers more than 400 KB of data. This behavior is far more disciplined than a typical browser tab left open pattern; the zero jitter gives the impression of an automation that looks as if it came straight out of a cron table. Consequently, the engine formulates the same hypothesis once again: `Long-running, low-jitter, automated beacon pattern - strong suspicion of C2 / background agent`. The analyst reading this line senses internally that this connection doesn't want to drop. It feels as though someone has quietly opened a remote console, and we're hearing its heartbeat.

- **10.0.0.30→93.184.216.34:80**
  This entry represents a flow where some periodic characteristics are present, but the statistical indicators suggest that the communication may be partially influenced by normal web activity. The following metrics summarize the timing behavior and traffic characteristics used by the analysis engine to evaluate this flow:
  - Score: 0.21
  - Number of events: 200
  - Average interval: 6.0 seconds
  - Jitter (`cv`): 0.24 (noticeable fluctuation)
  - Total duration: 1,195.0 seconds (approximately 19–20 minutes)
  - Total data: 1,635,800 bytes (~1.6 MB)
  - Profile: `WEB_LIKE_BEACON`
  - Human likelihood: 0.70

The third flow isn't as cold and mechanical as the first two. There's an average interval of 6 seconds, but the jitter rises to 0.24; the rhythm is still regular, yet irregularities resembling human interaction are observable. Moreover, the data volume is significantly higher, at approximately 1.6 MB.

This picture can easily be explained by scenarios such as heavy web usage, automatic page refreshes, or live content (feeds, live scores, dashboards). For this reason, the engine adopts a cautious stance and formulates its hypothesis as follows: `Although there is a certain regularity in the rhythm, the jitter and duration indicate that the traffic may be mixed with normal workstation web/noise traffic`. From an analyst's perspective, this flow belongs near the lower end of the suspicion list: It's noted peripherally, but it doesn't trigger the same deep-seated alarm feeling as the first two.

- **Silence on the DNS/proxy/firewall side**

  At the bottom of the screen, the following sections appear empty:

  `=== Proxy Anomalies ===`

  `=== Firewall Anomalies ===`

  This indicates that, in this run, only flow data derived from the PCAP was used, and DNS/proxy/firewall logs weren't incorporated. In other words, the correlation engine is currently reasoning solely based on rhythm. Once DNS metadata or firewall enforcement data are added to render the picture three-dimensional, the story behind these beacons will become much clearer.

This single snapshot represents that quiet moment at an SOC analyst's desk in which the logs aren't shouting, and the SIEM isn't turning red; yet the rhythm in three simple lines implies that an invisible agent is breathing somewhere.

The first and second flows are heartbeats pulsing with machine-like composure; the third flow is a weak rhythm that could be lost within human noise. This is where the power of CTI begins: in seeing these subtle differences, and in reading numbers as a behavioral map that touches the very emotion of the operation.

8.5 Monitoring Attacker Infrastructure and Lateral Movement

Once an attacker sets foot inside a network, the traces left behind are often invisible to the naked eye. Yet this invisibility doesn't mean the absence of traces, just that the correct perspective hasn't been applied. The fabric of the network, much like a piece of cloth, subtly deforms the moment it's touched; even the attacker's very first move creates a slight vibration in the natural flow of traffic. Sometimes this vibration is hidden within a tiny fraction of a second; at other times, it's concealed between the lines of log files.

Monitoring attacker infrastructure requires going beyond the surface-level visibility of IP addresses or domain chains. This is because an attacker's strength lies not only in the infrastructure they use, but in how they use it. True monitoring doesn't focus on the attacker's technical components alone; it seeks to comprehend the operational logic behind them.

For this reason, the scope isn't limited to tracking C2 endpoints. Only when all of the following are considered together does the attacker's full picture emerge:

- Rhythm of C2 beacons
- Selection of pivot points within the internal network
- Direction and speed of lateral movement chains
- Method of establishing covert communication channels
- Degree of protocol abuse
- Domain → NS → ASN behavioral chains within the external infrastructure

These elements may represent technical steps for an attacker, but for an analyst, they are clues that reveal the intent embedded within behavior. No movement inside a network is ever completely silent. Even in scenarios where the attacker actively attempts to remain hidden, silence itself is a behavior.

Silence may manifest as the following:

- Abnormally low bandwidth utilization
- Pauses within specific time windows
- Unnecessary brief contacts between internal nodes
- Repetitive communication patterns with reduced entropy
- Suspiciously persistent connection durations, such as an "immortal" TCP session

Most of these anomalies may appear ordinary or insignificant at first glance. However, an experienced threat analyst knows that the more an attacker strives to become invisible within the network, the more artificial the regularity of their behavior becomes. We'll walk through a few methods analysts can utilize to look out for these anomalies in the following sections.

8.5.1 Intent-Based Monitoring

Legitimate user traffic contains randomness, irregularity, and organic variance. In attacker traffic, amateurism can't be concealed, while professionalism becomes excessively flawless. Both ultimately surface as anomalies. This is where monitoring attacker infrastructure isn't about searching for a trace; it's about questioning why a behavior exists. This questioning can take a few forms:

- Why is a connection so regular?
- Why does it always flow in the same direction?
- Why are packet sizes so similar to one another?

- Why does a machine whisper something outbound every 15 seconds throughout the night?
- Why do two servers that normally never communicate suddenly probe each other within the internal network?

The answer to these questions is usually a single word: *intent*. Intent is the attacker's true objective, and every technical trace carries the shadow of that objective. A beacon isn't merely a connection; it says, "I am ready." A pivot isn't just a transition; it says, "I want broader access." A covert channel isn't merely a hidden transport pipe; it says, "This network won't see me." Hearing these inner voices requires the ability to intuit attacker behavior.

The following sections discuss a few ways to uncover the attacker's intent.

Beaconing Detection

Beaconing is more than an exchange of packets; it's the rhythmic echo of an attacker communicating with their implant. Beacon traffic is often small, regular, unidirectional, and low-volume. However, for an advanced CTI analyst, beaconing can't be identified solely by time intervals because the microdynamics of the flow determine everything.

Critical indicators used in advanced beaconing detection are listed here:

- **Temporal regularity**
  Beacon intervals—whether 5 seconds or 120 seconds—are abnormally stable compared to human behavior. A user never breathes this regularly; malware does.
- **Directional silence**
  Outbound packets are observed, but inbound responses consist of only a few bytes. This indicates that the beacon is just polling for commands. Such behavior is far removed from real-world protocol patterns.
- **Microjitter patterns**
  Some implants believe they can conceal themselves by adding jitter; however, jitter values are typically pseudo-random, low variance, and mathematically dependent. Because this dependency doesn't produce human entropy, it's quickly exposed.
- **Burst-before-silence anomalies**
  Some advanced APT implants perform a small TCP warm-up activity immediately before sending a beacon. This appears in firewall and NetFlow data as a mini SYN flicker. It's a subtle technical subsignature that often goes unnoticed.

Every beacon is a message between the attacker and the C2 saying, "I am alive and ready." The rhythm of this message reveals the attacker's operational discipline.

Lateral Movement Detection

C2 connections established with the external world are often more visible; the real danger begins inside the internal network. *Pivoting* is the process by which an attacker moves from one machine to another to expand access. In this process, the attacker's goal is to remain silent. Yet the network carries the echo of even the quietest steps taken within it.

Advanced indicators of lateral movement detection include the following:

- **Kerberos and Windows New Technology LAN Manager (NTLM) behavioral anomalies**
  When an attacker uses pass-the-hash, pass-the-ticket, or overpass-the-hash techniques, the following are observed:
  - Kerberos timestamp deviations
  - Abnormal ticket-granting service (TGS) request density
  - Repeated LAU (Logon Type 3) events

  These behaviors are far removed from normal user interaction.
- **East–west traffic density**
  The sudden emergence of a connection between two servers that normally don't communicate indicates that the attacker is searching for a pivot point within the internal network. Such connections are typically short lived, low volume, and focused on target discovery.
- **RDP shadow patterns**
  If RDP is used during lateral movement, the characteristic frame burst behaviors of RDP expose the pivot activity.
- **DNS lateral beacons**
  Professional attackers discover pivot points via DNS. Consecutive anomalous internal DNS queries represent the reconnaissance phase of pivot movement.

When an attacker touches a node inside the network, it can never be fully concealed. The connection may be silent, but even the rhythm of silence leaves a shadow.

Covert Channel Identification

Covert channel detection is one of the most difficult areas of network forensics because the attacker deliberately constructs an out-of-protocol communication. HTTP, DNS, ICMP, and TLS can serve as a mask for the attacker. However, the mathematics of the behavior beneath the mask never aligns with what is natural. Some advanced covert channel identification techniques include the following:

- **Entropy behavioral drift**
  Traffic containing a covert channel exhibits entropy that deviates significantly from normal application traffic. For example, in a C2 channel embedded within ICMP, the entropy of the ICMP payload may be twice the normal level.
- **Header field modulation**
  Attackers may transmit data via TTL, window size, the Don't Fragment (DF) bit, padding, or even sequence numbers. These micromanipulations can only be detected through metadata fingerprinting.

- **Timing channels**
  Time-based covert channels use time gaps (e.g., 200 ms = 0, 400 ms = 1) rather than content to carry information. Such channels produce patterns in flow graphs that appear irregular yet are mathematically meaningful.
- **Protocol abuse patterns**
  QUIC, HTTP/2 push streams, DNS ANY query chains, and ICMP echo-reply flip patterns all represent a method attackers use to open channels by exploiting the dark corners of protocols. These behaviors don't appear in human traffic. This supposedly invisible channel, when viewed at the flow level, shines like a stable anomaly.

In a real incident, an SOC analyst notices unusual traffic on a quietly operating Windows server:

- Flow rhythm: A regular 27-second heartbeat.
- Pivot behavior: Short bursts from the same server to port 445 on two different internal IPs.
- DNS behavior: TXT queries arriving at three-second intervals.
- ICMP: Some echo packets carry payloads twice the normal size.

When viewed in isolation, nothing is definitive. But when read together, the picture is clear: An implant was pivoting inside the internal network and establishing a covert channel over ICMP and DNS.

This example represents one of those cases in which every whisper from every layer tells a single story while monitoring attacker infrastructure. Beaconing is the attacker's breathing. Pivot traces are their footsteps. The covert channel is the way they say what they don't wish to say. Together, they reveal the attacker's full operational model: rhythm, method, infrastructure quality, internal network behavior, level of concealment, and operational maturity.

Monitoring attacker infrastructure is about being able to hear every step the attacker takes. Sometimes even the smallest deviation is sufficient for the dark silhouette in the background to emerge. And this is why, in advanced CTI analysis, the primary objective is to capture the step before the path the attacker walks is visible.

8.5.2 C2 Pulse Mapping

An attacker's C2 infrastructure may appear, from the outside, as an ordinary HTTPS request, a simple DNS query, or a small TLS handshake. However, this is a surface-level smokescreen. The attacker's true intent is hidden within the operational rhythm between the implant and the command center. This rhythm is often the real fingerprint of C2 traffic; not the size of the packets, but the manner in which they are sent; not the content, but the breathing behind it.

C2 pulse mapping is an advanced technique aimed not only at detecting and interpreting this rhythm. The mathematical relationships within the rhythm can reveal the attacker's level of discipline, maturity of their framework, operational speed, working windows, and even whether the attacker was rushing before or after the intrusion. Let's dig into how C2 pulse mapping goes beyond packet analysis to deconstruct the attacker's behavioral architecture.

The Nature of the C2 Pulse

A C2 implant must communicate to exist. Some speak frequently: They send a small whisper every five seconds. Some are silent: They say nothing for one to four minutes and then suddenly produce a short burst. Others attempt to mimic human behavior using far more sophisticated jitter mechanisms. Yet the pulse always reveals itself. The C2 pulse has three fundamental components:

- **Temporal cadence**

  The communication cycle defined by the C2 implant may be fixed, jittered, or pseudorandom.However, in every case, it diverges sharply from natural traffic variance. Real-world user behavior is irregular, intensity-based, intermittent, and context dependent.

  A user doesn't send packets every 5 seconds while refreshing a browser page. A user doesn't necessarily touch the external world every 20 seconds. User traffic never operates like a metronome, but C2 implants do.

- **Structural pulse pattern**

  Both the timing of the pulse and its internal structure carry meaning:

  - Burst → silent → burst cycles
  - Low-volume yet regular handshake behavior
  - Outbound-heavy packet flows
  - Minimal command payloads in the inbound direction
  - Connection lifetimes remaining consistently within the same range

  Each of these represents the structural nature of a C2 channel.

- **Entropic coherence**

  Real traffic entropy is irregular and multimodal. C2 traffic entropy, especially in implants with preset beacon intervals, becomes overly regular. When these three components come together, the C2 pulse ceases to be invisible and becomes a rhythmic behavior.

The Deeper Purpose of C2 Pulse Mapping

The C2 pulse mapping is performed because the one thing an attacker's implant can't conceal is time. The Cobalt Strike, Sliver, Mythic, Meterpreter, and Lazarus-specific custom beacon frameworks each handle timing differently. Here are some examples:

- **22–25 second polling: Python-based weak implants**
  The beacon interval of these implants almost never changes. Time-series analysis reveals a collapse in variance.
- **5-second jitter + microbursts: QUIC-based C2s**
  The QUIC library produces small but characteristic bursts on each transmission.
- **60–75 seconds of silence: Stealth APT operations**
  This silence isn't true silence; it's an operational breathing window.
- **1–3 KB outbound bursts: Onset of exfiltration**
  These bursts most often appear after pivoting.
- **Fixed-size inbound command payloads: Low-profile C2 architecture**
  The attacker sends minimal data, often consisting solely of a command string.

C2 pulse mapping correlates these indicators to decode the operational logic of the threat actor.

Every attacker makes mistakes, even if infrequently. These mistakes aren't technical; they are rhythmic. During an operation, an APT operator may do the following:

- Configure faster beaconing
- Persistently attempt to establish connections
- Miscalculate jitter values
- Behave impatiently within the attack window

These rhythm disruptions are entirely human-driven. This is where microlevel emotional tone analytics comes into play. Some actors, such as Lazarus, employ rhythms with these characteristics:

- Intense at the initiation of the attack
- More cautious as the operation progresses
- Ultralow frequency after reconnaissance

This behavioral shift resembles the emotional state of an operation log. The C2 pulse represents not only a technical signal but an operational state of mind.

Real-World Example

In a real incident, the C2 pulse appeared as follows:

- Fixed 27-second interval
- ±1.8 second jitter
- 180–210-byte outbound beacons
- 90-byte inbound command responses
- Mini TCP warming (preheating) before each burst every 5 minutes
- Low-TTL renewal on the DNS side every 10 minutes.

These indicators were collected. Here are the resulting conclusions:

- Not Python-based because the jitter variance is higher.
- Not Cobalt Strike because the packet size range differs.
- Not Sliver because the burst density isn't constant.
- Not Lazarus-like because no ultra-silence was observed.

Conclusion: This was a custom-developed, QUIC-based operational C2 implant. This conclusion was derived not solely from flow entropy, but from the behavior of the pulse itself.

The C2 pulse can serve as a practical indicator for threat hunters by revealing timing patterns within C2 communication. Changes in beacon frequency, pauses in activity, or sudden increases in traffic may indicate shifts in attacker behavior, such as a failed beacon, the establishment of a pivot point, or the start of data exfiltration. For this reason, mapping C2 pulse patterns provides analysts with a straightforward way to observe whether an implant remains active and how its operational rhythm evolves over time.

8.5.3 Pivot Point and East-West Traffic Monitoring

Every step an attacker takes inside the network leaves a small rise in temperature along the horizontal veins of the network. The challenge is to notice that heat in time.

An attack campaign begins at a gateway facing the outside world; however, the real story is written inside the internal network, along the horizontal paths where east–west traffic breathes. The moment an attacker establishes the first pivot point, they become a silent traveler moving internally, changing direction, and seeking opportunity. Although the steps taken by this traveler are designed to go unnoticed, they still touch the network's natural ecosystem and therefore generate subtle behavioral fractures. A few questions are relevant:

- How are these fractures detected?
- What traces does an attacker leave while moving inside the internal network?
- How do these traces merge with external rhythms to form the complete picture?
- How can an analyst capture these invisible paths?

This section reveals the anatomy of these questions in detail.

Pivot Point

A *pivot point* is the springboard an attacker uses to move beyond the first compromised internal node. This point typically appears as a user workstation, an application server, a misconfigured service, a management interface protected by weak credentials, or an endpoint with escalated privileges. However, what truly defines a pivot point isn't a technical weakness, but the operational logic chosen by the attacker. In pivot analysis, we must always ask the following question: “Why did the attacker choose this particular node as the pivot?”

Because pivot selection is never random. The node’s position within the network, the distribution of roles, access rights, and the density of east–west traffic are all factors that play a role in the attacker’s strategic decision.

A good pivot point does the following:

- Makes it easier to remain invisible inside the network
- Enables silent proximity to high-value assets
- Creates lateral movement without generating anomalies
- Forms a reliable backbone for the lateral movement chain

The traces left by an attacker during pivoting are sometimes nothing more than millisecond-level increases in connection attempts. Sometimes they appear as a small fluctuation in Kerberos TGS request density. At other times, they reveal themselves as unexpected variance in internal DNS queries. These microsignals mark the beginning of the bigger picture.

East-West Traffic

Traffic that opens to the external world constitutes north-south traffic; however, when attackers expand within the internal network, the most critical element is east–west traffic—horizontal flows from machine to machine, service to service, and segment to segment.

East–west traffic is normally not high volume, and it’s characterized by stable source-destination motifs. It’s tied to specific workloads, limited in temporal variance, and largely predictable in structure.

Precisely for this reason, an attacker’s touch becomes easier to detect: The traffic they interact with spills beyond the bounds of its natural order. The types of these spillovers include the following:

- **Unexpected peer relations**
  If traffic suddenly appears between two servers that never normally communicate, this represents reconnaissance prior to pivoting.
- **Burst-based lateral probing**
  For target discovery within the lateral network, an attacker performs short-lived, low-volume, serial connection attempts. This leaves a micropulse signature in TCP SYN density graphs.
- **Protocol-inconsistent behavior**
  SMB traffic isn’t limited to file sharing; it’s also a gateway to malicious lateral movement. Anomalous SMB session setup patterns are an early indicator of pass-the-hash attacks.
- **Authentication spike patterns**
  A sudden increase in NTLM or Kerberos activity indicates that a lateral movement chain has begun. east–west traffic analysis reveals how these anomalies are interconnected.

The Microtrace of a Pivot

A pivot movement typically unfolds as follows:

1. The attacker gains initial access to the internal network. This access is often low-privileged.
2. Credential harvesting then begins by targeting Local Security Authority Subsystem Service (LSASS), Security Accounts Manager (SAM), or keychains.
3. Using the acquired credentials, east–west reconnaissance is performed.
4. The most suitable target is identified.
5. The attacker establishes a short-lived, low-volume connection.
6. This connection feeds the next pivot in the chain.

Pivot traffic often appears invisible because the attacker communicates solely between internal segments. However, assuming it's invisible is a misconception. Pivot traffic carries the following signatures:

- Stable yet short flow durations
- Low packet size variance
- High destination diversity
- Deterministic connection attempts
- Anomalous authentication sequence behavior

None of these signatures alone indicate that an attacker was here, but when viewed together, the path the attacker walked becomes clear. Ask yourself this: Is this traffic a requirement of an application, or an invisible hand trying to move through the network?

Real-World Example

In a real incident, combined pivot and east–west analysis produced the following picture:

- A user device sent SMB requests at three-second intervals to three file servers it normally never communicated with.
- None of these requests involved file reads; they remained solely at the session setup level.
- Ten minutes later, the same device attempted an RDP handshake with a SQL server.
- During this period, Kerberos TGS activity increased, yet the user didn't appear to be actively logged in.
- Fifteen minutes later, the same device conducted east–west reconnaissance by following a port sequence such as 445 → 5985 → 135.

When these small anomalies were combined, it became clear that the attacker was pivoting within the internal network and advancing toward the highest-privileged node.

This example demonstrates how east-west traffic, through seemingly insignificant movements when viewed individually, can form a complete threat chain.

Direction, Time, Behavior

Pivot analysis isn't just about looking at the target. It requires a three-dimensional thinking model:

- **Direction (directionality pattern)**
  - Who are the target servers?
  - Is there an organic relationship between them?
- **Time (temporal drift)**
  - Why do these connections occur within this exact time window?
  - What does the distribution of silence look like?
- **Behavior (operational consistency)**
  - Does this traffic align with legitimate business workflows?
  - If it doesn't, which phase of the attack is the adversary executing?

When these three dimensions converge, the nature of the pivot becomes visible: Is it reconnaissance, privilege escalation, or preparation for data exfiltration?

Attackers generally try not to draw attention externally; internally, however, they take more controlled, more deliberate, and more systematic steps. This systematic behavior, no matter how carefully it's concealed, leaves small yet consistent traces within the living fabric of internal network traffic. The pivot point marks the beginning of these traces. East-west traffic, in turn, is the map of the route the attacker follows inside the internal network. Reading this map is the most effective way to interrupt the attacker's journey before it's completed.

8.6 Summary

Modern CTI increasingly relies on behavioral analysis of network traffic rather than direct inspection of packet content. As encryption technologies such as TLS 1.3, HTTP/2, QUIC, and DoH conceal payload data, analysts must focus on observable characteristics such as timing patterns, packet size distributions, flow duration, and protocol behavior.

This chapter demonstrated how PCAP analysis, protocol fingerprinting (JA3/JA3S), and flow-based telemetry (NetFlow/IPFIX) can reveal attacker activity even when traffic is encrypted. Tools such as Wireshark help visualize timing structures and session behavior, while protocol-level artifacts, including TLS negotiation patterns and HTTP/2 frame sequences, can expose imperfect implementations used by attacker frameworks.

The chapter also emphasized the value of flow-level behavioral modeling, including entropy analysis, jitter patterns, and traffic density observations. These techniques allow analysts to distinguish between normal user activity, automated beaconing, and data exfiltration patterns. Examples such as DNS beacon detection, QUIC timing anomalies, and TTL manipulation further illustrate how subtle statistical deviations can expose attacker infrastructure or tooling.

Ultimately, the key principle is that encryption hides content but not behavior. By analyzing the structural and temporal properties of network traffic, CTI analysts can detect C2 activity, identify attacker operational patterns, and derive actionable intelligence even in environments where packet payloads remain inaccessible.

In the next chapter, we'll switch gears to host-based forensic analysis, with a focus on Windows telemetry.

Chapter 9
Host-Based Forensic Analysis and Windows Telemetry

When you examine a Windows system through a forensic lens, you quickly realize that behind an ordinary desktop lies a vast story. Every step taken by the user, every reaction of the system, and every trace an attacker attempts to conceal persist in the silent memory of the endpoint. Host-based forensic analysis is the art of interpreting this silent memory, and cyber threat intelligence is the upper layer that transforms this meaning into operational intelligence.

On the surface, Windows appears to be a neatly functioning operating system (OS); yet the depth of its event records, the layered nature of the Windows registry structure, the transient but critical traces in memory, and the surgical precision of Windows System Monitor (Sysmon) telemetry ensure that every step taken by attackers survives far longer than they expect. Sometimes an Event ID 4688, a directory trace hidden within ShellBags, or a fragment of a thread lost in memory can each become an independent clue sufficient to unravel an entire campaign.

Endpoint visibility can be challenging for many analysts at first due to the large volume of data, the abundant noise, and the complex traces. Over time, however, a rhythm begins to emerge within this chaos, much like distinguishing a familiar set of footsteps within a dense crowd. Every threat actor group, every malware family, and even every campaign leaves behind unique and recurring behavioral patterns. The value of cyber threat intelligence (CTI) is hidden in the ability to recognize these patterns.

At the core of advanced threat hunting, solving a real advanced persistent threat (APT) campaign, and performing behavior-based attribution lies the same principle: The endpoint never truly goes silent; it just waits for the right questions to be asked. For this reason, this chapter brings together all layers of host-based telemetry—event logs, registry, memory artifacts, Sysmon, and Event Tracing for Windows (ETW)—under a single framework. From this, you'll understand how to collect data and in which context data gains value, how it exposes attacker tradecraft, and why incident response remains incomplete without endpoint visibility.

9.1 Role of Host-Based Forensics in CTI

Threat intelligence often focuses on the big picture: campaigns, infrastructures, actor groups, and tactic, technique, and procedure (TTP) models. However, the true face of an attack is most often hidden deep within the endpoint. This is because every action taken by an adversary, whether a silent PowerShell command or an invisible dynamic link library (DLL) injection, first comes to life at the host level. For this reason, a fundamental building block of CTI is host-based forensic analysis.

Some analysts say that host forensics is noisy, which is true. But much like picking out a specific set of footsteps on a crowded street, an attacker's process creation habits, the command-line arguments they favor, the subtle markers left in the registry, the timing of modules loaded into memory all form a behavioral signature. Capturing these signatures in CTI does more than solve a single incident; it enables us to anticipate the same actor's future campaigns.

For this reason, the host-based forensic analysis role in CTI is critical as it performs the following:

- Reconstructs adversary behavior
- Provides solid evidence for attribution efforts
- Exposes the operational routines of APT group
- Fills visibility gaps in endpoint detection and response (EDR) or security information and event management (SIEM) environments
- Transforms threat hunting from an intuitive exercise into an evidence-based discipline

In this section, you'll see that endpoint telemetry is more than a repository of logs; it's a behavioral microscope that reveals an attacker's true identity, intent, and methods. When this microscope is used correctly, no attacker ever truly remains invisible.

9.1.1 The Strategic Value of Endpoint Telemetry

An adversary's true intent—what they actually do, and how they do it—takes place on the host. Network records may sometimes fall silent, proxy layers can obscure traces, and command-and-control (C2) infrastructures can be changed rapidly. However, the data recorded by the endpoint most often comprises footprints that the attacker leaves unintentionally. So, what turns these traces into strategic value?

First, endpoint telemetry enables us to reveal adversary behavior at the TTP level. The recurring behavioral motifs (what they did, how they did it, in what sequence they progressed, which tools they preferred, and so on) can be specific to a particular adversary group. It's no coincidence that some APT groups consistently favor specific PowerShell parameters or repeatedly use the same technical variant for injection. These habits appear in host telemetry as behavioral signatures. Even the attacker's personality can sometimes be hidden among the logs.

Endpoint telemetry is also the most reliable source for behavioral modeling with MITRE ATT&CK. This is because MITRE ATT&CK isn't a theoretical framework; it's a categorized representation of real techniques observed in the field. The process creation chain of an Event ID 4688, Sysmon's network connection events, persistence keys in the registry all align directly with a technique in the MITRE ATT&CK matrix. This enables the analyst not only to resolve a single attack but also to understand the adversary's entire operational strategy. Host telemetry makes the entire behavioral map visible, including which steps the attacker followed throughout the kill chain, which alternative techniques they preferred, where they remained quiet, where they rushed, and so on.

But there's another factor that truly makes telemetry strategic: reality checking. CTI analysis largely relies on collected data and inferences; however, if those inferences aren't validated, the analysis is always incomplete. This ground truth—that is, the actual behavior observed in the field—is where endpoint data comes into play. Ground truth may refer to which user account a phishing campaign truly affected, whether an attacker actually executed a specific command (not just in theory), whether an exfiltration attempt that appeared in logs left any trace on the system, and so on. The most accurate answer comes from the endpoint. Data doesn't conceal, embellish, or fabricate scenarios: It simply states what happened.

Endpoint telemetry is the dimension through which CTI gains reality. It measures, validates, and interprets the adversary's behavior in the field. Perhaps most importantly, it provides an honest answer to the following question: "What did the attacker actually do?"

Table 9.1 summarizes the three main points we examined within a consistent and professional framework: TTP tracking, MITRE ATT&CK modeling, and ground-truth validation.

Strategic Element	**Added Value Provided by Endpoint Telemetry**	**Impact Within CTI**	**Analyst's Focus**
Monitoring adversary behavior at the TTP level	Makes behavioral motifs visible through process creation, registry changes, memory artifacts, and Sysmon Event IDs	TTP-based campaign analysis, extraction of adversary habits, behavioral signature generation	Tools used by the attacker, command templates, injection behaviors, persistence methods

Table 9.1: The Strategic Role of Endpoint Telemetry Within CTI

Strategic Element	Added Value Provided by Endpoint Telemetry	Impact Within CTI	Analyst's Focus
Behavioral modeling via MITRE ATT&CK mapping	Provides evidence that directly maps to MITRE ATT&CK techniques, concretizing the technical level	Reconstruction of the attack chain, identification of weak defensive points, development of behavior-based detections	Correctly establishing the relationship of the telemetry → MITRE ATT&CK technique → TTP chain
Using host data as ground truth	Represents field reality in cases of log manipulation or network visibility gaps	Validation of CTI inferences, elimination of false positives, clarification of operational reality	Confirming whether an event actually occurred, checking telemetry consistency

Table 9.1: The Strategic Role of Endpoint Telemetry Within CTI (Cont.)

9.1.2 The Limits of Endpoint Visibility

Endpoint telemetry is powerful, but it's never flawless. No matter how many traces you collect on a Windows system, there will always be an unseen area, a dark gap, a behavior that remains silent. Understanding these gaps is one of the most important responsibilities of a CTI analyst because an analyst who doesn't understand the limits of visibility is prone to making incorrect inferences or even missing the adversary's primary move altogether.

So why isn't the endpoint fully visible? Where do these limits originate?

The Kernel-User Mode Separation

The Windows OS operates on a complex architecture in which two distinct worlds are intertwined: kernel mode and user mode. What most of us refer to as telemetry (Sysmon records, ETW events, process creation traces, registry changes) exists only in the upper layer of this architecture, namely the *user-mode* space. This is the domain where applications run, user actions are recorded, and system behavior is often observable. For this reason, most telemetry sources appear easily accessible, readable, and analyzable. However, this visibility doesn't reflect the system's true depth; it shows only what surfaces to the top.

The real issue begins in the shadow of *kernel mode* because adversaries are well aware that leaving traces at the user-mode level is risky. A single log entry, a Sysmon Event ID, or a simple process chain can expose their operation. Therefore, more sophisticated actors prefer to move themselves into deeper layers of the system; namely, the kernel level. This is where things change. Kernel mode is a structure that controls the core of the system, communicates directly with hardware, and operates as the most privileged domain of the OS. Many activities that occur at this level are, by their nature, not visible to user-mode telemetry.

Techniques such as direct kernel object manipulation are designed specifically to deliberately bypass the limits of visibility. When an attacker directly manipulates kernel objects, there's very little left for user-mode tools to observe. When a rootkit positions itself at the kernel level, the standard process list still appears clean. While System Service Descriptor Table (SSDT) hook manipulations silently redirect system calls (syscalls), Sysmon doesn't produce direct or reliable telemetry unless additional kernel-level or behavioral visibility is available. In even more advanced scenarios, an attacker who infiltrates the hypervisor layer can create a second shadow reality above the OS so that while the system reports its own behavior with misleading clarity, a much darker process is actually underway.

Personal Experience with a Kernel Mode Rootkit

I had read many times how these techniques work on paper, but the impact of encountering a kernel mode–based rootkit for the first time was entirely different. The logs were pristine. The process list was surprisingly orderly. There wasn't the slightest anomaly in Sysmon's records, and yet the system's behavior felt slightly off. Something wasn't as it should have been. At that moment, I felt the full weight of the gap between what telemetry shows us and what it doesn't. It was as if the system was saying on the surface, "Everything is fine," while whispering, "But be careful—something is hidden."

This experience made me understand just how critical the invisible wall between kernel and user mode truly is. User-mode telemetry is an excellent starting point for tracking adversary traces, but it never provides a complete picture. Sometimes, we approach the truth only by noticing the gaps. Silence in telemetry is often the loudest signal. When a rootkit removes itself from listings, it actually produces a strong "I am avoiding observation" indicator. This behavior alone constitutes intelligence.

It's no coincidence that adversaries choose to establish themselves at the kernel level. It's a dark space where one can operate while remaining silent. Moreover, most standard telemetry tools observe this darkness with closed eyes. For an analyst, the real challenge lies in interpreting this silence because a system doesn't always speak, but it does act. A thread disappearing in an unusual manner, a handle being created at unexpected moments, a slight alteration in interrupt handling structures (e.g., interrupt descriptor table [IDT]-related mechanisms) don't appear in Sysmon, but they don't need to. These anomalies, when examined carefully enough, are clear enough to signal an operation unfolding at the kernel level.

To better structure these visibility boundaries and their analytical implications, Table 9.2 summarizes how different system layers influence what can and can't be observed through standard telemetry. It also highlights how adversaries leverage each layer's inherent opacity and how analysts must adapt their reasoning accordingly.

Layer	What Is Visible? (With User-Mode Telemetry)	What Is Not Visible? (With Default Tools)	Adversary's Advantage	Analyst's Interpretation/ Inference
User mode	Process creation (4688), registry modifications, Sysmon Event IDs, ETW provider outputs	Microtraces of certain injection techniques, short-lived threads, ephemeral handle activities	Easy manipulation, log bypass, user-mode hook cleanup	High telemetry visibility but high manipulation risk
Kernel mode	Not directly visible: Direct kernel object manipulation (DKOM), SSDT hooks, kernel callback manipulations not reported in user mode	Rootkit behavior, object manipulation, kernel call redirection, hidden processes and drivers	Persistence + invisibility; antivirus (AV)/EDR evasion	Silence = signal; interpreting anomalous behavior via CPU, handle, and thread patterns
Hypervisor (ring -1)	Not directly observable in OS; user-mode and kernel-mode telemetry appear as normal	Virtual machine (VM)–exit manipulations, shadow page table tampering, context-switching obfuscation	Deceiving the OS's own perception of reality; invisible control	Overly clean telemetry suspicion indicates need for hypervisor-level threat analysis
Firmware/ Unified Extensible Firmware Interface (UEFI) (pre-boot/ pre-OS layer)	No telemetry produced at the OS level	Bootkits, UEFI manipulation, diskless persistence	Execution power before the system loads	Lack of telemetry indicates suspicion of physical/ firmware-based threats

Table 9.2: The Impact of the Kernel Mode/User Mode Separation on Forensic Visibility

While the discussion so far has focused primarily on user-mode and kernel-mode visibility, it's important to recognize that modern adversaries increasingly operate beyond these layers. At the hypervisor level (ring -1), attackers can manipulate the execution environment itself (intercepting context switches, altering memory mappings, or controlling VM exit behavior) while allowing the OS to continue reporting seemingly legitimate activity. In such scenarios, both user-mode and kernel-mode telemetry may appear internally consistent yet fundamentally deceptive. This creates a dangerous analytical condition where the system appears "too normal," requiring the analyst to question both anomalies and the absence of anomalies.

Even deeper, at the firmware and Unified Extensible Firmware Interface (UEFI) level, adversaries can establish persistence mechanisms that execute before the OS is initialized. These techniques leave no direct trace within OS-level telemetry, effectively bypassing all traditional logging and monitoring mechanisms. As a result, the absence of visibility itself becomes the primary indicator, forcing analysts to extend their investigation beyond software artifacts and into hardware trust boundaries, boot integrity validation, and firmware state analysis.

In this layered model, visibility isn't simply reduced as we move downward—it's fundamentally transformed. Each layer introduces new attack surfaces and new forms of analytical uncertainty. Understanding these transitions is essential for interpreting what telemetry doesn't show and for recognizing when the observed reality may itself be manipulated.

Anti-Forensic Techniques

In the cyber threat ecosystem, perhaps the reality that adversaries understand best is that leaving no trace is far easier than erasing traces that have already been left. This is one of the core philosophies of attack operations. For this reason, anti-forensic techniques aren't just post-attack cleanup tools, but an integral part of the attack chain itself. In fact, many APT actors turn maintaining silence from the very beginning to the very end of their operations into a specialized skill.

Although most of the telemetry sources available to us today are designed to capture traces left by attackers, the methods used by adversaries have become faster, smarter, and more invisible. Therefore, the issue is more than just capturing traces. We're seeking to answer the question: Why are there no traces? This is where telemetry gaps become critically important.

The following anti-forensic techniques may appear simple, yet their impact is surprisingly profound:

- **Event ID 1102: Log cleared**
  An attacker may clear the entire security log. This is a classic move. However, this action itself carries a strong signal. The sudden reset of logs generates an alert in most SIEM systems. In other words, while trying to remain silent, the attacker actually creates even louder noise.
- **Windows Management Instrumentation (WMI) repository manipulation**
  The WMI structure stores critical datasets such as process information, service status, and event sources. An attacker may corrupt the WMI repository, selectively delete it, or manipulate log collection mechanisms by creating custom event consumers. This technique is frequently used, especially by fileless malware families.

- **ShimCache/AmCache timestamp manipulation**

  These records, which reflect Windows program execution history, are among the first places analysts examine. By manipulating timestamps in these areas, an attacker can render traces of an executed file invisible within the timeline.

  So, is the problem solved? No. For an analyst, such manipulations themselves constitute a finding. No matter how skillfully timestomping is performed, time inconsistencies across the rest of the system eventually begin to speak.
- **In-memory execution (reflective loading) instead of disk-based artifacts**

  Modern malware is often loaded into memory without touching the disk. Techniques such as reflective DLL injection or Portable Executable (PE)–less execution allow code to be executed without writing anything to disk. As a result, disk forensics is rendered ineffective. Unless an analyst captures a memory dump, the behavior remains completely invisible.
- **Sysmon manipulation or disablement**

  Deleting, uninstalling, corrupting the configuration file, overriding the event provider, or masking the service appearance of Sysmon are all well-known anti-forensic tactics.

  Some APT actors don't remove Sysmon; instead, they make it appear to be functioning normally but it doesn't actually trigger critical Event IDs. On the surface, there seems to be no issue; in reality, telemetry has been deliberately silenced.

While basic techniques are commonly observed, silence is achieved through far more sophisticated means in the operations of truly advanced actors. For example, credential harvesting from the Local Security Authority Subsystem Service (LSASS) process by an APT group shouldn't be thought of as dumping. Some actors inject threads directly into `LSASS.exe`, keep the code in memory for only milliseconds, and exfiltrate credential data via an ephemeral pipe. This pipe has the following characteristics:

- Doesn't use the file system
- Produces no event logs
- Isn't directly detected by Sysmon under default configurations
- Is likely gone by the time a memory snapshot is taken

Although the behavior occurred, the observer tools were too late to see it. This is where the telemetry gap emerges: The event happened, but the trace would only have been visible if it had been observed at that exact moment.

Can CTI Be Derived from Anti-Forensic Silence?

CTI can be derived from this silence because what an attacker tries to hide is often where they are most vulnerable, for example:

- An attacker who disables Sysmon is likely attempting to evade process creation-based detections, which indicates a TTP preference.

- An actor who manipulates ShimCache is likely attempting to avoid leaving persistent disk traces, increasing the likelihood of fileless behavior.
- A group performing memory-only credential extraction is focused on operational speed and stealth, which indicates technical maturity.
- An attacker who corrupts the WMI repository has most likely established persistence via WMI event consumers, which is a strong indicator of MITRE ATT&CK T1546.003.

Anti-forensic techniques, in a sense, reveal an attacker's personality, fears, areas of expertise, and operational habits. For this reason, from a CTI perspective, even silence is data.

Logged Evidence Versus Actual System Behavior

Does everything that happens on a system get reflected in logs? No. Is everything that appears in logs an exact representation of what actually occurred? Again, no. Logs are a reflection of reality: sometimes accurate, sometimes incomplete, and sometimes a reflection manipulated by the adversary. Here are a few examples:

- **Event logs/Sysmon**
  If an attacker executes a command via `powershell.exe`, Sysmon Event ID 1 records it. But what if the same command is executed by injecting it into `dllhost.exe`? Sysmon detects this only if it's properly configured. If it's misconfigured, the behavior passes silently.
- **Registry**
  A persistence key may be created. But what if the attacker performs this within a transactional registry session and never commits it? In that case, the change remains as a partial trace that can be observed only by specific forensic tools.
- **Memory**
  A module may be loaded into memory without ever being written to disk. In such a case, disk artifacts reveal nothing about the incident.

All of this reminds us that logs don't tell the full story; we must interpret the story. The critical question is how an analyst can cope with these limits. The answer is both technical and psychological:

- Accept telemetry gaps.
- Use alternative sources of evidence to fill those gaps.
- Interpret the silent spaces between signals.

Example: Identifying the Adversary Through Signal Loss

Consider the following characteristics of a specific APT group:

- Doesn't use well-known tools such as `procdump.exe` to dump `LSASS.exe`

- Patches the `MiniDumpWriteDump` application programming interface (API) to avoid triggering Sysmon
- Transfers the dump through a pipe in RAM and never writes it to disk
- Terminates all threads at the end of the operation using a silent-kill technique

The result is that there's almost no telemetry, but there's an unusual increase in the number of LSASS handles on the system. A very brief CPU spike appears. A millisecond-scale gap is noticed between log entries.

This is where the limits of visibility turn into an advantage. Even the effort an attacker makes to leave no trace becomes a signal. That effort tells us about the attacker's technical level, operational security (OPSEC) discipline, and tactical priorities.

The endpoint visibility limits are a roadmap, not an obstacle. If you know where the system falls silent, you can better understand where it speaks. If you know what logs don't show, you can interpret what they do show more deeply. And, if you recognize telemetry gaps, you can follow even the shadow of the adversary.

9.2 Advanced Configuration of Event Logs and Audit Policy

Every forensic analysis journey conducted on a Windows system eventually leads to the same destination: event logs. Logs are the system's memory. Sometimes they are honest, sometimes incomplete, and sometimes a memory subtly manipulated by an attacker, but they remain one of the most powerful traces that tell us about the past.

Figure 9.1 illustrates the advanced audit policy configuration within the Windows**Local Group Policy Editor** screen, specifically under **System Audit Policies - Local Group Policy Object**. At this level, granular audit categories such as **Account Logon**, **Detailed Tracking**, **Object Access**, and **Privilege Use** can be configured to generate security-relevant telemetry. The screenshot highlights that all categories are currently set to **Not Configured**, indicating the absence of enhanced auditing at the system level. Without proper configuration, critical security events, such as process creation, privilege escalation, and object access, may not be logged, significantly limiting forensic visibility and detection capabilities. This reinforces the importance of enabling advanced audit policies to support effective digital forensics and incident response (DFIR) and behavioral analysis.

A well-configured audit policy makes it possible to read system behavior as if deciphering the notes of an orchestra. Every Event ID has a rhythm: 4624 tells you a door has been opened, 4625 says a door has been forced, 4688 reveals that a process has been born, and 1102 whispers that memory itself is being erased. Sometimes this rhythm is calm, sometimes it accelerates, and sometimes it becomes inconsistent enough to betray the presence of an attacker. This is where the analyst's internal intuition comes into play: "Is this log behaving the way it should?" Audit policies are an invisible framework that defines the quality of logs.

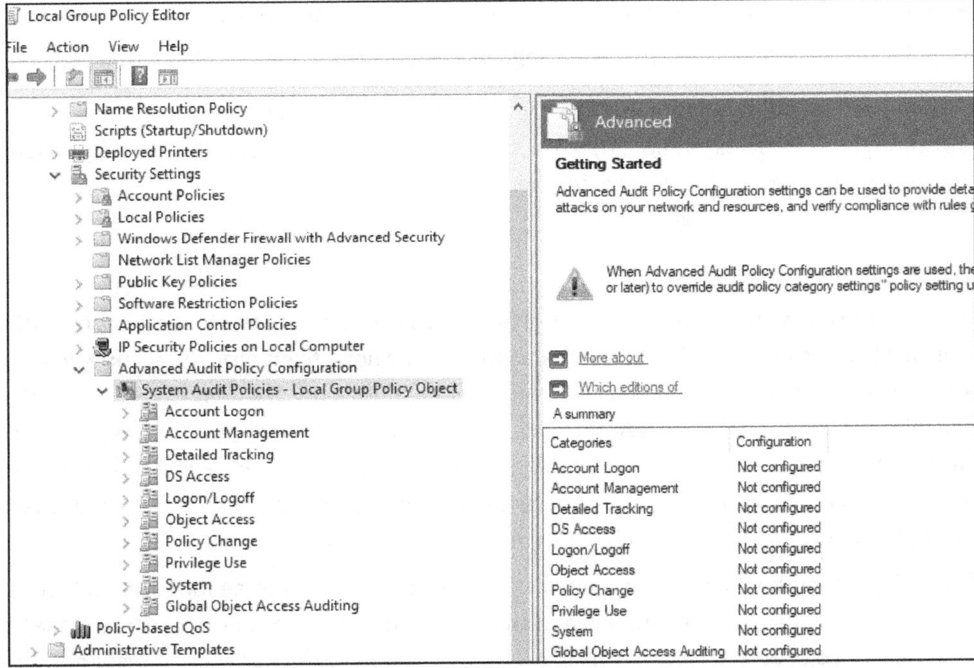

Figure 9.1: Advanced Audit Policy Configuration Interface in Windows Local Group Policy Editor

Real-World Example

I experienced the importance of audit policies firsthand in the field and was genuinely surprised. During a penetration test, there were almost no traces on the target system. Attacker activities such as file creation and command execution were all real, yet nothing had been recorded because the audit policy had been left in its default state. The system wasn't speaking, or rather, it had never been given instructions to speak. At that moment, I realized the true technical meaning of "What you can't see can't protect you."

From the perspective of modern CTI processes, there's another critical aspect of event logs: correlation. A single event may appear harmless on its own; however, a properly configured audit policy enables hundreds of events to come together and produce meaning. A single failed logon attempt is insignificant, but hundreds of them in succession indicate a brute-force attack. A single PowerShell command is ordinary, but when combined with a registry modification, it may indicate persistence. Audit policy makes these relationships visible because without visibility, correlation is impossible.

Advanced configuration doesn't merely determine which events are recorded; it reveals which behaviors are truly meaningful, for example:

- Without process creation (4688), there's no command-line analysis.
- Without logon/logoff events (4624/4634), account misuse can't be detected.

- Without object access auditing, access to sensitive data remains invisible.
- Without Sysmon events in the 1–11 range, behavioral modeling can't be performed.

In this section, we'll explore the fundamental logic of the log ecosystem, the role of advanced audit policies within CTI, the behavioral meanings of Event IDs, and why proper configuration serves as a lifeline for forensic analysis.

9.2.1 Depth of the Security Log

The security log is one of Windows' oldest yet still one of its most valuable sources of verifiable truth. If Sysmon is a microscope that clarifies behavioral telemetry, the security log is a rhythm ledger that records the system's fundamental reflexes. Data on who touched the machine, what they did, how they did it, and which privilege level they used while doing it is all hidden in the depths of these logs. It's like watching a star-filled sky: complex at first glance, cold from a distance, yet a coherent whole that gains meaning when you focus on the right point.

These logs form the backbone of forensic analysis, especially in three fundamental areas: logon behavior, process creation events, and log-clearing attempts. Each of these makes a different stage of an attack chain visible, provided you have the patience to recognize the trace you're looking for and the depth to derive meaning from it.

In the following sections, we'll cover the most relevant events from a CTI perspective and then walk through different ways they can be manipulated.

Understanding Events

First, let's examine authentication behaviors. The 4624 (successful logon) and 4625 (failed logon) events record a system's "door movements," such as which user arrived, with which session type they logged on, from which computer they came, when they came, and so on. Sometimes these events appear entirely ordinary. At first glance, they may not even create the impression of an attack, but an experienced analyst knows that logon events are the attacker's breathing rhythm.

> **Real-World Example**
>
> Once, while examining a security log, I sensed that consecutive chains of 4625 events weren't an innocent case of mistyping a password. The rhythm was different. Timing inconsistencies in the intervals, variations in the account names, and the fact that the source wasn't a single IP were all signs not of brute-force noise, but of a subtler discovery effort. In moments like these, the security log turns small details into warning lights.

One of the most frequent mistakes attackers make is their inability to imitate authentication behavior in a way that looks natural. A user doesn't attempt to log on from three different workstations within a minute. A server doesn't endure a bombardment of failed logons for minutes. The same account doesn't generate interactive and service-type logons only a few seconds apart. Such behaviors leave the attacker's shadow between the lines of the logs.

The 4688 (process create) events, in particular, carry a distinct depth. A record of the moment a process is born is often the first concrete evidence of attacker behavior. It reveals the command line, parent process information, token usage, and the existing integrated security context.

To be blunt, Event ID 4688 is more than a process create record; it often bears the fingerprints of the methods attackers prefer. A PowerShell invocation triggered via `cmd.exe`, a Living Off the Land Binary (LOLBin) executed with unusual arguments, or a logical inconsistency in a parent-child relationship are never coincidental.

Real-World Example

During an analysis, I noticed a .NET assembly behavior triggered via `rundll32.exe` on an otherwise ordinary Windows server. At first glance it was a small detail; yet the Base64 blob in the command-line arguments, the token information, and the timing of the parent process clearly showed that this wasn't an ordinary execution. A few lines of information provided by the security log were enough to expose the entire structure of this silent attack.

Token abuse behaviors are especially important because attackers often use stolen tokens or duplicate-token mechanisms to conceal privilege escalation activities. Although the security log may not always capture these transitions, small inconsistencies in parent-child relationships draw the analyst's attention.

In the security log, Event ID 1102 (log clearing) appears on the surface as nothing more than information that the log was cleared. But for an analyst, this event is one of the moments when the system screams. Because no reasonable user, no legitimate operation, and no routine maintenance process resets the security log, seeing a 1102 event is often the technical equivalent of saying, "Someone wanted to erase the traces."

Log clearing attempts are directly related to an attacker's understanding of OPSEC. Some attackers behave crudely and clear all logs at once. Others act with surgical precision, deleting only incriminating events and leaving the rest as they are. Both approaches vary depending on the attacker's technical maturity, the criticality of the incident, and the objective of the operation.

Yet whether crude or refined, every log-clearing attempt produces a clue. An attacker trying to hide traces can't completely eliminate the act of hiding itself. For this reason, Event ID 1102 and related anomalies are among the gold-standard indicators in forensic analysis.

When the security log is cleared, Windows automatically records this action with Event ID 1102. This event contains critical metadata that provides answers to questions such as these:

- Who cleared the log?
- Under which user context was the action performed?
- Which system component triggered it?
- What occurred immediately before the clearing?
- Did the log channel return to normal operation after the clearing?

Important points to consider include the following:

- **The 1102 event itself can't be deleted.**
  Windows writes it to the log channel using a special security mechanism. However, if an attacker completely replaces the event log file (EVTX) or manipulates the ETW provider, this behavior will manifest differently.
- **The 1102 event doesn't always occur after an attack.**
  Some actors clear logs in the middle of the attack chain to both break the timeline and render subsequent actions invisible.
- **The absence of 1102 is itself a finding.**
  If specific events have disappeared from the security log but no 1102 event is present, this is often an indicator of a selective deletion technique.

Table 9.3 provides a detailed summary of the different event types we've discussed, including their value from the CTI perspective.

Event Type	Event ID	What Does It Indicate?	Attacker's Typical Tactics	What the Analyst Should Pay Attention To	Value Produced from a CTI Perspective
Logon behaviors	4624/4625	Successful or failed authentication attempts	Password spraying, brute-force attacks, account discovery for lateral movement, service logon masquerading	Logon type, `TargetUserName`, `SourceIP`, number of failed attempts, rapid repetition intervals	Analyzes the attacker's initial access and account discovery methods

Table 9.3: Critical Security Log Event Types and Their Contributions to Forensic Analysis

Event Type	Event ID	What Does It Indicate?	Attacker's Typical Tactics	What the Analyst Should Pay Attention To	Value Produced from a CTI Perspective
Process creation (behavioral traces)	4688	Recorded when a new process is created; makes the command line and parent-child relationship visible	LOLBin abuse, token impersonation, script-based attacks, command obfuscation	Consistency of `ParentProcessName`, suspicious patterns in the command line, timing correlations	Behavior-based TTP extraction, MITRE ATT&CK mapping, reconstruction of the attack chain
Token abuse/ privilege manipulation	4688 + security context fields	Token information used during process creation	Privilege escalation using stolen tokens, creation of duplicate tokens	`TokenElevationType`, restricted admin behaviors, `LocalSystem` impersonation	Identifies the attacker's privilege escalation techniques; infers the actor's maturity
Log-clearing attempt	1102	Cleared security log	Trace removal, timeline manipulation, selective deletion	Gaps in log flow after event 1102, timestamp inconsistencies, anomalies in Event ID counts	Determines the presence of anti-forensic techniques; key signals regarding the attacker's OPSEC discipline
Selective log manipulation	Doesn't produce a single Event ID	Partial deletion or modification of security log content	Removal of sensitive behaviors, fabrication of a false timeline	Log file size anomalies, changes in event frequency	Analyzes the quality of the attacker's OPSEC strategy and target priorities

Table 9.3: Critical Security Log Event Types and Their Contributions to Forensic Analysis (Cont.)

Manipulation Techniques

Advanced APT groups and some red team specialists don't clear logs in their entirety. Doing so is too obvious, too noisy, and leaves traces. Instead, they selectively delete the following one by one:

- 4624/4625 events that reveal the attack chain
- 4670/4672 events that carry traces of privilege escalation
- 4688 events that show the process creation chain
- 4698 scheduled task records that prove persistence attempts

In this case, Event ID 1102 isn't triggered because the security log hasn't been completely cleared.

Direct manipulation is performed on the event log file itself (EVTX binary patching). An analyst can detect this silence only through the following three techniques:

- **Event frequency analysis**
  The normal volume of specific Event IDs is compared with the current volume. For example, when there are thousands of 4688 records, but there are no events during several critical minutes, this indicates a gap.
- **Timestamp anomaly**
  An unrealistic time gap between log timestamps, such as
  - 10:00
  - . . . no records . . .
  - 10:17

  is typically the moment when a selective deletion operation was performed.
- **Channel integrity value**
  Windows maintains integrity metadata for each log channel. If this value has changed, the log file has been manipulated.

EVTX Binary Patching

EVTX binary patching refers to the direct manipulation of the Windows event log file format at the binary level, without using standard logging APIs. Instead of clearing logs through native system functions, which would generate artifacts such as Event ID 1102, the attacker modifies the underlying `.evtx` file structure itself. This may involve altering record headers, adjusting offsets, corrupting or recalculating integrity-related fields (e.g., checksums), or selectively invalidating specific event entries. As a result, targeted events become unreadable or effectively invisible to standard log viewers and many EDR solutions, while the overall log file appears intact and operational. This technique allows adversaries to remove or conceal specific forensic evidence with minimal noise and without triggering conventional logging alerts.

The techniques attackers use to interfere with logs aren't limited to EVTX deletion or patching. More sophisticated methods include the following:

- **ETW provider tampering**
  ETW plays a critical role in event generation. An attacker may disable, hook, or override ETW event providers. In this case, no events are generated at all, meaning nothing appears, including Event ID 1102.
- **Kernel-mode manipulations**
  Kernel-level rootkits can directly affect the event log mechanism:
  - Log-writing functions are hooked.
  - Audit API calls are blocked.
  - Event ID generation is selectively suppressed.

  In such cases, the logs don't appear wrong, but reality is diminished.
- **EVTX structural modification**
  Some actors modify the EVTX file at the binary level:
  - Invalidating event records without deleting them
  - Manipulating cyclic redundancy check (CRC) fields
  - Breaking chunk alignment to render specific event blocks invisible
  - Rewriting deleted record headers

  Such manipulations may not be detected by EDR solutions, but deep forensic tools (EVTX recovery tools, hex-based parsers) can identify them.

For an analyst, log clearing attempts both indicate a threat and serve as clues about what the attacker was afraid of. These questions are the key to understanding the attacker's intent:

- Why were the logs deleted?
- Why was that specific moment targeted?
- Why were only certain Event IDs removed?
- Why was the timeline manipulated?

Here are some examples:

- If the attacker deleted only 4688 events, they left a process-focused trace. This indicates an attempt to conceal the tool execution chain.
- If 4624/4625 records were removed, the attacker was likely trying to hide the lateral movement phase.
- If the 4698 scheduled task was destroyed, the attacker was attempting to conceal the persistence mechanism.

Every manipulation reveals what the attacker wanted to hide, which is invaluable information from a CTI perspective.

Security Log Advice

The security log is valuable, but it's never absolute. The better you know it, the faster you notice its silences. As an analyst, my own approach is as follows:

- When you see a 4624 record, don't ask, "Who logged in?"; instead ask, "Is this logon a normal behavior?"
- When analyzing a 4688 record, don't focus solely on the command line; examine the parent-child relationship and token usage.
- When you encounter a 1102 event, don't panic; immediately compare the time periods just before and just after it.

This seemingly simple approach can make even an attacker's most concealed moves visible. The depth of the security log doesn't come merely from the number of Event IDs it contains, but from the story these logs collectively tell. These logs best reveal the interaction an attacker establishes with the system, their silent preferences, tactical mistakes, and hastily taken steps.

Example: Behavioral Privilege Escalation Detection Using Python

In our example shown in Figure 9.2, the behavioral detection engine named Privilege Escalation Behavioral Detection Engine v4 performs analysis on a normalized JSON event stream derived from Windows logs. The engine aggregates privilege escalation behaviors that attackers typically prefer within a 15-minute time window, generates a risk score, and automatically produces an analyst interpretation.

```
C:\Users\admin\PycharmProjects\Books>python privilege_escalation_evtx.py test_security_events.jsonl

{
    "entity": "H.Yener ARICI-0xABC",
    "score": 30,
    "risk": "Low",
    "behaviors": [
      "SeDebugPrivilege assigned (4672)"
    ],
    "mitre": [
      "T1068",
      "T1548",
      "T1134"
    ],
    "timestamp": "2025-01-26T10:00:10Z",
    "analyst_note": "In the context of account "H.Yener ARICI" and logon context '0xABC', multiple privilege-
     escalation-related signals were observed.
     Assignment of "SeDebugPrivilege". This privilege allows reading/writing other process memory, manipulating running
     processes and interacting only to a very limited set of accounts; seeing it on an unexpected identity is a strong
     indicator of privilege escalation preparation. While this behavior set does not by itself prove a completed
     attack, it warrants contextual review, especially if the account plays a critical role."
}
```

Figure 9.2: Behavioral Privilege Escalation Detection Output Highlighting SeDebugPrivilege Assignment and Contextual Risk Scoring

In the shared output file, the engine detected a noncritical yet notable `SeDebugPrivilege` assignment within a session (`0xABC`) associated with the user `H.Yener ARICI`. Although this signal alone doesn't constitute a definitive indicator of an attack, it's noteworthy from the perspective of attacker behavioral models.

The output shown in the image contains the following core fields:

- **`entity:`** `"H.Yener ARICI-0xABC"`
  Indicates under which user and logon ID context the analysis was collected.
- **`score:`** `30`
  The risk score generated solely by the `SeDebugPrivilege` assignment.
- **`risk:`** `"Low"`
  Classification of the chain as low risk because it's still an immature privilege escalation attempt at this stage.
- **`behaviors:`** `"SeDebugPrivilege assigned (4672)"`
  One of the strongest signals of privilege escalation behaviors.
- **`mitre:`** `["T1068", "T1548", "T1134"]`
  The techniques with which the event is associated according to the MITRE ATT&CK framework.
- **`timestamp:`** `"2025-01-26T10:00:10Z"`
  The moment the signal was generated.
- **`analyst_note:`**
  A long explanation in which the engine interprets and contextualizes the raw data entirely in analyst language.

Let's expand the analyst interpretation produced by the engine from a deeper digital forensics perspective:

- **Context of the signal**
  The session context is specified in the output:

  ```
  User: H.Yener ARICI
  Logon ID: 0xABC
  ```

  The behavioral analysis engine correlated only limited activity within the 15-minute window. However, the detected activity is a classic sign of low-intensity privilege escalation preparations.
- **Critical behavior detected**
  `SeDebugPrivilege`, associated with Event ID 4672, is one of the most critical keys for an attacker in the Windows ecosystem. This privilege enables the following:
  - Accessing the memory of other processes
  - Injecting code into a process
  - Reconfiguring services
  - Exercising full control over sensitive security processes such as `LSASS.exe`

Therefore, its appearance on the system outside of default administrative accounts is highly anomalous.

- **Risk score and rationale for the assessment**
  The reasons the engine classified this situation as low risk are listed here:
  - The chain still consisting of only a single behavior
  - The absence of other supporting signals such as service creation (7045), impersonation, `runas.exe`, or token enumeration
  - The low event volume

  However, the key point here is that every privilege escalation attack begins with low intensity. We shouldn't miss the first signal or wait for the chain to mature. Therefore, even if the danger level appears low, the event shouldn't be ignored entirely.
- **MITRE ATT&CK mapping**
  The engine's technical mapping is as follows:
  - `T1068`: Exploitation for Privilege Escalation: Preparations for exploitation aimed at privilege escalation.
  - `T1548`: Abuse Elevation Control Mechanisms: Manipulation of user rights.
  - `T1134`: Access Token Manipulation: Indicates attackers' attempts to operate through token-based techniques.

These technical correlations demonstrate that even a single signal may potentially represent the beginning of a much broader attack chain.

In the analysis note, we can see that although this is a singular signal, this privilege is one of the mechanisms most frequently abused by attackers in the Windows OS. The behavioral detection engine provided the analyst with operational insight by interpreting both the raw visibility of the event as well as its possible position within a broader attack chain. This example illustrates why, in modern security analysis processes, holistic interpretation of behaviors is far more critical than analyzing individual events in isolation.

9.2.2 Design of an Advanced Audit Policy

In a Windows environment, an audit policy can be considered a memory setting that defines what the system will remember. Its default state is often silent, reluctant to record events and insufficient for understanding threat behaviors. For this reason, designing an advanced audit policy isn't merely about generating more logs; it's about ensuring that the system provides meaningful information to analysts, researchers, and CTI teams.

It's tempting to think that increasing logging is the path to an effective audit policy: The more records, the better the visibility, right? But this isn't how things work in the field. At a certain point, the volume of logs increases so much that meaning is lost. It's like adding one more drop to the ocean—nothing changes.

This section explains how a system can be made to record events and, more broadly, danger itself. Advanced audit design rests on three fundamental concepts: a profile aligned with attack surface reduction, differentiated visibility across domain and standalone systems, and log enrichment principles for EDR/SIEM integration.

An Audit Profile Aligned with Attack Surface Reduction

Protecting a system against attacker behavior doesn't end with updating antivirus signatures or deploying EDR agents. True security begins with understanding the system's own internal discipline and keeping it under observation. The audit policy is the visible manifestation of this discipline. The system doesn't inherently know what is important; we teach it what should be recorded.

At this point, the attack surface reduction (ASR) approach becomes a security philosophy: Reduce the paths an attacker can take, and illuminate the remaining paths with a spotlight.

The logic of ASR is to tighten defenses and align telemetry with attacker behavior at an engineering level. For this reason, advanced audit profiles in modern environments specifically monitor the behaviors that attackers most frequently touch throughout the kill chain. The following behavior categories are the operational comfort zones that attackers have historically preferred:

- **Privilege escalation attempts**
  Privilege escalation is an indispensable step in most attack chains. An attacker gains access to a system as a low-privileged user but must climb higher to achieve their objective. This is where the audit policy must come into play. Examples include the following:
  - Visibility of processes attempting to obtain `SeDebugPrivilege`
  - Unusual process chains running with a `LocalSystem` token
  - Inconsistent logon types triggered via `runas.exe`
  - Privilege escalation performed through service creation (Event ID 7045)

  These are classic examples, but there are also lesser-known, advanced signals. If an attacker accesses the `GetTokenInformation` API via PowerShell with unusually high frequency, this is typically preparation for token structure enumeration. Many EDR solutions treat this behavior as noise, but in reality, it's a strong indicator of the reconnaissance phase preceding privilege escalation.

- **Token abuse and impersonation**
  One of an attacker's favorite techniques is impersonation, that is, advancing within the system by effectively claiming, "I am actually another account." The technical traces of this include the following:
  - Creation of duplicate tokens (very short-lived handle spikes)
  - Silent takeover of another process's token
  - Transfer of service account or `LOCAL SYSTEM` tokens into user-mode processes
  - Increased calls to `CreateProcessWithTokenW`

These behaviors are typically visible in the extended fields of Event ID 4688; however, most defensive environments don't enable these fields. Therefore, a well-designed audit profile should activate all subfields that contain token information.

Some APT groups use thread-level impersonation during token abuse. In this case, the token change occurs at the thread level rather than the process level. Event ID 4688 doesn't capture this; however, ETW's `ThreadSetToken` event provider can. An advanced audit profile can enable this provider.

- **Remote execution patterns**

  The paths attackers use for lateral movement are limited, but powerful: PsExec, WMI, Windows Remote Management (WinRM), and remote command execution over Server Message Block (SMB) sessions.

  Each of these tools has its own distinctive, fingerprint-like telemetry signals, for example:

  - PsExec-based remote execution is a widely used technique that enables remote command execution over SMB by creating a temporary service on the target system. This activity leaves recognizable traces in Windows logs. Specifically, Event ID 4624 (Logon Type 3) indicates a network-based authentication using valid credentials, which is typically followed by Event ID 5145 showing access to administrative shares such as `ADMIN$` or `IPC$` used to transfer and execute the PsExec service binary. When these events occur in close temporal proximity (particularly involving administrative shares), they strongly suggest remote execution via PsExec or a PsExec-like technique.
  - WMI-based execution allows remote command execution through remote procedure call (RPC) mechanisms, often without producing clear process creation logs. Event ID 5858, recorded in the `Microsoft-Windows-WMI-Activity/Operational` log, represents a WMI operation failure and is commonly associated with RPC communication issues or access constraints. While not inherently malicious, repeated or contextually unusual occurrences of this event, especially during lateral movement, may indicate failed or partially successful remote execution attempts. In successful cases, related events (5857–5861) can reflect query execution and provider activity; however, advanced attackers may disable or evade this logging. Therefore, analysts should correlate such events with behavioral indicators, such as short-lived spikes in `WmiPrvSE.exe` activity, rather than relying solely on log presence.
  - WinRM-based remote execution is a legitimate administrative method that enables remote command execution over HTTP or HTTPS. In this context, Event ID 4648 indicates that explicit credentials were used to initiate a logon session, often reflecting remote authentication with supplied credentials. This is followed by Event ID 4624 (Logon Type 3), confirming a successful network logon. The sequence of Event ID 4648 followed by 4624 (Logon Type 3), particularly when associated with processes such as winrm.exe or wsmprovhost.exe, serves as a strong indicator of remote command execution. Because WinRM is commonly used for legitimate administrative purposes,

distinguishing malicious use requires contextual analysis, including unusual source hosts, timing anomalies, or atypical account usage patterns.

- SMB session hijacking introduces subtle yet highly indicative anomalies within the authentication and session lifecycle. Instead of establishing a clean and fully traceable logon sequence, the attacker leverages an already authenticated SMB session or reuses captured credentials to execute actions without generating a corresponding, consistent authentication trail. As a result, the event flow may exhibit short-lived or incomplete credential mappings—such as transient Logon Type 3 sessions that lack a clear initiation context, abrupt session terminations, or inconsistencies between logon events and subsequent resource access (e.g., file share interactions without an expected preceding authentication pattern). These irregularities are often brief and easily overlooked, but when correlated across logon sessions, share access events, and account usage patterns, they form a distinctive behavioral signal of session hijacking rather than legitimate administrative activity.

Command execution via WMI (MITRE ATT&CK T1047) is often visible through Event IDs 5857–5861 under the `WMI-Activity Operational` log. However, APT actors may disable this log. Even so, very short-lived handle spikes in WMI provider host processes reveal the presence of this behavior. This signal becomes visible only with a well-tuned audit policy.

- **Persistence mechanisms**
  If an attacker has gained access to a system and intends to remain there, persistence is inevitable. It's one of the first areas an audit policy must monitor. Critical points include the following:
  - Scheduled tasks (Event ID 4698)
  - Registry run keys (Sysmon Event ID 13 Security Log Object Access)
  - Service creation (Event ID 7045)
  - COM hijacking (less known but effective)
  - WMI event consumer persistence (Event ID 5861 signal chain)

  COM hijacking attempts usually appear not as command execution in Sysmon or the security log, but as changes in DLL load patterns. In other words, the key indicator isn't the persistence record itself, but the timing anomaly that occurs in the behavior of loaded modules. An advanced audit policy should be configured to capture module load telemetry as well.

- **Script-based attacks**
  PowerShell, WMI, `rundll32.exe`, `mshta.exe`, and so on are the quietly operating engines of the modern attack ecosystem. An audit profile must make the following behaviors visible:
  - Base64-encoded PowerShell commands
  - Use of the `-nop`, `-w hidden`, and `-encodedcommand` arguments
  - Remote JavaScript/HTML loading via `mshta.exe`

  - Code execution via `rundll32.exe` outside of exported functions
  - Remote script execution via WMI

  Many APT groups don't use PowerShell Antimalware Scan Interface (AMSI) bypass to evade command-line logging; instead, they rely on inline .NET reflection. In such cases, the command line appears benign, but a reflection invoke chain forms within process memory. Monitoring the command line alone is insufficient; script block logging, module logging, and deep command capture must be enabled in the audit profile.

If an audit policy doesn't monitor these areas, CTI analyzes only the output of the attack. But a correctly designed audit policy achieves the following:

- The attacker's TTPs become visible.
- Behavioral analysis is performed accurately.
- MITRE ATT&CK mappings become clear.
- Incident response takes minutes rather than hours.
- Lateral movement chains can be broken.
- Even anti-forensic attempts turn into detectable signals.

Different Audit Scopes in Domain and Standalone Systems

One of the most common misconceptions I encounter when performing log analysis in the Windows world is that a Windows system is a Windows system: They all behave the same. However, once you begin working in the field, you quickly realize how far this assumption is from reality. A domain environment and a standalone machine may run the same OS on paper; yet from a forensic telemetry perspective, they are like two different universes.

I felt this difference during a real incident investigation: Moving from the complex, echoing log streams of a domain environment to the singular and clear signals of a standalone machine felt like stepping from a crowded avenue into a quiet street. Both have advantages and disadvantages, and understanding this distinction forms the cornerstone of a correct audit policy.

Domain Environments

A domain structure is like a large living organism in which hundreds of systems are managed from a single center and whose behaviors affect one another in a chain. The brain of this organism is the *domain controller (DC)*. Every authentication, every access request, and every service call passes through its filter first.

The value of Event IDs 4768 and 4769 stems from their position at the foundation of inter-system trust. These logs don't just say that a user attempted to log on to the system; they state all of the following in a quiet yet definitive manner:

- Where the user came from
- Which service account was targeted
- Through which service principal name (SPN) access was requested
- Which protocol variant was used

Real-World Example

I once witnessed an incident in which an unusual SPN request within a 4769 chain revealed the entire attack sequence. A single small record told a great deal.

In domain environments, threats spread laterally. Once a machine is compromised, the attacker typically moves on to another target. Here, the values of 4624 logon types function almost like a road map. If you want to track an attacker's movements, you must understand how specific logon types reflect different forms of remote access and interaction within the environment:

- **Logon Type 3**
  Represents a network-based authentication, typically associated with SMB access, WMI execution, or tools such as PsExec, and is often observed during lateral movement between systems.
- **Logon Type 10**
  Corresponds to remote interactive access via Remote Desktop Protocol (RDP), indicating that the attacker has established a direct interactive session on the target machine.
- **Logon Type 11**
  Reflects cached interactive logons, which occur when previously used credentials are reused locally in the absence of a DC, and may indicate offline authentication scenarios or credential reuse patterns.

Real-World Example

On one occasion, I realized that an attacker had begun using PsExec solely by noticing a rhythm anomaly in the Logon Type 3 chain. At first glance, it looked ordinary, but the time intervals weren't human. Logs sometimes whisper to us to "Pay attention, this step isn't real."

Because domain environments are the home of centralized management, *Group Policy Object (GPO)* determines the fate of audit settings. If all systems receive the same policy, telemetry remains consistent. Otherwise, the following happens:

- Critical logs may be missing on some machines.
- Unnecessary noise may appear on others.
- SIEM correlation becomes fragmented.
- Reconstructing the chain link by link during incident response becomes more difficult.

For this reason, consistency of audit policy in a domain environment is a fundamental requirement for forensic visibility. When GPOs are misconfigured, blind spots emerge where attackers can effectively hide.

Finally, domain admin privileges are like a master key that opens every door in the network. For this reason, the appearance of domain admin tokens is always a red flag. It's not normal for a domain admin account to do the following:

- Log in to a regular workstation
- Execute commands via `cmd.exe`
- Run arbitrary queries using PowerShell
- Establish sessions in the middle of the night

In fact, such behaviors often reflect an attacker's OPSEC weakness. Understanding token usage patterns can sometimes change the fate of an incident. On one occasion, we detected a credential harvesting operation targeting the entire domain solely due to a slight increase in the frequency of token elevation events.

Standalone Systems

A standalone machine isn't as complex as a domain environment, but this simplicity makes it more readable. Every behavior occurs locally, every anomaly is more clearly visible, and each step taken by an attacker leaves a more pronounced trace.

A standalone system has the following characteristics:

- The number of accounts is limited.
- Privileges are more constrained.
- Logon patterns are more consistent.

For this reason, even a single chain of failed logons stands out. A behavior that would be lost in a domain environment appears like a shining indicator on a standalone system.

On a standalone machine, persistence mechanisms are far more visible. The following traces all stand out easily:

- Run keys added to the registry
- Local administrator abuse behaviors
- Service creation records (Event ID 7045)
- Scheduled task activities

In a domain environment, minor persistence activity may be lost in background noise; on a standalone system, it practically declares, "I am here."

In standalone systems, most services such as RDP, SMB, or WinRM are either disabled or tightly restricted. For this reason, events such as the following carry significant importance:

- 4672 (special privileges assigned)
- 4624 Logon Type 10 (remote interactive logon)
- 4648 (explicit credential usage)

Even the appearance of a single one of these events at an unusual time generally indicates the presence of an attacker.

In domain systems, threats move laterally from one machine to another or from one account to another. The log patterns are complex, yet highly meaningful. In standalone systems, however, threats grow vertically: privileges are elevated on a single machine, persistence is established, and a silent dominance is achieved. The traces are clear, and the signals are strong.

An audit policy must understand this distinction. A poorly designed policy may generate excessive noise in one environment while creating critical silence in another. As an analyst, you know this well: Too much noise is just as dangerous as too much silence. What truly matters is learning to hear the unique rhythm of each environment.

Table 9.4 provides a detailed comparison of standalone systems versus domain environments.

Category	Domain Environment	Standalone System	Meaning for the Forensic Analyst
Authentication	Kerberos-centric; the 4768/4769 Ticket-Granting Ticket (TGT)/Ticket-Granting Service (TGS) cycle produces critical signals.	Windows New Technology LAN Manager (NTLM) and local account behavior is dominant; logon patterns are simpler and more readable.	Domain: Complex chain analysis is required. Standalone: Single anomaly detection is easier.
Lateral movement	Threats spread laterally; 4624 Logon Type 3/10/11 is critically important.	Lateral movement is limited; most attacks grow vertically (local privilege escalation).	Domain: Connection mapping is required. Standalone: Privilege escalation-focused analysis is performed.
Audit policy management	Centralized distribution via GPO; inconsistent GPOs lead to complex telemetry gaps.	Local audit settings are controlled individually, which provides more predictable log generation.	Domain: Policy consistency is critical. Standalone: Misconfiguration is detected more quickly.

Table 9.4: Differences in Audit Scope and Forensic Visibility in Domain vs. Standalone Systems

Category	Domain Environment	Standalone System	Meaning for the Forensic Analyst
Token usage	Domain admin tokens carry high risk across the network; the token abuse chain is a very strong signal.	Local admin privileges are pronounced; impersonation attempts are more clearly visible.	Domain: Token usage pattern analysis is essential. Standalone: A small anomaly may indicate a major compromise.
Persistence mechanisms	Persistence may blend into noise due to GPOs, software, and domain services.	Persistence is far more pronounced; run keys, Event ID 7045, and scheduled tasks provide clear signals.	In standalone environments, persistence detection is faster and sharper.
Remote execution	Tools such as PsExec, WMI, WinRM, and SMB are frequently used; event chains are complex.	Remote access channels are limited; Event ID 4672 + Event ID 4624 Type 10 are the most critical signals.	Domain: Execution pattern analysis is required. Standalone: A small movement creates a large echo.
Telemetry complexity	Many systems → many logs → much noise: Without proper correlation, traces may be lost.	One system → one context → low noise: Anomalies emerge more quickly.	Domain: Advanced SIEM correlation is mandatory. Standalone: Even manual analysis yields quick results.
Attacker behavior profile	APT groups prefer lateral spread; DC-targeted discovery and token rehearsal signals are observed.	Typically, the privilege escalation + persistence pair is dominant.	Scenario recognition capability differs; the analyst must view the two environments through different lenses.

Table 9.4: Differences in Audit Scope and Forensic Visibility in Domain vs. Standalone Systems (Cont.)

Log Enrichment Principles for EDR/SIEM Correlation

Modern security architectures derive meaning not from standalone components, but from an ecosystem in which components continuously supply data to one another. Within this ecosystem, the audit policy serves as a raw data provider. However, raw data alone isn't usable. Without proper enrichment, logs quickly degrade into isolated and meaningless aggregates within SIEM platforms. For example, a standalone 4688 (process creation) event may simply indicate that `cmd.exe` was executed; however, without additional context, it's impossible to determine whether this activity is benign or part of an attack chain. When enriched with its parent process (e.g., `winword.exe` spawning `cmd.exe`), command-line arguments (e.g., encoded PowerShell execution), and user context (e.g., a nonadministrative user invoking administrative tools), the same event becomes a clear indicator of suspicious behavior. Similarly, a 4624 (Logon Type 3) event on its own may represent routine network activity, but when correlated with 5145 (share access) and followed by a 4688 execution on a remote host, it forms a recognizable lateral movement pattern such as PsExec or SMB-based execution.

For a log to truly generate value, three principles must be applied:

- **Contextual enrichment**
  A 4688 process creation record conveys attack-related meaning only when combined with the following information:
  - Parent process
  - Command line
  - Token information
  - User SID
  - Comparison against baseline behavior

  These elements form the foundation for building a behavioral model within a SIEM.
- **Consistency of correlation keys**
  The fields that establish links between events must be in a consistent format:
  - `TargetUserName`
  - `LogonID`
  - `ProcessID`
  - `WorkstationName`
  - `IpAddress`

  Consistency across these fields enables SIEM platforms to automatically group events into coherent attack chains.
- **Time synchronization (time fidelity)**
  Even if the audit policy is correctly configured, incorrect system clocks will corrupt all telemetry. Attack chains become convoluted, and timelines fragment. For this reason, timestamp accuracy is of critical importance from a forensic perspective.

An audit policy is more than a configuration; it's a perspective. Producing too many logs makes you blind. Producing too few logs makes you deaf. The correct balance can only be achieved through experience. Every system has a different threat model. Every environment has different workflows, user behaviors, and risk profiles. Accordingly, the audit policy must be shaped by the identity of the environment itself.

Sometimes, a single 4670 record on a server that normally operates in complete calm, a seemingly ordinary 4625 chain, or a slightly delayed 4688 distribution can signal that something is wrong. When the audit policy is correctly designed, the system whispers the truth to you.

9.2.3 Behavioral Monitoring with Event Tracing for Windows

In Windows, every call, every operation, and every kernel transition flows through ETW in its rawest form. If Sysmon is a microscope, ETW is truly the system's pulse. Once you learn how to read that pulse correctly, you can sense even the most subtle movements of an attacker. This section explains why ETW is indispensable for behavioral monitoring, how it works, and why even the most effective defense mechanisms can become blind if ETW isn't properly configured.

One of ETW's greatest strengths is its ability to capture behavior not at the API level, but at the syscall level. Even if an attacker attempts to bypass user-mode APIs, low-level calls made to the kernel (i.e., a syscall) can't escape detection, for example:

- An unusually high frequency of `NtCreateFile` calls
- `NtAllocateVirtualMemory` indicates preparation for process hollowing
- `NtWriteVirtualMemory` indicates shellcode injection
- `NtCreateThreadEx` indicates remote thread creation
- `NtQuerySystemInformation` indicates credential discovery phase

None of these activities are directly visible in the event log. Sysmon may capture them, but only depending on specific configurations. ETW, on the other hand, collects these calls directly from their source: the kernel itself. Some APT groups perform injection operations using chunked write techniques specifically to evade certain ETW providers. However, when syscall frequency changes, ETW detects it. This discrepancy has no corresponding log entry; it emerges solely within ETW's silent observation.

Through ETW, you can observe how an attacker interacts with a process, what they manipulate and for what purpose, and even disruptions in the rhythm of syscall frequencies.

ETW is also very well known to attackers. Neutralizing such a powerful monitoring mechanism is a natural part of the attack strategy of advanced actors in particular. Some malware samples are listed here:

- Disable ETW providers
- Manipulate ETW buffers
- Force events written to ETW to be dropped
- Disrupt Sysmon's ETW-based event stream
- Silence all telemetry through kernel-mode ETW hooks

These techniques may sound complex, but they all have been repeatedly observed in real-world operations.

Real-World Example

An attacker had disabled ETW's `Microsoft-Windows-Threat-Intelligence` provider. According to the logs, there were no anomalies. Even Sysmon was silent. However, buffer fragmentation was observed in another ETW provider. This small inconsistency exposed the attacker's ETW manipulation.

Sometimes, it's not the system's silence itself, but the cracks within that silence that reveal the truth. If an attacker attempts to disable ETW, they are actually telling you something very important: "I am afraid of being seen here." This fear constitutes a critical behavioral signal from a CTI perspective.

ETW's strength, generating an enormous volume of data, is also its weakness. Monitoring a single kernel call is useful, but processing thousands of them per second is a stream that's difficult to manage. For this reason, ETW-based monitoring necessarily requires optimization:

- Unnecessary providers should be disabled.
- Buffer sizes should be adjusted according to system load.
- Event throttling mechanisms must be correctly configured.
- The balance between kernel providers and user-mode providers must be maintained.
- ETW events forwarded to SIEM platforms must be properly categorized.

We can liken this process to a photographer adjusting focus. If the focus is too wide, everything appears blurred. If it's too narrow, you miss the big picture. Proper ETW optimization makes both the attacker's behavior and the intent behind that behavior visible.

For example, an increase in `NtProtectVirtualMemory` calls within a specific time window may indicate that malware is preparing for anti-debugging or anti-instrumentation techniques. Some advanced EDR solutions also collect ETW events themselves; however, if the provider subscription method isn't properly configured, an attacker may render the EDR's telemetry subscription entirely ineffective, even without explicitly disabling ETW. This detail is often overlooked by many analysts.

An important lesson is that if other log sources have gone silent, ETW may still be speaking. An attacker may delete the security log and disable Sysmon, and nothing appears in Windows Event Viewer. Yet in the lower layers of ETW, there may still be a slight increase in calls, a subtle buffer anomaly, or an out-of-rhythm syscall sequence that all tell the truth.

For this reason, when investigating an attack, don't abandon logs, but don't blindly trust them either. When deep telemetry is required, treat ETW not as a tool, but as an ally. Those small signals the system whispers to you can sometimes illuminate the entire attack.

To better illustrate how ETW extends visibility beyond the limitations of traditional logging mechanisms, Table 9.5 provides a structured overview of the key behavioral domains that can be observed through ETW telemetry. As you can see, ETW enables analysts to detect activities that remain invisible to standard event logs, including direct syscall invocation, in-memory payload execution, and attempts to interfere with telemetry pipelines themselves. The table also emphasizes an important analytical principle: When expected signals are absent or distorted, this absence may itself represent a meaningful indicator of adversarial intent. In this sense, ETW doesn't just provide additional data, it reveals the gaps, inconsistencies, and behavioral rhythms that expose otherwise hidden attacker activity.

Category	Behaviors Monitored by ETW	Techniques Used by the Attacker	Why They Aren't Visible in Normal Logs	Forensic Value Provided by ETW
Syscall-based suspicious activities	`NtAllocateVirtualMemory`, `NtWriteVirtualMemory`, `NtCreateThreadEx`, `NtProtectVirtualMemory`, `NtQuerySystemInformation`	Process hollowing, shellcode injection, credential discovery, anti-debugging techniques.	API bypass, Sysmon configuration gaps, and user-mode limitations of the event log prevent visibility.	Captures raw behavior at the kernel level, which reveals the attacker's true interaction footprint.
User-mode API bypass behavior	Syscall sequences recorded by ETW even when Win32 APIs are bypassed.	Direct syscall invocation, syscall stomping, chunked writes.	Disabling API hooks misleads the event log.	Defeats the attacker's attempts at API invisibility.
ETW anti-forensic behaviors	Provider disable attempts, buffer corruption, subscription manipulation.	Disabling ETW providers, breaking EDR ETW subscriptions, kernel-mode ETW hooks.	EDR/Sysmon see only their own event streams; ETW manipulation is rarely reported.	Produces a suspicious silence signal that indicates the attacker's fear of visibility.
Memory manipulation behavior	Sudden memory allocation + write + protect changes.	Reflective DLL loading, encryption stub preparation, payload unpacking.	With no disk artifacts, the event log assumes nothing happened.	Reveals the behavioral trace of transient but critical in-memory operations.
Rhythm disruption in kernel behavior	Syscall frequency anomalies, irregular thread movements.	Kernel exploit preparation, stealth implant deployment.	Kernel behavior remains below normal logging thresholds.	Exposes the rhythm change during the attacker's preparation phase.

Table 9.5: Critical Areas of Behavioral Monitoring with ETW and an Interpretation Guide for Analysts

Category	Behaviors Monitored by ETW	Techniques Used by the Attacker	Why They Aren't Visible in Normal Logs	Forensic Value Provided by ETW
High-frequency telemetry optimization	Provider-based event density, buffer overflow signals.	Detection evasion through telemetry noise generation.	Normal logging mechanisms drop events under load.	Makes even telemetry saturation manipulation visible.
AV/EDR evasion attempts	ETW subscription manipulation, ETW consumer blocking.	Disabling the EDR telemetry pipeline.	The EDR may silently fail if it can't ingest ETW data.	Marks the moment the attacker's invisibility strategy fails.

Table 9.5: Critical Areas of Behavioral Monitoring with ETW and an Interpretation Guide for Analysts (Cont.)

9.3 Windows Registry Forensic Analysis

When you think about the registry, several things may come to mind: a branching structure, nested keys, values, and timestamps. But for a forensic analyst, it's a vault of memory—silent, deep, and layered. Very often, this vault is where an attacker's most persistent traces remain hidden the longest because the registry maintains a behavioral map of the system. What did the user do? Which program ran? When? Was a persistence mechanism established? Every action has a corresponding trace here, no matter how small—or at least, it should. The absence of such a trace is itself a significant signal at times.

This section approaches registry forensics not just as reading keys, but as a structured analysis of behavioral evidence. The discussion is organized around four core areas: user interaction artifacts, which reflect user activity through most recently used (MRU) lists and navigation traces; persistence mechanisms, which reveal how attackers establish long-term footholds within the system; artifact consistency, which focuses on temporal and structural relationships between registry keys; and program execution evidence, which captures traces of executed applications through artifacts such as ShimCache and AmCache. Together, these areas provide a framework for interpreting registry data as a coherent behavioral narrative rather than isolated entries.

9.3.1 Understanding the Registry

At first glance, the registry appears static. Hive files, defined structures, `NTUSER.DAT`, `SYSTEM`, `SOFTWARE`, and so on are all foundational components of the system, yet there's an anomaly. The registry is a dynamic structure; hives are updated as user sessions change, applications are installed, and system configurations evolve. This variability sometimes challenges the analyst. At other times, it becomes the indicator that reveals the most critical truth.

You might first hear the "voice" of the registry in an incident investigation, perhaps with a strange feeling at a moment when something seems missing. Logs fail to tell a complete story, Sysmon remains silent, ETW exhibits a broken rhythm but an unidentifiable vibration, and then you turn to the registry. With a single click, you notice it: a value that shouldn't be there, an incorrect timestamp, or a subkey that looks new but behaves as if it were old.

Real-World Example

The first time I experienced this feeling, I noticed inconsistent access times in the ShellBags records within `NTUSER.DAT`. The file system was pristine—too pristine. But the registry revealed that a folder within the user profile had been viewed at an unusual hour. The attacker had succeeded in cleaning the traces, but could not erase the registry's memory. That was the moment I truly understood the real power of registry forensics.

So, what do we look for in the registry?Not everything. Because searching for everything means finding nothing. Registry forensics proceeds through four core behavioral signals:

- **User interaction (user activity evidence)**
  MRU lists, ShellBags, `TypedPaths`, and `OpenSaveMRU` artifacts indicate where the user has navigated. Sometimes these traces appear ordinary, but a single recently opened file record can reveal the starting point of an entire attack chain.
- **Persistence (persistence mechanisms)**
  If an attacker has touched any of the `Run`, `RunOnce`, `Services`, Image File Execution Options (IFEO), and `AppInit_DLLs` areas, traces will remain—sometimes inadvertently. A malicious scheduled task may have been removed, but an access control list (ACL) change beneath the registry can persist.
- **Artifact consistency (temporal and structural consistency)**
  The registry is logical. Dates, timestamps, and parent-child relationships are expected to be consistent. Attackers often fail to replicate this logic. If one key appears new while the adjacent key appears old, the question is simple: "Why?"
- **Program execution evidence**
  MUICache, RecentApps, AppCompatCache (ShimCache), and AmCache provide a snapshot of executed programs. AppCompatCache may provide approximate timing, while AmCache is more precise. Using them together allows the attacker's behavior to be reconstructed chronologically. The registry may sometimes appear disordered, but at that moment, you're on the right path, for example:
  - ShimCache records are updated, yet the file system appears clean. The attacker deleted the file; the application executed, and the trace remained.

- An AmCache record exists, but the hash doesn't match. The attacker replaced the binary; persistence continues.
- The timestamp of a Run key differs from the system clock. Timestomping may have been applied.

These contradictions are a fascinating aspect about the registry. They resemble a small, silent game of chess between the attacker and the analyst.

The registry is like the inner voice of a Windows system where logs can be shaped, Sysmon can be silenced, and ETW can be manipulated. The registry can also be manipulated, but interventions in the registry are usually more costly and tend to leave behind traces such as timestamps, parent-child inconsistencies, hive integrity issues, or cross-artifact mismatches.

The registry is a place most attackers prefer to forget, and forgotten places speak the loudest. If an attacker is careless when adding persistence, timestamps expose them. If they are meticulous, the structure of the parent key exposes them. If they are professional, they may still forget something. Humans forget; machines are honest.

The registry is a powerful artifact for the following reasons:

- Even after an incident, a large portion of the registry remains preserved without being affected by RAM.
- Compared to the file system, manipulating the registry requires higher operational cost and often produces forensic inconsistencies.
- Even when attackers believe they have cleaned the registry, shadow records remain.
- It reveals the evolution of system behavior over time.

Registry forensics is the notebook in which an attacker's story is written silently. When you learn to read this notebook correctly, you can assemble a complete picture that includes everything the logs didn't say, the moments Sysmon missed, and the behaviors ETW pointed to but could not resolve.

To translate these behavioral signals into actionable forensic analysis, Table 9.6 presents a structured mapping of critical registry artifacts, their investigative meaning, and the ways in which attackers attempt to manipulate them. Rather than treating each artifact in isolation, the table emphasizes how persistence mechanisms, execution traces, and user interaction records intersect to form a coherent behavioral narrative.

As illustrated in Table 9.6, artifacts such as `Run` keys, `Services`, IFEO configurations, and AmCache entries provide complementary perspectives on attacker activity—ranging from persistence establishment to execution validation. At the same time, the table highlights a key analytical principle: no single artifact should be considered definitive on its own. Instead, forensic value emerges through correlation—cross-checking timestamps, validating paths and hashes, and identifying inconsistencies between registry data and other system artifacts such as file system records or event logs. This multiartifact approach enables analysts to move beyond isolated indicators and reconstruct attacker behavior with higher confidence.

Artifact/Location	What Does It Indicate?	Forensic Significance	Attacker Manipulation	What the Analyst Should Check
Run/RunOnce Keys				
HKCU/HKLM...\Run	Programs executed at user logon	Most common persistence point	Timestomping, fake paths, key addition/removal	Parent key timestamp, consistency with ShellBags and AmCache
Services				
HKLM\SYSTEM\CurrentControlSet\Services	Newly created or abused services	Silent and persistent backdoors	Binding a malicious binary to a legitimate service name	Event ID 7045 + AmCache + file timestamp analysis
IFEO				
HKLM...\Image File Execution Options<exe>	Process hijacking/debugger redirection	High-impact persistence technique	Adding a malicious debugger path	Consistency of the IFEO Debugger value with other artifacts
AppCompatCache (ShimCache)	Possible program execution traces	Postmortem execution analysis	ShimCache cleaning, timestamp manipulation	Not definitive evidence, so verify with AmCache
AmCache				
C:\Windows\AppCompat\Programs\Amcache.hve	Hash, path, and time information of executed programs	Strong source for validating execution relationships, though not guaranteed to generate records for every execution scenario	Binary replacement, hash collision	Time correlation with Run keys, Services, and ShimCache
MUICache				
HKCU...\Shell\MUICache	Applications executed by the user	UI-based interaction evidence	Adding fake display names	Whether the EXE name matches the actual path
RecentApps				
HKCU...\Search\RecentApps	Recently executed applications	User behavior analysis	Rarely manipulated	Timeline mismatch verification

Table 9.6: Advanced Registry Artifacts and Their Forensic Value

Artifact/Location	What Does It Indicate?	Forensic Significance	Attacker Manipulation	What the Analyst Should Check
ShellBags				
HKU<SID>...\BagMRU	History of visited folders	Can retain traces of even deleted folders	Leaving ghost folders	Inconsistency between the file system and ShellBags
UserAssist				
HKU<SID>...\UserAssist	Counters for executed programs	Reveals the rhythm of user behavior	Counter reset	Irregularities in counter increase patterns
TypedPaths/RunMRU/OpenSaveMRU	Recently used command and file paths	User/attacker path activities	MRU overwrite, leaving empty lists	Whether the list is unusually clean
WMI Persistence Artifacts				
HKLM\Software\Microsoft\Wbem\CIMOM	WMI event consumer persistence	Critical for diskless persistence	WMI repository manipulation	Correlation with WMI-Activity logs
Remote Desktop Protocol (RDP)/Terminal Services Client (TSClient) Artifacts				
HKU<SID>...\Terminal Server Client	RDP history	Lateral movement detection	Cleaning RDP history	Correlation with security log 4624 (type 10)

Table 9.6: Advanced Registry Artifacts and Their Forensic Value (Cont.)

9.3.2 Reconstructing User Behavior Through the Registry

One way to understand an attack is to examine what the malicious code did. But there's a second method that's even more effective: examining what the *user* did. Because user behavior is the fundamental rhythm of everything in a system, the attacker must disrupt that rhythm, rewrite it, or insert their own melody into it. This is exactly where registry forensics comes into play.

The registry contains the forensic reflections of a person's working habits, file preferences, navigation routes, and even, at times, the hasty mistakes made in the rush of the moment. We'll take a close look at user behavior traces in the following sections, including MRU artifacts, relationships between executed applications, and profile migration.

MRU Artifacts

MRU artifacts may appear at first glance to be a simple usage history, but for a forensic analyst, they are like a specialized window that brings together the psychological, behavioral, and operational traces of what occurred on the system. MRU keys in the registry don't just list recently opened files; they carry the silent yet meaningful reflections of the user's mental flow, curiosity, haste, and even behaviors manipulated by an attacker because these records provide a unique context both for understanding user habits and for understanding how an attacker infiltrated those habits. When you look at MRU values, you don't just answer the question "Which files were opened?"—you begin to construct a cognitive model that addresses much deeper questions:

- Why were the files opened?
- Is the timing logical?
- Is the rhythm of the sequence natural?
- To whom does this behavior belong?

MRU content forms a microtimeline of the user's interaction with the file system. A file being opened twice within the same minute may not be normal user behavior; it may relate to the attacker testing a payload, executing a command incorrectly, or evaluating the compatibility of a target binary before establishing persistence. A rarely mentioned fact in DFIR literature is that repetitive yet irregular accesses in MRU time series often reveal the noise generated by automated scripts, particularly the pre-execution tests of dropper- or loader-type malware. These tests are often missed by Sysmon because command-execution chains sometimes manipulate the parent-child relationship. MRU, however, records only the behavior; intent isn't questioned, just recorded.

Figure 9.3 demonstrates the use of PowerShell to extract `OpenSavePidlMRU` artifacts from the registry under `HKCU\Software\Microsoft\Windows\CurrentVersion\Explorer\ComDlg32`. The command enumerates registry subkeys and retrieves their associated properties, revealing MRU entries related to files accessed through common dialog boxes (e.g., **Open**/**Save** dialogs).

The output shows multiple indexed binary values along with `MRUListEx`, which defines the order of recent access. Although the entries are stored in binary format, they contain encoded references to file paths and user interaction history. These artifacts provide valuable insight into user activity, including recently opened files and directories, even in cases where corresponding file system traces may have been deleted.

From a forensic perspective, `OpenSavePidlMRU` entries are particularly useful for reconstructing user behavior and identifying access patterns. When correlated with other artifacts such as ShellBags, AmCache, and file system timestamps, they can reveal inconsistencies, hidden activity, or traces of attacker interaction that aren't visible in standard logs.

```
PS C:\Users\admin>

PS C:\Users\admin> Get-Childitem "HKCU:\Software\Microsoft\Windows\CurrentVersion\Explorer\ComDlg32\OpenSavePidMRU"
>> ForEach-Object {
    >> Get-ItemProperty $_.PSPath
>> }
MRUListEx  : {20, 0, 31, 80...}
0            {4, 0, 31, 80...}
1            {2, 0, 31, 80...}
2            {2, 0, 31, 80...}
3            {2, 0, 31, 80...}
4            {4, 0, 31, 80...}
5            {2, 0, 31, 80...}
6            {4, 0, 31, 80...}
7            {2, 0, 31, 80...}
8            {2, 0, 31, 80...}
9            {2, 0, 31, 80...}
10           {4, 0, 31, 80...}
11           {2, 0, 31, 80...}
12           {2, 0, 31, 80...}
13           {4, 0, 31, 80...}
16           {4, 0, 31, 80...}

PSPath       : Microsoft.PowerShell.Core\Registry::HKEY_CURRENT_USER\Software\Microsoft\Windows\CurrentVersion\Explorer\
PSParentPat  : Microsoft.PowerShell.Core\Registry
PSChildName  : *
PSProvider   : Microsoft.PowerShell.Core\Registry::HKEY_CURRENT_USER\Software\Microsoft\Windows\CurrentVersion\Explorer\

MRUListEx  : {20, 0, 31, 68...}

PSPath       : Microsoft.PowerShell.Core\Registry::HKEY_CURRENT_USER\Software\Microsoft\Windows\CurrentVersion\Explorer\
PSParentPath: 7z
PSChildName  : 7z
PSProvider   : Microsoft.PowerShell.Core\Registry::HKEY_CURRENT_USER\Software\Microsoft\Windows\CurrentVersion\Explorer\
```

Figure 9.3: Extraction of OpenSavePidlMRU Registry Artifacts via PowerShell

In MRU records, what truly matters is the context in which a file was opened. For example, if an EXE file located in a directory that hasn't been accessed for a long time suddenly appears in the MRU list—especially if `LastWriteTime` has been manipulated in the file system—this may indicate that, despite the attacker's timestomping, a trace was still left within the registry's nonlinear time model. This is a nuance that only advanced analysts are likely to notice: timestamps in MRU sequences don't always align exactly with file system timestamps. Therefore, MRU is a powerful counter-validation source for detecting an attacker's temporal manipulations.

Similarly, a technical detail known by very few DFIR experts is that some path values stored in MRU may belong to deleted or moved files. In such cases, MRU may inadvertently expose the attacker's attempt to clean the environment because even if the file is deleted, the MRU key doesn't forget it. This is invaluable information, especially for detecting staging files that C2 tools briefly open before relocating. In fact, incidents exist in the literature where certain APT groups were exposed because, after deleting their dropper files, they failed to notice the remnants in MRU. MRU records are this powerful because some operations triggered through Windows APIs, even when well-hidden by the attacker, still interact in the background with Explorer shell components. MRU silently records that interaction.

Another advanced detail to examine during MRU analysis is the entropy of directory access patterns. A normal user accesses files and folders in a meaningful logical order: project file → images folder → report → email attachment, for example. In attacker behavior, however, this rhythm breaks down; accesses become more fragmented, more random, and more target-focused. Sensing this rhythm reflects the forensic analyst's experience. For example, in one incident, the MRU list contained three folder names consecutively, and none appeared to share anything in common. Yet the star-shaped pattern in the access order (three touches arriving at short, long, and very short intervals) indicated that an attacker had triggered a tool that rapidly scanned specific folders. Sysmon didn't detect this because the process had been triggered through the legitimate `Explorer.exe`. MRU, however, recorded the pure behavior. This is where the registry's greatest strength becomes visible: Even the attacker's technique of hiding noise within legitimate processes can be caught in MRU's behavioral traces.

The most difficult part of MRU analysis is drawing the invisible line between natural human behavior and attacker behavior. Humans are indecisive; sometimes they open a file three times, then suspect something, and close it. But the attacker's rhythm is different: It's mechanical, targeted, and short-lived. An analyst's greatest strength is the ability to sense these rhythm disruptions in MRU content. Even a single MRU record, especially if the file path doesn't align with normal usage habits, may indicate the starting point of the attack, the loading phase, the discovery stage, or even preparatory activity prior to persistence.

Perhaps the most valuable aspect of MRU artifacts is that regardless of the attacker's intent, the registry leaves at least a small trace, such as a value, a path, or a timing inconsistency. Sometimes, it's a typographical error, an empty slot, or an odd proximity between two values. These aren't often visible in logs, filtered by Sysmon, or lost within noise in ETW. But the MRU structure records behavior in a pure form, above all of these systems.

ShellBags

ShellBags artifacts store detailed information about folder navigation within the Windows Explorer interface. Unlike MRU artifacts, which primarily reflect file access, ShellBags record how users interact with directory structures, even when no files are opened. This makes ShellBags particularly valuable in forensic scenarios where attackers attempt to remove file-based evidence. Even if a folder or file has been deleted, ShellBags may still retain traces of its existence, including folder paths, view settings, and access patterns.

From a behavioral perspective, ShellBags answer a critical question: Where did the user (or attacker) navigate within the system?

Key forensic characteristics include the following:

- Persistence of deleted folder traces
- Reconstruction of directory traversal behavior
- Detection of anomalous or nonhuman navigation patterns
- Identification of reconnaissance activity (browsing without execution)

Attackers rarely prioritize cleaning ShellBags due to their complexity and distributed structure across user hives. As a result, ShellBags often act as a residual memory layer, preserving activity that other artifacts fail to capture.

UserAssist

UserAssist artifacts track application execution from a user interaction perspective, recording how frequently programs are launched and how user activity evolves over time. Unlike MRU, which reflects access, UserAssist provides insight into execution frequency and behavioral rhythm. Entries are stored in an encoded format and include execution counters that increase as applications are used. This allows analysts to confirm execution and evaluate whether the pattern of usage aligns with natural human behavior.

From a forensic standpoint, UserAssist answers the following questions: How did execution occur, and does it reflect normal user interaction?

Key analytical value includes the following:

- Detection of abnormal execution frequency (automation, scripts)
- Identification of counter resets (possible anti-forensics)
- Differentiation between human-driven and automated behavior
- Correlation with MRU and ShellBags for execution validation

Here are some examples:

- A rapidly increasing counter in a short time window equals automation
- A reset counter indicates possible tampering
- Execution without MRU trace indicates programmatic execution

UserAssist transforms execution artifacts into behavioral signals, making it essential for distinguishing attacker activity from legitimate use.

Relationships Among Recently Executed Applications

The registry may not directly state which application a user executed and when; however, it preserves a substantial amount of indirect evidence related to application usage:

- **MUICache**
  Displays names of applications with which the user interacted.

- **RecentApps**
  Provides a consistent list of recently executed programs.
- **UserAssist**
  Provides the number of times each application was executed.

When these three artifacts are examined together, it becomes possible to reconstruct user behavior statistically, logically, and chronologically.

Have you ever noticed that the rhythm of UserAssist counter increases shows significant differences between a user and an attacker? Here are a couple of examples:

- A user may open Excel 15 times in a day, but an attacker can't trigger the same file at millisecond intervals.
- A user may use `cmd.exe`, but the `cmd.exe` instances opened by an attacker are usually short-lived.

These small differences is where the subtlety of forensic analysis truly resides.

Real-World Example

In one incident, I observed that the counters of several applications under UserAssist had been reset. At first glance, this appeared benign; perhaps the user profile had been refreshed? But ShellBags records were still retaining old folder navigation traces. The inconsistency was clear: The attacker had cleared the counters but forgotten ShellBags.

Profile Migration Traces

User profiles may be moved to different locations over time, for example, during a domain transition, a new machine deployment, or a change of username. During such processes, the registry doesn't completely erase traces belonging to the old profile.

As a result, small "shadow rooms" emerge where an attacker may hide or where forgotten data may persist. Indicators of profile migration within the registry commonly include the following:

- Unexpected SIDs under `HKU`
- Version differences among `NTUSER.DAT` files
- Run/MRU/MUICache records pointing to old profile paths
- Folder references under ShellBags that no longer exist
- GUIDs in UserAssist that don't belong to the current user

Real-World Example

In one case, I observed that a new user profile had been created, yet ShellBags records belonging to the old profile had been carried over. This wasn't an ordinary system migration; the attacker had created a new profile to obscure file traces but had failed to account for ShellBags' migration behavior. The registry doesn't forget its former owner. This may sound somewhat dramatic, but while an attacker can change their profile, they can't change the registry's memory.

Reading user behavior from the registry can reveal what a profile has experienced, how a user actually used their computer, and even whether the attacker was hurried or professional. MRU, ShellBags, and UserAssist are more than technical objects; they are the fingerprints of digital behavior.

When examining registry artifacts, don't focus on a single record. Each artifact is like half of a sentence; to understand the full sentence, you must listen to them all.

Remember!

Sometimes, the registry doesn't tell you about an attack directly, but it tells you about the rhythm of the hands that touched it: the haste, the attempt to conceal, the forgetfulness. Often, this is even more valuable than the attack itself.

Cross-Correlation Analysis

While each registry artifact provides valuable insight on its own, their true forensic strength emerges when they are interpreted together. Up to this point, MRU artifacts have been examined in detail as indicators of file access and user interaction. However, MRU alone can't fully represent user behavior, as it lacks context regarding how folders were navigated or whether execution actually occurred.

To address this limitation, two additional artifacts must be considered alongside MRU. ShellBags provide visibility into folder navigation patterns, revealing which directories were accessed, even if no files were opened or if traces have been removed from the file system. UserAssist, on the other hand, captures execution-related interaction by tracking how frequently applications are launched and how user-driven activity evolves over time.

Together, these three artifacts—MRU, ShellBags, and UserAssist—form a complementary triad. MRU answers what was accessed, ShellBags explains where the user navigated, and UserAssist reflects how interaction and execution occurred. When correlated, they allow the analyst to distinguish between genuine user behavior, automated activity, and attacker-driven manipulation with significantly higher confidence.

Table 9.7 shows how these artifacts should be interpreted in combination, which patterns indicate normal or adversarial behavior, and which signals provide the most reliable basis for forensic conclusions.

Analysis Component	MRU	ShellBags	UserAssist	Forensic Meaning (Cross-Signal Interpretation)
1. Verified file opening behavior.	File path appears in MRU.	The same folder is shown as browsed in ShellBags.	The counter value of the related application has increased.	Alignment of all three signals: Genuine user behavior; high-confidence execution evidence.
2. Deleted file/ cleaned system trace.	The path appears in MRU but the file no longer exists.	ShellBags may still contain the folder.	The execution record may be absent in UserAssist.	Attacker deleted the file; MRU left a residual trace: Indicator of an anti-forensic attempt.
3. Suspicious short-interval access.	The same file appears repeatedly at short intervals.	ShellBags show nonrhythmic, random folder transitions.	UserAssist counter increase doesn't match natural human rhythm.	High likelihood of payload testing, script-based access, or loader behavior.
4. Fake user activity (decoy activity).	A usage sequence appears normal in MRU.	Folders were never actually opened in ShellBags.	UserAssist counters are unnatural or reset.	Attacker may have populated MRU to disperse traces: Behavioral inconsistency is a red flag.
5. Profile migration/user change traces	Paths belonging to an old profile appear in MRU.	ShellBags contain gaps or references to nonexistent folders.	The UserAssist `GUID` points to a different user.	Possible migration of old profile data: Attacker-driven profile creation or masking attempt.
6. Suspicious file appearing only in MRU.	The file is visible in MRU.	There's no corresponding entry in ShellBags.	There are no execution traces in UserAssist.	File may have been opened programmatically (dropper, macro loader, injection chain).

Table 9.7: MRU + ShellBags + UserAssist: Cross-Correlation Analysis Table

Analysis Component	MRU	ShellBags	UserAssist	Forensic Meaning (Cross-Signal Interpretation)
7. Folder appearing only in ShellBags.	The folder is not present in MRU.	ShellBags show folder navigation.	Inconsistent counters are used in UserAssist.	Attacker may have browsed files without executing an application: Reconnaissance phase.
8. UserAssist counter increase anomaly.	Application paths are inconsistent with MRU.	The folder structure may or may not align with ShellBags.	The counter increased very rapidly or was reset.	Attacker may have used automation, brute-force GUI interaction, or test-driven executable triggering.
9. Completely inconsistent trio.	MRU appears normal.	ShellBags are abnormal.	The UserAssist rhythm is disrupted.	Human vs. non-human behavior model: Access sequence inconsistent with normal user behavior; high likelihood of attacker activity.
10. Silence across all three artifacts (gap).	There's an empty list and minimal records.	ShellBags show almost no activity.	UserAssist counters are near zero or wiped.	Attacker attempted profile cleaning, MRU wipe, or UserAssist reset: Silence itself is evidence.

Table 9.7: MRU + ShellBags + UserAssist: Cross-Correlation Analysis Table (Cont.)

9.3.3 Making Persistence Mechanisms Visible

An attacker's greatest objective isn't to gain entry; it's to remain inside. For this reason, persistence is both the most frequently used and the most deliberately concealed phase of the attack chain. The registry, meanwhile, is the quietest yet most stubborn witness in this game. Because no matter how much the attacker obscures their traces, somewhere deep within Windows, there will remain a small key, a forgotten value, or a mistimed write operation. Making persistence mechanisms visible isn't only a matter of an analyst's technical skill, but also of intuitive sensitivity: the courage to ask, "Why is this key here?"

Persistence is achieved in several ways:

- **Service-based persistence**

  Run, RunOnce, and service-based persistence are classic yet still among the most effective methods in the persistence landscape. They appear simple and are often among the first avenues an attacker will try. But beneath this simplicity lies a major advantage: The registry doesn't like to keep such attempts hidden. A value added to a Run key will invoke a malicious binary at every reboot; service-based persistence, meanwhile, turns into a background service that controls the system.

> **Real-World Example**
>
> In one incident, I noticed a minor spelling mistake in the description field of an ordinary Windows service. No one paid attention to it. But that mistake was the first signal revealing that the attacker had embedded a malicious binary inside a legitimate service.

- **Winlogon-based persistence**

  Winlogon-based persistence is a far more insidious game. The attacker injects themselves into the heart of the Windows logon process by modifying critical registry values associated with user session initialization. Specifically, keys such as userinit, which defines the programs executed immediately after a successful logon; Shell, which determines the user's desktop environment (typically explorer.exe); and legacy mechanisms, such as Graphical Identification and Authentication (GINA) DLL, which historically controlled the interactive logon process, can all be abused to trigger attacker-controlled code at logon time. Even a small manipulation in these values can cause malicious code to execute the moment a user signs in.

  This technique can be hidden in a single line within an otherwise ordinary registry view, but its timing is never natural. In a normal system, the Shell value is explorer.exe; if an attacker replaces it with a chained execution command, the registry's behavioral rhythm breaks. An analyst senses this disruption because Windows doesn't treat such deviations as normal system behavior.

- **Persistence through AppInit_DLLs**

  AppInit_DLLs is the gateway to another world: DLLs that every graphical application loads at startup. When an attacker adds a single path here, the malicious DLL is carried into memory alongside every application across the system. Moreover, when this type of persistence remains silent on the file system, it becomes difficult to detect. But the registry doesn't lie: The AppInit_DLLs value is empty—it must be. In a system where that value is populated, either a security product is running or an attacker is. One of these two possibilities must be true.

- **Persistence through IFEO hijacking**

  IFEO hijacking is one of the most creative forms of persistence. A debugger intercepts the execution of an ordinary application, yet the debugger actually triggers the attacker's payload. This method is both persistence and an execution flow manipulation behavior. In the registry, the presence of a `Debugger` value under an application's IFEO is a red flag by itself because, in legitimate systems, this mechanism is almost never used.

Real-World Example

In one case, the attacker was triggering malicious code by adding a `Debugger` to `notepad.exe`. When Notepad opened, nothing seemed suspicious. The button sound and window behavior looked natural. But in the background, the IFEO debugger was triggering a second process. Sysmon had failed to capture this relationship, but the registry captured it.

- **Persistence through TaskCache**
  Registry-based scheduled task configurations represent a more advanced dimension of persistence. Cleaning Task Scheduler XML files is easy, but TaskCache under the registry still whispers the existence of tasks. If the `GUID` of an old or deleted scheduled task remains in the registry, the attacker performed an incomplete cleanup.

Real-World Example

In one forensic examination, an orphaned `GUID` under `TaskCache\Tree` provided the strongest clue that explained the entire timeline of the attack. The task no longer existed in the file system, but the registry hadn't forgotten it. Windows' memory is sometimes far stronger than attackers take seriously.

When making persistence mechanisms visible, an analyst's greatest ally is intuition. The registry isn't always orderly; sometimes it's complex with old and new values interwoven. Yet there's a very subtle difference between an attacker's touch and the system's natural behavior. Attackers are often hurried: Sometimes they add a value but forget to update the timestamp of the adjacent key. Sometimes they try to hide a DLL path, but a minor spelling mistake in a service description undermines the entire game. Sometimes they delete a scheduled task but forget to clean TaskCache. All of these small details are an analyst's greatest advantage.

In persistence analysis, the registry isn't just a data source; it's like a witness. It doesn't speak emotionally—it speaks consistently. No matter how professional an attacker may be, they can't escape the microtraces in the registry as long as there's patience to read them, intuition to follow the rhythm, and attention to interpret the silence.

Persistence mechanisms are always hidden within behaviors that appear legitimate. The registry also preserves those behaviors, but it never forgets. The attacker's aim is to remain inside; the analyst's aim is to make that persistence visible. More often than not, the answer is hidden in a small value deep within the registry.

To consolidate these persistence mechanisms into a structured analytical view, Table 9.8 provides a comparative overview of the most common registry-based persistence techniques, their operational behavior, and the forensic signals they produce. As shown, different persistence techniques (such as `Run` keys, service-based execution, Winlogon manipulation, and IFEO hijacking) operate at distinct stages of system initialization and user interaction. Despite these differences, they share a common goal: maintaining execution continuity while minimizing visibility. The table emphasizes that each technique introduces subtle inconsistencies, whether in execution paths, parent-child relationships, timestamps, or configuration values. For the analyst, the key identifying the persistence point and, more importantly, validating its legitimacy through cross-artifact correlation, behavioral consistency, and alignment with expected system baselines.

Persistence Type	Registry Location	Behavior/What It Does	Attackers' Purpose	Forensic Detection Clues for the Analyst
`Run/RunOnce` persistence	`HKCU/HKLM\Software\Microsoft\Windows\CurrentVersion\Run`	Automatically launches specific executables at user logon	Fast, low-profile persistence; blending into user processes	Parent key timestamp inconsistencies, unexpected paths, timestomping, suspicious binaries
Service-based persistence	`HKLM\SYSTEM\CurrentControlSet\Services`	Establishes persistence by creating or abusing a service	Creating a silent and continuously running backdoor in the system	Event ID 7045 + service description analysis + executable path inconsistencies; binary hash history
Winlogon persistence (`Shell/Userinit`)	`HKLM\Software\Microsoft\Windows NT\CurrentVersion\Winlogon`	Adds malicious execution to the logon phase	Automatic execution upon user login	Red flag if `Shell` value not equal to `explorer.exe`; chained execution commands
`AppInit_DLLs` persistence	`HKLM\Software\Microsoft\Windows NT\CurrentVersion\Windows\AppInit_DLLs`	Loads DLLs into all GUI applications at startup	System-wide propagation via global DLL injection	Value must be empty; if populated, indicates either a security product or an attacker; DLL signature must be analyzed

Table 9.8: Comparative Analysis of Registry-Based Persistence Mechanisms

Persistence Type	Registry Location	Behavior/What It Does	Attackers' Purpose	Forensic Detection Clues for the Analyst
IFEO (debugger hijacking)	`HKLM\Software\Microsoft\Windows NT\CurrentVersion\Image File Execution Options<exe>`	Triggers a different program when an executable is launched	Executing malicious code under the guise of a legitimate application	Debugger value unnatural in almost all systems; parent-child process chain inconsistencies
Scheduled task persistence (registry artifacts)	`HKLM\Software\Microsoft\Windows NT\CurrentVersion\Schedule\TaskCache\Tree`	Retains traces of task definitions even after deletion from disk	Concealing persistence and triggering tasks even after artifact cleanup	Tasks present in TaskCache but missing on disk indicates incomplete attacker cleanup
Shell extensions persistence	`HKLM\Software\Classes*\ShellEx`	Binds malicious code to shell extensions triggered by Explorer	Creating a hidden execution surface	Whether Extension `CLSID` values match legitimate Component Object Model (COM) objects
`Active Setup` persistence	`HKLM\Software\Microsoft\Active Setup\Installed Components`	Executes scripts when a user profile is created	Automatic malicious installation on new user profiles	Consistency of GUIDs under `Installed Components` with the system image
WMI event consumer persistence	`HKLM\Software\Microsoft\Wbem\CIMOM`	Executes malicious code when a WMI event is triggered	Diskless persistence, stealth execution	Anomalies when WMI repository and registry don't align; Consumer + Filter correlations

Table 9.8: Comparative Analysis of Registry-Based Persistence Mechanisms (Cont.)

9.3.4 Example: Behavioral Analysis of Registry-Based Startup Persistence Using Python

In Windows environments, one of the persistence points most frequently preferred by attackers is the `HKCU\Software\Microsoft\Windows\CurrentVersion\Run` key. This is because this area triggers applications that run at every user logon, and persistence mechanisms executed under the user profile are often less conspicuous. But is every `Run` entry truly suspicious? This question must be approached from both a technical and an intuitive perspective.

The example shown in Figure 9.4 presents the forensic evaluation of two entries obtained during live system analysis. The analysis was performed by querying the `HKCU\Software\Microsoft\Windows\CurrentVersion\Run Registry` key using a Python-based inspection script (`persistence_hunter.py`) in live mode. The script enumerates `Run` key entries; extracts associated metadata such as executable paths, value names, and last write timestamps; and applies a lightweight behavioral scoring logic based on factors including execution path location, naming patterns, and known software baselines. Each entry is then enriched with contextual interpretation, such as whether it resides in a standard installation directory (e.g., `Program Files`), whether the naming aligns with legitimate software, and whether its behavior is consistent with expected startup activity—allowing the analyst to distinguish between benign auto-start entries and potentially malicious persistence mechanisms.

```
C:\Users\admin\PycharmProjects\Books>python persistence_hunter.py --mode live
{
    "type": "Run",
    "hive": "HKCU",
    "key": "Software\\Microsoft\\Windows\\CurrentVersion\\Run",
    "value_name": "Icecream_Screen_Recorder_New_Auto_Start",
    "value_data": "C:\\Program Files (x86)\\Icecream Screen Recorder 7\\recorder.exe -tray",
    "last_write": "2025-11-27T17:06:08.056026",
    "suspicion_score": 1,
    "analyst_note": "Icecram_Screen_Recorder_New_Auto_Start' is launched via Run key, pointin
    a Program Files directory. Persistence anomaly detected."
},
{
    "type": "Run",
    "hive": "HKCU",
    "key": "Software\\Microsoft\\Windows\\CurrentVersion\\Run",
    "value_name": "EPSDNMON",
    "value_data": "C:\\Program Files (x86)\\Epson Software\\Download Navigator\\EPSDNMON.EXE",
    "last_write": "2025-11-27T17:06:08.056026",
    "suspicion_score": 1,
    "analyst_note": "EPSDNMON" is launched via Run key, pointing to a Program Files directory."
},
```

Figure 9.4: Behavioral Analysis Output of Registry-Based Startup Persistence via HKCU Run Keys

Let's take a closer look at both entries:

- **`Icecream_Screen_Recorder_New_Auto_Start`**
  This entry represents a completely legitimate record created by the Icecream Screen Recorder software installed on the user's system to launch itself in the system tray at startup. It contains the following information:
  - Registry location: `HKCU\Software\Microsoft\Windows\CurrentVersion\Run`
  - Executed application: `C:\Program Files (x86)\Icecream Screen Recorder 7\recorder.exe -tray`
  - Suspicion score: 1

Given that it resides under the `Program Files` directory and belongs to a signed and widely used application, an analyst wouldn't expect any anomaly in this entry. Nevertheless, the analyst still asks the following question: "Is this application's execution at startup consistent with user behavior, or could an attacker have been attempting to collect information via screen recording?"

In this case, considering the location, digital signature, and intended usage, the entry demonstrates consistency, and no forensic risk is identified.

- **`EPSDNMON` (Epson Download Navigator)**

  This entry is one of the update and notification components within the Epson printer software ecosystem. It contains the following information:

  - Registry location: `HKCU\Software\Microsoft\Windows\CurrentVersion\Run`
  - Executed application: `"C:\Program Files (x86)\Epson Software\Download Navigator\EPSDNMON.EXE"`
  - Suspicion score: `1`

  Its execution from a standard installation directory under `Program Files`, its inclusion in the manufacturer's own updater mechanism, and the presence of a logical auto-start usage scenario place this entry firmly in the benign category.

At this point, an analyst should ask, "Does this application genuinely belong to the manufacturer, or has a fake updater with a similar name been placed on the system?" Digital signature verification of the binary provides critical guidance here. In this example, both the location and behavior are entirely legitimate, and no abuse of persistence is observed.

The shared conclusion drawn from both cases is as follows:

- Not every `Run` entry is suspicious.
- However, every `Run` entry deserves a question.
- These questions form the analyst's intuitive validation framework:
  - Is the location legitimate?
  - Is the application's execution at startup consistent with user behavior?
  - Is the binary's digital signature trustworthy?
  - Does this entry coincide with an unusual period in the timeline?

This approach goes far beyond simple indicator of compromise (IOC) detection; it's the ability to sense the misalignment between the natural operational rhythm of Windows and an attacker's intervention. In these two specific entries, no indicators of malicious persistence were identified, and a startup profile consistent with the system's natural operation was observed from a behavioral perspective.

9.3.5 The Forensic Value of ShellBags, ShimCache, and AmCache

Every attack leaves a trace, sometimes very apparent and sometimes so subtle as to be nearly imperceptible. However, three specific components of the registry—Shellbags,

ShimCache, and AmCache—function for an analyst as three powerful windows through which traces, time, behavior, and intent can be forensically reconstructed. These artifacts may not tell a story on their own, but when brought together, they quietly but clearly delineate an attacker's presence on a system. In fact, in many cases, when logs fall silent, Sysmon misses activity, and ETW remains obscured, these three artifacts illuminate the true flow of events.

ShellBags record the traces left by a user or an attacker while navigating files and directories. For this reason, ShellBags represent one of the strongest pieces of forensic evidence of navigation history: If a user opened a folder—even if that folder was later completely deleted—ShellBags declare, through their remaining shadow records, "We saw it."

Real-World Example

I first understood the power of ShellBags during a ransomware incident. The attacker had encrypted files and then cleaned up the directories. The file system was entirely silent. Yet ShellBags still retained the names of the deleted folders, along with their pre-attack access timestamps. This small trace enabled the reconstruction of the entire attack chain.

Another powerful aspect of ShellBags' forensic value lies in their ability to reveal the rhythm of behavior. A user typically navigates folders in a logical order: top to bottom, left to right, following a coherent pattern. An attacker, by contrast, is often fast, erratic, and goal oriented. A folder is opened and immediately closed, followed by a sudden jump to another area of the system. These abrupt directional shifts are reflected in the rhythm captured by ShellBags. Sometimes, a sequence of actions spanning just three to five seconds can reveal the full character of an attacker's reconnaissance phase.

ShimCache, on the other hand, tells a different story by maintaining a list of applications that were possibly executed. The most critical word here is *possibly* because ShimCache isn't definitive execution evidence. Nevertheless, it contains the names, paths, and certain timing information of files that were executed or evaluated for execution. The forensic power of ShimCache becomes particularly apparent during disk-cleaning attempts. An attacker may have deleted a malicious EXE, leaving the file system seemingly pristine; yet the ShimCache entry may still list that file. At that precise moment, a small but valuable window opens for the analyst: "The system was aware of this file."

Real-World Example

I recall an incident in which a Windows server appeared completely clean. Neither event logs nor Sysmon revealed anything suspicious, but a strange EXE entry in ShimCache (located in an unusual directory) exposed the entire structure of the attack. The binary had long been deleted. Timestamps had been manipulated. But ShimCache preserved that small, forgotten line. Attackers forget, but the registry doesn't.

AmCache is considered the gold standard of forensic analysis because along with indicating that a program was opened, it also records hash information, the full file path, the initial execution time, and even publisher information in some cases. As such, AmCache serves as a source that verifies the executed binary with pinpoint precision. Where ShimCache says "This probably ran," AmCache can say "Yes, it ran, and it had this hash." In forensic analysis, the mutual corroboration of these two artifacts establishes a chained model of evidentiary confidence.

AmCache plays a particularly critical role in the staging and execution chains of APT actors. Even after an attacker loads and deletes a payload, the AmCache record often remains. The hash itself can reveal a great deal about the attacker's operational discipline: Was it randomized? Did the same binary appear with the same hash across different systems? Such behavior can sometimes expose the identity of an entire campaign.

When these three artifacts are used together, their forensic power increases exponentially. ShellBags tell you where navigation occurred, ShimCache indicates what kinds of files were interacted with, and AmCache confirms when and which binary was executed. Sometimes, solving an attack is like merging three distinct perspectives into a single narrative.

ShellBags show folder movement → ShimCache points to a suspicious EXE within that directory → AmCache confirms the execution time. This triad of signals reveals an attacker's footprints on the system not only technically, but behaviorally as well.

Analysts should never examine these three artifacts in isolation. Each is like a different word in the same sentence—meaning only emerges when they are read together. The silence of ShellBags sometimes gains meaning through the shout of ShimCache, and the definitive record left by AmCache sometimes completes the rhythm observed in ShellBags. Other times, all three point in the same direction, and the attacker's presence on the system emerges with full weight.

Remember!

An attacker can delete a binary, manipulate timestamps, and wipe logs, but they can't completely erase the memory of ShellBags, ShimCache, and AmCache. Somewhere, a fragment always remains, and that fragment is often enough to solve the entire attack.

No matter how strong the forensic value provided by these three artifacts may be, the analyst must pay close attention to certain sensitive details to interpret them correctly. A misread ShellBags record, an insufficiently evaluated ShimCache entry, or an AmCache record taken out of context can cause the entire investigation to be diverted in the wrong direction. Therefore, when approaching each artifact, it's essential to consider both the system's natural behavior and the attacker's potential manipulation techniques.

Before examining specific examples, it's important to highlight several common pitfalls and nuanced behaviors that analysts must consider when interpreting these artifacts:

- **There's a time travel effect with ShellBags timestamps.**
  ShellBags timestamps don't directly correspond to the actual timestamps of files or folders. This is because ShellBags record only the moment when a user viewed a folder, not the folder's last modification time. Here are a couple of points to consider:
  - Even if a folder is deleted, the ShellBags record remains; this is normal behavior.
  - ShellBags timestamps may not align with New Technology File System (NTFS) timestamps.

  The registry timestamp reflects the time the folder was opened by Explorer, not the folder's actual modification time. A common analyst error is interpreting the mismatch between a deleted folder and a ShellBags timestamp as attacker manipulation. This isn't always correct.

  From an analytical perspective, ShellBags should be interpreted as indicators of user behavior, whereas file system timestamps represent object-level temporal data; correlating these two perspectives provides a more accurate and contextual understanding of system activity.
- **ShimCache isn't a definitive execution record—it can be misleading.**
  One of the most common mistakes in the DFIR world is treating ShimCache as a definitive program execution record. In reality, the following is true:
  - ShimCache indicates that a file was evaluated by Windows, not that it was executed.
  - Some entries may be created without execution during metadata access.
  - Timestamps may be inconsistent depending on the Windows version.

  The key question the analyst must ask is "Was this EXE actually executed, or did the system merely evaluate the file?" The answer is usually obtained through correlation with AmCache.
- **AmCache hash information is powerful, but not recorded for every drive.**
  Although AmCache is a highly valuable artifact, it has certain limitations:
  - Programs executed from external USB drives may not always be written to AmCache.
  - Executables launched directly from compressed archives (ZIP, 7z, etc.) may not generate records.
  - If the AmCache file is corrupted, the system—not the attacker—may be responsible.

Warning

The absence of an AmCache record isn't proof that execution didn't occur. AmCache can only provide strong evidence that execution did occur.

- **The temporal logic of the three artifacts is different, so they can't be directly compared.**
  One of the most frequent analyst errors is attempting to align ShellBags, ShimCache, and AmCache timestamps on a single timeline. However, their temporal logic differs:

  - ShellBags: Viewing time
  - ShimCache: Evaluation time (file load/execution check)
  - AmCache: First execution time

  Because these temporal models differ, direct equality shouldn't be expected. Instead, logical coherence should be sought:

  - ShellBags shows that a folder was opened.
  - ShimCache shows that an EXE in the same folder was evaluated.
  - AmCache shows that the EXE was executed.

  If these conditions are met, the behavioral chain is complete. Exact timestamp alignment isn't required and is normal.

- **Anti-forensic manipulations are often performed through silence.**
  The following conditions may indicate attacker manipulation:

  - MRU completely empty while ShellBags is fully populated indicates a profile deletion attempt.
  - AmCache entirely cleared indicates a registry restore/rollback of artifacts.
  - ShimCache entry present but binary missing indicates a file wipe + metadata spoofing.
  - Corrupted ShellBags structure means there was manual registry hive manipulation.

  In particular, corruption of ShellBags is a strong indicator of registry editing by an attacker. Normal user behavior doesn't corrupt ShellBags.

- **The concept of ghost folders within ShellBags must always be examined.**
  A ghost folder is a folder that no longer exists on the file system but still appears in ShellBags. It has the following forensic significance:

  - The attacker deleted the folder.
  - A file wipe was performed.
  - A user profile was migrated.
  - Time manipulation occurred.
  - Exfiltration directories were cleaned.

  These traces are critical artifacts that only a small number of analysts examine. A Ghost Folder combined with the same directory appearing in ShimCache may raise suspicion of exfiltration preparation.

- **ShimCache alone is weak without cross-validation with UserAssist.**
  Many analysts make decisions based solely on ShimCache; however, ShimCache must be interpreted together with UserAssist. In UserAssist counter rhythms, sudden spikes, resets, and millisecond-interval increments indicate automation or an attacker command chain.

  If ShimCache shows a record, UserAssist either confirms or refutes the execution rhythm of that application.

- **Hive migration (profile migration) can be misinterpreted.**
  The presence of old profile SIDs under `HKU` doesn't always mean that an attacker created a new profile. It may result from domain exit, profile migration, or username changes.
  However, the following scenario is a critical red flag:
  - ShellBags retains old profile folder references.
  - UserAssist counters don't align with the new profile `GUID`.
  - AmCache contains binary records belonging to two different profiles.

  This threefold combination indicates profile manipulation or OPSEC-driven profile copying by an attacker.
- **Windows version–specific artifact behaviors must be known.**
  The behavior of ShimCache and AmCache varies by Windows version:
  - Windows 7 ShimCache has more limited timestamp logic.
  - Windows 10+ AmCache includes `Publisher` information and `Program ID`.
  - Windows 11 ShellBags structures contain more complex nested hierarchies.

  Failure to understand these version-specific differences can lead to incorrect conclusions:
  - ShellBags alone is limited.
  - ShimCache alone is potentially misleading.
  - AmCache alone is incomplete context.

  But when all three are combined, a single analysis can answer the following:
  - Where navigation occurred
  - Which file was seen
  - Which program was executed
  - Which file was deleted
  - Which timestamps were manipulated
  - Whether the activity was triggered by a user or an attacker

This triad functions as the "holy triangle" of registry forensics. When interpreted correctly, it allows the analyst to reconstruct an attacker's steps on the system not only technically but behaviorally as well.

9.4 Memory Acquisition and Memory-Based Forensic Analysis

RAM reflects the system's closest-to-the-moment and least-filtered operational state. A kind of digital stream of consciousness. When a good analyst learns to read this stream, the attacker's true face beneath the mask often begins to faintly emerge. So how much of this volatile universe called memory are we really able to capture?

A system's RAM, when examined up close, carries traces of adversary behavior: loaded modules, invoked APIs, suspicious thread chains, the seemingly random entropy of shell-code, and sometimes even a command sequence that clearly looks hastily written. The analyst must find the right path among all of these clues. The active execution phase of an attacker's behavior always begins in RAM, and it most often leaves its most forensically meaningful traces in RAM as well.

When we take an instantaneous snapshot of memory, however, are we truly capturing a reality that we "nail to time," or are we moving through a scene manipulated by the attacker? The following sections seek to address this question by exploring the importance of memory acquisition, the various collection methods, and how to detect anti-forensic techniques. We'll also examine some important memory forensics frameworks to help examine memory.

9.4.1 Why Is Memory Acquisition Critical?

Memory is critical because attackers are becoming increasingly more volatile. Fileless operations that leave no disk trace, PowerShell injections that appear and disappear, and fragments of code scattered within WMI are things we can only catch in RAM. If an attacker has built themselves a home in process memory—and most do—scanning the disk is of no use. RAM is like the heartbeat of forensic analysis; even the slightest arrhythmia carries meaning.

Sometimes, during an analysis, you open the RAM image and the module list looks perfectly clean. Everything seems by the book. But you dig a little deeper, scan the lower layers, and you find a 12-byte shellcode embedded in a forgotten page frame. That's the moment you feel the attacker is there. It's as if they were in a hurry; the code's entropy is high, but not flawless. There's a break in the rhythm.

Inevitable questions arise during memory acquisition:

- Does this memory image truly have integrity?
- Did a process kill itself during acquisition?
- Did a hypervisor-based rootkit distort visibility?
- Did an implant operating via direct memory access (DMA) manage to keep itself outside the snapshot?

Every question gives rise to a suspicion; every suspicion demands examining more layers. Capturing memory isn't enough—you must understand memory because RAM often carries the shadow of the attacker's real identity, briefly, in fragments, as pieces of behavior that contradict one another.

Memory analysis isn't linear; it advances through rhythmic fluctuations. First, the coarse structure is extracted: the process list, modules, open file descriptors. Then, the rhythm breaks; the analyst descends into the lower layers: handle tables, the Virtual Address Descriptor (VAD) tree, thread context blocks, and traces of API hooking. Somewhere, an unexpected turn occurs—an anonymous memory region attached to the middle of a legitimate DLL in the VAD tree. That small inconsistency is the attacker's true presence. Most of the time, while looking at RAM, an analyst finds themselves on the edge of a sentence like this: "This memory region should not have been here, but why is it here?" At that moment, the analysis ceases to be an abstract problem and turns into an intuitive struggle.

Partial Entropy

You observe high-entropy traces in RAM, that is, small islands that are compressed, encrypted, and perhaps XORed. These are often the heart of the implant. However, entropy isn't always high across the entire region; the attacker's code sometimes leaves flaws behind: a small low-entropy sequence, an inadvertently left function signature, or perhaps a half-finished mutex name. These minor drops are small advantages that fall into an analyst's hands. It's as if the attacker, while taking a breath, momentarily broke the rhythm through a brief lapse of attention.

The greatest misconception during memory acquisition is the thought, "I have taken an image; I am safe now." Yet the manipulations that occur in memory-injective techniques, process hollowing, unlinked modules, and hidden threads are all embedded within the snapshot. The question is, have you developed the perspective required to unravel this embedded structure?

Sometimes you look at a memory image and feel that something is wrong, but you can't say exactly what. That feeling is the moment when the intuitive gateway of a true forensic analyst begins to open.

9.4.2 Memory Acquisition Methods

Obtaining a memory image sounds like a single technical operation, but it's not. In reality, the way memory is captured varies according to the system's physical, logical, and even behavioral state at that moment. Each method leaves a direct imprint on the texture of the evidence that reaches the analyst.

In a cold memory acquisition, integrity stands like a wall; in hypervisor-assisted acquisitions, invisible layers reveal themselves; and in live acquisition, the pulse of RAM is captured at the moment it beats the fastest. But which one is closer to the truth? Which one captures the breath the attacker is trying to hide?

The three methods discussed in the following sections are among the most powerful tools of modern digital forensics, and at the same time, its greatest traps.

Cold Memory Acquisition

Cold memory acquisition is a RAM image obtained when the system has been completely powered down. It sounds paradoxical: How can memory be acquired from a system with no power? The real secret lies here: Dynamic RAM (DRAM) retains its contents for seconds after power loss—and under proper temperature conditions, even for minutes. Cold acquisition techniques (cold boot attacks) operate during this natural period of data inertia. This method is valuable for the following reasons:

- Implants running in memory can't kill themselves.
- Anti-forensic mechanisms can't be triggered.
- Active rootkit mechanisms and runtime hooks can't continue to function once power is cut.
- Hypervisor-based deceptions shut down.
- User-mode manipulations freeze.

No matter how sophisticated the attacker is, when power is cut, everything is laid bare. However, the lesser-known detail is that the decay time of DRAM cells isn't uniform across all regions. Data retention time in DRAM cells varies depending on temperature, hardware characteristics, and cell structure, and some memory regions may degrade later than others.

What does this mean? Consider the following:

- Implant code, due to its intensive activity, may remain in RAM longer.
- Unused system modules may decay before the attacker's artifacts.

Some analysts refer to this behavior as an entropic memory shadow. As memory decays, the attacker's traces are erased later than those of normal processes. This is the most ironic advantage of cold memory: Decay brings the attacker to the forefront.

Real-World Example

In an experiment conducted in a university laboratory, a process performing Advanced Encryption Standard (AES) encryption was halted, the device was powered off, and cold acquisition was performed within four seconds. The first stable blocks observed in the decay map were the AES round key memory regions. In other words, what the attacker wanted to hide survived longer due to the laws of physics.

Hypervisor-Assisted Memory Acquisition

Hypervisor-based memory acquisition refers to the process of capturing a VM's memory from the hypervisor layer, rather than from within the guest OS itself, typically by creating a snapshot or directly accessing the VM's memory space (e.g., via VMware, Hyper-V, or Kernel-based Virtual Machine [KVM]). Because this method operates outside the guest OS, it reduces the risk of interference, tampering, or visibility limitations imposed by user-mode or kernel-mode malware, and is therefore generally considered one of the most reliable approaches for obtaining a consistent memory image. Yet behind the scenes, an incredibly complex dance takes place:

- The guest OS perceives its own reality.
- The hypervisor holds a broader reality.
- The attacker most often hides within the guest OS.
- But some advanced implants can reach the hypervisor level and establish cross-layer camouflage.

Note

Many modern hypervisors perform memory snapshot operations through hardware-assisted memory virtualization layers such as extended page tables (EPT)/nested page tables (NPT).

This leads to three consequences:

- **Memory remapping traces become visible.**
  Microdifferences left by user-mode hollowing techniques in the hypervisor's lower layers appear as geographical anomalies within the snapshot.
- **Timestamps are inconsistent even if all DMA activities are masked.**
  DMA-based implants may appear passive at the moment of the snapshot; however, the mapping anomalies they leave in memory maps can be revealed through analysis.
- **Rootkits can't hide from the hypervisor.**
  This is true because the hypervisor is a layer above the kernel.

The hypervisor isn't as neutral as we assume. The following mechanisms affect the memory image in a way similar to how biological tissue is distorted under a microscope:

- **Memory ballooning**
  A dynamic memory management technique in which the hypervisor reclaims unused memory from the guest OS by inflating a "balloon" driver inside the VM, potentially causing portions of memory to be paged out or rearranged, which can lead to incomplete or noncontiguous memory acquisition.
- **Transparent page sharing**
  A memory optimization mechanism that identifies identical memory pages across VMs and consolidates them into a single shared page, making it difficult to determine the original ownership and context of certain memory regions during forensic analysis.
- **Snapshot compression**
  The process of compressing memory pages when creating snapshots to reduce storage overhead, which may alter the raw structure of memory data and require decompression, potentially introducing artifacts or obscuring fine-grained forensic details.
- **Host kernel caching**
  A mechanism where the host OS temporarily caches memory pages or disk-backed data for performance optimization, which can result in discrepancies between the guest's perceived memory state and the actual data captured during acquisition.

For this reason, RAM images acquired from a hypervisor aren't as "sterile" as commonly believed. Professional analysts refer to these effects as *HV distortion noise*.

Real-World Example

During an incident response engagement at a financial company, a RAM snapshot was taken from a server running on VMware ESXi. During Volatility analysis (see Section 9.4.4), the following was observed:

- A memory region appeared twice in the VAD tree.
- Only one instance was accessible from the guest OS.

It was determined that the attacker had performed process hollowing. The hypervisor's replication mechanism (page deduplication) had unintentionally exposed this manipulation.

Live Memory Acquisition

Live acquisition, that is, capturing RAM while the system is running, is both the most dangerous and the most necessary method in digital forensics for the following reasons:

- A significant portion of volatile artifacts can only be captured while the system is operational.
- All fileless attacks become visible at this stage.
- If the attacker is active, their behavior may change before the snapshot.
- The memory image carries the system's living tissue.

This is where the difficulty begins. During live acquisition, memory defends itself. Yes, you read that correctly. Memory, physically, reacts to the snapshot process in the following ways:

- New pages are loaded.
- Page faults occur.
- Processes rearrange their own memory regions.
- Memory traffic increases.

As a result, live acquisition can fragment attacker artifacts, create new traces, or trigger existing ones. Therefore, some anomalies in memory images belong not to the attacker, but to your own intervention.

A little-known risk in live acquisition is that the memory acquisition tool itself (e.g., Windows Physical Memory [WinPMEM], Magnet RAM Capture, Belkasoft RAM Capturer) loads its own drivers and ring-0 handlers. During this loading process, the following occurs:

- A block is allocated from kernel memory.
- Certain `EPROCESS` structures are re-referenced.
- Handle tables may briefly form a double image.

This phenomenon is rarely mentioned in the literature, yet in real-world cases it's one of the most frequently encountered types of distortion that complicates analysis.

Example

You're performing live acquisition from a machine with your incident response team. Everything seems normal, when in Volatility, you observe thread contexts associated with two different PIDs, yet sharing the same address space.

You might assume this to be a hyperjacking attempt by the attacker, whereas in reality, the following is true: The kernel driver of the RAM acquisition tool ITSELF Created a short-lived refcount jitter during thread enumeration.

This is where the analyst's intuition comes into play. Not every anomaly is malicious, but every anomaly must be explained.

Method Comparison

Now that we have an understanding of each memory acquisition method, let's compare the methods in Table 9.9.

Method	Attacker's Opportunity for Manipulation	Data Integrity	Visibility Level	Risk
Cold memory	Very low	Highest	Physical level	Dependent on hardware conditions
Hypervisor-assisted	Low	Medium–high	Cross-layer	Hypervisor distortions
Live acquisition	Medium–high	Variable	Widest	Artifact corruption

Table 9.9: Comparison of Memory Acquisition Methods

Cold acquisition is the closest to reality, live acquisition is the richest in information, and hypervisor-assisted acquisition is the most balanced method.

9.4.3 Detecting Anti-Forensic Techniques in Memory

Memory is the mirror attackers fear the most. Disks can be cleaned, log chains can be broken, and even the registry can be silenced, but RAM is where attackers are most vulnerable. For this reason, advanced threat actors develop various anti-forensic techniques to conceal their traces in memory. Interestingly, no matter how skillfully these techniques are applied, they often leave behind an irregular rhythm, a logical shadow, or an entropic trace. The analyst's task is to notice these silent distortions.

How can an anti-forensic behavior be recognized in memory? When we see nothing in RAM, is a great deal actually happening? Or conversely, when we see a great deal, are we in fact seeing nothing at all? This section stands at the center of these questions.

Practical Access and Validation of Memory Artifacts

While the concepts just described focus on behavioral interpretation, the analyst must also understand how to practically access and validate these structures within a memory image. Without direct interaction with memory analysis frameworks, these indicators remain purely theoretical. In practice, memory artifacts such as Process Environment Block (PEB) structures, `EPROCESS` objects, and VAD trees aren't accessed manually, but through specialized forensic frameworks that reconstruct OS abstractions from raw memory.

The first step in this process is establishing a reliable view of active processes. Tools such as Volatility or Rekall achieve this by enumerating processes through multiple perspectives, including active process lists (e.g., `PsActiveProcessHead`) and pool scanning techniques designed to detect hidden or unlinked processes. The analyst must compare these different views, identifying discrepancies between processes officially linked within kernel structures and those that exist only in raw memory. At this stage, the fundamental question becomes the following: Does every process observed in memory exist within the kernel's official process list?

Following this, attention shifts to the user-mode perspective through analysis of the *Process Environment Block (PEB)*, mentioned previously. The PEB provides insight into loaded modules, process parameters, and execution context as perceived by the process itself. By enumerating modules through PEB structures and comparing them with VAD-backed memory regions and kernel-level module tracking, the analyst can identify inconsistencies. Modules that exist in memory but are absent from PEB listings, or discrepancies between reported and actual execution flow, may indicate deliberate manipulation. This leads to a critical question: Is the process reporting its own structure truthfully?

The next phase involves reconstructing the previously mentioned *Virtual Address Descriptor (VAD) tree*, which defines how memory is allocated within a process. Through forensic tools, analysts traverse VAD nodes to identify memory regions, their base addresses, sizes, protection flags, and whether they are file-backed or private allocations. Particular attention is given to executable regions that aren't backed by files, missing VAD entries corresponding to active memory regions, and gaps between the logical VAD map and the actual contents of memory. These inconsistencies often indicate injection or anti-forensic manipulation. The guiding question here becomes the following: Does every executable region have a corresponding VAD entry?

However, advanced analysis can't rely solely on reconstructed OS structures. It requires direct interaction with raw memory. By scanning memory for PE headers, high-entropy regions, and opcode patterns, the analyst can identify artifacts that exist physically but aren't represented in logical structures such as the VAD tree or PEB module lists. This cross-validation reveals orphaned executable regions—areas of memory that the OS doesn't acknowledge but which still contain active code. At this stage, the analyst must ask the following: What exists in memory that the OS doesn't acknowledge?

Thread analysis then provides insight into execution behavior. By enumerating threads within each process and validating their count and start addresses, the analyst can detect inconsistencies between user-mode and kernel-level views. Threads that point to nonmodule memory regions or high-entropy areas, as well as mismatches in reported thread counts, may indicate hidden execution contexts. This leads to another key question: Where is execution actually happening?

Finally, kernel-level integrity must be evaluated. This involves traversing core structures such as `EPROCESS` linked lists, handle tables, and callback mechanisms to identify broken pointer chains, misaligned structures, or inconsistent offsets. These anomalies signal that the internal logic of the kernel has been disrupted. The analyst must therefore determine the following: Is the kernel's internal structure still consistent?

Across all of these stages, a single analytical principle remains constant: No single view of memory can be trusted in isolation. True forensic insight emerges only through cross-validation, that is, comparing PEB structures with VAD mappings, VAD representations with raw memory contents, process lists with pool scan results, and thread structures with actual execution contexts. Anti-forensic techniques don't eliminate artifacts; rather, they desynchronize them. Within these inconsistencies are where the most valuable signals emerge for the analyst.

Process Hollowing/Unmapped Hollow Regions

Process hollowing is an attack technique that many analysts interpret simply as an implant not visible on disk. Yet its reflections in memory are far more complex. A normal process progresses through standard sections such as the following:

- PE header
- `.text`
- `.data`
- `.rdata`
- `.idata`

In a hollowed process, however, the following are observed:

- Memory regions that appear empty in the VAD tree but carry `EXECUTE` permissions
- Regions marked as `Private` yet not associated with any module
- Memory blocks that exhibit high entropy when dumped

These blocks are sometimes extraordinarily silent with no import table, no relocation area, and no proper PE structure. It's as if someone has removed the heart of a module and replaced it with something else.

Detection indicators include the following:

- `MEM_PRIVATE` + `PAGE_EXECUTE_READWRITE` combination in the VAD tree
- Randomness spike in the dumped block (entropy > 6.5)

- Presence of sequences resembling PE offsets despite the absence of a PE signature
- Thread context diving into this hollow region

The analyst's internal question is "There's code, but there's no form, so why is this void executable?"

Unlinked Modules

Some implants remove the loaded module from the PEB list structures:

- `LDR_INLOAD_ORDER`
- `LDR_INMEMORY_ORDER`
- `LDR_ININIT_ORDER`

A DLL removed from these lists becomes invisible to normal analysis tools, yet it continues to reside in memory. This is an illusion of absence. The module may have been deleted from disk, may not be recorded in hash tables, or even have part of its name corrupted, but the VAD tree is honest enough that it can't hide an unlinked module.

Detection indicators include the following:

- A module region present in the VAD but absent from the Loader (LDR) lists
- A valid PE structure appearing when the region is dumped
- Jumps or inconsistent alignment in `VirtualAddress` fields
- Missing imports in the table, yet function pointers visible in memory

> **Real-World Example**
>
> In an APT analysis, an unlinked module was identified as a PE region of size 0x10000 visible in the VAD tree. There was no trace on disk. However, the timestamp in the PE header didn't match any legitimate process running on the system that day. The attacker had failed to erase the time trace.

Hooking Anti-Forensics

One of the most insidious forms of memory-based anti-forensics is masking the hooking structure. An attacker may manipulate layers such as these:

- **Import Address Table**
  Redirecting imported function pointers within a process so that calls to legitimate APIs are transparently routed to malicious code.
- **Export Address Table**
  Altering exported function mappings within a module to hijack how other processes resolve and invoke those functions.

- **SSDT hooks**
  Modifying kernel-level syscall dispatch entries to intercept or alter syscall behavior before it reaches the intended kernel routine.
- **Inline hooks**
  Overwriting the beginning of a legitimate function (e.g., inserting a jump instruction) to redirect execution flow to attacker-controlled code.
- **Hypercall intercepts**
  Interfering with hypervisor-level communication mechanisms to monitor or manipulate guest-to-hypervisor interactions, often used in advanced or virtualized environments.

However, what distinguishes advanced implants is that the hooks aren't visible—yet they are there. How can this be recognized? There's a difference in instruction density between the hooked function's machine code and the original code. Consider the following:

- **If a kernel API begins with a 15-byte stub: suspicion**
  In normal conditions, kernel APIs follow a consistent and expected instruction pattern at their entry point. An unusually long or atypical stub (e.g., ~15 bytes) at the beginning of a function may indicate that the original instructions have been overwritten to redirect execution flow. This often suggests inline hooking, where the attacker inserts a trampoline or jump sequence to transfer control to malicious code before returning to the original function.
- **If a `jmp rax` or `call rax` appears in the prologue: more suspicion**
  Function prologues are typically predictable and structured (e.g., stack setup instructions). The presence of indirect control transfer instructions such as `jmp rax` or `call rax` at the very beginning of a function is highly unusual and often indicates redirection to dynamically resolved or attacker-controlled memory regions. This pattern is commonly used in stealthy hooks to bypass static analysis and conceal the true execution target.
- **If short sequences resembling a return-oriented programming (ROP) chain are present: much more suspicion**
  ROP-like instruction sequences consist of small, chained instruction fragments designed to manipulate execution flow without relying on traditional function calls. The presence of such short, dense instruction patterns within a function, especially in unexpected locations, suggests deliberate manipulation of control flow. This is particularly suspicious because it indicates advanced evasion techniques aimed at avoiding detection while maintaining execution capability.

Partial entropy comes into play here. A hooked region typically has the following characteristics:

- Higher entropy than the original code
- Shorter but denser opcode sequences
- Disruptive of logical integrity while attempting to conceal itself

Kernel-Level Artifact Shadowing

Kernel anti-forensics is quieter, more subtle, and more destructive than user-mode tricks. Hiding processes in user mode is a kind of stage performance; hiding artifacts in the kernel is akin to rewriting the entire stage. For this reason, some advanced implants directly attack the kernel's structure to conceal themselves:

- They hide or re-chain the `EPROCESS` structure.
- They remove themselves from thread lists, appearing nonexistent in the scheduler's official view.
- They rewrite handle tables, breaking tracking chains.
- They delete or render kernel callbacks invisible.
- They manipulate global lists such as `PsActiveProcessHead`.

We must not forget this truth, however: The kernel can never be perfectly concealed. The Windows kernel operates with a strong expectation of internal consistency; when you disrupt this fabric, a small "tear" inevitably forms somewhere. This is where the analyst must look for the following symptoms:

- **Corruption in `EPROCESS` sequences**

  Under normal conditions, an `EPROCESS` structure is created for every process in the system, and these structures are linked to one another through specific linked lists. The doubly linked list that begins with `PsActiveProcessHead` functions as a kind of map of currently living processes in the system.

  Within this map, between `EPROCESS` structures, we can expect the following:

  - A consistent structure size
  - A predictable offset layout
  - A logically ordered sequence within the list

  Implants that employ kernel anti-forensic techniques typically do one of two things to remain invisible: completely remove their own `EPROCESS` from the list (unlink) or replace the real one with a fake, manipulated `EPROCESS`.

  In practice, the resulting phenomenon appears as follows:

  - Both the expected address ranges and the structural alignment between `EPROCESS` structures are disrupted.
  - Some `EPROCESS` structures are positioned in an unusually distant manner relative to others.
  - In some cases, residual pointers belonging to a "dead" `EPROCESS` remain within the list.

  This `EPROCESS` drift, that is, the "shift" of `EPROCESS` structures, isn't detected by most conventional tools. However, when an analyst deeply parses the memory image and traverses `EPROCESS` structures sequentially in raw memory, they eventually think, "Why is the address of this process structure so far offset compared to its neighbors?" This question may be the first clue of an implant shadowed at the kernel level.

- **Thread count mismatch**

  Every process reports the number of threads visible in user mode through structures such as the PEB. Task Manager, Process Explorer, and most EDR tools all reference this high-level view. However, when you descend into the kernel, the picture sometimes changes.

  Suppose that you examine the PEB and the process appears to be running with six threads, but when you inspect the kernel thread lists, you find only five threads associated with that same process. In this case, the question that should come to mind is simple yet unsettling: "Where is that missing thread?"

  Possible explanations include the following:

  - Anti-forensic code has removed its critical thread from the lists.
  - The thread structure still exists in memory but is no longer linked to global lists.
  - The thread context is being executed as a kind of proxy through another process.

  Kernel-based implants typically perform critical functions such as the following through a single hidden thread:

  - Receiving commands
  - Running the main loop
  - Timer-based triggering

  Therefore, even a difference of one thread can be a strong signal to an experienced analyst that something is going on in this system. It's a small numerical difference, but a great deal of noise.

- **VAD Inode/VAD self-deletion**

  For every user-mode process, there exists a VAD tree that defines that process's memory regions. In simple terms, the VAD says, "These address ranges are used for these purposes with these permissions." When malicious code performs injection or hollowing and then cleans itself out of the VAD structure after completing its task, an interesting scene emerges: There are still anonymous, executable (`EXECUTE`) blocks in memory, yet no node exists in the VAD tree to describe these blocks. The resulting situation includes the following:

  - Meaningless gaps in the VAD tree
  - Breaks in parent-child relationships
  - Inconsistencies in page protection information

Example

For example, according to the VAD, the range 0x7ff60000–0x7ff80000 appears to contain nothing. But when you scan raw memory, you find high-entropy code within this range. Moreover, the instruction pointer (relative instruction pointer/extended instruction pointer) of a thread points precisely to the middle of this region.

From a forensic analysis perspective, this means the attacker removed their trace from the VAD tree, but memory is still speaking. Most automated tools, relying solely on the VAD structure to produce a clean memory map, simply ignore such regions. This is why these gaps are a "silent scream" to an analyst:

- The map says, "There's nothing here."
- Raw memory shouts, "There's a lot here."

True forensic skill lies in hearing the discord between these two voices.

Kernel anti-forensics fundamentally plays on a single concept: amplifying the gap between what is visible and what is real:

- **EPROCESS drift**
  The geometry of the process list is distorted.
- **Thread mismatch**
  Statistical balances are shifted.
- **VAD self-deletion**
  A distance is introduced between the memory map and the actual memory contents.

Through these manipulations, the attacker does the following:

- Breaks the process-based correlation mechanisms of EDR solutions
- Invalidates the default assumptions of tools such as Volatility and Rekall
- Repaints the patterns the analyst's eye is accustomed to

Yet every manipulation simultaneously creates a structural imbalance. This imbalance manifests itself in offsets, pointer chains, tree structures, and sometimes solely in entropy distributions. The essence of kernel-level artifact shadowing is that the attacker doesn't remove objects—they sever their connections. The analyst, in turn, doesn't search for the missing object itself, but for the trace left behind by the broken connection.

Memory Artifact Fragmentation

This technique is employed when the attacker realizes that they can't hide their payload in memory as a single contiguous piece. The shellcode or implant module has the following characteristics:

- Split into small fragments of 32–64 bytes
- Interleaved with sequences that appear legitimate
- Arranged in a way that doesn't disrupt the overall entropy balance

These fragments deceive classical signature-based hunting approaches.

But there's one fact they overlook: Fragmented code in memory always contains a higher number of pointers. Pointer density should raise suspicion of fragmentation.

If, within a module, you observe an unusually high number of the following, then there's malicious activity present (but well hidden):

- Jump tables
- Return addresses
- Unaligned opcodes
- Disconnected strings

Anti-Forensic DMA Tactics

DMA-based implants (modern techniques used for EDR bypass) access memory through devices rather than through the CPU. Therefore, they don't trigger page faults, they don't generate kernel hooks, and they don't create threads.

From the outside, it appears as though nothing is happening. However, DMA isn't truly invisible. DMA implants have the following characteristics:

- Exhibit abnormal density in the physical address map
- Create additional mapping layers in the EPT/NPT tables
- Introduce microscopic deviations in cache flush behavior

These deviations can be revealed through time-based analysis; specifically, high-frequency translation lookaside buffer (TLB) patterning. A DMA implant introduces inconsistencies in the TLB by causing irregular patterns in address translation behavior, such as unusually frequent TLB misses, rapid invalidation (flush) cycles, or nonuniform access timing across memory regions. Under normal conditions, TLB activity follows relatively stable and predictable patterns aligned with process execution and memory access locality. However, when DMA-based access bypasses the CPU's normal memory management flow, it disrupts this consistency.

High-frequency TLB patterning refers to the analysis of these rapid, low-level fluctuations in TLB behavior over time, where subtle anomalies (often invisible at higher abstraction layers) can indicate external or non-CPU-driven memory access. This type of signal isn't observable through standard Volatility modules, but can be identified using low-level profiling, hardware performance counters, or specialized memory analysis techniques.

The Greatest Flaw of Anti-Forensics

No matter how sophisticated they are, all anti-forensic techniques share a common weakness: They disrupt the natural rhythm of the system. Under normal conditions, memory does the following:

- Switches pages at predictable intervals
- Follows a consistent module loading order

- Exhibits a predictable thread distribution
- Maintains a monotonic EPROCESS chain

When anti-forensic techniques are activated, the following happens:

- The module order shows asynchronous jumps.
- Thread contexts fail to mirror one another.
- Memory permission levels become inconsistent.
- The entropy map fractures like the sound of a cracked consonant.

9.4.4 Memory Forensics Frameworks

Memory forensic analysis is one of the most powerful tools targeting a large portion of modern threat actors' concealment strategies. This is because an attacker can manipulate the file system, logs, and the telemetry layer; yet it's difficult to fully control the instantaneous and raw nature of RAM. Memory is an unfiltered, uncensored snapshot of what is happening in the system. Memory forensics frameworks exist precisely to examine this snapshot.

Volatility, Rekall, and entropy-based module analyses all approach the same question from different angles: What truly happened in this system's memory? Seeking the answer to this question is critical because memory is where the attacker may commit their greatest mistake.

Behavioral Process Analysis with Volatility 3

Volatility 3 is a framework that goes far beyond the simplistic logic of "take pslist, look at pstree, mark the suspicious process." Yes, it still enumerates process lists, shows handles, and lists DLLs, but its real power comes from the ability to test the behavioral consistency behind these structures. What makes Volatility 3 valuable can be summarized with this idea: A process is betrayed not by its signature, hash, or name, but by its behavioral rhythm.

This behavioral rhythm manifests itself in a memory dump along three primary axes: parent-child relationships, handle/thread topology, and the traces left by "missing" things. Each matters on its own, but when read together, they become a behavioral model that reveals the attacker's story:

- **Parent-child inconsistencies: Cracks in the process lineage tree**

  In a normal Windows system, the process lineage tree repeats certain patterns. A `cmd.exe` spawned under `explorer.exe`, a `powershell.exe` launched from within it, and so on. User interactions typically follow a logical flow. Therefore, when the parent-child relationship is disrupted, this is more than a technical anomaly, but a behavioral alarm. Volatility 3 steps in here by doing the following:

  - Shows process hierarchy through `pstree`-like outputs
  - Validates lineage through parent PID and GUIDs

- Reconstructs the "true" lineage tree using `EPROCESS` relationships in memory structures

A `powershell.exe` process within a user session is most often born under `explorer.exe`. But if, in a memory dump, the parent GUID points to a service process, for example, an `svchost.exe` or an unrelated `spoolsv.exe`—this isn't ordinary. This may indicate several things:

- The attacker disrupted the parent reference after process hollowing.
- A legitimate process chain was imitated, but the GUID assignment doesn't match the real hierarchy.
- As a result of process injection, new threads were placed inside a fake process.

Volatility 3's advantage is that it reads such relationships not merely as a list, but through structural linkage. A process tree that looks normal to the naked eye can suddenly turn into a story full of inconsistencies when Volatility scans internal structures.

If the parent-child chain doesn't flow in a memory dump, there's a point somewhere in the system that an attacker has touched. The first place to look to find that point is GUID matches.

- **Abnormal handle and thread relationships**

  A process list shows you the shell, whereas handle and thread lists show you the internal organs. Advanced actors prefer manipulating the inside of a process rather than changing its outward appearance:

  - They create new threads after injection.
  - They move these threads into the address space of other processes.
  - They run malicious code inside an innocent process via remote threads.
  - They touch the memory space of other processes through handles.

  With Volatility 3, you can observe the following:

  - Which objects and processes a process is connected to through handles analyses
  - The origin of threads via threads and `thrdscan`
  - Which module controls which thread through stack traces

Real-World Example

In a memory analysis case involving a suspected process injection scenario, while examining a RAM dump, I observed this:

- Visible process: `svchost.exe`
- Thread owner: `svchost.exe`
- Stack trace: Entirely filled with `cmd.exe` and subsequently invoked script engine functions

The thread was saying, "I am `svchost`," but the inside of its brain smelled like `cmd`. This small inconsistency revealed an injection mechanism aimed at command execution running in the background.

Volatility 3 shows such structures—where the inside doesn't match the outside—very clearly. From a behavioral perspective, none of the following is normal:

- If a thread's start address is outside the boundaries of loaded modules
- If handles extend to an unexpectedly large number of different processes
- If the thread count is far above normal limits for the process type (e.g., a `rundll32.exe` with hundreds of threads on an otherwise calm system)

What we call an anomaly is sometimes a single-line handle record.

- **Memory artifacts of absence: The trace left by what isn't there**
  Classical forensic approaches typically focus on the following: "What exists, what is running, what is loaded?" Yet in memory forensics, what isn't there can also be evidence. With Volatility 3, you can see components that are expected—but missing—within a process's DDL lists, module maps, and VAD structures, for example:
  - The absence of certain system DLLs that should normally exist in every `explorer.exe` process
  - A process appearing as though it never loaded a DLL it imported
  - Regions in the VAD map that contain a PE header but don't appear in the module list

All of these can be indirect indicators of techniques such as process hollowing, unmapped execution, and reflective loading. Volatility enables you to ask: "This process should have had this module, but it doesn't. Why?" Volatility's true magic lies in this sentence: Nothing in RAM can be completely hidden; even when something is being concealed, the manner in which it's removed leaves a behavioral trace. Removing a DLL from a process, cleaning the module list after injection, and fragmenting a PE header and presenting it as anonymous memory are all cleanup for the attacker and abnormality for the analyst.

A small but important tip for behavioral analysis is that when working with Volatility 3, it's highly valuable to think of processes not as single photographic frames, but as connected scenes:

- Is the parent-child chain normal?
- Is the thread/handle topology compatible with the process type?
- Does the DLL/VAD structure match expected system behavior?

If the rhythm breaks between these three links, Volatility 3 shows you what happened and what is happening even though it shouldn't. Most of the time, in memory forensics, the real breaking point occurs here: It's not a process's appearance, but its inconsistency that speaks.

Table 9.10 presents a comparative view of how normal process behavior differs from attacker-induced anomalies when examined through Volatility 3. Volatility 3 enables analysts to detect not only what is present in memory but also what is missing, inconsistent, or structurally misaligned. Each category highlights a specific layer of behavioral validation, demonstrating how subtle discrepancies, such as broken lineage chains, abnormal thread execution contexts, or mismatches between VAD regions and module lists, can reveal underlying anti-forensic techniques. In this sense, the table serves as a practical interpretation guide, helping analysts translate low-level memory artifacts into meaningful indicators of attacker activity.

Behavioral Analysis Category	Normal System Behavior	Typical Indicators of Attacker Behavior	Anomalies Detectable with Volatility 3	Forensic Significance/Analyst's Interpretation
Parent-child relationships	Processes form a hierarchy consistent with natural user workflows (`explorer.exe` → `cmd.exe` → `powershell.exe`).	Parent GUID mismatch, parent PID manipulation, fake process chains, interactive programs spawned under service processes.	`pstree`, `pslist`, `psscan`, `EPROCESS` parent pointer analysis.	A broken chain is an early indicator of process hollowing, injected processes, or masquerading attempts.
Handle/thread topology	Threads produce stack traces consistent with their process modules; handle lists are limited and predictable.	Handles opened to memory spaces of other processes, remote thread creation, abnormally high thread counts, thread start addresses outside module boundaries.	`handles`, `threads`, `thrdscan`, `vadinfo`, stack trace analysis.	Strong correlation with injection, reflective loaders, hollowing, or command execution frameworks (C2 implants).
Artifacts of absence (missing modules)	Each process loads a DLL set appropriate to its environment; critical system DLLs are always visible.	Missing DLLs, PE files hidden from module lists, modules present in VAD but absent from loader lists, corrupted PE header regions.	`dlllist`, `ldrmodules`, `vadyarascan`, VAD–LDR comparison techniques.	Reveals reflective loading, mapped–unmapped tricks, and anti-forensic module hiding techniques.

Table 9.10: Comparison Table of Volatility 3 Behavioral Process Anomalies

Behavioral Analysis Category	Normal System Behavior	Typical Indicators of Attacker Behavior	Anomalies Detectable with Volatility 3	Forensic Significance/Analyst's Interpretation
Memory structure inconsistencies	VAD, handle, thread, and module structures form a coherent system graph.	Random-looking VAD boundaries, overlapping memory regions, unnamed high-entropy segments.	`vadtree`, `malfind`, entropy analysis, detection of unbacked memory regions.	Indicates that malicious payloads are resident in memory in compressed or encrypted form.
Process lifecycle anomalies	Process lifetimes align with system events and user actions; PID reuse follows expected patterns.	Artificially prolonged lifetimes, orphan processes (no parent), exit code manipulation, misuse of suspended states.	`psscan`, `psinfo`, `lifetimes`, detection of orphaned `EPROCESS` structures.	Red flags for fileless malware, in-memory C2 implants, and stealth execution techniques.

Table 9.10: Comparison Table of Volatility 3 Behavioral Process Anomalies (Cont.)

Detection of Injection, Hollowing, and Reflective Loading with Rekall

Rekall is the surgical scalpel of the memory forensics ecosystem. Thanks to its stable architecture and deep plugin support, it's extraordinarily effective at uncovering process manipulation techniques.

Techniques such as process hollowing, reflective DLL loading, and thread injection form the core of fileless attacks. What Rekall does best is capturing the anomalies left behind by these techniques:

- **Hollowing indicators**
  The original EXE header remains in memory, but the process image region is entirely filled with a different PE. This is Rekall's primary red flag.
- **Reflective loading indicators**
  If the loaded module has the following characteristics, then Rekall treats it as a "patched PE" and flags it immediately:
  - A misaligned PE header
  - Inconsistent section offsets
  - An import table smaller than expected

  Reflective loading bypasses the file system and is therefore detected late by many EDR solutions, but memory forgives nothing.

- **Remote thread injection detections**
  If a thread's start address lies outside the boundaries of any loaded module, this strongly indicates injection or reflective loading. Rekall compares these boundaries to deliver fast and decisive results. Sometimes a single line in a report is enough to map the entire attack.

Entropy-Based Module Detection

In memory forensics, the concept of entropy is an indicator that most analysts initially underestimate, yet it can be lifesaving against advanced attacks. Entropy is a behavioral indicator that marks points of deviation from memory's natural rhythm. And memory is the most honest witness of what is happening in a system: No matter how hard an attacker tries to conceal a module, the density trace the module leaves behind doesn't vanish completely. The entropy of a normal Windows DLL typically falls within the 6.5–7.5 range because a DLL consists of a statistical mixture of code, data, metadata, the PE header, and import tables. This mixture is predictable and orderly. Therefore, as entropy rises, you know that something has disrupted the natural flow.

But if entropy exceeds 8.0, there's nothing natural there. Entropy at this level most often carries three critical meanings:

- **Encrypted payload**
  Advanced malware often resides in memory in encrypted form and is decrypted only immediately before execution. The reason is very simple: EDR solutions look for signatures, YARA rules look for strings, and AV engines look for known sequences. However, encrypted data resembles none of them.

  Encrypted payloads have the following characteristics:
  - Create sharp entropy spikes
  - Appear independent of the code region
  - Reside as anonymous memory in the VAD structure
  - Are unrelated to the import table

  For this reason, entropy analysis can detect encrypted modules before YARA does. A memory segment that an EDR labels as clean often rises in the entropy graph like a red column.

- **Packed/compressed shellcode**
  Many implants, especially APT toolchains with loader architectures, are written into memory in compressed form. These packing techniques may not be as obvious as Ultimate Packer for eXecutables (UPX), a widely known and easily identifiable executable packer commonly used to compress binaries; most often, a custom packer is used. Packed shellcode exhibits high entropy for the following reasons:
  - Compression algorithms equalize data distribution.
  - They eliminate the natural separation between code and data.

- They completely erase traces of the PE structure.
- They form a homogeneous mass of noise in memory.

Sometimes, observing this density is like suddenly hearing a low-frequency hum within an otherwise normal silence in the forest. It's invisible, but it's there. The analyst's task is to hear that hum.

- **Memory filling for anti-forensic purposes**
  Some attackers are surprisingly creative: By filling certain regions of memory with meaningless noise, they both mislead analysis tools and hide their real payloads inside a complex crowd.

  Such filling techniques are implemented through infrastructures such as these:
  - Random byte patterns
  - XOR with a fixed key
  - Noise generation via pseudo-random generators

  Here, entropy becomes a kind of noise-density map.

Entropy analysis is so effective because an attacker may hide an invisible module, delete its header, corrupt the PE structure, and manipulate the DLL list, but they can't conceal it without altering the statistical nature of the data. Even if an EDR can't interpret behavior, entropy can see the mathematics of behavior.

Example: A Silent 64 KB Density Region

In one memory analysis, the Module List appeared clean. Module names were normal, paths were normal, and there was no unusual DLL. But there was a small detail: A certain segment of memory had an entropy value of 8.1. What did this mean?

- No PE structure
- No imports
- No strings
- No name
- No path

But there was an overly orderly randomness. When the entropy scan flagged this region, it was as if the system whispered, "This silent area is full of noise... perhaps you should take a look." When that 64 KB segment was unpacked, what emerged was a compressed Cobalt Strike Beacon. For this reason, always remember that sometimes the loudest thing is the thing that appears the quietest.

Traditional security systems verify module names, perform hash comparisons, and may apply whitelisting controls. But they miss this: If attacker behavior is packed, names and hashes no longer carry value. Behavior can't be packed-only deferred. Entropy analysis, however, captures the intent inside the package. The EDR says clean, YARA stays silent, and Sysmon sees nothing, but entropy simply stands quietly and never lies.

Analysts shouldn't rely solely on module names, treat hashes as definitive indicators, or consider the DLL list sufficient on its own—all must be validated through cross-artifact correlation.

First, derive the density profile of the memory map because not every attack leaves a trace, but density always does. Entropy-based analysis is the greatest enemy of modern fileless attacks—and one of the greatest allies of a memory analyst.

The entropy value of each module in memory reveals its statistical nature and behavioral identity. The classification in Table 9.11 provides a clear, rapid, and reference-grade framework for anomaly detection when analyzing a memory dump.

Entropy Range	Category	Probable Content	Occurrence in Normal Systems	Forensic Risk Level	Analyst's Interpretation/ Notes
0.0–4.0	Low entropy	Empty memory, text blocks, null padding, low-diversity data	Frequently observed (especially empty pages and padding areas)	Low risk	Generally insignificant; however, abnormally low entropy may indicate an anti-forensic overwrite attempt.
4.0–6.0	Low-medium entropy	PE header, metadata, import table, static data sections	Common in DLL/EXE structures	Low–medium risk	Normal PE structures fall within this range. However, an entire module remaining at this level may be suspicious (except for very small modules).
6.0–7.0	Normal entropy	Standard Windows DLLs, mixed code and data	Very common and expected	Normal behavior	An average DLL typically resides in this range. However, persistent entropy increases above 7.2 in critical system processes may constitute an anomaly.
7.0–7.5	High-normal	Mixed code segments, optimized modules	Common	Requires attention	Most normal DLLs in memory fall within the 7.0–7.5 range. However, these values pose risk in unnamed or unbacked segments.

Table 9.11: Module Classification by Entropy Value (Memory Forensic Analysis)

Entropy Range	Category	Probable Content	Occurrence in Normal Systems	Forensic Risk Level	Analyst's Interpretation/ Notes
7.5–7.9	Abnormally high	Packed code, compressed segments, encrypted metadata	Rare in normal modules	Medium-high risk	This range is frequently observed during fileless payload staging or reflective loading. If no PE header is present, detailed analysis is mandatory.
8.0–8.3	Suspicious density zone	Encryption, custom packers, compressed shellcode	Not normally observed	High risk	Entropy values of 8.0 and above-when interpreted in context-strongly suggest hidden or encrypted payloads.
8.3–8.6	Stealth payload indicator	Cobalt Strike Beacon, Meterpreter, custom remote access trojan (RAT) modules	Very rare/ abnormal	Critical risk	This range often indicates embedded, encrypted, or compressed attack code. Should be correlated with behavioral scanning.
8.6–8.9	Anti-forensic noise	Randomly filled memory blocks, noise masking	Not seen in normal systems	Extremely critical	The attacker may have used random-fill techniques to conceal modules. Unbacked memory plus this entropy level almost certainly indicates malicious activity.
8.9–9.0+	Dangerous anomaly	Cryptographic material, dense encrypted blobs, polymorphic shellcode	Seen only temporarily during cryptographic operations or active crypto API usage	Highest risk	A persistent entropy region at or above 9.0 typically contains a fileless implant, C2 beacon, or self-modifying code. This is an immediate investigate now flag.

Table 9.11: Module Classification by Entropy Value (Memory Forensic Analysis) (Cont.)

Example: Detecting Memory Gaps as Anti-Forensic Zero-Fill Behavior Using Python

The memory analysis output shown in Figure 9.5 was generated by parsing a raw memory dump using a Python-based analysis script designed to scan virtual address space regions page by page. The script iterates through memory in 4 KB aligned segments, calculates entropy values for each page, and classifies them based on content characteristics (e.g., zero-filled, high-entropy, or executable patterns). During this process, consecutive pages composed entirely of 0 bytes (0x00) were identified and labeled as `"ZERO_FILL_PADDING"`, indicating a potential anti-forensic memory manipulation pattern. This condition led the detection tool to classify the relevant memory region as `"ZERO_FILL_PADDING"`.

```
C:\Users\admin\PycharmProjects\Books>python memory_padd

{
    "offset": "0x100000",
    "size": 4096,
    "entropy": 0.0,
    "classificationn": "ZERO_FILL_PADDING"
},
{
    "offset": "0x101000",
    "size": 4096,
    "entropy": 0.0,
    "classificationn": "ZERO_FILL_PADDING"
},
{
    "offset": "0x102000",
    "size": 4096,
    "entropy": 0.0,
    "classificationn": "ZERO_FILL_PADDING"
},
{
    "offset": "0x103000",
    "size": 4096,
    "entropy": 0.0,
    "classificationn": "ZERO_FILL_PADDING"
},
{
    "offset": "0x104000",
    "size": 4096,
    "entropy": 0.0,
    "classificationn": "ZERO_FILL_PADDING"
},
{
    "offset": "0x105000",
    "size": 4096,
    "entropy": 0.0,
    "classificationn": "ZERO_FILL_PADDING"
```

Figure 9.5: Detection of Zero-Filled Memory Pages Indicative of Anti-Forensic Memory Padding

The key points inferred from the data are as follows:

- **Repeating `ZERO-FILL` pattern across consecutive pages**
  The offset of each page follows the form:
  - `0x100000`
  - `0x101000`
  - `0x102000`
  - `0x103000`
  - `...`
  - `0x105000`

  These addresses demonstrate a linear progression in the Virtual Address Space (VAS) with natural 4 KB page alignment. This indicates that the memory wasn't cleared or filled randomly, but rather in contiguous blocks.

  Analyst interpretation: Filling more than five consecutive pages with identical content ceases to be normal application behavior and strongly increases the likelihood of an anti-forensic action. In areas such as the heap, stack, or code region, zero-fill behavior at this density isn't normally observed.

- **Entropy value: 0.0 → completely synthetic void**
  The entropy value of each page is `"entropy": 0.0`. This indicates that the blocks contain no random data whatsoever and are composed entirely of 0x00 bytes.

  Analyst interpretation:
  - Observing an entropy value of 0.0 in normal process memory is extremely rare.
  - During memory allocation, the OS may occasionally provide zero-initialized pages; however, this doesn't occur in such large, contiguous blocks.
  - Attackers deliberately fill memory blocks with zeros to complicate memory forensic analysis, erase artifacts, or create a "noise curtain" within the dump.

  Therefore, an entropy value of 0.0 doesn't merely indicate an empty page, but rather a potential signal of manipulation.

- **`ZERO_FILL_PADDING` classification**
  The tool's classification field is `"classification": "ZERO_FILL_PADDING"`. This classification indicates the following:
  - The memory page contains no meaningful data.
  - It was most likely subjected to an anti-forensic padding operation.
  - It may represent artifact cleanup, runtime manipulation, or an attempt to conceal code prior to the memory dump.

  Analyst interpretation: Real-world malicious implants (e.g., Cobalt Strike, Sliver, Lazarus-style actors) do the following when they detect that a memory dump is imminent:
  - Encrypt their code.
  - Fill memory regions.

- Remove VAD entries.
- Overwrite pages entirely with zeros to eliminate residual traces.

This structure produces a fingerprint that closely matches those behaviors.

- **Offset distribution and VAD anomaly**
Because the pages are contiguous, this region can be described as forming a single block segment in memory, as shown in Table 9.12.

Start	End	Page Count	Size
0x100000	0x105000	6	24 KB

Table 9.12: Example Memory Block

Analyst interpretation:

- A completely empty 24 KB region isn't natural.
- This size range is commonly preferred for shellcode cleanup operations.
- If this segment appears as `PrivateMemory` or `Unknown` in the VAD tree, the anti-forensic suspicion increases significantly.

What does this data represent within the dump? The data shown in Figure 9.5 may indicate scenarios such as these:

- Cleared traces during the dump process
- Removed payload regions
- Shellcode deallocation leaving only zeros behind
- Noise-generation behavior by malware attempting to evade analysis

Such zero-fill segments typically appear in two primary scenarios:

- **Post-execution wiping**
The malware zeroes its memory before terminating.
- **Anti-forensic memory noise padding**
A silent ghost block is created to confuse EDR or forensic tools.

In both cases, such segments must always be analyzed in conjunction with other memory regions, code permissions (`RX`/`RWX`), VAD structures, and indicators of attacker behavior.

The consecutive zero-fill memory segments observed in this output are a typical indicator of memory wiping behavior frequently used in modern anti-forensic techniques. An entropy value of `0.0` confirms that the pages are entirely zero-filled, while their large and contiguous nature supports the conclusion that this isn't random, but a deliberate manipulation. Such a structure is a strong signal that the actual malicious code in memory was either wiped prior to the dump or concealed using padding intended to hinder analysis.

9.5 Summary

We'll wrap up this chapter with Table 9.13, which recaps the core Windows telemetry sources used in host-based forensic analysis that we've covered in depth, comparing which attacker behaviors are detected by which source, the signals analysts should monitor, and their value from a CTI perspective.

Telemetry Source	Captured Artifacts	Detected Behaviors/ Signals	Value for CTI	Key Considerations for the Analyst
Event logs (security, system, application)	Logon/logoff records, process creation, policy changes	Failed authentication attempts, privilege escalation, service manipulation	Cornerstone of chain reconstruction; backbone of the timeline	Focusing on critical Event IDs such as 4624/ 4625, 4688, 1102, and 7045
Registry (`HKCU`, `HKLM`, ShimCache, AmCache, ShellBags)	User activities, persistence mechanisms, file open/delete history	Persistence, lateral movement, application execution history	Reveals attacker behavioral patterns	Verifying whether key timestamps have been manipulated
Memory artifacts (RAM dump)	Active processes, modules, unencrypted credentials, injection traces	Process hollowing, DLL injection, C2 threads, malware behaviors	Most reliable signal source despite anti-forensic tactics	Applying correct techniques to prevent integrity loss during memory acquisition
Sysmon (Event ID 1–26)	Process creation, network connections, file changes, driver loading	C2 communication, LOLBins activities, data exfiltration	Most critical telemetry for behavioral detection and campaign correlation	Using a well-crafted Sysmon configuration to reduce noise
ETW	Kernel-level API calls, process/ thread events, network functions	AV/EDR evasion techniques, malicious API usage patterns	Exposes attacker tradecraft at the lowest level	Noting that even attempts to block ETW constitute an attacker signature

Table 9.13: Windows Telemetry Sources: Forensic Capabilities

Telemetry Source	Captured Artifacts	Detected Behaviors/ Signals	Value for CTI	Key Considerations for the Analyst
File system artifacts (MFT, USN Journal)	File creation/ deletion operations, timestamps, change logs	Timestomping, data manipulation, forensic laundering	Critical for data integrity analysis	Correlating MFT with registry and Sysmon
Network Telemetry (Windows Filtering Platform/Sysmon/ ETW-Net)	Outbound connections, port usage statistics	C2, beaconing, lateral movement	Indispensable for APT behavioral analysis	Applying time-series analysis to detect anti-beaconing techniques

Table 9.13: Windows Telemetry Sources: Forensic Capabilities (Cont.)

Now, let's shift our focus to the incident response process in the next chapter.

Chapter 10
Integrating CTI into Incident Response

While monitoring logs, you may encounter hundreds of alerts—but which constitute a real threat? The incident response cycle—from detection to containment, from eradication to feedback—only becomes valuable when it progresses with accuracy. A team lost among false signals may realize the real threat only when it's already too late.

Cyber threat intelligence (CTI) plays a critical role in the incident response process. Rather than a meaningless string of numbers and a behavioral pattern, indicators of compromise (IOCs) and indicators of attack (IOAs) are both windows into the attacker's mind. When these elements are combined with live telemetry, the effect is like a foggy photograph coming into focus: Boundaries become clear, threats fall into place, and dark areas are illuminated.

Sometimes, on a security information and event management (SIEM) screen, you see nothing more than an IP match—dry, colorless, silent information. But CTI reminds you of which campaigns that IP has been used in before, which infrastructure cluster it belongs to, and what kind of habits the attacker exhibits. A single line of log data suddenly appears before you with its full context. At that moment, the analyst's perspective changes; they move beyond just seeing an alert and begin to read an entire attack narrative.

This is where SIEM integration and rule tuning become critical. Without CTI-driven enrichment, detection rules tend to operate in isolation, producing alerts based on static thresholds or generic patterns. However, when CTI is embedded into SIEM logic, rules evolve from simple match conditions into context-aware detection mechanisms. Indicators are no longer evaluated solely by their presence, but by their relevance, recency, associated threat actors, and observed behavior patterns. Rule tuning, therefore, isn't just about reducing false positives, but about aligning detection logic with the current threat landscape. A well-tuned SIEM, informed by CTI, doesn't just detect anomalies—it prioritizes what truly matters.

Security orchestration, automation, and response (SOAR), on the other hand, is the mechanism that translates this depth provided by CTI into action. It reduces to seconds the correlation steps that would otherwise take an analyst hours; it flags high-risk assets, initiates necessary isolations, and even anticipates attacker intent based on known behavioral patterns in some cases. Yet effective automation isn't achieved through rigid playbooks alone.

When CTI is integrated into SOAR workflows, playbooks become adaptive: Enrichment steps dynamically pull intelligence from multiple sources, decisions are guided by threat confidence and context, and response actions are adjusted based on the nature of the adversary. In this sense, automation becomes intelligence-driven orchestration.

The power of automation can at times be sharp enough to humble human intuition. Yet the source of this sharpness is the layer of meaning provided by CTI, interpreting, comparing, and placing data within the correct historical and operational context.

The value CTI adds to the incident response process can be summarized in a single phrase: transforming the right information, at the right time, into the right response. This transformation represents a critical inflection point in every organization's security journey. Because in a world where threats are multiplying, attackers are accelerating, and infrastructures are becoming increasingly complex, simply detecting is no longer sufficient. It's necessary to understand what happened, why it happened, who did it, and what was targeted. That is why CTI exists: to provide incident response with data, meaning, logic, and an intuitive sense of direction.

You should ask yourself: "If CTI didn't exist, how long would it take to understand what this alert means?" In many cases, the answer is simple: too long. CTI is central to modern incident response because it accelerates understanding and enables timely, informed decisions.

Now, let's examine the core components of this chapter, including how CTI enhances detection through SIEM integration and rule tuning, how CTI drives automated response through SOAR playbooks, and how indicators and behaviors are contextualized to support accurate analysis, prioritization, and decision-making within the incident response lifecycle.

10.1 The Role of CTI in Incident Response

Incident response is often perceived as a firefighting reflex: An alarm is triggered, analysis begins, and the team is mobilized. Yet in the cyber domain, fires are invisible; there's no smoke and no heat, and sometimes you don't even realize that an attack has begun. It's at this point that CTI comes into play as an unseen guide that shapes every phase of incident response.

In the following sections, we'll see how CTI impacts each phase of incident response, with a particular emphasis on preparation, detection, and analysis.

10.1.1 CTI in the Complete Incident Response Cycle

To understand the role of CTI within the incident response process, we must step back and observe the entire cycle: preparation, detection, analysis, containment, eradication, and lessons learned. At first glance, these phases may appear to form a straight line. In practice, however, incident response is often a complex loop that repeatedly pulls the analyst backward, pushes them forward, and continuously forces reassessment. CTI serves as the most stable and consistent reference point within this loop. It reminds you where you are, what you're facing, and where you need to go.

During the *preparation phase*, CTI strengthens the organization against future storms. By describing the habits of threat actors, the paths they use, and the assets they target, CTI allows you to hear the footsteps of an attack that hasn't yet begun. When writing a playbook, always place the risk profile provided by CTI on the very first page because without that profile, every decision you take remains incomplete.

In the *detection phase*, the role of CTI becomes even more pronounced. Distinguishing the real adversary from hundreds of blinking alarms on a SIEM screen can't be achieved through technical capability alone; it requires context, source credibility, and an understanding of historical patterns. At this stage, CTI transforms a single-line alarm description into an entire attack narrative. An IP address is no longer just a number; it becomes the trace of a campaign, the signature of a threat group, or perhaps the continuation of another incident.

In the *analysis phase*, CTI acts like a thread running through all findings. It explains why the small anomalies you encounter in memory dumps, Domain Name System (DNS) queries, or process trees are there in the first place. Analysts often feel like detectives, and CTI is the magnifying glass that reveals the invisible connections between the files.

During the *containment and eradication phases*, the impact of CTI becomes more operational in nature: Which system should be isolated first? Which account must be suspended immediately? Which persistence technique represents the attacker's return path? Answering these questions correctly requires the technical details of the incident and an understanding of the attacker's character. CTI defines that character; it reveals intent, objectives, and behavioral habits.

Finally, *lessons learned* refers to the moments after every incident when we stand in the silence of the security operations center (SOC) and look back: Which steps were correct, which were lacking, and which behaviors could we have recognized earlier? In this phase, CTI functions like a teacher. It collects every piece of data, every anomaly, and every behavioral fragment derived from the incident, and strengthens institutional memory. It reduces future response delays when the same attack reappears and eliminates recurring mistakes.

In short, CTI is a strategic layer that supports all phases of the incident response lifecycle. It provides the context and direction needed to transform response efforts into an effective, intelligence-driven defense.

10.1.2 Establishing an Intelligence-Driven Defense Architecture

The preparation phase is the foundation of the incident response lifecycle, directly shaping the effectiveness of all subsequent actions. At this stage, CTI doesn't just present past threats; it transforms defense architecture by aligning it with the operational realities of the current threat landscape. CTI enables the design of a security infrastructure that's both predictive and context-aware by combining organizational risk profiles; sector-specific threat patterns; adversary tactics, techniques, and procedures (TTPs); and historical campaign behaviors.

What may appear as routine configuration is, in practice, a structured engineering process. CTI informs which behaviors detection systems should prioritize, how thresholds should be calibrated, and which telemetry sources must be collected. Adversary techniques, such as specific command-and-control (C2) patterns or identity-focused attack methods, directly influence monitoring strategies, correlation logic, and detection sensitivity. In this way, CTI is translated into the operational logic of the system.

One of the most critical outcomes of this phase is optimized logging. Effective incident analysis depends on data collected before the incident, accurately, consistently, and with sufficient depth. CTI identifies which artifacts are meaningful within attack chains, guiding the selection of relevant log sources such as Windows events, Event Tracing for Windows (ETW) providers, and network telemetry. This targeted approach reduces noise while preserving high-value signals, preventing the data gaps that often hinder analysis.

CTI also plays a key role in positioning defensive controls. By revealing commonly used persistence methods, lateral movement techniques, and evasion strategies, it enables prioritization tailored to the organization's threat exposure. As a result, security controls shift from generic implementations to targeted mechanisms aligned with real adversary behavior.

Although often invisible, the preparation phase determines how effectively an organization can perceive and respond to incidents. Weaknesses at this stage (such as missing telemetry, misaligned detection logic, or irrelevant indicators) emerge as operational failures during active incidents. In contrast, a CTI-driven preparation phase establishes both technical capability and a form of analytical readiness, allowing teams to understand not just what is happening, but why and what may follow.

Ultimately, this phase represents an engineering process in which CTI embeds intelligence into the defense architecture-enhancing detection accuracy, refining data collection, and aligning controls with threat reality. The strength of incident response depends largely on the robustness of this foundation, continuously shaped and updated by CTI.

We'll explore strategies to achieve this threat landscape modeling in the following sections.

Advanced TTP Frequency Analyses: Beyond Ordinary Distributions

Conducting standard TTP frequency analysis on MITRE ATT&CK is common; however, an advanced analyst derives the real value by examining changes in TTP density over time, variations tied to campaign periods, and shifts in technical preferences driven by adversaries' operational pressures, for example:

- **The change rhythm of subvariants of T1078 in the financial sector**
  In certain financial centers, it has been observed that the T1078 (Valid Accounts) technique intensifies in 60-day cycles. The reason is that some advanced persistent threat (APT) groups tend to use stolen credentials before account cleanup is performed. This pattern is rarely discussed in open sources, yet it holds significant value within internal CTI operations.
- **The use of T1190 with a hit-and-pivot model in the healthcare sector**
  In the healthcare sector, T1190 (Exploit Public-Facing Application) isn't used just as an external entry point; in some campaigns, following exploitation, adversaries pivot to internal services within a very short time frame (on the order of seconds to minutes). This ultrashort pivot behavior is unique due to the high connectivity density of healthcare infrastructures.
- **Linguistic profile shifts in T1566 (Phishing) variants within government institutions**
  Some threat actors enhance phishing content by performing organizational jargon modeling that almost exactly replicates official correspondence patterns used in specific countries. This technique can only be detected through linguistic analysis and completely bypasses standard IOC scanning.

Such advanced analyses determine which behavioral patterns an organization's SIEM correlation engines should focus on. By moving beyond standard technical distributions, they establish an intuitive framework for understanding the attacker's cognitive map.

Critical Asset-TTP Mapping

Although asset-TTP mapping may appear on the surface to determine which asset is being targeted, in reality, it's a far more complex analytical process. This mapping combines assets' technical characteristics, data flow structures, identity relationships, and adversaries' campaign habits into a unified matrix.

The following original examples illustrate how advanced TTP mapping is applied in practice:

- **Building a T1552 (Credential Dumping) risk model through an endpoint identity correlation chain**
  In some organizations, certain endpoints become indirect targets because they operate within the same session as critical service accounts. The adversary doesn't consider this endpoint valuable in itself; however, by performing credential dumping, they can pivot to the next link in the chain. This represents a relationship that is invisible in traditional asset classifications yet extremely decisive in threat modeling.
- **Silent risk generation of management networks in data centers with respect to T1021 (Remote Services)**
  Management virtual local area networks (VLANs) often aren't perceived as threats because they are accessible only to administrators. However, certain APT groups employ

complex lateral movement chains under T1021 in which Remote Desktop Protocol (RDP)/Windows Remote Management (WinRM) pivots specifically target these network segments. Revealing this information enables management networks to be remodeled according to their true threat surface.

- **Misplacement of T1610 (Deploy Container) technique variations in container-based environments**
  It's known that in some campaigns, attackers exfiltrate data by launching ephemeral containers after gaining access to container orchestrators. This behavior is often absent from traditional asset maps because ephemeral containers aren't persistent assets. However, CTI-supported asset-TTP mapping makes the role of these assets within the attack chain visible.

Such advanced mappings play a critical role in determining which servers an organization should isolate, which network segments should be hardened, and which identities should be positioned within high-risk categories.

Threat landscape modeling is one of CTI's most strategic functions and goes beyond determining who is attacking or which technique is being used. Through detailed TTP frequency analyses, intra-campaign behavioral rhythms, sector-specific technical variations, and advanced asset-technique mappings, the defense architecture transforms into a structure that is threat-shaped, intuitive, forward-looking, and contextually optimized.

Table 10.1 systematically addresses which TTPs different asset types are exposed to.

Asset Type	Relevant TTP	Adversary Objective	Risk Rating	CTI Recommendation
Endpoint + privileged account	T1552 (Credential Dumping)	Transition to the next account in the chain	High	Local Security Authority Subsystem Service (LSASS) protection, session isolation
Management network (mgmt VLAN)	T1021 (Remote Services)	Lateral pivot + access deepening	Very high	Segment hardening, access restriction
Ephemeral container	T1610 (Deploy Container)	Rapid operations for data exfiltration	Medium	Container runtime telemetry, ephemeral log consistency

Table 10.1: Asset-TTP Mapping Matrix

This mapping does more than determine which asset is risky; it enables the anticipation of the adversary's operational intent, target selection logic, and the likely directions of progression within the attack chain. Let's break down these mappings in more detail:

- **Endpoint + privileged account: T1552 (Credential Dumping)**
  The use of endpoints by adversaries to access privileged accounts represents one of the quiet yet most effective steps within the attack chain.

  An endpoint may not always appear to be a primary target; however, credentials stored in memory (LSASS, Data Protection API [DPAPI], browser credential caches, token artifacts) constitute an ideal springboard for transitioning to the next link in the chain. It's dangerous for the following reasons:
  - The compromise of an endpoint is often perceived by SOC teams as an isolated incident.
  - In reality, the adversary's true target is the account itself, and not the system.
  - Once a privileged account is compromised, the balance of the entire defense architecture is disrupted.

  When credential dumping behaviors aren't detected at the TTP level, the adversary can advance through the chain without visibility. LSASS protection, ETW-based credential access monitors, and session isolation—recommended by CTI—are critical controls that enable the chain to be broken.
- **Management network (mgmt VLAN): T1021 (Remote Services)**
  Management VLANs are often invisible sources of risk because access is limited to administrators, leading to the assumption of a low threat surface. For many APT groups, however, the opposite is true: The management network is viewed as a pivot point and a means of access deepening.

  The following shows the adversary's behavioral logic:
  - A single identity that gains access to the management network is sufficient.
  - Due to the nature of this network, the assets accessible through it are critical: domain controllers (DCs), orchestrators, management consoles, backup servers, and so on.
  - T1021 techniques (RDP, Secure Shell [SSH], WinRM) attract almost no suspicion in this environment.

  From a CTI perspective, the very high risk rating is based on the technical impact of the attack and the potential to establish sustainable persistence. Segment hardening—defined as the strict isolation and control of network segments through access restrictions, protocol limitations, and monitored communication paths—is therefore recommended by CTI for these reasons:
  - Narrows access paths
  - Reduces the maneuvering space for lateral movement
  - Forces the threat actor into more predictable behaviors
- **Ephemeral container: T1610 (Deploy Container)**
  Ephemeral containers are often disregarded from an operational risk perspective because of their short lifespan. However, some advanced threat actors leverage this short-lived nature as an advantage.

Following is the adversary's operational technique:

- After gaining access to the orchestrator, the adversary can launch an ephemeral container within seconds.
- This container may be used for data exfiltration, establishing C2 tunnels, or executing diskless payloads.
- When the container is removed, almost no trace of the incident remains; log consistency is disrupted.

Enforcing container runtime telemetry makes the lifecycle of ephemeral containers visible. Ephemeral log consistency, in turn, ensures that every container (however temporary) leaves a telemetry trace. Without this configuration, container-based attacks become invisible micro-incidents.

This matrix shows that threat modeling goes beyond asset classification to understanding adversary behavior. It helps explain why a target is chosen, which techniques are used, and how the attack is likely to progress. Accordingly, asset–TTP mapping becomes a key enabler of proactive defense within CTI.

Technical Profiling of Adversary Habits

Building on this asset–TTP mapping, the next step is to translate these associations into a deeper technical understanding of how adversaries actually operate in practice. CTI extracts the following technical parameters related to the operational behaviors of threat actors:

- **C2 protocols and packet-level characteristics of traffic**
  For example, an APT group's use of jitter-based beaconing at 60-second intervals enables network intrusion detection system (NIDS) signatures to be redesigned accordingly.
- **Exploit frameworks and payload types in use**
  Is it a memory-only implant or dynamic link library (DLL) sideloading? This information directly guides which modules should be activated within the endpoint detection and response (EDR).
- **Technical fingerprints of persistence mechanisms**
  Technical patterns are identified, ranging from registry manipulation to Windows Management Instrumentation (WMI) event subscriptions.

Beyond static indicators, CTI enables a deeper analysis of C2 behavior at the temporal and protocol levels. Rather than relying solely on fixed patterns, analysts must examine how traffic evolves over time and how subtle variations can reveal adversary intent. The following considerations highlight key dimensions of this behavioral analysis:

- **Analyzing beaconing behavior by periodicity and via variance and deviation analysis**
  It's not sufficient for C2 traffic to exhibit a 60-second beacon interval. The analyst should also examine the following:

- Beacon jitter percentage (% jitter)
- Intra-period delay deviations
- Rhythms of change in Transmission Control Protocol (TCP)/User Datagram Protocol (UDP) options
- Transport Layer Security (TLS) fingerprint variations (JA3/JA3S)

Some APTs alter the beacon interval every four to five cycles to hinder detection. This behavior can't be identified by observing the average period alone.

- **Applying packet-level timing analytics for C2 traffic**
  Most adversaries are detected not by traffic volume but by traffic rhythm:
  - Anomalies in packet length distributions
  - Burst packet behaviors
  - Signature-evasive variations of keep-alive packets
  - Maximum transmission unit (MTU) manipulations
  - Low byte-density queries used for DNS tunneling

  These signals allow a C2 channel to be distinguished from normal user traffic.
- **Differentiating exploit frameworks at the operational footprint level**
  Exploit frameworks have distinctive traces. Analysts should pay attention to the following:
  - Return-oriented programming (ROP) chain patterns used in memory (sequences of chained return addresses that enable code execution without injecting traditional shellcode)
  - Stack misalignment indicators created by the exploit (unexpected changes in stack structure that often signal low-level exploitation activity)
  - Payload staging behavior (single-stage vs. two-stage), where single-stage payloads execute immediately, while two-stage payloads first establish a loader that retrieves or decrypts the final payload
  - Expected post-exploitation process injection paths (which processes are typically targeted for code injection after initial access)
  - Application programming interface (API) call sequences used by shellcode (NT Layer DLL [NTDLL]–based patterns, where low-level Windows system calls (syscalls) are invoked directly to evade higher-level API monitoring)

  These traces are vital for profiling which actor or ecosystem the exploit is associated with.
- **Categorizing payload types into behavioral classes**
  An analyst shouldn't classify a payload solely as "memory-only" or "sideloaded DLL"; instead, a technical profile should be constructed using the following criteria:
  - The payload's anti-analysis capabilities
  - The encryption/obfuscation scheme employed

- Memory allocation strategy (e.g., use of `VirtualAllocEx` with read, write, and execute (RWX) permissions, where memory is allocated as readable, writable, and executable to enable in-memory code execution, or heap spraying techniques that distribute payload fragments across memory to improve reliability and evade detection)
- Anomalies created within the parent-child process chain
- Entry point behavior (e.g., creating a suspended process and injecting code via a remote thread, allowing the payload to execute within a legitimate process context while bypassing standard execution flows)

These details enable a technical comparison of the adversary's maturity level.

- **Identifying recurrent "wet fingerprint" values of persistence mechanisms**
  Each threat actor's persistence behavior is unique. Analysts should focus on the following:
  - Common key-naming conventions used within Windows registry subpaths
  - Trigger types in WMI event subscriptions (interval versus system event)
  - Folder naming patterns within scheduled tasks
  - Binary path manipulation during service creation
  - Preferred directory structures during DLL search order hijacking

  An actor's persistence fingerprint often remains consistent even across subsequent campaign variants, making it a powerful matching mechanism for CTI.
- **Deriving the adversary's operational tempo behavior**
  This is a frequently overlooked yet highly valuable indicator. A CTI analyst should answer the following questions:
  - Does the adversary move quickly, or slowly and cautiously?
  - What is the time gap between initial access and lateral movement?
  - At what intervals are privilege escalation attempts repeated?
  - Is persistence established early or late within the attack chain?

  This temporal model plays a critical role in determining the actor's maturity level.
- **Examining the side-effect artifacts of APT groups**
  Even advanced adversaries leave traces within the attack chain; these traces are often observed not at the file or network level, but in system behaviors. Side artifacts to be examined include the following:
  - Increases in specific `WmiPrvSE.exe` process instances
  - Changes in LSASS handle access patterns
  - Abnormal ETW provider activity
  - Spikes in kernel call patterns (especially syscalls)
  - Traces of log deletion attempts for anti-forensic purposes

  These artifacts enable deeper refinement of the technical profile.

- **Recalibrating security controls according to CTI**
  During the preparation phase of the defense architecture, technical configurations are performed in an intelligence-driven manner:
  - Rewriting SIEM correlation rules according to TTPs: For example, if the intensity of T1053 (Scheduled Task/Job) is high, the correlation window for task scheduler events is narrowed.
  - Adjusting the telemetry level of EDR sensors based on the threat profile: Some actors heavily utilize process injection, causing memory scanning levels to increase. Others prefer PowerShell, making script block logging mandatory.
  - Adding behavior-based signatures to network intrusion detection system (NIDS)/network intrusion prevention system (NIPS) systems: Instead of traditional IOC-based detection, C2 behavioral rhythms are processed.
- **CTI-supported design of the logging strategy**
  The most common technical mistake during the preparation phase is focusing on log volume; whereas what truly matters is the value of the logs.

  A CTI-supported logging strategy includes the following:
  - Prioritizing telemetry points preferred by APTs: Sysmon Event IDs 1, 7, 10; DNS Client logs; and ETW providers
  - Ensuring timestamp integrity, which is a critical step for detecting attackers' timestomping behaviors.
  - Using correlation windows because some actors move slowly, while others move very quickly. CTI determines whether this window should span minutes or hours.
- **Proactive hardening of the defense architecture**
  CTI data produces direct technical counterparts within defense configurations:
  - Tightening zero-trust policies according to the threat profile, especially in sectors where identity-focused attacks are prevalent
  - Redesigning network segmentation with APT lateral movement paths in mind
  - Planning honeypot, honeynet, and deception infrastructures based on threat actor behaviors, for example, specialized Kerberos traps for groups inclined toward credential harvesting
- **CTI-based development of incident response playbooks**
  Playbooks are more than templates; they must align with the technical capabilities of threat actors. CTI directly contributes content in the following areas:
  - Prioritized asset lists
  - Potential attack chain steps
  - Applicable persistence methods
  - Likely evasion techniques
  - Expected artifacts and telemetry structures

The preparation phase is the point at which CTI is blended with technical knowledge and forms the most robust foundation of defense. Every configuration performed here—every rule adjustment and every telemetry decision—directly determines the speed, accuracy, and operational effectiveness of the incident response team when an incident occurs.

10.1.3 Context-Oriented Alert Validation Mechanisms

The detection process is perhaps one of the most paradoxical moments in cybersecurity. Every alert on your screen represents a potential attack; yet at the same time, many of them are nothing more than meaningless noise. It's easy to become lost in this noise, especially if you're working in a SOC environment where the alert stream grows like an avalanche within minutes. It's at this point that CTI quietly comes into play; like an invisible hand, it provides an additional layer of context that strengthens an analyst's decision-making muscles. The signal-to-noise ratio transforms into a kind of intuitive filtering mechanism.

From experience, there are moments when a SIEM alert conveys nothing on its own. Your eyes fixate on a single line on the screen: "Process Injection Detected" or "Suspicious Network Connection." Without context, these phrases feel empty. In one such instance while investigating an alert, a small note from the CTI platform caught read: "This IP is a variant of the C2 infrastructure used in the X campaign two weeks ago." Sometimes, a single line changes the emotional state; uncertainty gently recedes, replaced by focused attention. This is why context-oriented alert validation serves as a support system that alleviates the analyst's cognitive load.

Context gives meaning to data. However, the context referred to here isn't superficial supplementary information; it's a multilayered fabric composed of the technical convergence of adversary behavior models, infrastructure chains, campaign periods, and historical incidents. To understand an alert, it's necessary to see both the present moment and the adversary's historical rhythm. Knowing that some APT groups operate late at night, that some intensify activity only during specific holiday periods, and that others have used nearly the same process tree for years without significant change is the true power of the context provided by CTI.

Alerts generated by EDR and SIEM systems are often one-dimensional signals. CTI renders these signals three-dimensional. It doesn't view an alert merely as suspicious; it ties the alert to the adversary's habits, references previous campaigns, and assigns weight based on the prevalence of the TTP in use. Even the timing of an alert can sometimes be a critical indicator; for example, it has been observed that the beaconing behaviors of certain known threat actors recur within specific time windows. Such timing information can provide invaluable clues regarding how frequently correlation engines should perform analysis.

Of course, behind this process lie robust technical models. The alert validation mechanism is fed by the following core components:

- Lifespan of the IOC and its campaign context
- Sectoral frequency of the TTPs

- Current state of the adversary's preferred infrastructure chain
- Behavioral anomaly scoring
- Expected movement model (e.g., the likelihood of lateral movement)
- Similar traces observed in the adversary's past campaigns

When these elements converge, an alert becomes a powerful signal that generates a clear prediction about where the attack chain may currently reside.

From an operational perspective, CTI-supported validation acts as a consistent analytical layer for the analyst. It captures details that may be overlooked, maintains reliability under time pressure or fatigue, and reduces the risks associated with rapid decision-making. Table 10.2 presents the key parameters used to enrich alerts with CTI, showing how each contextual layer contributes to improving accuracy and prioritization.

Contextual Layer	Description	Impact on the Alert	Typical Source
IOC lifespan	Validity window of the IOC	Reduces false positives	Threat feeds, MISP Threat Sharing (MISP)
Campaign context	History of similar attacks	Strengthens the alert's linkage to a real attack	CTI database
Infrastructure chain	C2 domain/IP clusters	Enables C2 correlation	Passive DNS (pDNS)
TTP frequency	Sector-specific technique distribution	Improves determination of alert severity	MITRE ATT&CK
Behavioral model	Adversary's rhythmic behaviors	Enables timing-based validation	Network detection and response (NDR)/endpoint detection and response (EDR) telemetry
Anomaly score	Difference between expected and observed behavior	Clarifies alert prioritization	Machine learning/user and entity behavior analysis (UEBA) systems

Table 10.2: Alert Contextual Enrichment Matrix

To effectively apply this enrichment model in practice, analysts should follow several key principles when interpreting alerts within a CTI-driven context:

- **An alert should be evaluated both as a SIEM/EDR output and as a potential stage within the attack chain.**
  Every alert may represent the manifestation of a real TTP. Therefore, the analyst should

mentally visualize where the alert might fall within the chain (IA → execution → C2 → lateral movement → exfiltration). Without context, such positioning can't be performed.

- **Before trusting an IOC, its temporal value (freshness) must always be verified.**
  Outdated (stale) IOCs, especially those from context-poor or low-quality feeds, can significantly degrade alert validation quality by increasing false positives. Analysts shouldn't interpret an alert without examining the IOC's last-seen date, variant history, and whether the same IOC has been reused within the same campaign.
- **Campaign context should be used both for matching and for understanding adversary intent.**
  If a campaign is financially motivated, the impact of an alert differs; if it's espionage-driven, the priority level rises dramatically. The CTI campaign profile directly guides how urgently an alert should be investigated.
- **A network-based alert shouldn't be interpreted without infrastructure chain correlation.**
  A single IP address or domain means nothing on its own. C2 cluster analysis must be performed by checking the same registrar, Autonomous System Number (ASN), certificate fingerprint, and infrastructure lineage (variant chain).

  If this step is skipped, critical C2 relationships may be overlooked.
- **TTP frequency can be misinterpreted without sector context.**
  For example, while T1078 may be a critical indicator in the financial sector, the same behavior may be far more commonplace in other sectors. Therefore, analysts must normalize MITRE ATT&CK technique frequencies according to their own sector to avoid misjudgment.
- **Without behavioral model inference, the timing of an alert can't be interpreted correctly.**
  Some APTs exhibit specific operational rhythms:
  - Fixed beacon intervals
  - Connections at specific hours
  - Intraday intensity fluctuations

  Analysts must verify where the alert's time window falls within these rhythms.
- **Using the anomaly score as a sole decision-maker is risky; it must be evaluated together with context.**
  A high anomaly score doesn't necessarily indicate an attack, and a low anomaly score doesn't fully exclude an attack. The anomaly score is a behavioral clue and is incomplete without CTI context.
- **The accuracy level of the alert source (source confidence level) must always be evaluated.**
  Some EDR engines are highly sensitive to process injection, while certain SIEM rules tend to generate noise. Therefore, the historical false-positive rate of the alert source should be taken into account.

- **Variants of the TTP that generated the alert must always be examined.**
  T1059 (Command Execution) is a broad category; its variants (e.g., PowerShell, CMD, WMI, Python; and in macOS environments, AppleScript) create entirely different risk profiles. Analysts shouldn't determine the risk level of an alert without identifying the correct variant.
- **The reliability of the CTI source used for alert validation (intel quality scoring) must be verified.**
  Errors in the intelligence source can skew the entire validation process. Therefore, the following parameters must be checked:
  - Confidence level
  - Reliability of the source
  - Context completeness
  - Integrity of STIX/TAXII
- **While validating an alert, the adversary's expected next step must be anticipated.**
  An alert indicates not only the current event but also the most likely next step in the chain:
  - If credential dumping has occurred, privilege escalation is expected.
  - If a C2 beacon exists, tooling download is expected.
  - If lateral movement traces are present, domain enumeration is expected.

  This foresight dramatically increases incident response speed.
- **Even if a context-enriched alert turns out to be a false positive, it still produces value; relationships learned during contextual processing must be fed back into the defense architecture.**
  While extracting context, regardless of whether the alert is real, analysts may identify the following:
  - A new C2 variant
  - A new TTP chain
  - A new anomaly pattern

  This information must be updated within the CTI repository.
- **Decisions made without context increase the analyst's cognitive load; context-based validation removes decision fatigue from the process.**
  Well-structured context frees the analyst's mind from unnecessary evaluation burden and enables focus on critical alerts. For this reason, context is an operational efficiency tool.
- **Every alert should be validated not with data alone, but with a narrative.**
  Without this sequence, an alert remains just a line of text: alert → context → behavior → chain → narrative

10.1.4 Integrating Technical Findings with the Threat Model

The analysis phase is a critical stage in incident response, where fragmented technical evidence must be transformed into a coherent attack narrative. Individual findings such as anomalies in memory, network telemetry, or file system artifacts often appear insignificant in isolation. However, when combined with CTI, these elements gain context and begin to reveal patterns that indicate adversary behavior and intent.

At this stage, the analyst's role extends beyond observation to correlation. Technical findings provide concrete evidence of activity, while the threat model (particularly frameworks such as MITRE ATT&CK) offers a structured understanding of how and why adversaries operate. When these layers are aligned, isolated signals can be mapped to specific stages of the attack chain. For example, process injection artifacts in memory, combined with C2 patterns in network traffic and persistence indicators in the file system, allow the analyst to position the adversary within the lifecycle with greater confidence.

CTI further accelerates this process by linking observed behaviors to known campaigns, infrastructure patterns, and adversary tendencies. A technical anomaly is no longer evaluated in isolation; it's assessed in relation to known threat activity, enabling faster and more accurate interpretation. This integration also supports forward-looking analysis, helping analysts anticipate likely next steps based on adversary behavior, tempo, and historical patterns.

Ultimately, the convergence of technical findings and the threat model enables a more complete understanding of the incident: its scope, progression, and potential impact. This approach transforms analysis from a reactive task into a structured, intelligence-driven process.

The following sections examine how this integration can be applied across different data sources and analytical layers.

Multilayer Evidence Integration

Multilayer evidence integration is one of the most critical methods used during the analysis phase of incident response. The primary objective is to transform technical artifacts obtained from different sources into a coherent whole within the same attack chain. Modern attacks don't leave traces on a single surface; instead, they generate simultaneous and fragmented signals across memory, network, file system, or ETW channels. For this reason, when an analyst focuses on only a single data category, the full scope of the attack can't be observed.

True attack detection becomes possible when correlations are established across layers:

- **The position of memory artifacts within the attack chain**
  Technical findings observed in memory dumps (`memdump`) contain the earliest and most raw evidence of the attack chain:
  - T1055 (Process Injection) behaviors (`RemoteThread`, `CreateRemoteThread`, `QueueUserAPC`, `NtMapViewOfSection` indicators)

- Suspicious heap allocations (RWX memory segments)
- Portable Executable (PE)–less shellcode signatures resembling reflective loader structures

These artifacts indicate actions performed by the adversary very close to the execution phase. Detection of injection alone doesn't reveal the adversary's intent; however, when combined with network telemetry and file system traces, the position within the attack chain becomes clear.

- **Analysis of C2 behavior through network telemetry**
  Network telemetry represents the surface through which the adversary communicates with the environment. Critical signals in this layer include the following:
  - 60-second beaconing
  - Low jitter percentages
  - Application-layer protocol variants used for C2 (T1071)
  - TLS fingerprint mismatches (JA3/JA3S)
  - Regular size patterns in outbound packets
  - Domain activity matching pDNS history

  These indicators validate whether the injection artifact observed in memory represents merely a local anomaly or is generated by an implant actively maintaining C2 communication. The existence of an attack often emerges from the alignment of memory and network telemetry sets.
- **File system traces and persistence behavior**
  File system artifacts typically indicate the consolidation phase of the attack chain. Examples include the following:
  - T1547 (Startup Registry/Service Persistence)
  - Unauthorized DLL chains (DLL Search Order Hijacking)
  - Timestomping (T1070) variants
  - Unexpected PE timestamp values

  In particular, persistence mechanisms demonstrate that the adversary doesn't intend to execute code only once, but rather aims to remain in the environment for an extended period. Persistence mechanisms indicate that the adversary's activity isn't limited to a single execution but is part of a broader, staged attack progression. When analyzed together, different data sources reflect distinct phases of this progression: Memory artifacts point to execution activities, network telemetry reflects C2 communication, and file system traces reveal persistence mechanisms. Taken together, these layers provide a structured view of how the attack evolves across the environment.
- **Gaining behavioral depth through ETW logs**
  ETW is a critical signal source, especially in advanced attack analysis. The following ETW providers supply valuable data for multilayer integration:

  - `Microsoft-Windows-Kernel-Process`: Process creation/injection patterns
  - `Microsoft-Windows-Security-Auditing`: Token manipulation/privilege escalation
  - `Microsoft-Windows-Kernel-Network`: Network behavior deviations
  - `Microsoft-Windows-PowerShell`: PowerShell-based reconnaissance and execution

  ETW signals function like connective tissue between behaviors observed in memory and their consequences on the network.

- **Detecting metric inconsistencies through user behavior analytics (UBA)**
  UBA adds an additional layer that strengthens the context of technical artifacts. The following anomalies may indicate that a T1055 injection isn't the result of a software fault, but rather an attack technique employed for potential credential access:
  - A user session lasts longer than normal.
  - The same user performs activities on multiple machines within a short time span.
  - Login patterns deviate during nighttime hours.
- **Constructing the TTP chain as a model**
  After integrating technical findings, each data layer can be directly associated with a specific stage of the attack. Memory artifacts often indicate process injection activity (T1055), reflecting execution on the host. Network telemetry typically reveals beaconing behavior (T1071), representing C2 communication. File system traces, in turn, point to persistence mechanisms (T1547), showing how the adversary maintains long-term access. This chain directly corresponds to the sequence commonly observed in modern APT operations: execution → C2 → persistence.

  Through integration, the analyst can clearly answer the following questions:
  - At which stage of the chain is the adversary?
  - Is there a risk of lateral propagation?
  - Is privilege escalation expected?
  - Are supporting TTPs aligned sequentially within the chain?

  These answers directly determine how the incident should be handled.

Why is this integration essential, from a technical perspective? An anomaly detected in a single layer may represent a software error, a side effect, benign behavior, or a false positive. However, consistent signals recurring across three layers are rarely coincidental, for example:

- Injection in memory may be incidental.
- Outbound jitter may be normal traffic.
- Persistence DLL chain may be a software update.

Yet if consistent signals appear across all three layers, the likelihood of adversary behavior increases significantly in practice and should be prioritized.

Multilayer integration improves analysis by reducing false positives, clarifying the attack chain position, and enabling more accurate assessment of adversary behavior and risk.

Resolving Adversary Intent Through Technical Traces

Adversary intent is one of the most critical analytical layers in modern incident response processes because where an attack is attempting to go is just as important as where an attack originates from or which tools it uses. While technical findings indicate the adversary's current position within the attack chain, intent analysis provides a strong forecast of how that chain is likely to progress. This foresight directly determines the speed of detection and containment actions.

The reason incident response analysts transform technical signals into intent is straightforward: Every technical artifact is a reflection of the adversary's strategic thinking. When the following traces are interpreted correctly, they reveal the adversary's operational objective:

- **Reconnaissance intent**
  The types of behaviors an adversary exhibits in the initial phase typically indicate reconnaissance motivation. These signals are largely oriented toward understanding system structures, user relationships, and network paths. Technical indicators include the following:
  - High volumes of WMI query calls (e.g., WMI command-line [WMIC] process list, WMIC user account)
  - Increased activity of the `Microsoft-Windows-WMI-Activity` provider within ETW
  - Lightweight Directory Access Protocol (LDAP)–based domain enumeration attempts
  - PowerShell command sequences such as `Get-NetUser` and `Get-NetGroup`
  - Elevated Address Resolution Protocol (ARP)/Network Basic Input/Output System (NetBIOS) broadcast activity

  These findings indicate that the attack chain is still in the reconnaissance phase. CTI context enables prediction of which TTPs may follow in later stages. For example, if it's known that a specific APT group conducting reconnaissance typically shifts toward credential access within 10–15 minutes, the analyst can take preventive action accordingly.
- **Credential access intent**
  LSASS access, token manipulation, or Security Accounts Manager (SAM) dumping attempts indicate that the adversary is preparing for the next stage: privilege escalation. Technical indicators include the following:
  - `OpenProcess/ReadProcessMemory` access targeting `lsass.exe`
  - Suspicious intensity of `SeDebugPrivilege` usage
  - DPAPI blob access observed in `Procmon` logs
  - Credential tool-chain behaviors observed via `Microsoft-Windows-Security-Auditing` (Event ID 4688) events in ETW
  - `MiniDumpWriteDump` calls triggered from abnormal processes

This behavior constitutes a very strong signal of the adversary's intent to progress along the attack chain. CTI's contribution is particularly evident here: While some APT groups perform credential dumping to test initial access, others initiate lateral movement within minutes after dumping credentials. If the analyst interprets intent correctly, the next link in the chain can be transformed into a proactive incident response action.

- **C2 stabilization intent**

  The behavior of the C2 channel reveals the adversary's level of control and the maturity of the operation. Technical indicators include the following:

  - Reduced beacon jitter variance (indicating a transition into a stable interaction mode)
  - Fixed-interval beacon flows
  - Repeated TLS fingerprint matches (JA3)
  - Pipelining or chunking behavior in C2 traffic
  - Transition to privacy-enhancing transport protocols such as DNS over HTTPS (DoH)

  These signals indicate that the adversary intends to remain in the environment for an extended period. Accordingly, incident response priorities shift: The focus is no longer solely detection, but the speed of isolation.

- **Lateral movement intent**

  Attempts to expand access correspond to one of the most dangerous stages of the attack chain. When this intent is detected, the incident response team effectively races against time. Technical indicators include the following:

  - Server Message Block (SMB) session establishment anomalies (`TreeConnect`, `SessionSetup`)
  - Unexpected connections initiated via WinRM
  - `wmic/node:`-based calls
  - Increased Kerberos pre-authentication request volume
  - Behaviors consistent with tools such as PsExec
  - Suspicious write-access attempts on `Admin$`

  This set of behaviors indicates that the adversary is preparing to transition from credential access to lateral movement. CTI clarifies which APT groups exhibit this specific rhythm and which techniques typically follow.

When intent analysis is performed correctly, the following occurs:

- Containment times are reduced (because the targeted stage is known).
- Root cause analysis is more accurately positioned (the analyst rapidly identifies the beginning of the chain).
- Blocking can occur before lateral movement (the chain is broken).
- Adversary maturity level is assessed (targeted campaign vs. opportunistic attack).
- Playbook activation is optimized (incorrect actions are avoided, and the correct playbook is triggered).

At its core, the process follows a clear progression in which technical findings are interpreted as observable behaviors, these behaviors are then analyzed to infer adversary intent, and that intent ultimately drives the strategic decisions taken during incident response. This chain determines the success of incident response. Adversary intent represents far more than a single log entry or a simple anomaly; it's the technical surface reflection of the adversary's mental model. An analyst who can read these signals can anticipate an attack, intercept it, slow it down, and often suppress it before the chain is completed.

When adversary intent analysis is combined with the context provided by CTI, it becomes one of the sharpest decision-making mechanisms within the incident response process.

10.2 Detection and Validation with IOCs and IOAs

The role of indicators of compromise (IOCs) and indicators of attack (IOAs) in detection and validation processes constitutes one of the most decisive thresholds of modern cybersecurity. At a critical moment within the incident response chain, a single IOC can sometimes change the entire direction of an analysis; at other times, an IOA can reveal the malicious intent hidden behind behavior that appears ordinary on the surface. These two concepts are never just technical indicators—IOCs and IOAs are carriers of a subtle yet extremely critical link established between the tangible traces left by the adversary in the environment and the defender's experience and intuition.

The detection process often begins with a simple alert. A short line appearing in a SIEM log, a low-priority warning displayed on an EDR console, or a single network connection attempt. At first glance, each of these signals may seem routine, even insignificant. What is decisive at this point, however, isn't the signal itself, but the context in which it appears. In the construction of context, the correct interpretation of IOCs and IOAs plays a central role. The same IOC can take on a completely different meaning when observed in a different time window or on a different system. Similarly, behavior that appears benign in isolation can become the harbinger of a serious threat when combined with other links in the attack chain.

IOCs represent the most tangible traces left behind by an adversary after making contact with a system. A file hash, an IP address, a domain name, a registry key, or a specific file path are often perceived as cold and mechanical data points at first glance. Yet every IOC is, in fact, the frozen state of a specific malicious action—the projection of a behavior that occurred in the past. For this reason, IOCs show what an attack has left behind. However, these indicators lose value over time; variants change, and depending on the adversary's tactics, they can rapidly lose their validity. This reality removes IOCs from being absolute sources of truth on their own and renders them dependent on context.

IOAs, on the other hand, represent the living side of an attack. The IOA approach, which focuses on behavioral indicators, aims to understand not only what the adversary has left behind but what they are attempting to do. An anomalous process injection attempt, an unexpected PowerShell execution, repeated privilege escalation attempts within a short period, or a rhythmic beaconing pattern indicate that adversary intent is still active and that the attack chain may be progressing. For this reason, IOAs are more dynamic than IOCs and often carry clues about what may come next.

True detection and validation power emerges at the point where IOCs and IOAs are considered together. While an IOC tells the analyst that a similar trace has been observed in the past, an IOA whispers whether the same threat is becoming active again today. This dual reading enables the analyst to evaluate the attack's progression over time instead of focusing on a single moment. As a result, it becomes clearer which stage the attack is in, what may be targeted next, and which systems are at risk.

In modern security environments, this approach is vital. Today's SOCs are surrounded by high volumes of noise: false positives, context-free alerts, and thousands of warnings that contain no real threat. Within this noise, extracting the signal requires both technical tools and the analyst's reasoning capability. For this reason, the validation phase isn't a simple "matched/not matched" check. On the contrary, it's a multilayered process that includes evaluating the validity period of an IOC, understanding the position of an IOA within the attack chain, and correlating all of these findings with threat intelligence.

The following sections examine how IOCs and IOAs are applied in practice, including their lifecycle and reliability, their role in behavioral detection and validation, and how they are combined with CTI to support accurate analysis and decision making within the incident response process.

10.2.1 The Lifecycle of IOCs and Their Operational Value

An IOC, at first glance, appears to be nothing more than a small and technical indicator: a hash, an IP address, a domain name, or a certificate fingerprint. However, this surface-level appearance is deceptive. In reality, every IOC is an external manifestation of a time-spanning relationship formed with the adversary's operational habits.

Some IOCs are inherently transient. They lose their meaning within hours, are replaced by new ones, and leave behind only a brief trace of contact. Others remain active for months; they reappear repeatedly across different incidents, different victims, and different time frames. Rarer still are those that become the invisible backbone of a threat campaign. Even as years pass, they persist unchanged, silently carrying the adversary's operational identity. This diversity removes the IOC concept from being a static indicator and transforms it into a temporal, contextual, and behavioral analytical object.

For this reason, the value of IOCs can't be explained solely through the concept of matching. The real issue is being able to read how long an indicator persists, in which contexts it appears, and with what degree of consistency. From a cyber defense perspective, this is far more than a technical skill. It's an art of reading against time. When an analyst learns to focus not on the indicator itself, but on its behavior over time, its repetition patterns, and its resistance to change, IOCs cease to be mere alarm-generating signals and instead become clues that reflect operational character.

Real-World Example

I remember the day I first realized the importance of an IOC's lifecycle. There was a hash match flashing on the SOC screen—a bright, clear, aggressive red icon. Even the tone of the alert seemed to shout, "Found it—this is it!" But a few hours later, when I saw that the adversary had simply repackaged the same file and completely changed the hash, I felt a slight sense of disappointment. A signal had suddenly become meaningless. That day, I understood that not every IOC ages at the same speed, and an analyst must learn to distinguish what is valuable by observing its behavior over time.

To better understand this variation, it's useful to examine how different types of IOCs behave over time and how their lifecycles influence detection and validation decisions:

- **Hash IOCs: A short-lived spark**
  Hash values are like the fastest-fading flower within the IOC family. They are powerful, sharp, and striking, but short-lived because modern adversaries can produce thousands of different hash variants of the same malware in the following ways:
  - Inserting a small whitespace into the code
  - Changing compilation settings
  - Using a packer or a crypter

  For this reason, hash IOCs offer high accuracy but only for a very limited period of time. If an analyst clings to a hash, they may often be relying on a trace the adversary abandoned hours ago. This awareness determines the direction of the defense.
- **IP IOCs: Medium lifespan, variable silence**
  IP addresses can, in some campaigns, be observed for longer periods compared to hash IOCs; however, due to modern infrastructures, dynamic hosting models, and rapid rotation techniques, they are often extremely fragile and volatile. Some actors change IPs within days, while others broadcast C2 traffic from the same IP for weeks. The most interesting aspect here is that the lifespan of an IP reflects the adversary's operational discipline, for example:
  - APT groups targeting the financial sector may sometimes use the same ASN blocks for years.
  - Ransomware gangs, on the other hand, tend to increase IP rotation to remain hidden.

  Observing an IP IOC that remains unchanged for a long period may, in certain scenarios, indicate that the adversary is using a stable infrastructure; however, this must be evaluated together with alternative explanations such as intermediary services, shared hosting, or network architecture constraints.

When examining IP IOCs, you may be curious to get answers to "How long has this address been alive? And why has it not changed yet?" These answers can help clarify the adversary's intent.

- **Domain IOCs: A thin line between ephemerality and persistence**
  Domain IOCs are among the most complex members of the IOC ecosystem because a domain's lifespan varies dramatically depending on the DNS model used, the hosting service, the adversary's strategic planning, and the nature of the campaign.

  Some domains operate for only a few hours and disappear like a shooting star (phishing, one-time C2 footholds). Others serve the same campaign family for months or even years (large-scale, state-sponsored infrastructures). Domain IOCs become exceptionally valuable, especially when they have the following characteristics:

  - Registrar structures repeat.
  - Name Server (NS) records exhibit similar patterns.
  - Subdomain templates are distinctive (e.g., `cdn-update-win32[.]com`).
  - pDNS history is rich and consistent.

  Domain IOCs are like the handwriting of the adversary. Letters and words may change, but the rhythm of the handwriting often remains the same.

- **Certificate fingerprints: Relatively persistent and high-value traces**
  Certificate fingerprints (especially SHA-1/SHA-256 fingerprints) are among the most resilient IOCs. This is because certificate management requires operational care and consistency, particularly when the adversary directly controls the infrastructure. While some actors prefer to reuse certificates for long periods, others may rapidly change these traces through automation or intermediary services.

  It has been observed that APT groups using TLS-based C2 communication sometimes don't change certificate fingerprints for months (or, in some cases, even years). For this reason, a certificate IOC is worth its weight in gold when it comes to linking a campaign chain.

  When analysts see a fingerprint match, the thought is usually something like, "This can't be a coincidence. There's a story here."

Why Is Lifespan So Important?

The lifespan of an IOC directly affects the analyst's speed and accuracy of decision-making:

- If a hash is aging, focus on behavior.
- If an IP is changing, follow the infrastructure chain.
- If a domain is consistent, the campaign may still be ongoing.
- If a certificate fingerprint consistently reappears, in most cases, it indicates continuity associated with the same operational infrastructure, campaign line, or actor; however, this linkage must be validated by considering infrastructure ownership and intermediary layers.

An IOC's behavior over time sharpens detection accuracy. It makes understanding the chain easier, eliminates noise, and grants the analyst deep intuition. The lifespan of IOCs reveals the operational character of the threat actor to the defender. Some indicators disappear quickly; others are resilient. These differences resemble a threat philosophy. Over time, you learn that some IOCs point only to a single location while others reveal an entire journey.

For an analyst, the true value lies in recognizing this distinction and using it to uncover broader adversary behavior. At that moment, a seemingly minor trace can reveal a much larger operational pattern.

10.2.2 Strengthening IOA-Based Behavioral Detection Logic

IOA-based behavioral detection represents the most vivid, most dynamic, and most human-scented side of modern cyber defense because here, there's no longer the frozen face of a hash, nor the cold silence of an IP address. Instead, there's movement, intent, rhythm, habit, and often the feeling that something is going wrong. IOA reads every step taken by the adversary in behavioral form before it ever appears as a technical trace. For this reason, sometimes a small deviation in a process tree, a sequence of syscalls, an unexpected handle opening, or a single anomalous connection echoing in the silence of the night can reveal the entire chain.

Behavioral detection focuses not on what the adversary does, but on how it's done. This distinction is one of the most powerful differentiators shaping the future of cyber defense. There's that moment during a night shift in the SOC when a slight drowsiness sets in; the glow of the screen presses heavily against your eyelids. Just then, an alert appears in the EDR: "Suspicious Process Tree Anomaly." At first glance, it seems like nothing's wrong:

```
chrome.exe → powershell.exe → rundll32.exe...
```

But there's a rhythm disruption in that chain. Something isn't flowing naturally, making the next quesion, "Why would Chrome start PowerShell under normal circumstances?" This is exactly what we call an IOA: the scent of behavior changes, the tone of the flow becomes distorted.

Behavioral detection interprets adversary behavior through three primary structures, which we'll discuss in detail in the following sections:

- Syscall sequences
- Process tree anomalies
- Unusual changes in access patterns

Each of these generates a strong signal on its own; but together, they reveal the adversary's mental model.

System Call Sequences

Syscalls expose the most intimate behaviors of a malicious payload. Because regardless of which language, framework, or packer an adversary uses, they ultimately reveal themselves through the calls they make to the operating system, for example:

- The sequence `NtAllocateVirtualMemory` → `NtWriteVirtualMemory` → `NtCreateThreadEx` represents a classic injection chain.
- The sequence `NtOpenProcess` → `NtReadVirtualMemory` is an early indicator of credential-dumping behavior.
- An unusually high volume of `NtQuerySystemInformation` calls reveals the reconnaissance phase.

For example, when a normal Word document is opened, it generates approximately 120–180 syscalls. If that same Word document contains an exploit, the syscall sequence suddenly flows with an entirely different rhythm. This is where that subtle inner stir within the analyst is born: "These calls don't belong to the same melody." IOA makes these microscopic rhythm deviations visible.

Abnormal Process Tree Structures

Every process tree tells a story. Under normal conditions, it flows like this:

```
explorer.exe
└── chrome.exe
        └── adobeupdater.exe
```

But with an adversarial touch, it looks like this:

```
chrome.exe
└── powershell.exe
        └── rundll32.exe (remote thread)
```

When an analyst sees this, internal questions arise: "What is the purpose of this chain? Why does it go from Chrome to PowerShell?" This is where the value of process tree–based IOAs lies: They make unnatural relationships visible.

Real-World Example

In a financial institution, there was a PowerShell command that appeared benign in the EDR. The process tree was as follows:

```
excel.exe
└── powershell.exe
        └── certutil.exe
```

Individually, none of these processes appeared malicious, but when combined, they formed an almost "whispering" IOA: This chain reflected the adversaries' Living Off the

Land Binaries (LOLBins) tactic. Although each process may appear legitimate on its own, their combined sequence reflects a coordinated execution pattern that aligns with known adversary tradecraft, making the activity detectable only when evaluated as a behavioral chain rather than as isolated events.

The result was an intrusion attempt contained within five minutes. The power of IOA lies within an early intuition that an IOC could never achieve.

Unusual Access Patterns

One of the most critical IOAs domain is the way an adversary accesses resources:

- From where?
- At what time?
- In what order?
- How frequently?

An unexpected pattern is often the first signal of an approaching attack, for example:

- A single user account accessing five different servers simultaneously at 03:40 at night
- An unexpected handle being opened to the LSASS process
- An increasing scanning rhythm in SMB connections at short intervals
- WMI queries repeating in 1-second cycles

These behaviors are interpreted at the IOA level, and the analysis focuses on determining which stage of the adversary chain is currently active.

Real-World Example

Before lateral movement began at a customer environment, there was only a single sign—the silent repetition of the `net.exe` command on specific systems. No IOC generated an alert, but the IOA revealed the intent: "The adversary is performing reconnaissance." The action we took at that moment broke the chain before it could even be established.

IOAs are critical because they capture behavioral patterns that are far more difficult for adversaries to alter than static indicators. While hashes, domains, and IPs can be easily changed, underlying behaviors (e.g., discovery, credential access, and lateral movement) remain relatively consistent. By focusing on these patterns, IOAs provide a reliable way to identify adversary intent.

The Analyst's Road Map

We now have an understanding of the relevant structures, but the key challenge is how to operationalize them in real-world analysis. Moving from theory to practice requires a structured approach that translates behavioral insights into actionable detection logic. For analysts, this means recognizing patterns and consistently applying them to interpret activity, assess intent, and guide response decisions. This section provides a road map for strengthening IOA-based detection logic.

IOA-driven systems require three core capabilities:

- **Behavioral model construction**
  The normal rhythm of ordinary behavior is defined. Every deviation becomes a potential IOA.
- **Intent-focused evaluation**
  Which stage of the attack does a behavior indicate? An analyst who can answer this question can anticipate the incident chain.
- **Multilayered context**
  Syscall analysis + process tree + access patterns + CTI intelligence provides the highest level of detection accuracy.

In behavioral detection, most analysts look at large indicators: Is the process tree broken? Are WMI queries excessive? Has SMB scanning increased? Yet true mastery lies in capturing the microdeviations that the adversary is often not even aware of.

The first microdeviation category is *timing anomalies* (subsecond timing deviations). Normal user behavior is organic. The human hand has a rhythm; machines do not, for example:

- When a user opens and closes files, delays typically fluctuate between 300 and 900 ms.
- A bot-based adversary, however, produces a perfectly repeating access rhythm such as 4 ms → 4 ms → 4 ms → 4 ms.

This small but sharp anomaly isn't processed by most EDR solutions. But an IOA engine perceives that the rhythm isn't human.

Real-World Example
I felt the same thing one night while examining Sysmon logs in the SOC: "This click doesn't belong to me. This rhythm wasn't produced by a living person."

Next is the *microjitter pattern (MJP)*. An adversary applies jitter while attempting to conceal C2 traffic. But jitter is never perfect. The adversary's framework in use, scripting language, local system clock, and command execution all leave traces at a microlevel. This trace is unique, almost like the adversary's breathing pattern. This microjitter enables identification of the adversary even after a version update.

Other microdeviations are part of the adversary's *motor signature*; an advanced concept that most analysts have never encountered. In some APT groups, a distinctive coding style leaves traces in behavioral detection:

- **Thread management characteristics**
  Some groups create a large number of short-lived threads (thread bursts). Others operate using a single long-running thread. This preference remains unchanged for years. It's not a signature, but a character trait.
- **Memory management routines**
  Some adversaries prefer this chain: `VirtualAlloc` → `memset` → `memcpy`. Others frequently use this one: `NtAllocateVirtualMemory` → `RtlMoveMemory`.

  A behavioral engine can perceive this preference as an IOA. This is an operational habit that remains unchanged even if the hash changes, the domain changes, or the IP changes.

Sometimes, an adversary executes no explicitly malicious process. They drop no malware file and establish no C2 connection. So how are they detected? Detection comes through *side effects*. Adversarial behavior unintentionally produces small tremors:

- **Kernel-mode timer drift**
  Some implants hook timers, causing microscopic drift in the system clock. The user never notices this. But the behavioral engine does.
- **Handle leak patterns**
  Poorly written malicious implants leave behind thread or handle leaks. Four leaks within five minutes isn't normal. This leak is a behavioral trace of the adversary's code quality.
- **Cache placement anomalies**
  When an adversary uses a memory-resident implant, tiny waves form in the CPU cache lines. This wave creates only a few milliseconds of deviation in performance counters, but for a sufficiently capable IOA system, even this is a signal.

Psychological IOA

Yes, you read that correctly: Psychological IOA is a real concept. It refers to the adversary's mental state being reflected in their technical behavior.

The impatient adversary moves quickly, spawns many processes, breaks log consistency, and exhibits irregular lateral-movement rhythm. An impatient adversary usually can't remain in the network for long.

The patient APT, on the other hand, does the following:

- Performs only reconnaissance for three days
- Advances laterally in pinpoint steps rather than all at once
- Repeats access patterns at the same hours of the night
- Generates extremely low syscall noise

This adversary's IOA is silence. Sometimes IOA isn't noise, but silence.

Going beyond the microdeviations, analysts should keep the following considerations in mind to make the most of IOAs:

- **Look for the hidden signal within the noise.**
  An adversary's behavior may not be meaningful on its own. But when clustered, an adversary profile emerges, for example:
  - Syscall deviation: 1%
  - Process tree anomaly rate: 3%
  - Access pattern disruption: 2%

  Individually, none of these trigger an alert. But together, they form a 6%–7% behavioral cluster that is statistically unique. Like a distinct echo within the noise.
- **Use IOA to distinguish tools.**
  IOA can reveal not only the adversary but even the tool the adversary is using, for example:
  - Cobalt Strike: Specific syscall jitter model + pipe access pattern
  - Sliver: More regular thread management + TLS handshake variants
  - Mythic: Stable beacon period + Python-based syscall sequences
  - Havoc: High memory stability + low handle leakage after injection

  This type of detection is something almost no IOC can achieve.
- **Beware of silent adversary behavior.**
  Some adversaries exert extraordinary effort to avoid producing IOAs. But they forget one fact: Zero IOA is also an IOA.

  A human user does the following:
  - Makes mistakes
  - Misclicks
  - Opens and closes processes
  - Runs scripts
  - Gets lost between tabs

  The adversary, on the other hand, tries to remain perfectly clean.

  This overly clean behavior is, in fact, the strongest signal—much like the suspicion raised by an excessively orderly desk that appears never used.
- **Aggregate your findings into a profile**
  Advanced analysts can aggregate adversary behaviors and construct a profile:
  - Slow-moving?
  - Aggressive?
  - Access-centric?
  - Tool-dependent?
  - Reconnaissance-focused?

- Multi-thread-oriented?
- Low syscall noise?

This profile enables identification of the adversary even while the chain is still in progress, which is the greatest strength of IOA:

- Identifying the adversary before they are visible
- Detecting without file artifacts
- Reading intent even if the infrastructure changes
- Breaking the chain before it's established

IOCs indicate what has occurred, while IOAs provide insight into ongoing activity and likely next steps. Behavioral detection focuses on identifying patterns that reveal adversary intent, such as syscall anomalies, thread deviations, or unusual process relationships. By interpreting these signals, analysts can move beyond isolated observations and develop a clearer understanding of the attack's progression.

10.2.3 Multilayer Correlation with Live Telemetry

In multilayer correlation with live telemetry, reliance is no longer placed on a single signal alone; network flows, EDR events, DNS records, file integrity checks, and process behaviors all come together and are placed on a common timeline. In a sense, the original timeline of a cyberattack is reconstructed.

This section describes a larger structure built upon IOA-based behavioral detection: the effort to turn live telemetry into a single breath, a single rhythm. This section explores how multilayer correlation is operationalized in practice, including the alignment of telemetry across time; the integration of network, endpoint, and DNS data; and the identification of behavioral patterns that emerge only when these layers are analyzed together. It also examines how correlation strengthens detection accuracy, reduces ambiguity, and enables analysts to reconstruct the sequence and causality of an attack with greater confidence.

The Foundation of Multilayer Correlation: Time Itself

In cyber incidents, time is an invisible compass; however, for this compass to function correctly, time must first be calibrated (NTP synchronization, clock drift tolerance, time zone normalization, and log collection delays). Correlation performed without this calibration can misrepresent the sequence of events and causality.

Real-World Example

I felt this very clearly during a SOC night shift. Network telemetry was showing a mysterious outbound spike. In the same minute, the EDR reported that a process had been injected. On the DNS side, there was a suspicious domain resolution. The file integrity module flagged a DLL modification 90 seconds later.

> Individually, each had low contextual value; however, in an environment where time synchronization was verified, their convergence within the same short window (e.g., a few minutes) significantly increased the likelihood of shared causality. At that moment, the following thought crossed my mind: "This many coincidences can't be a coincidence." The power of correlation begins there.

To operationalize this approach in practice, analysts must treat time alignment as a structured preprocessing step rather than an implicit assumption. Here's the process flow:

1. Verify time synchronization across all telemetry sources by comparing NTP status, identifying clock drift (e.g., ±2–5 seconds tolerance for endpoints, tighter for network devices), and normalizing all timestamps into a common reference (typically UTC). SIEM pipelines should enforce this normalization during ingestion, ensuring that logs from EDR, DNS, network sensors, and file integrity systems are directly comparable.
2. Once temporal consistency is established, begin correlation by defining an appropriate time window. In practice, analysts typically work with sliding windows (e.g., 1–3 minutes for high-speed attacks, 5–15 minutes for slower, stealth-oriented activity). Events are then grouped based on proximity within this window, allowing patterns to emerge across layers. For example, a process injection event in EDR can be correlated with outbound network connections and DNS resolutions occurring within the same interval, forming a candidate activity cluster.
3. Perform cross-layer validation. Analysts shouldn't rely on a single telemetry source, but instead confirm whether multiple layers support the same hypothesis. This involves mapping each event to a functional category—execution (process activity), communication (network/DNS), and persistence or modification (file system changes)—and verifying whether these categories align in sequence. SIEM correlation rules or query-based approaches (e.g., joining logs on timestamp and host identifiers) are typically used to automate this step.
4. Filter noise before aggregation for effective correlation. Baseline behavior must be established to distinguish expected system activity from anomalous patterns. This can be achieved by profiling normal process trees, typical DNS query frequencies, and standard network flows. Events that deviate from these baselines—especially when they appear simultaneously across multiple telemetry sources—should be prioritized for deeper investigation.
5. Validate causality rather than mere coincidence. Temporal proximity alone isn't sufficient; supporting indicators such as shared host context, process lineage, or known CTI-linked infrastructure must reinforce the relationship. For instance, if a suspicious DNS query resolves to an IP associated with a known campaign, and this aligns with a process injection event on the same host, the confidence level of correlation increases significantly.

By following this structured approach (time normalization, window-based grouping, cross-layer validation, noise reduction, and causality verification), multilayer correlation becomes a repeatable analytical process rather than an intuitive exercise. This is the point at which correlation moves from observation to actionable intelligence.

Correlation with Network Flows

Network telemetry tells us where the attacker comes from, where they go, and the rhythm with which they communicate, for example:

- Regularly spaced (e.g., ~60-second) connection attempts or short-lived sessions over port 443 (taking jitter tolerance and the possibility of legitimate automation traffic into account)
- An ETW anomaly that began 30 minutes earlier on the same host
- A newly registered domain resolution observed in DNS logs

Each of these signals alone produces limited evidence; when evaluated together, they form a high-confidence hypothesis of active C2 communication and trigger validation steps (proxy logs, TLS fingerprinting, process-socket correlation, EDR network telemetry, DNS responses/time to live (TTL), and pDNS comparison). Network flows are particularly useful in answering the following questions:

- Are the connections occurring at expected times?
- Is the traffic volume normal?
- Has the destination IP ever been seen before?
- Which model do the packet lengths conform to?

Attackers may sometimes drop no files, perform no process injection, and even evade EDR entirely. Completely hiding network rhythm is difficult; however, attackers can weaken this trace through jitter, proxy chaining, content delivery network (CDN)/legitimate service impersonation, and timing diversification. For this reason, rhythm isn't an indicator that delivers judgment on its own; it's one that gains strength when combined with other layers.

In practice, correlating network flows follows these main steps:

1. Begin with normalizing and aggregating telemetry from sources such as NetFlow, Zeek, and firewall logs within a common time window. Analysts typically start by filtering for anomalous patterns (e.g., periodic connections, rare destination domains, or unusual port usage) and then pivot to host-level context by matching source IPs with endpoint identifiers from EDR data.
2. Perform session-level analysis by examining connection frequency, duration, byte distribution, and packet size patterns to distinguish automated beaconing from legitimate traffic. Tools such as Zeek logs or enriched NetFlow records can be used to identify periodicity, jitter, and burst behavior. At this stage, TLS fingerprinting (e.g., JA3/JA3S) and certificate metadata should be correlated to determine whether encrypted traffic aligns with known benign or suspicious profiles.

3. To further strengthen correlation, link network activity with process-level telemetry. Analysts should verify which process initiated the connection (e.g., via EDR or Sysmon Event ID 3) and whether this behavior is consistent with expected application activity. For example, a browser process initiating outbound HTTPS traffic may be normal, whereas a system process or scripting engine (e.g., PowerShell) establishing similar connections may indicate C2 activity.
4. Incorporate infrastructure context by checking domain age, pDNS history, ASN ownership, and known associations with threat campaigns. If multiple signals (e.g., periodic beaconing, uncommon infrastructure, and suspicious process origin) converge within the same time frame, the analyst can elevate confidence and proceed with containment or deeper validation steps.

By following this workflow—pattern detection, session analysis, process correlation, and infrastructure validation—network flow analysis becomes a repeatable method for identifying and confirming adversary communication.

The Role of EDR Events

EDR telemetry is one of the most critical sources of IOA-based detection; however, the level of visibility may vary depending on product configuration, sensor coverage, and event generation policies. Process injection, thread creation, handle opening, and token manipulation all carry the steps taken by the adversary. EDR often provides the signal that a meaningful execution or privilege/access behavior has occurred on this host; however, due to certain techniques (telemetry gaps, kernel/user-mode visibility limitations, allowlist/policy effects), part of the event may not be visible at all.

Network flow, on the other hand, says "And this activity communicated with the outside." DNS logs add "And to this address." When viewed together, the picture becomes clear.

Real-World Example

In a real incident, I observed telemetry as follows:

- 14:02:31 Suspicious thread creation/remote thread creation (remote thread start) behavior (EDR; in some products reported under low-level call names such as `NtCreateThreadEx`, while in others under a technical label)
- 14:02:36 Outbound traffic to a third-party domain (network)
- 14:02:40 New subdomain resolution (DNS)
- 14:03:15 File integrity alert (FIM)

This was more than a single process tree. This was an operational chain.

In practice, working with EDR events follows these main steps:

1. Validate the telemetry coverage and event fidelity. Analysts should first confirm whether key behaviors (e.g., process injection, remote thread creation, and token manipulation) are consistently logged and whether any gaps exist due to sensor configuration or policy constraints. This step is critical, as missing or partially captured events can lead to incorrect assumptions during analysis.
2. Correlate EDR events with host and process context. Analysts should examine parent-child relationships, process lineage, command-line arguments, and associated handles to determine whether the observed behavior aligns with legitimate activity. For example, a remote thread created within a browser process may indicate injection, but its significance depends on which process initiated it and under what context.
3. Extend correlation beyond the endpoint. EDR events must be aligned with network and DNS telemetry using a common timeline. In practice, analysts match process execution timestamps with outbound connections (e.g., via EDR network telemetry or firewall logs) and verify whether the same process initiated external communication. This process–network linkage is essential for confirming whether execution behavior is connected to C2 activity.
4. Validate the external infrastructure involved to strengthen confidence. This includes checking domain age, pDNS history, ASN ownership, and known associations with threat campaigns. If a suspicious process behavior, outbound connection, and domain resolution all align within a short time window and share contextual indicators, the likelihood of malicious activity increases significantly.
5. Reconstruct the sequence as an operational chain rather than isolated events. By ordering EDR, network, DNS, and file integrity events chronologically, it becomes possible to identify execution, communication, and persistence phases within a single timeline. This structured reconstruction allows analysts to move from detection to decision, determining whether containment, escalation, or deeper investigation is required.

The Power of DNS Logs

DNS, in the eyes of many analysts, is merely a service that resolves domain names. Yet from the attacker's perspective, DNS is the gateway to invisibility because its lightweight, fast, quiet, and ignored by most SIEM rules. For this reason, DNS logs are the most critical yet least appreciated component of a multilayer correlation structure.

During a real attack investigation, understanding that DNS logs are often the first telemetry to speak and the last to fall silent helps you see how deep a story this layer carries. The power of DNS lies not in the data itself, but in the rhythm of its progression, the inconsistencies in its age, and the kinship relationships within the topology.

The following three signals make DNS logs the most powerful layer capable of tracking the adversary's breath:

- **Domain lifeline**
  A domain's age is a strong contextual indicator in suspiciousness assessment; however, it's not decisive on its own and must be interpreted together with registrar/NS changes, hosting infrastructure, pDNS history, and traffic behavior.

  Newly created or newly activated domains increase the likelihood of abuse, including C2 and phishing; however, legitimate scenarios such as campaigns, products, CDNs, or corporate integrations must also be considered. Most attackers register domains immediately before launching an operation. This registration often has the following characteristics:

  - Aligns with operational hours
  - Exhibits similar naming patterns
  - Is kept for a very short duration (one year or less)
  - Has no pDNS history or only a very weak one

  In some campaigns, it can be observed that the infrastructure is created very close to the attack window (on the scale of hours to days); such a fresh infrastructure signal, when combined with pDNS history and registration/NS changes, produces a strong alert. A domain this fresh almost shouts that a new operation is coming.

  Long-standing domains often carry a higher likelihood of legitimacy; however, due to scenarios such as expired-domain takeover, hijacking, or purchase, registrar/NS changes, certificate/hosting changes, and sudden breaks in traffic patterns must always be checked.

  A long-standing domain is often legitimate. However, there are exceptions: Some APT actors purchase old and abandoned domains and use them as C2. Therefore, rather than looking only at age, the following three parameters are examined:

  - Has the registrar changed?
  - Have NS records been updated in recent months?
  - Has the domain's traffic volume suddenly increased?

  If even one of these questions is answered "yes," the domain may be suspicious despite its age.

  Domain age analysis enables the prediction of adversary infrastructure before a campaign begins. It's akin to listening to the attacker's footsteps in advance.

- **Resolution rhythm**
  The rhythm of DNS traffic is one of the clearest indicators distinguishing human behavior from machine behavior. Adversary implants often exhibit a low-variance, periodic, or weakly jittered resolution rhythm (e.g., recurring queries at 30–90 second intervals). However, because modern malware can employ jitter and randomization, rhythm analysis must be performed together with variance/distribution metrics (standard deviation,

coefficient of variation [CV], burst behavior). Human behavior, by contrast, is organic and irregular. Users may open a page, pause, click on something else, then go somewhere else, and subsequently do nothing for several seconds or minutes.

This natural irregularity is the hardest behavior for an attacker to imitate. Rhythmic DNS resolutions may indicate the following:

- Beacon-based C2 communication
- Command exchange concealed via DoH
- A malware implant's heartbeat function
- Exfiltration preparation
- Red team tools (Sliver, Cobalt Strike, Havoc, etc.)

Sometimes a single rhythm is enough to expose the adversary. When I first noticed this, the following thought crossed my mind: "This query couldn't have come from a human finger."

■ **Domain-cluster correlation**
One of the most powerful yet least utilized aspects of DNS logs is their ability to reveal infrastructure kinship. This approach doesn't focus on whether a domain is suspicious on its own; it examines its "family" to see if a domain matches the following:

- The same registrar
- The same ASN
- The same country location
- The same subdomain pattern
- The same pDNS siblings
- A similar Secure Sockets Layer (SSL) certificate fingerprint
- A similar Who Is (WHOIS) pattern

These similarities significantly increase the likelihood that the domain belongs to the same campaign/infrastructure family; for certainty, they must be validated through pDNS matches, certificate chain/fingerprint correlation, hosting artifacts, and the observed TTP/IOA chain. This makes the infrastructure topology that the adversary attempts to conceal visible.

Example

The following domains belong to the same campaign:

- `cdn-updatewin32[.]com`
- `cdn-winupdate32[.]net`
- `update-win32cdn[.]org`

All of them have the following traits:

- Hosted on the same ASN
- Registered using the same registrar
- Registered within the same period
- Carry the same SSL fingerprint

The moment you recognize this cluster, the picture becomes clear: This isn't a single domain; it's an operation. DNS logs break their silence at that point and tell the analyst: "This isn't an isolated indicator. This is part of a chain, and there are other links in the chain waiting for you."

Most cyberattacks don't generate noise. An adversary may drop no files, evade EDR, and change IP addresses, but the rhythm, age, and kinship relationships within DNS traffic flows are the behaviors that attackers find the hardest to conceal. When an analyst learns to read DNS logs, they begin to see both the queries and the intent behind them, the rhythm, and the quietly woven infrastructure topology. That small creak can sometimes reveal the entire attack.

File Integrity Traces

Attackers aim to operate invisibly; however, when they touch the file system, they leave traces. File integrity monitoring (FIM) is invaluable particularly in the following cases:

- When a persistence mechanism has been added or modified at the file/folder level
- When unexpected changes or timestamp deviations occur in critical directories
- When DLL replacement or binary patching is observed

Note

Registry-based triggers aren't within the scope of FIM; they are covered by registry auditing, EDR, Sysmon, or dedicated registry telemetry.

When combined with IOA, network flow, and DNS, FIM completes the entire picture. In an example where there was only a small DLL modification, EDR showed no process activity, the network was silent, and DNS appeared normal, FIM indicated a minor deviation. The timing of the DLL modification coincided with the same time window as an LSASS access attempt. This correlation alone isn't definitive; however, when the process tree, signature verification, the DLL's hash/signer information, the loading chain, and the LSASS access technique are validated together, the hypothesis of a credential access operation is strengthened.

Correlation Along the Time Axis

Examining all telemetry sources one by one is like flipping through the pages of a book. Placing them on a time axis, however, is reading the entire story. Time-based correlation asks the following:

- Is this event related to the one I saw before?
- At which moment of the chain does this behavior occur?
- Is this small signal the opening of the door to a larger attack?

When different layers generate signals within the same time window, the likelihood of false positives decreases and the probability that the event represents a chain increases. However, each layer must still be independently validated to ensure accuracy and to account for potential gaps in telemetry, scope limitations, or ingestion delays. In practice, this involves confirming reconnaissance activity through the correlation of DNS queries and ETW signals, validating injection behavior using EDR events supported by kernel-level indicators, assessing C2 establishment by combining network flow analysis with DNS activity, and verifying persistence mechanisms through FIM alongside registry telemetry.

The clustering of these signals within a short time window (e.g., 3–7 minutes) may indicate the early phases of an attack chain; however, it shouldn't be used as a definitive judgment without excluding scenarios such as legitimate automation or patch processes.

The Hidden Power of Multilayer Correlation: Making Silent Events Audible

The most difficult challenge in cybersecurity is detecting silent attacks. If an adversary can move through a system without writing files, changing IP addresses, generating hashes, or triggering EDR, then you need IOA + live telemetry correlation to break the silence.

We can do this because no adversary can perfectly conceal all of the following at once:

- Timing
- Rhythm
- Access sequence
- Kernel-level deviation
- DNS micropatterns
- Anomalous handle graphs
- Microscopic inconsistencies in process relationships

This "noise within silence" becomes audible through multilayer correlation. IOCs are like photographs. IOAs are video. Live telemetry correlation? That's the director's chair overseeing the entire video stream. When you build a system that combines all telemetry sources and aligns time, recognizes relationships, suppresses noise, and reveals intent, the adversary is no longer invisible.

10.2.4 Intelligence Enrichment Techniques for Reducing False Positives

Reducing false positives is an exhausting battle in cybersecurity. For an SOC analyst, it's sometimes not the alert itself that is tiring, but the uncertainty implied by the alert. An IP match, an unknown domain, and an unusual process behavior all may appear as threats, yet most of them are just noise. This noise accumulates in an analyst's mind; each alert creates a small vibration, and each vibration thickens the fog of uncertainty a little more. For this very reason, one of the most striking truths of threat hunting is that real threats are most often silent, while false alarms are surprisingly loud. The attacker moves forward with invisible steps, saying nothing; yet a misconfigured system, an application in its testing phase, or an ordinary update service produces particularly conspicuous sounds.

Data gains meaning only when combined with context. Without context, every alert appears equally dangerous; as context is added, some quickly fade, while others suddenly turn into a bright red point. WHOIS records, ASN information, historical attack campaigns, and threat actor profiles are the unseen yet most powerful tools for making this distinction. For this reason, most analysts eventually realize that the value of an alert doesn't arise from the alert itself, but from the contextual lights added around it. Context transforms a point into a map. It turns a single IP into a story, a single domain into a history, and a single behavior into an intention.

The following sections examine these enrichment techniques in detail, focusing on how specific data sources contribute to reducing false positives in practice.

WHOIS Records: The Biography of a Domain

Years of experience in SOC environments revealed that what enables an alert to be understood instantly isn't just technical knowledge, but intuition—the trace left by experience. The raw form of a signal never creates an emotional response in the analyst's heart, but when the signal merges with context, that inner feeling emerges: "Yes, this could be real." Sometimes, simply looking at a WHOIS query is enough to understand whether an alert is dangerous. It's easy to distinguish the sharp scent of a newly registered domain from a domain that has been in the hands of the same owner for years. Hidden within WHOIS records is a timeline: the moment a domain is born, the moment it changes hands, and the long periods during which it sinks into silence. Each of these whispers something small to the analyst during threat hunting: "Stay calm, this is an old friend," or "Be careful, this is the shadow of something new."

A WHOIS record is like a domain's identity card with its name, date of birth, owner, and historical changes written there. Answers to the following questions play a major role in filtering out false positives and bringing real threats to the surface:

- Why is WHOIS so critical?
- Is the domain newly created?
- Are the ownership details real or concealed?

- Has the registrar changed suddenly?
- Have the NS records been moved to suspicious providers?

Real-World Example

During a SOC shift, constant "risky domain" alerts were being generated. Everyone was concerned. I checked the WHOIS record: The domain had belonged to the same company for 14 years. At that moment, the meaning of the alert became clear. This wasn't a threat, but a misconfigured update service. WHOIS is one of the fastest ways to reduce false positives.

ASN Information: The Neighborhood Where an IP Address Lives

ASN information carries the spirit of the neighborhood in which an IP resides. Some neighborhoods open onto dangerous streets; others have been clean and orderly for years. Knowing which ASN an IP comes from is like understanding that IP's character. Some ASNs have long been associated with malicious infrastructures; traffic originating from them naturally raises suspicion. Other ASNs belong to well-established, reputable providers; a momentary anomaly from such a source is often nothing more than a technical mistake. Once an analyst learns this distinction, a large number of false positives is filtered out mentally before an alert is even opened.

Real-World Example

At a bank, an IP that appeared benign in outbound traffic became instantly suspicious through ASN analysis. The ASN check revealed AS9009. The sentence that passed through my mind at that moment still stays with me: "No good news comes from this neighborhood." Within minutes, a Cobalt Strike C2 was identified. ASN information reveals the signal hidden within the noise.

An IP address isn't merely a sequence of numbers; it has a neighborhood that is defined by ASN. But why is it important? Some ASNs host well-known malicious infrastructures: bulletproof hosting providers, botnet C2 centers, network blocks with persistent abuse reports, and so on. Following are a couple of examples:

- Certain hosting/virtual private server (VPS)/carrier ASNs may appear more frequently in abuse cases; however, an ASN alone isn't proof of maliciousness. ASN is a contextual layer that increases risk weighting and must be evaluated together with pDNS, certificates, traffic rhythm, behavioral telemetry, and in-incident correlation.
- Traffic observed from such ASNs may require faster triage; however, decisions must be made using multiple signals.

Historical Attack Campaigns

Historical attack campaigns are a poetic aspect of intelligence enrichment because attackers' histories tend to repeat themselves. A behavioral atlas is gradually formed by a domain pattern used by an actor years ago, beacon intervals rotating with the same rhythm, ASNs sourced from the same geographies, and so on. This atlas contains both the attacker's past steps and the potential steps they may take in the future. Looking at campaign data shows the analyst whether an alert is just a trailer or a full scene from a movie. Sometimes a domain that appears new is simply the continuation of an old threat with only two letters changed. In such a moment, the feeling inside the analyst is indescribable: "I have seen you before. And this time, I didn't hesitate to recognize you."

Most attackers reuse their behaviors repeatedly. Developing new techniques is labor-intensive; advancing with old habits is easier. For this reason, examining past campaigns is extremely valuable in determining whether a new alert is actually the continuation of an old story.

Campaign data reduces false positives via these questions:

- Has the subdomain pattern been seen before?
- Is the same ASN + registrar combination recurring?
- Does the beacon interval resemble the rhythm of a previous campaign?
- Is the targeted sector the same?

Real-World Example

A new domain detected in an organization closely resembled a domain from an APT campaign in 2020:

- 2020: `cdn-update-ms[.]com`
- 2023: `cdn-update-ms[.]net`

Only three letters were different. Such a similarity couldn't be a false positive.

Threat Actor Profiles

Threat actor profiles are the final layer of color that completes this picture. Every APT group has its own rhythm: Some are impatient, some patient; some aggressive, some subtle. If an analyst knows the actor's rhythm, they can quickly assess the authenticity of an alert. If the attack behavior doesn't align with the actor's historical sequence of tactics, the likelihood of a false positive increases. If the alert overlaps with the actor's known TTP chain, the alert suddenly gains seriousness. This knowledge creates a quiet confidence in the threat hunter's mind, much like recognizing a familiar face in a crowd.

Actor profiling is an excellent guide for interpreting alerts. A profile tells us the following:

- Preferred tools
- Active hours

- TTP sequence progression
- Frequently used infrastructures

Real-World Example

An SMB brute-force activity had emerged. The CTI team said, "This behavior can't belong to the APT-X group. They usually use RDP. This is most likely a false positive." Thirty minutes later, it was confirmed as an error from a monitoring agent. Profile knowledge eliminates false positives and makes the remaining signals far more meaningful.

Passive DNS Analysis

pDNS is a memory book that shows which IPs a domain has pointed to in the past, which subdomains it has appeared with, and when changes occurred. It's valuable because even if attackers change the domain, its history can't be erased.

pDNS reduces false positives in the following ways:

- If a domain has pointed to known legitimate IPs for years, a sudden change raises suspicion.
- If a domain has frequently been associated with malicious IPs in the past, risk increases.
- If a domain exhibits fast-flux behavior, the likelihood of botnet infrastructure increases.

Real-World Example

A domain appeared suspicious at a customer environment. It was newly registered but looked very clean. When we examined pDNS, we saw that the domain had moved across eight different IPs within 24 hours. Because this behavior could align with scenarios such as fast-flux, dynamic infrastructure, or CDN redirection, it must be evaluated as a risk signal after legitimate possibilities are excluded. The probability of a false positive dropped dramatically. pDNS is the invisible link between an attacker's intent and a domain's history.

SSL/TLS Certificate Enrichment

SSL certificate fingerprints are among the most difficult components of attacker infrastructure to change because they aren't cheap, they are labor-intensive, and frequent renewal carries risk. For this reason, certificate information serves as an excellent filter for reducing false positives.

Certificate information generates signals as follows:

- The recurrence of the same fingerprint across different domains confirms an infrastructure chain.

- Very short certificate lifetimes (1–3 months) are consistent with operational C2 infrastructures.
- A suspicious certificate authority (CA; not common providers like Let's Encrypt, but exotic CAs) increases the risk signal.
- Known signatures of malicious tools may be observed in certificate parameters (e.g., Cobalt Strike JARM fingerprints).

Real-World Example

Alerts were constantly being triggered for an IP; the IP itself appeared ordinary. However, the TLS fingerprint matched exactly a variant previously used in an APT campaign. At that moment, my mind clicked: "Alright, I know you." This small indicator enabled us to unravel the entire incident.

URL and Domain Pattern Analysis

Attackers change domain names, generate variants, and apply obfuscation, but one thing they often repeat unconsciously is their linguistic patterns. Consider this example:

- `cdn-update-win32.com`
- `ms-win32-update.net`
- `update32-cdn-windows.org`

These all look different, but the patterns are the same. The linguistic fingerprint of a threat actor is revealed when these patterns are analyzed:

- Domain naming structure
- Subdomain hierarchy
- URL path technique
- Parameter usage

This method is unique in reducing false positives because attackers may change their infrastructure, but they can't change the way they think.

Reverse Infrastructure Correlation

An advanced method for reducing false positives is as follows:

- The IP/domain used to generate the alert isn't examined first.
- Known attacker infrastructure clusters are identified beforehand.
- The values observed in the alert are scanned against these clusters.

This method interprets the value of the signal first through the attacker.

Example

All ASN, registrar, and subdomain patterns used by APT-28 over the past five years are extracted. If the incoming alert matches this cluster, the probability of a false positive drops dramatically.

For an analyst trying to understand attacker behavior, this technique evokes the following feeling: "I've learned their breathing rhythm, and this alert doesn't resemble that rhythm at all." Thus, false positives are rapidly eliminated.

Threat Intelligence Confidence Scoring

Not every intelligence source is equal. Therefore, when validating an alert, the reliability level of the source must always be evaluated. Reliability is measured by the following:

- Source's historical accuracy rate
- Completeness of context
- Freshness duration of the IOC
- Consistency of actor correlation
- Intelligence acquisition method of the source (sensor-based vs. analyst-based)

Example

If an IOC appears in three different low-confidence feeds and is absent from any reputable sources, the alert is most likely a false positive. Poor intelligence increases noise; high-quality intelligence cuts through noise.

Technique Comparison

Reducing false positives is the art of transforming an alert from a solitary data point into a meaningful event. Table 10.3 summarizes and compares the techniques we've discussed to give a complete picture.

Enrichment Source	Elements Examined	Primary Purpose	Impact on False Positives	Example Use Case
WHOIS	Domain age, ownership information, registrar, NS changes	Assessing the domain's legitimacy history and stability	Reduces unnecessary alerts for old and stable domains	Domain owned by the same company for 14 years: Indicates a misconfigured update service

Table 10.3: Intelligence Enrichment Sources and Their Impact on False Positives

Enrichment Source	Elements Examined	Primary Purpose	Impact on False Positives	Example Use Case
ASN	Network (AS) associated with the IP, country, provider type	Determining the IP's neighborhood and risk profile	Prioritizes traffic originating from notorious ASNs	AS9009 (M247): Rapid identification of outbound traffic as Cobalt Strike C2
Historical attack campaigns	Domain patterns, beacon interval, target sector, tools used	Determining whether a new alert is a continuation of an old campaign	Reduces false positives by filtering behaviors inconsistent with known campaigns	`cdn-update-ms[.]net`: variant of the 2020 `cdn-update-ms[.]com` campaign
Threat actor profiles	Actor TTP sequence, active hours, preferred protocols and tools	Correlating the alert with the actor's typical behavior	Rapidly filters alerts that don't match the profile	APT-X doesn't use brute force: An SMB brute-force alert likely false
pDNS	Historical IP mappings, subdomain relationships, fast-flux behavior	Understanding a domain's movement over time and its intent	Lowers noise for domains with a long, clean history	8 IP changes within 24 hours plus operational fast-flux: False positive likelihood decreases
SSL/TLS certificate fingerprint	Certificate lifetime, fingerprint reuse, CA type, known C2 fingerprint matches	Revealing infrastructure continuity and campaign linkages	Distinguishes benign vs. malicious domains sharing the same fingerprint	Ordinary IP + same TLS fingerprint as an APT campaign: Dramatical increase in alert significance
URL/domain pattern analysis	Naming patterns, subdomain hierarchy, path/parameter structure	Identifying the attacker's linguistic fingerprint	Filters domains/URLs that don't match the attacker's style	Linking variant domains in the `cdn-update-win32` family to the same campaign
Threat intel confidence scoring	Source reliability, IOC freshness, context completeness	Measuring the signal quality of intelligence sources	Deprioritizes low-confidence, single-source IOCs	IOC seen only in three low-confidence feeds likely marked as noise

Table 10.3: Intelligence Enrichment Sources and Their Impact on False Positives (Cont.)

Example: Extracting Intelligence from TLS Fingerprints with Python

When attempting to understand the infrastructure hidden behind a domain name, sometimes the only truly reliable source is a TLS certificate that silently presents itself. A certificate, much like a person's handwriting, inevitably leaves behind certain clues about its owner, even unintentionally. The analysis conducted for `haydararici.com.tr` becomes a clear illustration of this as it included an innocently appearing domain, a certificate that leaves retrospective traces, a cryptographic identity whose validity is confirmed, and an operational time window. When all of these elements are examined together, even a standalone JSON output transforms into a meaningful intelligence artifact.

As shown in Figure 10.1, this output was generated using a custom Python-based TLS fingerprint analysis script developed for this study, which programmatically retrieves certificate metadata (e.g., validity period, issuer details, Subject Alternative Name (SAN) entries, and cryptographic parameters) directly from the target service over port 443. This approach ensures that the analysis is based on live telemetry rather than static or third-party data sources.

```
C:\Users\admin\PycharmProjects\Books>python tls_fingerprint_analyzer.py  --port 443
{
  "target": "",
  "port": 443,
  "cert_info": {
    "subject": {
      "CN": "*."
    },
    "issuer": {
      "C": "US",
      "O": "Let's Encrypt",
      "CN": "R13"
    },
    "not_before": "2025-11-07T11:53:19+00:00",
    "not_after": "2026-02-05T11:53:18+00:00",
    "cert_valid_days": 89,
    "san_dns": [
      "*.",
      ".com.tr"
    ],
    "signature_algorithm": "sha256",
    "key_type": "RSA",
    "key_size": 2048,
    "is_ca": false,
    "key_usage": [
      "digital_signature",
      "key_encipherment"
    ],
    "extended_key_usage": [
      "1.3.6.1.5.5.7.3.1",
      "1.3.6.1.5.5.7.3.2"
    ]
```

Figure 10.1: TLS Fingerprint Analysis Output Revealing Certificate Validity, Cryptographic Identity, and Operational Time Window

In this section, we'll see what the extracted data signifies, which questions it raises in the mind of a SOC analyst, and why these seemingly small technical findings carry significant threat intelligence value.

Now, let's walk through what the data obtained during the analysis signifies, which questions it triggers in the mind of a SOC analyst, and why these small technical findings possess genuine threat intelligence value:

- **Subject**

  The `"subject"` field of a certificate provides the identity of the asset. `CN=*.haydararici.com.tr` indicates that the target is most likely a corporate infrastructure. The meaning of the wildcard usage is clear:

  - Enabling the same certificate to apply to dozens of subdomains
  - Allowing internal and external services to operate under a single cryptographic umbrella
  - Simplifying update and maintenance processes

  From a threat intelligence perspective, wildcard certificates can be both good and bad news. They are common in most corporate environments and may serve as a signal in favor of legitimacy; however, attackers may also employ wildcards in suitable scenarios. For this reason, a wildcard isn't exculpatory evidence, but merely a weighted contextual indicator. Accordingly, the wildcard observed here produces a weak-to-moderate contextual signal in favor of legitimacy.

- **Issuer**

  The certificate being signed by Let's Encrypt reflects a healthy trend followed by nearly all modern web services. Let's Encrypt can also be used by advertising-oriented attack infrastructures; however, within the context of this analysis, the issuer behavior is closer to legitimate use.

  Nevertheless, there's a rule in threat intelligence: A CA may be legitimate, but the infrastructure may not be. For this reason, a CA is never evidence on its own. In this case, however, when evaluated together with other indicators, the issuer information strengthens the certificate's clean profile.

- **Validity**

  The certificate's validity period is 89 days. This is entirely normal within the natural policy framework of Let's Encrypt; however, this value is also a behavior frequently observed in C2 infrastructures. Operationally, a short lifetime has the following characteristics:

  - Infrastructure is rapidly renewed
  - Flexible and automation-oriented
  - Operates within a continuously updated structure

  For this reason, `short_validity = true`, as correctly flagged by the analysis engine, doesn't generate suspicion on its own, but—due to operational behavior similarity—it's a metric that must be flagged.

- **SAN**

  A clean and consistent SAN list indicates the following:

  - Domain shadowing isn't present in the infrastructure.
  - Subdomain diversity isn't under attacker control.
  - The certificate isn't tied to a shared or rented CDN.

This quiet yet critical detail points to a level of regularity that is rarely observed in attacker infrastructures. In most malicious infrastructures, the SAN list is either completely empty or contains illogical domain fragments.

- **Public key**
  RSA 2048 remains secure and standard by today's standards. However, from an intelligence perspective, the real issue is that malicious infrastructures often use minimal key requirements. Effortless deployment, automated certificate generation, and rapid distribution are the primary reasons. An RSA 2048 key points to an automated Let's Encrypt renewal infrastructure typical of a modern organization. This further strengthens the profile.
- **Extended key usage**
  There are two values in the `extended_key_usage` field:
  - `1.3.6.1.5.5.7.3.1` → `serverAuth`
  - `1.3.6.1.5.5.7.3.2` → `clientAuth`

  `serverAuth` is the correct and expected value for a web server role. `clientAuth` is normal in certain corporate configurations; used for user/machine authentication.

  If this were a C2 infrastructure, we would typically observe only `serverAuth` or no `EKU` at all. Therefore, this combination is consistent with a corporate architecture.
- **Basic constraints**
  Figure 10.2 shows the continuation of our output, starting with the `"basic_constraints"`. It's important to confirm that the server isn't a CA. CA misconfigurations frequently encountered in poorly configured systems (e.g., an internal CA certificate mistakenly assigned to a server) can create opportunities even for attackers. In this case, we have the following:
  - `ca` = `false`: Correct configuration
  - `path_length` = `null`: Expected condition

  Therefore, there's no suspicion.
- **Fingerprints**
  The SHA-1 and SHA-256 fingerprint values are akin to the certificate's fingerprints. These values play a critical role in IOC matching. The analysis engine reached the following conclusion: `known_c2_match` = `false`:
  - This fingerprint doesn't match any known C2 profiles in our current reference/IOC repositories. This is a signal that reduces suspicion; however, it doesn't clear it on its own.
  - This also doesn't match common C2 profiles such as Cobalt Strike, Sliver, or Mythic in our reference/IOC repositories.
  - There's no negative profile associated with it in the IOC pool.

  These are strong indicators that reduce suspicion from an intelligence perspective, but they aren't exculpatory evidence on their own.

```
    ],
    "basic_constraints": {
      "ca": false,
      "path_length": null
    },
    "subject_key_id": "26dafb878072033c30511aced927f0436af845f2",
    "authority_key_id": "e7ab9f0f2c33a053d35e4f78c8b2840e3bd69233"  ←
  },
  "fingerprints": {
    "sha1": "0ebde5b232781c04216bdccc886de14be8cc540d",
    "sha256": "56b44a97b6a315df7a53b9e2c5e32ed6af88b16db83621222cfbf9d283eacdad",
    "ja3": null,
    "jarm": null
  },
  "ioc": {
    "known_c2_match": false,
    "matches": []
  },
  "heuristics": {
    "short_validity": true,  ←
    "exotic_issuer": false,
    "issuer_name": "CN=R13, O=Let's Encrypt, C=US",
    "is_ca_cert_on_tls": false,
    "missing_server_auth_eku": false,
    "suspicious_score": 0.3
  },
  "tls_session": {
    "tls_version": "TLSv1.3",
    "cipher_suite": "TLS_AES_256_GCM_SHA384",  ←
    "cipher_protocol": "TLSv1.3",
    "alpn_protocol": null
  },
  "analysis_time_utc": "2025-12-10T08:24:27.680887+00:00"
}
```

Figure 10.2: Extended TLS Fingerprint and Heuristic Analysis Highlighting Certificate Lineage, Session Parameters, and Behavioral Risk Indicators

- **Heuristics**

  The score produced by the analysis engine is 0.3 (low risk). This score originates from a single factor: `short_validity = true`. However, due to Let's Encrypt's natural policy, this condition is expected and represents a contextual false-positive example, not a genuine suspicion. An intelligence analyst's correct interpretation of this value demonstrates the maturity of the system.

- **TLS session**

  Negotiated session parameters include the following:

  - `TLSv1.3`
  - `TLS_AES_256_GCM_SHA384`
  - `alpn_protocol`: None (likely consistent with HTTP/1.1 usage)

  The use of TLS 1.3 indicates that the organization follows modern security practices, the server software is up-to-date, and the legacy structures that would be vulnerable to third-party exploits aren't present.

  Attacker infrastructures may use TLS 1.2 or TLS 1.3. What is often distinguishing isn't the version itself, but the TLS fingerprint (JA3/JA4/JARM), certificate continuity, Application Layer Protocol Negotiation (ALPN) consistency, and traffic rhythm. For this reason, the TLS profile also remains on the clean intelligence side.

This analysis output provides a simple yet powerful lesson: A certificate is more than a cryptographic document in modern cyber intelligence; it's a behavioral indicator. The following indicators we've obtained, when considered together, point to a properly configured TLS infrastructure of a corporate service:

- Legitimate CA
- Well-structured SAN configuration
- Modern protocol support
- Consistent enhanced key usage (EKU) profile
- Clean fingerprints with no IOC history

From an intelligence perspective, this example is an ideal case study demonstrating how vital context is when filtering truth from noise.

10.3 Contextualization of Threats and Impact Analysis

Understanding a threat is about both identifying it and determining where it fits within the broader attack chain and what it truly affects. On their own, technical artifacts (e.g., hashes, IP addresses, or event logs) carry limited meaning. These elements begin to reveal their significance only when context is applied.

In practice, a single alert rarely tells the full story. Its meaning emerges when it's correlated with surrounding signals: prior activity, user behavior, infrastructure patterns, and related telemetry. An event that appears insignificant in isolation may, when aligned with other indicators, represent a continuation of an attack chain. Contextualization is therefore the process of connecting these elements to understand what happened, why it happened, and what may follow.

This distinction is critical because not every detected threat translates into meaningful impact. A technical anomaly may remain local and benign, while a seemingly minor signal—when combined with privileged access, critical systems, or related behaviors—can indicate significant risk. Without contextualization, it's difficult to distinguish between noise and events with real operational consequences.

At its core, threat contextualization answers three key questions:

- What does the event represent technically?
- Which assets does it affect?
- What impact may arise when it's combined with other signals?

When these questions are addressed, the threat becomes a measurable risk rather than an abstract observation.

Impact analysis builds on this foundation by translating technical findings into potential business and operational consequences. It enables analysts to assess both the current state of an incident and its likely progression and severity. This approach shifts analysis from isolated detection toward a structured understanding of risk.

The following sections examine how contextualization and impact analysis are applied in practice, including techniques for correlating signals, evaluating affected assets, and assessing potential impact across the attack lifecycle.

10.3.1 Interpreting Adversary Intent

Interpreting adversary intent is the most critical stage in transforming an attack from technical noise into a real narrative. Because what an attacker does is often visible; it manifests itself in logs, network traces, and process behaviors. But why it's done—that's what remains invisible. This is the darkest area of cybersecurity: the resolution of intent.

Interpreting adversary intent, at a certain point, ceases to be purely technical analysis and turns into a form of silent inner reading that develops within the analyst's mind. The steps taken by the attacker whisper not only what has been done but also what will be done. Every process call, every DNS resolution, and every failed authentication attempt appears, on the surface, to be nothing more than a single line; yet when read in the proper context, it becomes a small window into the attacker's internal motivation. This is why intent analysis often carries a more human and more intuitive dimension than other analytical processes. At times, the analyst senses a change in the rhythm of the attacker's behavior before even examining the logs. We realize that something is not progressing as it should, much like a single wrong note played in a very familiar melody.

The invisible aspect of intent is often hidden in the attacker's patience. For example, some threat actors behave aggressively as soon as they initiate the attack chain, rapidly move into lateral movement, and attempt to obtain high privileges in a short time. In such attackers, the rhythm is intense; it resembles a runner gasping for breath. Yet there are others who, after gaining access to a system, don't produce a single line of log for days. They simply sit, wait, and observe. This silence, in fact, conveys a great deal. When the rhythm of that silence begins to break, the feeling within the analyst also changes: "This patience isn't meaningless. There must be a significant objective behind this waiting."

Real-World Example

A user account had been performing only low-privilege queries for weeks: a few LDAP requests, one or two WMI calls, and a quiet file-listing routine. All of it looked so innocent that most analysts attributed the behavior to routine activity. However, the timing of these queries disturbed me; such a homogeneous and orderly pattern often points to automation or a systematic intent, and no definitive judgment can be made without excluding the possibility of legitimate (benign) automation. When I saw on the CTI platform that a similar rhythm matched the long-term positioning phase of an APT group, I felt a cold silence somewhere inside me. That silence was the harbinger of an approaching attack. The intent was unmistakable: The adversary was settling in, learning, and waiting for the optimal moment. Behaviors that appeared harmless to an external observer revealed, when viewed in the proper context, how dangerous the attacker's patience could be.

Intent analysis also shines in the areas where the attacker makes mistakes. No matter how professional attackers may be, they leave behind small inconsistencies that reflect their character at some point in the operation. For instance, a financially motivated attacker typically seeks rapid return; therefore, reconnaissance behavior is hasty, and log rhythms are fragmented. Strategic actors, by contrast, are more controlled; their steps are orderly and their rhythms are smooth.

Real-World Example

I'll never forget a case in which I noticed these two behaviors intermingled within a single attack: the adversary was leaving both financial and strategic traces. This contradiction revealed that the attack was, in fact, a hybrid operation—money was being targeted, but the internal structure of the organization was also being mapped. Such hybrid intents become visible only in the hands of an analyst with strong CTI context.

In some cases, intent becomes far clearer than the targeted asset itself. For example, a seemingly ordinary PowerShell command may appear completely benign when viewed in the wrong context; yet when it occurs in close proximity to a critical server, it may signal the attacker's strategy of moving inward. The same command may mean nothing when executed on an end-user system, but when executed near a DC, alarm bells begin to ring. Here, the technical signal doesn't change; what changes is the context, and context makes intent visible.

Real-World Example

On one occasion, what alerted me wasn't a single call, but the fact that `NtQueryInformationProcess` formed a recurring pattern within a specific process tree and together with specific adjacent signals (e.g., unusual handle access, suspicious memory reading or dump preparation, and privilege indicators). This call can normally be used in many scenarios within a system. But when it appears in the middle of a particular process tree, within a specific time window, and in conjunction with a specific user, it suddenly gains meaning. This pattern bore strong similarity to certain behavior clusters previously observed in credential access operations; however, the final interpretation was validated not by a single call, but by the combination of process tree, privilege context, and adjacent telemetry. The attacker had said nothing. Yet the sequence of calls spoke its intent: "I am preparing."

Intent analysis can't be considered independently of impact. Some attackers focus on data, some on systems, and others on identities. Without understanding a threat actor's intent, it's nearly impossible to estimate the magnitude of the impact they may create. For example, a small manipulation within authentication systems may indicate that a financially motivated group is laying the groundwork to drain accounts. The same behavior, if it points to a strategic actor, may aim to undermine the overall trustworthiness of the system or serve a long-term espionage objective. In other words, the same signal can reflect entirely different intents in different contexts.

Throughout this process, the analyst focuses not just on raw technical data, but on the psychological model underlying that data because an attacker's intent is often reflected in behavioral rhythm. A rushed attacker disrupts the rhythm. A panicking attacker makes unexpected mistakes. A patient attacker becomes almost invisible. Reading these rhythms represents the intuitive side of threat hunting. A small emotional reaction often precedes technical analysis. You see a signal, and an unexplainable tightness forms inside you. That tightness is often the moment when the attacker's intent "touches" you. As the analyst matures, they learn to place these emotional microsignals into a technical context.

Ultimately, intent analysis defines the attacker not only by what they do but why they do it. An attack becomes a real threat only when its intent is resolved; or conversely, through context, it becomes clear that it's a false alarm. Intent is the invisible heart of an attack; it determines its rhythm, draws its direction, and reveals its target. When an analyst understands intent, what stands before them is no longer an indistinct shadow. That shadow takes shape, is named, and becomes manageable. At that moment, the analyst thinks to themselves: "Now I see you. Now I know what you want."

Table 10.4 shows the different types of intent, their indicators, characteristics, target, and potential impact.

Intent Type	Technical/ Behavioral Indicators	Timing and Rhythm Characteristics	Typical Targets	Impact Potential
Financial (financial gain)	Credential harvesting, RDP brute-force, proximity to payment systems, injection attempts against web applications (web inject), use of malware	Relatively rapid progression, short dwell time, aggressive trial-and-error	Payment systems, ERP, accounting, online banking	Direct financial loss, fraud, account draining

Table 10.4: Adversary Intent Types and Typical Indicators

Intent Type	Technical/ Behavioral Indicators	Timing and Rhythm Characteristics	Typical Targets	Impact Potential
Strategic (espionage/geopolitical)	Long-term reconnaissance, low-noise access, custom toolchain, targeted data exfiltration	Silent presence lasting weeks/ months, low-frequency but consistent activities	Government institutions, defense, energy, critical infrastructure	Information leakage, reputational damage, strategic position weakness
Reconnaissance	Network mapping, Active Directory (AD) enumeration, port scanning, patterned SMB/ LDAP/WMI queries, user/group discovery	Slow but persistent queries, typical mapping rhythm	Entire network, critical server segments, DCs	Laying the groundwork for the next attack phase
Destructive (destructive/ wiper)	Volume shadow copy deletion, backup targeting, mass deletion, master boot record (MBR)/file system manipulation, disabling security components	Intense activity in a short time, high noise, sudden burst	Backup systems, file servers, hypervisors	Difficult-to-recover data/ system loss, prolonged outage
Hybrid (mixed/ multistage)	Both financial and strategic traces, both reconnaissance and destruction preparation, combined use of different toolsets	Rhythm changes between phases, "mixed" traces in behavioral profile	Cross-system targets, both data and financial assets	Loss of money, information, and reputational/ operational damage

Table 10.4: Adversary Intent Types and Typical Indicators (Cont.)

Let's break down the adversary intent types in more detail:

- **Financial intent**

  Financially motivated threat actors design their operations with a focus on rapid return. In other words, the attacker's patience cycle is low. This characteristic is reflected directly in technical behavior. Technical indicators typically concentrate around systems that directly interface with monetary flows:

  - Compromise of user credentials (especially when multifactor authentication [MFA] is weak)
  - Attempts to gain remote access through RDP brute-force attacks
  - Web injection attempts in areas close to payment APIs
  - Payloads aimed at collecting session tokens on e-commerce platforms

  The rhythm of such actors is distinctly fast. Because there's no objective of long-term persistence, dwell time remains short, and aggressive trial-and-error behavior is observed. Targets are generally areas where financial assets reside:

  - Banking interfaces and accounting applications
  - Credit card payment systems
  - Payroll/payment modules within enterprise resource planning (ERP) systems

  The impact is direct financial damage. Even a single access can result in account draining or the redirection of payment instructions.

  The most distinctive characteristic of financial actors is their operational impatience. When an analyst observes rapidly repeated login failures on EDR telemetry or unusual traffic patterns around payment systems, the analysis often points toward this intent type.

- **Espionage/geopolitical**

  Strategically motivated actors are the most patient, systematic, and controlled group within the attack landscape. As with APTs that are aligned with nation-states, the objective isn't money, but information superiority, geopolitical advantage, or manipulation of critical infrastructure.

  Technical indicators are quiet and designed to leave minimal traces:

  - Reconnaissance lasting days or even months
  - Lateral movement with very low signal frequency
  - Use of signatureless, custom toolchains
  - Minimal yet carefully executed exfiltration aimed at specific files

  The rhythm of such attackers is like a whisper. Their concern for remaining undetected is so strong that behaviors reflected in logs may resemble random system activity. Targets are critical and strategic entities:

  - Government institutions
  - Energy companies

- Defense industry
- National infrastructures

The impact is direct data leakage or erosion of stability at a state level. These attacks don't produce immediate impact, but if they remain unnoticed, they lead to strategic losses. To understand strategic actors, an analyst often needs to focus on the calmness of the signal. The quieter it is, the more serious it may be.

- **Reconnaissance**

  Reconnaissance is the first and inevitable phase of the attack chain. For this reason, reconnaissance behaviors are often confused with false positives because they cause no apparent harm on the surface. Yet for the attacker, reconnaissance is the stage of drawing the map; the fate of the attack is determined here. Technical behaviors include the following:

  - Systematic scanning of network segments
  - Extraction of user/group/permission topology within AD
  - Regular queries against services such as SMB, LDAP, and WMI
  - Tests aimed at understanding how protected each system is

  The rhythm is slow but persistent. Attackers conducting reconnaissance don't rush; they attempt to measure how the system responds to them.

  The target is the entire network, but the following in particular:

  - DCs
  - Critical servers
  - Segments hosting sensitive data

  The impact isn't direct damage; rather, it indirectly increases the success rate of the primary attack. If an analyst misses reconnaissance behavior, the likelihood of stopping the next phase decreases significantly.

- **Destructive/wiper**

  Destructive threat actors don't seek gain. A malicious motivation is their signature: The objective is to cause damage, create disruption, and generate chaos that is difficult to reverse. The technical indicators of such attacks often begin with attempts to prepare the environment:

  - Deletion of volume shadow copies
  - Neutralization of backup systems
  - Disabling security processes
  - Mass file deletion or file system manipulation
  - Registry-level preparation for wiper activity

  The rhythm is extremely fast. When the moment of attack arrives, all actions intensify within minutes. Targets are assets with high operational value:

  - File servers
  - Backup infrastructure
  - Hypervisors and VM clusters

  The impact is the most severe: system collapse, data loss, and operational outages lasting days. An analyst's ability to recognize this intent early may determine an organization's survival.

- **Hybrid (mixed/multistage)**

  Hybrid intent represents a combined approach increasingly favored by attackers. Here, there isn't a single objective, but multiple goals, and the attack progresses in layers, for example:

  - Financial data is collected first.
  - Strategic documents are exfiltrated afterward.
  - In the final stage, a destructive action is executed to cover their tracks.

  Technical behaviors are diverse:

  - Use of tools resembling both APT and cybercrime groups
  - Mixed chains such as reconnaissance + financial data exfiltration + wiper preparation
  - Use of multiple toolchains against the same target

  The rhythm is variable. It may be quiet in one phase and aggressive in another. This variable rhythm is the most distinctive technical trace of hybrid attacks. Targets span different systems:

  - Financial data
  - Sensitive documents
  - Operational infrastructure

  The impact is threefold:

  - Financial loss
  - Information loss
  - Operational damage

These attacks are the most dangerous because they contain misleading traces and behavioral mimicry that can divert the analyst in the wrong direction.

To systematically interpret this complexity, analysts must move beyond isolated observations and evaluate intent through multiple contextual dimensions. Table 10.5 outlines the key factors, data sources, and guiding questions that support a structured assessment of adversary intent, helping translate diverse signals into a coherent understanding of attacker objectives.

Context Dimension	CTI/Telemetry Source	Question Analyst Should Ask	Contribution to Intent
Targeted system/ asset type	EDR, configuration management database (CMDB), asset inventory, AD data	What exactly is the attacker orbiting around?	Distinguishes whether the intent is financial, strategic, or random
TTPs used	MITRE ATT&CK mapping, EDR events, log correlations	Which actor profile and objective type is this TTP set closer to?	Clarifies the actor type and intent segment
Timing and continuity	SIEM timeline, beacon rhythm, login patterns	Is the rhythm of this activity patient, rushed, or irregular?	Strengthens the distinction between strategic and opportunistic attackers
Targeted data/function	Data loss prevention (DLP) logs, database access logs, file access records	Which data or business process is this behavior moving toward?	Reveals data-focused, identity-focused, or system-focused intent
Attacker's past campaign behavior	CTI platform, historical incident reports	What did the same actor want before, and what are they pursuing now?	Shows the continuity and evolution of intent
Organizational value of the target	Business impact analyses, critical asset inventory, business continuity management (BCM) documents	What would the real-world cost be if we lost this target?	Establishes the link between intent and potential impact
Position in the attack chain	Kill chain/MITRE ATT&CK phase mapping	Is this event at the beginning, middle, or end of the chain?	Sheds light on whether it's reconnaissance, exploitation, or destruction
Emotional/intuitive tone of the behavior	Analyst experience, prior case memory	Which past incident does this remind me of?	Adds an intuitive validation layer to technical data

Table 10.5: CTI Context in Intent Analysis and Key Questions to Be Asked

Let's examine these context dimensions in more detail:

- **Targeted system/asset type**
  One of the most fundamental CTI building block questions analysts must ask themselves is, "What is the attacker circling around?" Because the asset an attacker touches is often direct evidence of intent, EDR, CMDB, or AD data reveal which systems the attacker is currently interacting with, for example:
  - If activity revolves around ERP systems, this indicates financial motivation.
  - If behavior is observed near DCs, this indicates strategic escalation or reconnaissance.
  - If the attacker operates superficially only on endpoints without converging toward a clear target, this indicates an opportunistic, low-maturity, or poorly targeted threat actor profile.

  This context provides the following insight: Where the signal emerges, the intent is usually directed toward that location.
- **TTPs used**
  Every attacker has a distinctive behavioral fingerprint, which is revealed by the set of TTPs employed. MITRE ATT&CK mapping, EDR events, and log correlations allow the analyst to break the shell of this model and look inside.

  The analyst asks themselves this critical question: "Which actor's style does this combination of tools and techniques most closely resemble?" This question is critical for the following reasons:
  - Financial groups typically use commodity malware + credential dumping + rapid exfiltration.
  - APT groups exhibit custom loaders + slow lateral movement + targeted reconnaissance behavior.

  This comparison clarifies intent: Tools betray the attacker's personality.
- **Timing and continuity**
  Time is one of the strongest clues in understanding adversary intent. SIEM timelines, beacon intervals, and login patterns all reveal the attacker's level of patience. Analysts asks themselves, "Is this rhythm patient, rushed, or unstable?":
  - Patient rhythm: Strategic APT (weeks/months)
  - Rushed rhythm: Financial motivation or ransomware preparation
  - Irregular rhythm: Inexperienced threat actor relying on trial and error

  An attacker can't hide their rhythm, and rhythm is the mathematical equivalent of intent.
- **Targeted data/function**
  The types of files an attacker accesses, the business functions they touch, and the data groups they seek reveal the "content" of intent, for example:
  - Customer data appearing in DLP logs: Financial/personal data theft
  - Schema discovery behavior in database logs: Strategic espionage

- Backup-focused access on file servers: Destructive intent
- Movement near authentication services: Preparation for privilege escalation

Analysts ask themselves, "Which business process is this behavior moving toward?" The target always explains why the attacker came.

- **Attacker's past campaign behavior**
  CTI platforms, historical attack reports, and campaign analyses form a behavioral archive that shows how an actor has operated over time. Analyst ask themselves the following: "What did this actor want before, and what are they pursuing now?"

  If the TTP model, word choices, domain patterns, or IP/ASN relationships resemble past campaigns, the intent is progressing along a consistent objective. This enables the identification of threat trends:

  - If the same actor has targeted the energy sector for three years, a new signal likely points in the same direction.
  - If the same toolchain previously exfiltrated data, it may carry a similar mission again.

  Past behavior illuminates the future.

- **Organizational value of the target**
  This context is often overlooked by technical security specialists, yet it's one of the most critical components of threat intent. Analysts ask themselves this question: "What would the real-world cost be if we lost this asset?" The answer to this question changes based on the following:

  - Priority level of the signal
  - Severity of the intent
  - Urgency of response
  - Potential business impact of the threat

  Here are some examples:

  - If the attacker is browsing an ordinary user folder, this is low criticality.
  - If the same behavior appears at the core of a financial workflow, the intent may be serious.
  - Even a small anomaly on a DC may align with the most dangerous intent categories.

  The value of the target determines the weight of the intent.

- **Position in the attack chain**
  Through cyber kill chain or MITRE ATT&CK phase mapping, analysts ask themselves the following: "Where is the attacker right now?" This directly influences intent because each phase represents a different objective:

  - Reconnaissance: Collection/learning intent
  - Exploitation: Gaining access intent
  - Lateral movement: Privilege expansion intent

  - Exfiltration: Data theft intent
  - Impact/destroy: Destructive intent

  The same signal can take on different meanings depending on its position in the chain. For example, SMB queries may appear benign during reconnaissance, but when they move closer to backup servers, they may herald destruction. Position changes everything.

- **Emotional/intuitive tone of the behavior**
  This row is the most human part of the table. Over time, every analyst begins to associate certain attack rhythms with cases they have encountered in the field. Analysts ask themselves this question here: "Which incident I've experienced before does this remind me of?" This intuitive connection provides the following:
  - Early reflexes for attacks carrying similar rhythms
  - Intent signals that technical data alone can't explain
  - The grounding of the feeling "something about this behavior feels wrong" into technical context

  Intuition is often underestimated in cybersecurity, yet it plays a critical role. Some attacks leave no clear technical signal; however, the tone, rhythm, and even the silence of behavior can reveal intent. This layer represents analyst intelligence formed by the convergence of experience, memory, and intuition.

10.3.2 Impact Scope Analysis and Calculation of Propagation Potential

To understand how dangerous a piece of malware truly is, it's often necessary to look not at what it initially does, but at what it can do. Every attack appears small in its first step: a minor process anomaly, a slight DNS irregularity, a tiny log deviation, and so on. The first stone thrown by the attacker almost always appears innocent. Yet the real danger lies in that stone's potential to grow as it rolls.

In cybersecurity, impact scope analysis measures the following:

- How far can a piece of malware move?
- How widely can it spread?
- Which part of the identity chain can it touch?
- How fragile does each touched link make the entire chain?

These questions aren't purely technical; they also test the analyst's internal intuition for reading the future of a threat. A piece of malware may be silent in logs yet create noise within the chain, or it may behave visibly but fail to expand. Impact and propagation analysis is the science of understanding an attacker's potential.

The first indicator of an attacker's mobility emerges from the breadth of access within the compromised system. For example, malware running on an end-user workstation may not appear serious on its own. However, it ceases to be an ordinary incident and becomes a major propagation risk if that same malware does the following:

- Runs with local administrator privileges
- Encounters exposed credentials on the system
- Operates on a machine joined to the domain
- Has user accounts cached locally

Real-World Example

During a SOC shift, malware delivered through an apparently ordinary Chrome plugin discovered identity artifacts on the system, including cached credentials and traces indicating `runas` usage. This minor detail turned into a bridge that could allow the attacker to advance toward the DC. The initial signal was small, almost invisible; yet the propagation potential was enormous. At that moment, the impact analysis became clear: More than just a endpoint incident, this was the beginning of a chain that could spread across the entire organization.

One of the most reliable ways to understand a piece of malware's lateral movement capability is to evaluate the attacker's ability to touch the identity chain. Because more often than not, it's not systems that are compromised, but identities. Systems are merely platforms upon which identities operate.

The identity chain is like an invisible path extending between the following:

- User identities
- High-privilege accounts
- Service accounts
- Kerberos tickets
- Cached credentials
- Machine accounts

When an attacker compromises a single link in this chain, the entire chain becomes endangered. This is why an analyst must always ask the following question: "Which part of the identity chain can this attacker touch?"

This question is also the direct mathematics of propagation capacity. Threats that spread within a network typically exhibit three behavioral models:

- **Linear propagation**
  The attacker advances in a single direction, moving from one system to another, and then to the next. This behavior is usually indicative of an inexperienced threat actor or one seeking rapid results.

- **Branching propagation**
  Multiple branches emerge from a single point. After gaining access at one location, the attacker simultaneously targets several different machines. This explosive style of propagation is commonly observed in highly aggressive campaigns.
- **Silent perimeter expansion**
  The attacker selects a central point and quietly touches surrounding systems, then waits without moving far from that center. This behavior is typically characteristic of the patient reconnaissance phases of APTs. An analyst's ability to recognize these three models is critical for estimating how far a threat may spread.

Another key indicator of propagation potential is the network topology and the number of segments the attacker can reach. For example, in one case, an apparently ordinary PowerShell script was able to move—due to a misconfigured VLAN—from a financial segment to an internal test segment, and from there into the DC segment. The script itself appeared low-impact in isolation. The topology, however, was dangerous. Sometimes the threat isn't in the code, but in the organizational structure.

At this point, the analyst asks, "How many segments can this propagation connect?" The answer often determines the fate of the impact analysis.

The emotional dimension of impact scope analysis lies in the risk the analyst feels internally. Some incidents appear technically small, yet their propagation potential is so large that the analyst feels a tightening sense of unease: "This will grow... this will definitely grow." This intuition is often correct.

An experienced analyst sees the current incident and, importantly, what that incident may become three steps later. This isn't a mathematical analysis, but an intuitive projection shaped by experience.

The most critical (and often overlooked) aspect of impact scope analysis is accounting for both current conditions and for how an attacker can reshape the environment's weaknesses. Some threats don't just exploit opportunities; they create them. When calculating propagation potential, the analyst must internalize the following reality: Malware doesn't move solely between systems; it takes the shape of gaps in an organization's security model. If segments aren't properly isolated, malware finds the gaps between segmentation walls. If identity management is weak, malware discovers the fragile points of privilege. For this reason, impact scope analysis must measure current propagation, as well as the attacker's ability to generate new paths.

The risk created by an attacker isn't limited to the area in which they move; how they can use the information they access also determines the multiplier effect of propagation. For example, an attacker may not be able to access the DC, but if they can read a domain administrator's weekly script routines, they can later use that knowledge to launch a second attack on the identity chain. This is an invisible link in the chain and is often not detected in logs. For an analyst, such a situation represents a "silent link" ready to expand. Impact analysis remains incomplete if this link isn't considered.

When evaluating propagation potential, the attacker's operational creativity must also be taken into account. Many modern threats leverage configuration remnants that have gone completely unnoticed within the environment:

- An old service account believed to have been disabled
- Forgotten SMB shares
- A highly privileged credential left in a test environment
- Tokens stored in plaintext within continuous integration/continuous delivery (CI/CD) pipelines

Any of these can become a pivot point for the attacker. This situation is often excluded from technical reports because the attacker hasn't yet used them. However, during potential-state evaluation, the analyst's task is to imagine all paths that are unused but usable and add them to the risk multiplier. True impact analysis encompasses what has occurred and the worst-case scenario that could occur.

At a more advanced level of impact scope analysis, the attacker's ability to misdirect comes into play. Some threat actors deliberately leave false rhythms to make propagation potential difficult to assess. For example, by touching low-importance files, they lead security teams to conclude that it's merely a low-impact malware, and then trigger a second payload that targets privileged accounts. This dual architecture is particularly common in hybrid attack models. For an analyst to recognize this tactic, impact analysis must examine both the scale of the incident and the attacker's rhythm manipulation.

The area an attacker can propagate into isn't limited to technical topology; organizational behavior patterns can also alter propagation. For instance, a common mistake made by IT teams is granting a user account elevated privileges temporarily to resolve an issue, and then failing to revoke those privileges. This one-time temporary privilege becomes an enormous springboard for the attacker. Impact scope analysis must account for such behavioral security gaps because many attacks grow not from technical vulnerabilities, but from human habits.

Sometimes, what determines the fate of propagation isn't the account the attacker has accessed, but the accounts they can see but haven't yet accessed. For example, malware can map the identity topology of an environment by issuing only `Get-User` and `Get-Group` queries in AD. This map enables the attacker to understand which accounts to target and which privileges represent the critical links in the chain. True impact analysis considers the future damage such invisible reconnaissance can cause because even if the attacker hasn't yet taken a step, they have already planned where to go.

Real-World Example

In one case, our impact analysis resulted in a high score even though the attacker performed no lateral movement at all. The reason was that the attacker identified the name of a critical file server through only a few SMB queries and made a small, test-level call related to one of the Kerberos vulnerabilities. Many analysts might have dismissed this as unnecessary log noise. Yet these two small indicators showed that the attacker's intent wasn't ordinary; they were preparing for a much larger journey. Propagation capacity wasn't about the steps the attacker had taken, but about the direction those steps pointed toward.

The deepest point of impact scope analysis lies in the attacker's ability to create a chain reaction. An attack sometimes expands not only at the point of initial access but through the indirect effects triggered by that access:

- A credential is stolen, and access to the backup server is obtained using that credential.
- Backups are disabled, so ransomware preparation accelerates.
- Ransomware spreads, and business continuity is disrupted.
- Business continuity is disrupted, and financial losses increase geometrically.

This chain grows not only with the attacker's capability but also with the vulnerability of the environment. This is the purpose of impact analysis: to see every link in the chain before it breaks and to solve the mathematics of the risk's expansion.

Some threats don't spread; they don't need to. The impact they create is explained by the criticality of the point at which they reside. Even a very small script running on a DC can cause the collapse of the entire identity infrastructure without spreading at all. For this reason, when calculating propagation potential, the following question always arises in the analyst's mind: "How much damage does this attack cause even if it doesn't move?" This question makes even the quietest threats visible.

Impact analysis is one of the rare domains in which the analyst combines technical data with emotional intuition. Logs show only what has happened; intuition reveals what could happen. When these two layers work together, the threat is no longer just a data point; it has a route, a tendency, and a potential. When the analyst reads that route, the future of the attack unfolds like a lifting fog, and a clear realization emerges: "The danger is not now. The danger is what this may evolve into." This is the true power of impact and propagation analysis: seeing not the attack itself, but the future of the attack.

To translate this analytical perspective into a consistent and repeatable assessment, impact and propagation potential must be evaluated through a structured set of criteria. Table 10.6 presents a scoring model that combines technical indicators, environmental conditions, and attacker behavior to quantify how a threat may expand and what level of risk it represents to the organization.

Evaluation Criterion	Observed Condition/Metric	Score Range	Meaning of the Risk
1. Identity access level	User → Local Admin → Domain Admin → Service Account Chain	1–5	The more privileged the identity, the higher the propagation potential.
2. Credential state	Cached credentials, Kerberos tickets, `runas` artifacts, plain-text secrets	1–5	If the malware can pivot to new accounts, the risk increases.
3. Lateral movement capability	WMI, PsExec, SMB, RDP, WinRM, Active Directory Federation Services (ADFS) abuse, Kerberos manipulation	1–5	The more transition channels available, the faster the propagation speed.
4. Network segmentation status	Strict segmentation → Weak segmentation → Cross-VLAN traversal	1–5	If segment traversal is possible, the entire network may be at risk.
5. Attacker rhythm/ posture analysis	Patient → Silent → Wave-like → Aggressive → Explosive	1–5	Rhythm determines the pattern and duration of threat growth.
6. Criticality of targeted systems	Low-impact endpoint → File Server → DC → Backup node → Hypervisor	1–5	Contact with critical systems creates a multiplier effect.
7. Security control status	EDR active/passive, antivirus (AV) bypass attempts, logging health	1–5	Weak defenses directly increase propagation potential.
8. Malware technical capability	Persistence, credential dumping, privilege escalation, worming behavior	1–5	The capability profile defines the limits of the attack.
9. Vulnerabilities available for propagation	Unpatched Common Vulnerabilities and Exposures (CVEs), exploitable misconfigurations, legacy protocols	1–5	The number of environmental weaknesses is a propagation multiplier.
10. Potential chain reaction impact	Backup loss, business continuity disruption, data exfiltration, operational shutdown	1–5	Impact reflects both technical damage and business magnitude.

Table 10.6: Impact and Propagation Risk Scoring Table

Total score calculation is shown here:

- **10–19: Low propagation risk (contain straightforward)**
  At this level, the activity should be contained through routine triage and standard validation procedures. Analysts should confirm the alert's context, verify whether it aligns with known benign behavior, and monitor for any escalation. Immediate disruptive actions are typically not required unless additional indicators emerge.
- **20–29: Medium propagation risk (monitor + investigate)**
  At this stage, the activity requires active investigation and closer monitoring. Analysts should correlate related events across multiple telemetry sources, assess affected assets, and determine whether the behavior is progressing along the attack chain. Preventive measures (e.g., restricting suspicious connections or increasing logging) should be considered.
- **30–39: High propagation risk (contain + lateral block)**
  At this level, containment actions must be initiated promptly. Analysts should isolate affected systems where appropriate, block lateral movement channels (e.g., SMB, RDP, WinRM), and identify compromised credentials or access paths. The focus is on limiting the attacker's ability to expand within the environment.
- **40–50: Critical propagation risk (immediate isolation + incident response full activation)**
  This level requires immediate incident response activation. Affected systems should be isolated without delay, incident response procedures should be fully engaged, and all relevant teams (SOC, incident response, IT operations) must be coordinated. The priority is to stop propagation, protect critical assets, and prevent widespread operational or data impact.

The impact and propagation risk scoring table does more than determine whether a threat exists or not; it calculates where the threat may grow in the future, how fast it can propagate, and which systems it may bring down. What makes this table distinctive is that it frames the attacker's behavior within a technical structure while simultaneously making the potential evolution of the attack visible. Let's take a closer look at the criteria:

1. **Identity access level: Strength of the chain**
   The account an attacker gains access to determines the fate of the attack:
   - User level (Score 1–2): Propagation is limited.
   - Local administrator (Score 3): Intra-system manipulation is possible.
   - Domain Administrator (Score 5): This level immediately pushes the propagation analysis into the red zone.

   Identity is the "key" carried by the attacker. The larger the key, the more doors it can open.
2. **Credential state: Silent reservoir of risk**
   Credential remnants left on a system are often invisible, yet highly dangerous. If an

attacker can access the following remnants, the propagation risk increases geometrically:

- Cached credentials
- Token remnants
- `runas` secrets
- Kerberos ticket reuse
- Plaintext configuration secrets

The malware doesn't remain where it is; it discovers new doors.

3. **Lateral movement capability: Pivot capacity**
   An attacker's ability to use lateral movement techniques is a clear indicator of propagation potential. WMI, PsExec, SMB execution, RDP hopping, and WinRM abuse each adds a new link to the chain. An attacker capable of using five different lateral channels is far more dangerous than one using a single channel because even if the response team blocks one path, the attacker can slip through another.
4. **Network segmentation: Hidden speed multiplier**
   Some of the most devastating attacks aren't technically sophisticated, yet they cause massive impact because segmentation is poor. If VLAN transitions, firewall rules, or access control lists (ACLs) are lax, every system becomes a bridge to another, and each bridge multiplies propagation. For this reason, even low-capability malware can turn into a disaster in a misconfigured network.
5. **Attacker rhythm analysis: "How does it move?"**
   Rhythm reveals the attacker's character. The following behaviors can be observed and interpreted:
   - Patient indicates APT.
   - Aggressive indicates ransomware.
   - Irregular indicates inexperienced actor.
   - Silent indicates reconnaissance.

   Impact analysis is incomplete without considering rhythm because rhythm determines the speed and direction of propagation.
6. **Criticality of targeted systems: Detonation point**
   Some systems aren't sources of propagation, but multipliers of propagation. For example, a threat that touches one of these systems grows much faster than one that doesn't:
   - DCs
   - Backup servers
   - Hypervisor clusters
   - File server root shares
   - Identity provider (IdP) systems

7. **Security control status: Health of the defense**
   The malware may not be strong, but if the defense is weak, the threat appears far more dangerous. For example, the following conditions are systemic weaknesses that increase propagation risk:
   - EDR is disabled.
   - AV bypasses quarantine.
   - Logs are corrupted.
   - SIEM correlations aren't functioning.

   The weaker the defense, the faster the threat grows.
8. **Malware technical capability: What can it do?**
   Here, the current behavior and the malware's potential capability is evaluated, for example:
   - It contains privilege escalation exploits.
   - It has worming features.
   - Credential dumping modules are active.

   Even if the attack hasn't yet spread, the risk score increases because the potential to spread is high.
9. **Available vulnerabilities: Open doors of the environment**
   Sometimes, the threat itself isn't strong, but the environment is so weak that the risk escalates. In the following cases, the root of the risk isn't the attacker, but the environment itself:
   - Unpatched CVEs
   - Legacy protocols (Windows New Technology LAN Manager [NTLMv1], SMBv1)
   - Misconfigured administrative shares
   - Weak group policy object (GPO) implementations
10. **Chain reaction impact: Loss multiplier**
    Some attacks remain confined to a single system. Others grow like an avalanche. The impact isn't just technical if an attack has the following characteristics, it strikes the entire organization:
    - Can access backups
    - Involves critical data
    - Affects business continuity
    - Has the potential to damage the supply chain

    For this reason, the final row of the scoring table measures the real-world consequence of impact analysis.

10.3.3 Determining Operational Priority According to CTI

Operational prioritization is the heart of incident response because the hardest thing in defense is knowing which danger to deal with first. On an SOC screen, dozens—sometimes hundreds—of signals appear at the same time; each has a different tone, and each has an uncertain urgency. At times, analysts feel like they are standing in a train station watching trains pass nonstop and wondering which one should I catch, which one is going to hit us directly, and which one is just the noise of a distant line?

CTI steps in here and forms the invisible backbone of events. Because CTI doesn't determine what should be addressed first, which incident threatens a critical asset, and which risk the organization can tolerate. Determining operational priority is far more than the technical dimension of threat analysis; it gains meaning through the combination of strategic vision, institutional memory, and the analyst's intuitive sensitivity.

Sometimes, an alert appears to carry very high risk, yet in reality it doesn't touch the organization. Sometimes, a small, faint signal is contacting the most fragile link in the chain. Determining operational priority is the art of seeing the difference between these two extremes.

When establishing operational priority, the first question analysts ask themselves is, "Which asset is this incident close to?" Within every organization, there's an invisible hierarchy of assets: file servers, DCs, payment systems, customer databases, backup infrastructures, hypervisor clusters, and so on, and they don't all stand on the same line.

Even if the technique used by an attacker is the same, the criticality of the asset it touches acts like a compass that changes direction. For example, if a PowerShell anomaly is observed on an end user, it's only a limited incident. But if the same anomaly is observed on a DC, all priorities of incident response change; the dashboard suddenly turns red.

Real-World Example

One of the most unforgettable moments in my personal experience is how an apparently ordinary LSASS access request could be so small on a normal endpoint that no one would care, yet when the same behavior occurred on an IdP server, it made me jump up. The same technical behavior carries two different meanings at two different points. CTI makes this difference readable.

When determining operational priority, the second question analysts ask themselves is, "Where does this incident's risk level fall within the limit the organization can bear?" Every organization has an invisible risk tolerance. Some are accustomed to dealing with more aggressive threat actors; others hold sensitive data, so even the slightest breach can cause major destruction. CTI clarifies this tolerance line.

Risk level must be read together with its organizational impact, for example:

- A file deletion behavior carries low risk on a typical endpoint.
- But if file deletion occurs on a backup server, it produces destructive risk.

Here, CTI whispers the following to the analyst: "This isn't just an incident; its consequences can create a wide resonance." The breadth of that resonance determines the priority of the incident.

The third critical factor is how the incident is positioned in the context of the threat actor. The question that passes through the analyst's mind this time is, "Which actor's way of working is this behavior consistent with?" If the signal aligns with the typical reconnaissance chain of a known APT group, the priority rises immediately.

APT groups never finish an attack in a single step. They come, they settle in, and when the time comes, they touch the target. On the other hand, if a financially motivated attacker is detected, the priority should still be high, but this time, the urgency comes from the rhythm. Financial attackers are fast and impatient. Within a few minutes, they can steal data, drain accounts, or build a fraud chain.

For this reason, CTI both tells who the actor is and shows the source of urgency:

- APT: Long-term but critical threat.
- Financial: Short-term but fast threat.
- Script-kiddie: Low threat but broad noise.
- Destructive group: Fast and irreversible threat

Each occupies a different place on the operational prioritization schedule.

My favorite aspect of determining operational priority is that the analyst must blend technical knowledge with intuition. Sometimes CTI data says an incident is only at a medium level, but the analyst notices a disruption in the signal's rhythm—something is too regular or too random. These emotional microsigns can indicate that an incident may be more important than expected.

Real-Time Example

In cybersecurity, small intuitions sometimes prevent major disasters. One day at a customer, I saw an abnormal Kerberos request on an ordinary user account. Technically, it was a low-priority signal. But in CTI, it had been reported that week that a particular APT group had begun using the same TTP in Europe. Suddenly, I felt an indefinable unease inside. I escalated the incident. Hours later, it became clear that the attacker was silently preparing privilege escalation. This is where CTI's true power emerges when combined with the analyst's intuition.

In determining operational priority, the last—and perhaps the most important—question analysts ask themselves is, "Which other incidents can this incident trigger?"

Impact analysis and operational prioritization converge at this point. Some signals aren't momentary; they generate chain reactions:

- A credential dump indicates lateral movement.
- A lateral movement provides access to the DC.
- DC access puts the entire identity infrastructure at risk.
- Identity infrastructure attack risks the collapse of the entire organization.

An analyst who can see this chain selects the correct link to prioritize. Sometimes, to break the chain, it's necessary to target the first step; sometimes the third step; and sometimes the step that appears the quietest.

In summary, operational prioritization is the compass that determines the direction of incident response. CTI is the magnetic field of this compass—without it, the needle loses its direction.

CTI tells the analyst the following:

- Look at this first because the impact may be large.
- This signal is small for now, but its rhythm isn't right—pay attention.
- This attacker is patient; if not detected early, it will grow.
- This behavior is close to a critical asset; increase the priority.
- This incident may exceed the organization's risk tolerance; act quickly.

The analyst combines all of this information in their mind and makes the most critical decision: "What should I stop first, right now?" This decision isn't purely technical, but depends on understanding how a threat may evolve. CTI supports this by helping analysts prioritize alerts within a broader operational context, rather than evaluating them in isolation. As a result, prioritization focuses on both the severity of an alert and its potential implications across the environment.

The following factors should be considered when determining operational priority:

- **Signal inconsistency across systems**
  Threat actors rarely leave a single trace; they break their traces into small pieces and distribute them across different systems. For this reason, operational prioritization should consider both signal intensity and inconsistencies between signals. When one log entry overlaps with another log in a different system, when a small deviation in a process tree combines with a slight irregularity in network traffic, or when an authentication alert correlates with DNS resolutions, the priority level can suddenly change. This is often noticed through personal intuition. A connection lights up in the analyst's mind, creating a small vibration that says, "These two signals can't be a coincidence." That vibration is the early indicator that the incident actually has a higher operational importance.
- **Weak signal close to a critical asset**
  Another advanced reality that complicates operational prioritization is the deliberate use of deceptive behaviors by threat actors. In some advanced APT operations, attackers

generate low-severity, "bait-like" incidents to measure the organization's attention. For example, they may create a harmless file on a normal user system that resembles a known malicious hash. While the SOC team pursues this small signal, the attacker quietly touches the identity infrastructure. In such situations, the only way to determine operational priority correctly is to focus not on the noise created by the signal, but on the silence of the context. Sometimes, the most dangerous incident is the one that makes no sound; the true power of CTI lies in making sense of this silence.

- **High-noise signal on a low-criticality asset**
  In some environments, the attacker's target isn't directly critical assets; instead, they seek to find the weakest link in the chain within the organization and first establish a bridge there. This bridge is sometimes not a technical vulnerability; it may be a note left in a system administrator's routine script, an API token forgotten in an automation pipeline, an unnoticed SSH key, or environment variable values stored in CI/CD systems. Analysts often don't see these elements as threats because they aren't directly related to an attack. However, an advanced CTI approach knows how critical these weak links are for an attacker. Therefore, when determining operational priority, both the current position of the attack and the paths the attacker may establish in the future must be taken into account.
- **Signal made suspicious by timing**
  In some scenarios, operational prioritization requires understanding not the attacker's technical speed, but the intent in their timing. For example, a credential access incident occurring at 03:00 at night isn't in the same category as the same incident occurring during business hours. Because an access attempt carried out at night is often a conscious timing choice by the attacker; this timing is a sign of operational preparation. Even this small detail can elevate an incident's priority from a lower level to a higher one. The analyst's task is to decipher the purpose of the rhythm at the point where technical signals and timing signals intersect.
- **Anomaly coinciding with a critical business process**
  Sometimes operational prioritization intersects unexpectedly with the organization's own business processes. For example, an anomaly seen in file access logs during the finance department's monthly closing period has a much higher priority compared to the same anomaly occurring at another time of the year. Or authentication failures occurring during a critical tender period in government institutions may be far more meaningful than those occurring on ordinary days. One of CTI's most important roles is to establish a connection between technical incidents and business processes. When this connection is made, operational prioritization suddenly becomes much clearer.
- **Noisy signal with potential to be a decoy**
  Another unknown factor that increases operational priority is the attacker's withdrawal behavior. Some advanced threat actors quietly disappear when they realize they have been detected. Although this is often perceived as low priority, in reality, the danger is the opposite: this withdrawal may indicate the attacker's intention to reposition for a

larger operation. In such moments, the analyst should look not at signal intensity, but at signal absence because sometimes the most critical moment of an attack is the moment when the attacker does nothing.

- **Disruption in user behavior rhythm**
  A user has access habits that operate in the same rhythm for years; if one day they suddenly connect to a different server or become active in an unexpected time window, this may appear as a small incident on its own, but within the CTI context, it's evaluated as possibly the first breath of an account taken over by an attacker. This small behavioral change can rapidly raise operational priority because attackers often don't leave a large trace in their first touch; they only slightly disturb the rhythm of behavior.
- **Incident capable of triggering a chain reaction**
  Finally, operational prioritization is sometimes performed not with technical truths, but with the mathematics of the worst-case scenario. More important than an incident's actual risk is the potential risk it carries. If the attacker's next move is unknown, the analyst should ask themselves the following question: "In the worst-case scenario, where could this incident lead?" If the answer touches the possibility of organizational collapse, that incident is elevated to the highest priority, no matter how small its technical signal may be.

In the CTI approach, operational prioritization is a strategic decision that shapes the future of the attack. As the analyst walks among signals, they protect both the current threat landscape and the fate of the organization. Most of the time, the critical decision becomes, "Which small signal is the harbinger of a great silence?" The correct answer to this question forms the invisible shield that protects organizations from disaster.

Table 10.7 functions as an operational decision engine, enabling SOC analysts to classify and prioritize every signal they encounter within its context. Instead of evaluating a signal in isolation, it does so in combination with asset criticality, actor type, potential impact, and the organization's business realities.

Signal/Incident Characteristic	Example Scenario	Asset Criticality	Threat Actor Context	Potential Impact Level	Recommended Operational Priority
Weak signal close to a critical asset	Minor LSASS access on a DC, low-noise anomaly	High	Consistent with APT-like reconnaissance TTPs	Propagation into identity infrastructure	Critical (P1)
High-noise signal on a low-criticality asset	Aggressive malware activity on an end user, numerous AV alerts	Low/medium	Financially motivated, fast profit-oriented	Local damage, limited spread	Medium/High (P2–P3)

Table 10.7: CTI-Based Operational Prioritization Decision Matrix

Signal/Incident Characteristic	Example Scenario	Asset Criticality	Threat Actor Context	Potential Impact Level	Recommended Operational Priority
Signal made suspicious by timing	Failed Kerberos request at 03:00, singular but matching CTI-correlated TTP	Medium/ high	Correlated with a recently reported APT campaign	Medium-to-high spread risk	High (P2)
Anomaly coinciding with a critical business process	Abnormal query volume on accounting database during financial closing day	High	Suspicion of financial or hybrid intent	Direct financial/ reputational loss	Critical (P1)
Noisy signal with potential to be a decoy	File resembling a known malicious hash but appearing harmless, single alert on endpoint	Low	Possible distraction (decoy) behavior	Indirect risk (analyst resource consumption)	Low/Medium (P3–P4)
Correlated small anomalies in quiet segments	Minor DNS rhythm disruptions across separate segments + slight identity anomalies	Medium/ high	Possibility of silent expansion (silent perimeter)	Chain-propagation potential	High/Very High (P1–P2)
Attacker trace suddenly disappearing from a critical system	Suspicious process chain visible in EDR for a while, then completely silent	High	Advanced actor realizing detection	Repositioning/ preparation for major attack	High (P2)
Disruption in user behavior rhythm	User with years of stable behavior suddenly accessing a different server	Medium	Suspected account compromise	Entry point into the identity chain	Medium/High (P2–P3)
Incident capable of triggering a chain reaction	File deletion activity on a backup server, simultaneous reconnaissance indicators near the DC	High	Destructive or ransomware pre-stage	Organizational collapse scenario	Critical (P1)

Table 10.7: CTI-Based Operational Prioritization Decision Matrix (Cont.)

> **Note**
>
> The operational priority codes can be interpreted as follows:
>
> - P1: Hard-to-reverse risks such as organizational collapse and compromise of identity infrastructure, backup, or hypervisor layers. Immediate action required.

- P2: High likelihood of spread, or actor alignment with APT/critical TTPs. Rapid triage + containment required.
- P3: Locally impactful but noisy/automated activity. Standard incident response workflow required.
- P4: High likelihood of decoy/deception or low impact. Monitoring/validation required.

Following is a concise explanation of key prioritization patterns:

- **Weak signal close to a critical asset**
  Even low-level activity on high-value systems (e.g., DCs) should be treated as high risk, as it may indicate early-stage access or preparation for credential compromise.
- **High-noise signal on a low-criticality asset**
  Aggressive or noisy activity on low-value endpoints often reflects opportunistic or automated threats. While action is required, the overall impact is typically limited.
- **Timing-based anomalies**
  Unusual activity during low-activity periods (e.g., late-night hours) may indicate reconnaissance or preparation phases, especially when aligned with known TTPs.
- **Anomalies during critical business processes**
  Even minor technical deviations can become high-risk when they occur during sensitive operational periods, such as financial closing or critical transactions.
- **Potential decoy signals**
  Some noisy or suspicious artifacts may be intentionally generated to distract defenders, requiring cautious but lower-priority handling.
- **Correlated low-level anomalies**
  Multiple small signals across different systems, when combined, may indicate coordinated adversary activity typical of advanced threats.
- **Sudden disappearance of attacker traces**
  The abrupt loss of visibility may indicate that the adversary has detected monitoring and is repositioning for a subsequent phase.
- **User behavior deviations**
  Unexpected changes in established user activity patterns may signal account compromise or identity misuse.
- **Chain-reaction capable incidents**
  Certain events become critical due to their potential cascading impact, particularly when they affect recovery mechanisms or core infrastructure.

10.3.4 Examples: Threat Contextualization in Practice

The following examples illustrate how threat contextualization is applied in real-world scenarios. Rather than evaluating alerts in isolation, these cases demonstrate how combining technical signals with CTI, asset criticality, and operational context can significantly change prioritization decisions. Each example highlights how seemingly low-noise or routine activity may, when properly contextualized, reveal indicators of high-impact threats.

Silent Hypervisor Touch (No Noise, Loss of Ground)

One day, you see the following on the SOC screen: a short but consistent access attempt occurring outside of business hours to a vSphere/Hyper-V management interface that is normally accessed only during maintenance windows.

What is unusual is this:

- The account attempting access belongs to a legitimate system administrator.
- The IP originates from an expected internal management segment.
- The fail/success ratio is normal; there's no brute force, no burst activity.

At first glance, it's very easy to say, "The system administrator must have done this." But CTI whispers something else: In recent months, some APT groups have been conducting attacks targeting the hypervisor layer during the pre-ransomware phase for the following reasons:

- If the hypervisor is compromised, all VMs can be taken hostage in a single move.
- Backup and production environments can be held hostage together.
- Agent-based EDR solutions can be disabled.

At this point, the signal isn't high-noise in the classical sense; on the contrary, it's too orderly, too clean, and too natural. This is where operational priority should be determined not by classic log volume, but by the strategic value of the layer being touched. The hypervisor layer is the backbone of the organization's infrastructure; therefore, even the slightest behavioral deviation observed here should be handled in the P1–P2 range.

Consider the following points:

- Don't automatically assume legitimacy just because a real admin account is being used.
- Evaluate hypervisor access logs not only by login success but by time, target, and session duration rhythm.
- Correlate anomalies in backup systems, storage layers, and snapshot operations occurring in the same time window.
- If there are CTI warnings about current hypervisor-targeted campaigns, tools, or proof of concepts (POCs), immediately elevate the incident to a higher priority level.
- In such cases, take seriously both the technical evidence and the question, "Would this person normally be here at this time?"

Abnormal Noise from a Normal Log Source (SIEM's Prediction Error)

Consider a web application firewall (WAF) that has been generating logs in the organization for years and has been relatively trouble-free. Every day it produces dozens or hundreds of low/medium severity alerts: cross-site scripting (XSS) attempts, simple SQL injection (SQLi) attempts, bot traffic, and so on. The SOC team has grown accustomed to the noise of this source.

One night, the WAF generates another alert:

- The source IP belongs to a known cloud provider.
- The pattern looks like a classic SQLi attempt.
- The WAF marks it as "blocked."

However, in the CTI platform, the following information exists:

- That same IP was recently reported as a C2 server in attacks against several critical financial institutions.
- The IP was used not only for exploitation but also for post-exploitation data exfiltration.
- The traffic passed as legitimate-looking in some APT-like operations.

In other words, while the SIEM labels this event as just another WAF incident at a low level, CTI says, "This IP isn't an ordinary script-kiddie IP." Here, operational priority should listen not only to the SIEM rule set, but to the CTI-based overlay. This incident should move from the level of "The WAF already blocked it, ignore it," to the level of "If this IP succeeds, which data layer could it touch?"

Consider the following points:

- Don't treat the severity assigned by the SIEM as ultimate truth; reassess it by correlating with CTI.
- Check whether the same IP appears in DNS, proxy, VPN, RDP, and SSH logs across different time windows.
- Don't relax simply because the WAF blocked it; examine WAF upstream/downstream logs for bypass attempts and 500/503 error patterns.
- Avoid developing source-centric blindness with the excuse that the log source is noisy.
- When calculating operational priority, answer this question: "If this IP succeeds, where could it reach?" together with the business unit (customer data, payment integrations, APIs, etc.).

An External Actor That Appears Internal (VPN + Identity + Timing Triangle)

The organization uses VPN + MFA in its remote work infrastructure. There has been a pattern operating smoothly for years:

- User X typically connects between 09:00 and 18:00.
- The connection is made from within Turkey.
- The same device and the same IP block are always used.

One day, you observe the following pattern:

- The same user appears to open a VPN session first from an Istanbul IP, and 40 minutes later from a different country.
- Both sessions are approved via MFA.
- The user navigates internal systems, particularly around Role-based Access Control (RBAC), Identity and Access Management (IAM), and password vault interfaces.
- There's no aggressive activity; no file deletion, no scanning, and so on.

From a purely technical perspective, the SIEM might not even classify this as "medium." But on the CTI side, it's known that certain actors have recently been attempting MFA fatigue and session hijack techniques, particularly TTPs such as "stealing ongoing sessions" and "capturing token chains via reverse proxy." In other words, the signal here says, "There's something that looks like a user, but doesn't behave exactly like a user."

At this point, operational priority must go far beyond the classic login success/login failure distinction. Geographic jump, timing, accessed application set, user habits, and CTI TTP matches should be combined, and this incident should be elevated to at least the P2 level.

Consider the following points:

- Don't fall into the dogma of "We're safe because we have MFA"; remember that MFA itself is a targeted control.
- Check whether the device fingerprint associated with the same user (user agent, OS build, browser footprint, etc.) aligns with the normal pattern.
- Combine VPN gateway, SSO, and IdP (ADFS/Microsoft Entra ID/Okta, etc.) logs on a single timeline and look for rhythm breaks.
- If CTI reports campaigns related to TTPs such as "impossible travel," "MFA bypass," or "session hijack," place this incident directly into that context.
- Don't simply ask the user "Was it you?" and move on; examine where the account accessed after successful login, which objects were viewed, and how long the session remained active.

The common point of these three examples is that when viewed individually, the signals don't ring disaster bells. But when context, rhythm, and the CTI layer come into play, operational priority suddenly jumps up.

True maturity begins not with the question, "How many alerts are there?" but "What does this signal say in context, and what could it become in the future?" This is the art of producing the answer to that question through CTI-based operational prioritization.

10.4 Summary

This chapter demonstrated how CTI enhances incident response by adding context, meaning, and direction to technical signals. Rather than treating alerts in isolation, we examined how CTI enables analysts to interpret IOCs and IOAs, correlate multilayer telemetry, and reconstruct attack activity as part of a coherent chain.

We explored how CTI supports each phase of the incident response lifecycle by guiding logging strategies, improving validation accuracy, and strengthening decision making. Particular emphasis was placed on contextualization, showing how technical findings gain significance only when evaluated alongside asset criticality, attacker behavior, and operational impact.

Finally, we highlighted how CTI-driven approaches improve operational prioritization, allowing analysts to distinguish between noise and meaningful threats, assess propagation potential, and respond with greater precision.

In the next chapter, we'll extend these concepts by focusing on how intelligence-driven processes can be operationalized through structured workflows.

Chapter 11
Intelligence-Driven Proactive Threat Hunting

In the world of cybersecurity, threats no longer knock on the door; they look for ways to slip inside silently. Defensive architectures that for years formed the backbone of security through alert-generating systems are increasingly being buried under noise. Within this noise, identifying the truly dangerous signal is no longer possible through technology alone, but requires human intuition and intelligence.

Intelligence-driven hunting places even a simple log anomaly into context; it whispers where a behavior, an intent, or a step in the attack chain stands. It fills in the spaces between the lines of MITRE ATT&CK, enabling us to understand the "why" behind tactics, techniques, and procedures (TTPs), indicators of compromise (IOCs), and adversary models. Because modern attackers don't leave a single trace, they strategically distribute fragmented, irregular, and sometimes nearly invisible clues. On their own, these clues may be meaningless, but when combined with intelligence, they form the backbone of an attack narrative.

Proactive hunting abandons the reactive defense mindset of "I'll look when an alert arrives" and activates the defense itself. In a sense, instead of waiting for threats, it walks toward them. Over time, this becomes an intuitive discipline developed by every defense team. Sometimes it's a simple Domain Name System (DNS) request within the environment, and sometimes it's an anomalous PowerShell module load of small yet meaningful movements. Like the faint rustle of leaves heard in the darkness of night, it's insignificant on its own, but vital when placed in the right context.

This chapter approaches hunting as a continuously evolving analytical reflex. It explains how cyber threat intelligence (CTI) provides hunters with a sense of direction, how behavior-based approaches improve detection quality, and why custom detection logic must be built not only on rules but on narratives. It details how telemetry sources complement one another, why weak signals can be the most valuable signals, and how a hunt program can reach an institutional level of maturity.

The future of cyber defense is moving away from a structure that positions itself based on external alerts. A mindset is now prevailing that says, "Before you come, I'm coming." Threat hunting is the embodiment of this very mindset. For an analyst, perhaps the most exhilarating moment is when a hypothesis is validated and an invisible attack chain becomes visible. In that moment, a subtle micro-expression stirs within the hunter: "There it is." This feeling is that rare, sweet moment of success in the security operations center (SOC) world, and this chapter aims to pave the path toward those moments with a scientific, systematic, and intelligence-driven methodology.

11.1 What Is Threat Hunting?

In addition to tools and telemetry, threat hunting relies on the analyst's judgment. Beyond automated detections, it often begins with recognizing subtle inconsistencies, i.e., small deviations that don't fully align with expected behavior. At its core, hunting is the disciplined practice of identifying and investigating those anomalies.

For many years, cybersecurity was managed through a reactive reflex. An attacker entered through a door, the system generated an alert, and the analyst reviewed the alert. This model worked for years, until attackers learned how to bypass alarms. Today, a sophisticated adversary can move through an environment for days, even weeks, without generating a single alert. Sometimes the only indicator of that silent progression is the timing of a beacon packet, the length of parameters in a PowerShell command, or a small deviation in a process tree. In other words, threats no longer always shout; they choose silence.

At this exact point, hunting enables a redefinition of defense. The reactive approach carries water once a fire has broken out. *Proactive hunting*, however, senses the smell of smoke before anyone else notices it.

Real-World Example

My first hunting experience was marked by a similar feeling. There was a 4624/4672 login pair that I saw frequently; nothing appeared unusual except that the minute the login occurred fell within a time window that wasn't typically busy in the environment. The IP was familiar as well, yet something bothered me. After a few queries and a few correlations, eventually, I discovered that the attacker was in the preparation phase of a Kerberoasting attack. In that moment, I understood that threat hunting is the work of a mind that doesn't limit itself to reading logs alone. Sometimes a behavior speaks louder than numbers.

We'll explore why threat hunting is important in the following sections and examine the limits of reactive security versus the capabilities and components of proactive hunting. We'll also dive into the application of a hypothesis-driven approach and key tools for filling information gaps.

11.1.1 Importance of Threat Hunting

Modern attacks aren't designed for real-time detection; on the contrary, they are designed to avoid detection. Frameworks such as Cobalt Strike manipulate jitter by shaping beacon rhythms to resemble normal network jitter. Advanced persistent threat (APT) actors don't flood endpoint logs; they use system APIs that leave low signal noise. In cloud-heavy environments, attackers often behave in a legitimate-by-design manner, quietly leveraging all available legitimate infrastructures for their own purposes.

Therefore, threat hunting isn't merely a technical process; it's a defensive language developed against the enemy's silence. We'll discuss the core aspects that strengthen defense in the following sections.

Overcoming Alert Fatigue

Alert fatigue is one of the most insidious challenges of modern SOC environments. A security analyst tries to identify real danger within a flood of alarms streaming across the screen throughout the day, which is like trying to catch a whisper in a crowd. While thousands of alerts are shouting at once, a silently advancing attacker can move in the background without generating any noise. Alerting systems are, in essence, mechanical reflexes that react to predefined patterns; attackers, however, don't operate mechanically. They seek gaps, create gaps, and move through gaps.

This is why threat hunting steps outside the alarm-centric way of thinking. An alert is a labeled form of an event; hunting, however, focuses on the behavior, not on the label: With which commands was a process launched? Why does a recurring memory access pattern occur on the same machine at the same second? Why does a user account access certain resources only outside of business hours? These are questions that SIEMs can't automatically ask, but that a hunter naturally wonders about. Hunting attempts to understand behaviors that don't generate alerts, and more often than not, the true face of an attack lies not on the alert panel but within these silent behaviors.

Real-World Example

I first realized the danger created by alert fatigue through a small rhythm anomaly in an extraordinarily clean environment. The endpoint detection and response (EDR) system had not generated a single alert for days; the screen looked almost sterile. That calmness alone was suspicious, but the real clue was hidden in the Windows System Monitor (Sysmon) records. Every day, at exactly 02:13:07—down to the millisecond—a remote thread injection attempt occurred with the same timing—neither faster nor slower—like a Cobalt Strike Beacon with jitter tuning applied. Technically, no EDR rule was triggered; yet the behavior carried the pulse of a quiet command-and-control (C2) channel that didn't want to be noticed. The second-level alignment was a reflection of the attacker's attempt to remain hidden.

Here lies the importance of overcoming alert fatigue: attacks are no longer noisy. Modern implants are designed to operate with low noise; they are engineered to bypass EDR policies and remain below security information and event management (SIEM) thresholds. A chain of behaviors (e.g., minimal handle accesses to Local Security Authority Subsystem

Service (LSASS) at regular intervals, PowerShell commands composed of concatenated Base64 blocks, and recurring patterns of short-lived network sessions) only gains meaning through behavioral analysis. An alert is triggered only at a very late link in this chain. A hunter, however, can notice the very first link.

For this reason, overcoming alert fatigue is one of the most critical capabilities of threat hunting. Like seeing the single meaningful brushstroke within a complex painting, in the eyes of an analyst, an ordinary log line can suddenly transform into something else: a clue, an inconsistency, or the beginning of a story. More often than not, the attacker is caught while quietly writing that story.

Capturing Invisible Attack Phases

The most dangerous aspect of attacks is often that although they have already begun, they appear as if they haven't. Cyber threats usually render themselves invisible in the middle of the chain, and sometimes even at the very first step. Most initial reconnaissance commands, credential touches, and early preparations for lateral movement don't light up a single indicator on an SOC screen. Attackers, like someone moving silently through a dark room, take invisible steps and proceed to the next phase before their traces become distinct. For this reason, capturing invisible attack phases is one of a hunter's most critical skills. Only hunters notice an anomaly before it takes the shape of a threat and intuitively sense what a behavior may evolve into in the future.

This state of invisibility has become even deeper in modern environments. Cloud services, endpoints, identity providers (IdPs), and application programming interface (API)–based access models provide attackers with quiet spaces in which to roam. For example, an OAuth application consent created in a Microsoft 365 environment can open the door to manipulating the entire corporate email flow without contaminating a single endpoint. This action generates no alert on the EDR; in fact, most SIEM rules don't even see it. Yet a small line appears in the Graph API logs: "`consentType: AllPrincipals.`" This value, which appears on paper to be an ordinary permission parameter, is in fact the first knot of a massive attack. That quiet beginning may become the initial link in a chain that will later compromise the entire tenant.

Similar silent phases appear across different platforms. For instance, an attacker may make a small change in Windows Active Directory (AD)—so small that most security policies consider it normal. One day, a user account's logon hours are expanded to "All Hours." At first glance, this may look like an operational requirement. However, the attacker has quietly opened the path to persistent access to that account. The behavior itself appears innocent; yet when viewed in context—especially if the account has been inactive for a long time or if the last password change occurred at unusual times—this touch becomes the unseen starting point of the chain. It produces no alarm; but that faint "Something is wrong here" feeling inside the hunter makes this small change meaningful.

Another example emerges on the network traffic side. An advanced actor doesn't use high-volume scans during reconnaissance. Instead, they may establish a low-volume, microscopic scanning rhythm that lasts for hours in the form of a single SYN packet sent to a different machine every five minutes. SIEM categorizes this as noise, and EDR doesn't see it. Even the firewall often doesn't record it. Yet this ultra-low-tempo reconnaissance traffic pattern is silent proof that an attacker has begun mapping the environment. This scanning behavior indicates the phase in which the attacker quietly investigates configuration files, remote procedure call (RPC) endpoints, or exposed services without drawing attention. In other words, the most important stage of the attack lies in the invisible middle of the chain.

A third example is hidden in timing inconsistencies within telemetry. Suppose an attacker performs process hollowing on an endpoint to hide within a legitimate process. This technique is now generally known and often detected by EDR solutions. However, a capable adversary doesn't perform hollowing in the classical way; instead, they add only a small memory region to the target process, execute their shellcode at a very low frequency, and create CPU usage fluctuations too small to be noticed. While everything appears normal on the surface, a subtle pattern begins to emerge in Sysmon Event ID 10 records: periodic but very small memory accesses to the same process identifier (PID). This doesn't resemble the typical behavior of a legitimate application. While this doesn't result in alarm, it creates a faint vibration on the hunter's intuitive radar.

The common point of these examples is that attackers are now mastering the silent sections of the attack chain. This is where the importance of the hunting process begins.

Building Intelligent, CTI-Driven Defense

The true value of CTI isn't just on the pages of a PDF, but in how that knowledge is used in the field. Threat hunting is the process that takes threat intelligence out of the laboratory and applies it to real environments, transforming it into a living organism. It takes CTI and reshapes it through the context of the current environment; it analyzes which TTPs are truly applicable, which techniques are likely to leave traces in a given architecture, and where specific attack models may hide. For this reason, hunting is the actionable form of intelligence.

Modern threat intelligence often presents highly technical and specific behavioral patterns. For example, if a CTI report details APT29's long-lived beacon model using DNS-over-HTTPS (DoH), a hunter knows that this isn't just information; it's a new behavioral template to be sought within the environment. DoH traffic is generally legitimate and considered normal by most security devices. However, in long-lived beacon models, millisecond-level repeating request timings, specific hostnames repeatedly resolved by the same endpoint, or changes in the rhythm of Windows `Crypt32` API calls used by the DoH client may stand out. Accordingly, the hunter's chain of thought progresses as follows: DNS logs → endpoint DNS API calls → DoH client behavior → timing analysis → pattern extraction. This technical chain can reveal the true face of an implant embedded in the environment that is designed to avoid generating alerts.

This is where the core strength of CTI-driven intelligent defense becomes apparent: Intelligence describes a behavioral pattern; the hunter determines how that pattern would manifest within their own systems. This is the critical point at which theory turns into practice.

Let's explore a few examples to demonstrate:

- **CTI guidance in OAuth token manipulation**
  Suppose a CTI report states that in FIN7's recent campaign, the group targeted refresh tokens for token replay attacks within Microsoft Entra ID. The report specifies that the attacker particularly performed the following:
  - Silent brute-force attempts over Internet Message Access Protocol (IMAP)/Post Office Protocol (POP) in tenants with legacy authentication enabled
  - Long validity periods for compromised refresh tokens
  - Use of the refresh token replay (RTR) model to bypass multifactor authentication (MFA)

  This intelligence should translate into the following technical approach during the hunting process:
  - Correlation of `RefreshTokenAge`, `RiskDetail`, and `TokenIssuerType` within Microsoft Entra ID sign-in logs
  - Repeated but low-frequency failed login attempts from the same IP
  - Anomalies where `AuthenticationRequirement: singleFactorAuthentication` is observed despite MFA being enabled
  - Session sequences in which different device globally unique identifiers (GUIDs) are authenticated using the same refresh token

  This analysis chain makes a deliberately nonalerting identity threat visible.
- **Using CTI to hunt fileless malware behaviors**
  Another CTI report may state that a state-sponsored group uses the following TTPs in PowerShell-based fileless implants:
  - Script block minification via `System.Management.Automation.dll`
  - Low-volume Windows Management Instrumentation (WMI) event consumer beacons running every 45 seconds
  - Lightweight memory scraping modules targeting Data Protection API (DPAPI) master key extraction instead of LSASS
  - A tendency toward thread stack spoofing rather than classic process injection

  Upon receiving this information, the hunter's technical approach is shaped as follows:
  - Entropy analysis of "minified" script blocks in Event ID 4104 `ScriptBlockLog` records
  - Low-frequency, recurring event consumer triggers in WMI-Activity Operational logs
  - Loading of `System.Management.Automation` observed in Sysmon ID 7 (`ImageLoad`) without any parent PowerShell process

- Memory pattern analysis of processes accessing DPAPI keys without touching LSASS
- Creation of thread stack regions in unusual locations indicative of stack spoofing

This hunting approach can uncover a sophisticated fileless implant that triggers almost no classical EDR rules.

Defense driven by CTI is a mindset transformation that seeks to understand how threats will appear, what shadows they will cast in the environment, and which behavioral models they will distort.

Strengthening Defense Not Once, But Continuously

Strengthening defense once is never sufficient in the modern threat landscape; because attackers aren't static, defense must not be static either. Threat hunting discovers attacks that already exist and, importantly, exposes weak, fragile, or entirely unseen areas within the defensive architecture. In this respect, hunting is akin to a high-resolution X-ray applied to the environment to determine the following:

- Are telemetry pipelines functioning correctly?
- Are log sources flowing completely?
- Are correlation rules able to capture real behavioral flows?

The answers to these questions usually emerge only during active hunting.

The points where the defense is leaking include missing logging, incorrect Event ID coverage, weak correlation logic, and improperly normalized fields. Yet these issues go unnoticed during normal operations because everything appears to be working. Hunting removes this false sense of visibility. For example, while searching for C2 activity, the complete absence of Sysmon Event ID 3 (Network Connection) records may appear to be a simple configuration error; in reality, however, it represents a massive blind spot that renders all outbound connections entirely invisible to the defense. This blind spot is silent because the defense "can't see," not because there's no threat.

For this reason, continuous strengthening of defense is inherent to the hunting cycle. A hunter recognizes both where the attacker is and where the defense itself is weak. Sometimes, identifying how to improve the defense is more valuable than finding the attack itself.

We'll demonstrate with an example of revealing a weak telemetry pipeline. Consider a hunter whose task is to hunt Kerberos-based lateral movement techniques used by a specific APT group. CTI indicates that the group particularly employs the following behaviors:

- Kerberos Privilege Attribute Certificate (PAC) manipulation
- Abuse of unconstrained delegation in subordinate trusted domains
- Unusual timing patterns in Ticket-Granting Service (TGS) requests

Based on this intelligence, the hunter begins analyzing the following telemetry chain:

- Security Event ID 4769 (TGS Requests)
- Sysmon Event ID 10 (Process Access)
- Netlogon Operational log (MS-NRPC)
- Key distribution center (KDC) service debug logs

During the analysis, the hunter notices that only a portion of Event ID 4769 records appear in the Windows security logs. Some are entirely missing. At first glance, this may seem like low-volume traffic; however, when the hunter examines the Netlogon logs, they discover that all TGS requests are recorded there in full. When correlation is attempted, it becomes clear that the security logs fail to record nearly 60% of these requests. The following describes what this means:

- The audit policy on the domain controller (DC) is misconfigured: Kerberos Authentication Service is enabled only for success events, while failure events are disabled.
- Certain Kerberos events have been coalesced by the Windows Filtering Platform (WFT) as part of noise reduction.
- Relevant logs have not been fully and correctly ingested into the SIEM due to improperly normalized security identifier (SID) resolutions in Sysmon.

These weaknesses cause the attacker to become almost entirely invisible when employing techniques such as these:

- **Golden ticket**
  The forging of Kerberos ticket-granting tickets [TGTs] by leveraging a compromised KRBTGT account hash, allowing the attacker to generate arbitrary authentication tokens with custom privileges. This effectively bypasses domain authentication controls and enables long-term, stealthy persistence.
- **Silver ticket**
  The creation of forged service-specific Kerberos tickets (TGS) using the hash of a service account, enabling direct access to targeted services without interacting with the DC. This evades centralized logging and significantly reduces the detection surface.
- **Overpass-the-hash**
  Quieter method (aka pass-the-key) where an attacker uses a captured Windows New Technology LAN Manager (NTLM) hash to request legitimate Kerberos tickets via the authentication protocol itself, blending into normal Kerberos traffic and avoiding the need for plaintext credentials while maintaining a low forensic footprint.

In other words, while searching for the attack, the hunter actually discovers a vast dark area in which the attack can hide. This finding directly translates into a defensive hardening action:

- Audit policy is reconfigured.
- Sysmon configuration is updated.

- The SIEM normalization schema is corrected.
- Correlation rules are redesigned.
- The DC telemetry pipeline is tested end-to-end.

As a result, the defense is strengthened, and the environment regains its own threat-detection capability. Thus, hunting becomes not just a mechanism for detecting attacks, but an improvement process that eliminates the environment's own blind spots.

The Value of Human Intuition

Some traces can truly only be felt. An attacker's rhythm, a subtle imbalance in behavior, or a millisecond-level deviation in network traffic are often things noticed by the human eye and mind, not by machines. SIEM classifies such signals as "insignificant," security orchestration, automation, and response (SOAR) labels them as "noise"; algorithms perceive the mathematical dimension of behavior but often miss the story embedded within it. Human intuition enters at this point in the form of a small, indefinable vibration that doesn't feel right.

A hunter is someone who enters a dark environment with a small flashlight. The flashlight doesn't illuminate the entire room; it illuminates only a single point. At that point, you either see something or you don't—but most importantly, you feel that you must look there. Every hypothesis is a step, and every step opens a new window. Sometimes, when you look through the window, you see only darkness and you accept it. But sometimes, the beginning of the entire attack chain winks at you from behind just a few bytes of irregularity. That's when you truly understand the value of intuition.

From a technical perspective, the value of human intuition manifests in the following areas:

- Millisecond-level shifts in time series
- Unnatural delays within process start chains
- Rarely observed combinations of API calls
- Memory access patterns that don't align with normal usage models
- Microdeviations in network traffic packet size distributions that appear ordinary yet are statistically meaningful

Algorithms discard most of these signals as "insignificant variance." The hunter, however, follows rhythm—not variance. Here are a few real-world examples to demonstrate:

- **Intuitively sensing a millisecond break in beacon rhythm**
  While investigating suspicious outbound traffic in the environment, a beacon with jitter tuning was attempting to blend into normal HTTPS traffic. Packet lengths were consistent, the domain Server Name Indication (SNI) information matched a popular content delivery network (CDN) domain, and the EDR generated no alerts. Yet something was unsettling: The beacon's transmission intervals appeared mathematically random, but I noticed a rhythm that didn't align with the "naturalness of randomness."

Normal applications mimic human behavior in jitter variations: delays increase during peak hours and decrease during off-hours. The beacon, however, did the opposite. Packets were sent at more consistent intervals during the quietest hours of the night.

This anomaly is too subtle for a SIEM to detect. Sysmon or firewall logs don't flag it. But a hunter can sense this rhythmic inconsistency.

When I dug deeper, I noticed that the JA3 fingerprint within the Transport Layer Security (TLS) `ClientHello` deviated slightly from the typical JA3 values of CDN clients. This small difference (statistically significant, yet too minor for systems to label as suspicious) revealed the entire story of the attack: A long-lived, low-noise beacon was operating silently in the environment.

- **Unnecessary haste within a process chain**

  In another case, while working as a hunter, I noticed something unusual about the startup timing of PowerShell. Everything appeared normal: The `ExecutionPolicy` was bypassed, and the commands were Base64-encoded, which was already a well-known technique. What bothered me, however, was the timing of the process chain.

  The transition from `Winword.exe` to `powershell.exe` typically occurs with a delay of 200–400 ms in normal user behavior. A user triggers a macro, the macro calls PowerShell, and the process naturally includes a delay. Here, however, the transition occurred in 80 ms. This speed didn't mimic human interaction; it indicated an automated chain.

  This timing deviation was ignored by the SIEM. The EDR didn't even mark it as suspicious because the behavior didn't match signature-based rules. But as the hunter, I sensed that this speed wasn't natural.

  As I deepened the investigation, I found traces of proxy dynamic link library (DLL) injection in the process tree. The attacker was triggering PowerShell via a thread within the Word process rather than through a macro chain. That was precisely why the delay was abnormal. This subtle technical difference—only a 100–200 ms deviation—was the key indicator that made the attack visible.

The common point of these two examples is that algorithms measure *expected behavior*, whereas the hunter listens to the *spirit of behavior*. The goal isn't just to observe the signal itself, but to understand the intent behind the signal. Sometimes, that intent can only be captured through intuition.

11.1.2 The Limits of Reactive Security and Alert Fatigue

Reactive security was a revolution in the early years of cybersecurity. When a system detected an anomaly, it generated an alert; the analyst reviewed the alert, followed predefined procedures, resolved the issue, and operations returned to normal. For a long time, this mechanism was regarded as the backbone of security teams. At that time, attacks were noisy enough, techniques were simpler, log volumes were lower, and threat actors were far less sophisticated than they are today. Reactive defense worked in that world.

But today? In today's threat landscape, reactive security resembles attempting to put out a forest fire with a fire extinguisher. While the flames continue to spread, the only information available to teams may be a delayed alert. Moreover, this alert is often either a false positive or generated at a very late stage of the attack chain. In other words, by the time the system begins to scream, the silent mechanisms have already been operating for a long time, and the attacker has passed through many doors, making progress without leaving obvious traces.

Modern attacks advance passively. They often behave as if nothing is happening. Some use legitimate APIs; others mimic ordinary user behavior. The real threat moves further and further away from visibility. This is where call alert fatigue emerges. On an SOC screen, thousands of alerts accumulate every day. Some of these alerts are unnecessary, some are repetitive, some are caused by faulty logs, and some are truly critical. However, the human mind is limited in its ability to distinguish the critical ones among thousands of alerts. Attention fragments, visual fatigue increases, and that small red dot within piles of alarms is easily missed. Naturally, analysts tend to place less emphasis on threats that don't generate alerts. Yet that silent threat is the true starting point of the invisible chain that advances unnoticed.

Example

In a corporate environment, an EDR may generate no serious alerts for days. On the surface, this appears to be a good thing, and many teams interpret it as everything is fine. However, a small pattern emerging in Sysmon records (e.g., a memory access loop occurring at the same second every night) may actually aligns with the jitter configuration of a Cobalt Strike, such as an implant. SIEM classifies this as a low-priority pattern because the statistical variance appears low. But a hunter senses this rhythm. For a reactive system, this inconsistency is nothing; for hunting, it's a suspicious heartbeat.

Reactive security has another limitation: the "appears to be working" belief. Many SOCs behave as if everything is normal because systems generate noise when they break; yet stealthy attacks generate no noise at all. When a server's CPU usage rises abnormally, everyone pays attention. But if an attacker increases CPU usage by only 1%, this change triggers no alerts. That 1% CPU shift, however, may represent the regular beaconing activity of an in-memory implant. Reactive security doesn't notice it because reactive security listens only to signals that shout; it doesn't hear whispers.

Example

Consider an SOC analyst that has new alerts arrive on the screen every five seconds. Most are caused by misconfigured applications, noisy logs, or transient performance fluctuations. While the analyst works through this noise, the real attack may be occurring on an endpoint that generates no alerts at all. The attacker uses legitimate Windows APIs, minifies PowerShell commands, touches credentials only in a read-only manner, and completely blends behavior into normal traffic. Reactive security can't catch this because a reactive system looks for where the attacker is "shouting." A modern attacker doesn't shout.

For this reason, threat hunting becomes a critical discipline that fills the gap left by reactive security. Hunting doesn't ask if there's an alert. Instead, it asks the following:

- Why is there no alert?
- Why does this behavior look the way it does?
- Is this silence natural, or is something being concealed?

A hunter's intuition often notices what machines can't see: one log line is slightly different from the others, the rhythm of a packet chain fits the definition of "normal" but not its naturalness, or user behavior is legitimate but incompatible with its context. This is where alert fatigue is broken: by a human's skeptical gaze.

Ultimately, the limit of reactive security is embedded in its very nature. It's reactive. It responds. But most modern threats advance without giving an opportunity for reaction.

Therefore, overcoming alert fatigue requires more than rules and alerts; it requires human intuition, behavioral analysis, contextual awareness, and a proactive investigative mindset. In this sense, threat hunting is the quiet yet most powerful reflex of modern cyber defense because an unseen trace is more important than a shouting threat.

11.1.3 Core Components of Proactive Hunting

Proactive threat hunting is more than an effort to find attacks; it's the determination to systematically investigate all areas in which threats may emerge. This approach requires far more than a passive defensive mindset. Proactive hunting means acting when there isn't yet an alert, seeking meaning in silence, and uncovering truths hidden in the gaps between logs, network traffic, and behavioral patterns. For this reason, proactive hunting is both a mental discipline and a technical architecture.

Understanding the core components of this discipline makes it possible to recognize the essence of the value a hunter delivers to both themselves and their organization:

- **Hypothesis-driven thinking: The starting point of the hunt**

  The most distinctive characteristic of proactive hunting is the mindset of "first form a hypothesis, then search for evidence." This approach operates in the reverse of the reactive security cycle of "alert arrives → investigate." A hypothesis defines the possibility of an attack; this possibility may be based on technical knowledge, threat intelligence, the internal dynamics of the environment, or the analyst's intuition.

  Sometimes a hypothesis begins with purely academic reasoning: "If PowerShell-based fileless activity is increasing in the environment, the entropy of ScriptBlockLog ID 4104 should have changed."

At other times, the hypothesis is triggered by a far more human sensation, for example, that faint feeling gnawing at an analyst when a user account has been a little too quiet.

The most accurate hypotheses sometimes begin entirely with a sense of unease. Technical justifications come later; intuition often moves the first stone.

- **Behavior-based analysis: Intent behind the patterns**

  One of the strongest components of proactive hunting is behavior-based thinking. Modern attacks are designed to avoid generating signatures. In other words, the attacker uses legitimate APIs, communicates through well-known protocols, and interacts silently with system components. You might be asking several questions:

  - Which command was executed?
  - Which file was created?
  - Which process was triggered under whose identity?

  These questions are, of course, important. But for the proactive hunter, the more critical question is this: Is the behavior natural?

  Unnatural behaviors typically manifest in the following ways:

  - Timings
  - Access frequencies
  - Combinations of API calls
  - Methods of memory region allocation
  - Rhythm anomalies in network traffic

  For this reason, behavioral analysis is the true compass of a hunter. Even if an attacker conceals their actions, they can't conceal their intent.

- **Telemetry maturity: You can't hunt what you can't see**

  The foundation of proactive hunting rests on accurate and complete telemetry flow, as we discussed in depth in Chapter 9. If logs are missing, correlation is broken, or pipelines are inconsistent, the hunter is navigating in a blind area. The phrase "You can't catch what you can't see" may sound simple, but it's one of the greatest truths in cybersecurity.

  Misconfigured Sysmon, incomplete ingestion of DNS telemetry into the SIEM, EDR agents producing only partial events, or audit policy errors in Windows Security logs are each capable of completely altering the course of hunting. For this reason, proactive hunting continuously tests how the environment tells its own story:

  - Are all Event IDs functioning correctly?
  - Are log formats properly normalized?
  - Are telemetry timestamps consistent?
  - Are the behavioral signals produced by the EDR complete?

  Sometimes a hunter doesn't find an attack; instead, the silence in the logs reveals a weakness in the defense. This, too, is an important part of hunting.

- **Contextual awareness: A single data point can lie**
  Proactive hunting doesn't rely on a single piece of data. A single log, a single event, or a single anomaly has no meaning unless it's examined within context. Contextual information is what transforms clues into something meaningful for the hunter.

  The same event can carry two entirely different meanings in two different contexts:

  - An unusual PowerShell session on an administrative account indicates critical suspicion.
  - The same session occurring on an IT automation server is normal operation.

  During a hunt, keep this one principle in mind: Data is never read in isolation. It's read together with surrounding data: time, user, device, and historical behavior. The accuracy of the hunt depends on context.

- **Intuition: The microsignals that technology can't capture**
  This is the part of proactive hunting that can't be mechanized: intuition. Sometimes a faint vibration within the hunter signals the presence of an anomaly that doesn't shout but whispers. The things machines discard as insignificant variance (e.g., a millisecond shift in a beacon's timing, unnecessary haste within a process tree, and a strange combination in an API call), the hunter hears as disruption in the rhythm.

  This intuition is an inner voice shaped by experience. It's not entirely technical, nor is it entirely emotional. It stands somewhere in between.

- **Continuity: The hunt ends, hunting never does**
  Proactive hunting isn't a task; it's a process that's not done once, but continuously. Threats change, environments change, user behaviors change, and technology changes. For this reason, hunting is a neverending cycle:

  - Observation
  - Hypothesis
  - Testing
  - Validation
  - Learning
  - Improvement

  After a long hunt, it's not possible to turn off the tap and relax because the next day, another attack may be born in the same silence. Proactive hunting keeps defense alive, allowing it to breathe and building its resilience.

The subcomponents of proactive hunting are both technical steps and cognitive processes. The following elements are layers that complete the previous sections and constitute the true nature of hunting:

- **The art of learning an environment's normals: Knowing your own ecosystem**
  One of the most critical skills for a hunter is knowing what normal looks like because the path to identifying anomalies passes through internalizing normal behavior. Here's a

real-world example: I once noticed that SQL servers responded to queries with a delay of 3–4 seconds at certain hours. At first glance, this looked like a problem, but when I learned that intensive data backups were performed during those hours, I realized that this was normal behavior. This taught me that an anomaly is confusion for someone who doesn't know the context; for someone who does, it's a signal.

For this reason, the hunter's first task is to observe systems, become familiar with them, decode their rhythm, and understand their natural fluctuations. In doing so, the attacker's footsteps become far more distinct.

- **Thinking from the adversary's perspective: Understanding what the attacker would prefer to avoid**
  Hunters don't think only from the defender's perspective; they also look through the attacker's eyes. Questions such as these are normal fundamental reflexes of proactive hunting:
  - If I infiltrated this environment, where would I enter?
  - What would I abuse?
  - Which telemetry would I want to evade?

  Some attack techniques make sense only when you understand where the defense is blind, for example:
  - An APT attempting lateral movement without touching highly privileged accounts in an AD environment will often target service accounts.
  - Memory-resident implants that evade EDR detection hide behind low API usage profiles.
  - Cloud attackers target permission structures with tenant-wide impact because such actions blend into noise within session-based telemetry.

  A hunter continuously visualizes these possibilities in their mind—like searching for the missing piece of a puzzle.
- **Reading the time factor: Decoding the chronology of behavior**
  An important component of proactive hunting is time-sensitive thinking. *When* an event occurs can often be more critical than how it occurs, for example:
  - If the same user account normally logs in only between 09:00 and 17:00 on weekdays, a session occurring at 04:12 on a Sunday morning is a time anomaly.
  - If PowerShell commands never run outside business hours and a `ScriptBlockLog` Event ID 4104 appears at midnight, this is technically small but contextually significant.
  - If Kerberos TGS requests normally occur 3 times per minute and suddenly spike to 43 times, this may indicate brute-force activity or reconnaissance preparation.
- **Psychological resilience in hunting: Making peace with silence**
  Proactive hunting is sometimes hours of silence where nothing is found. An anomaly you've pursued for months may not be a real threat. This can be discouraging, but it's also the natural rhythm of the profession.

For example, during an anomaly analysis conducted on Sysmon ID 11 (`FileCreate`), I spent approximately 130 hours searching for small temporary files created by an APT group prior to ZIP-based dropper activity. For 130 hours, I found no malicious files. Yet that process taught me the full boundaries of normal behavior. The following month, what enabled me to identify a malicious payload—nearly identical to benign files but distinguished by a 23 KB size difference—was precisely that patience.

- **Investigative curiosity: The power of "What-if?" questions**
  Sometimes what redirects a hunter isn't a technical discovery, but curiosity. The question that fuels the spirit of hunting is "What if the attacker could do this?" This question does the following:
  - Generates new hypotheses
  - Leads to examining logs that were previously ignored
  - Changes the way telemetry is interpreted
  - Pulls the analyst out of standard thinking patterns

  Curiosity is often the only way to stay one step ahead of the attacker. Because attackers are creative; defense must be as well.
- **Post-hunt learning: Every hunt lays the groundwork for the next**
  Proactive hunting gains its value not from the result, but from the process. Even when a hunt appears unsuccessful, it reveals new insights about the environment:
  - Missing log sources are identified.
  - Performance bottlenecks are discovered.
  - User behavior patterns become clearer.
  - The attack surface is reduced.
  - New detection logic is designed.

  Each hunt makes the next one stronger. Every false trail becomes a reference for the next correct one, and every attempt trains the hunter's intuition algorithm a little more.

With these components in mind, let's think back to how this compares to reactive security. The difference between reactive security and proactive threat hunting illustrates how two distinct security cultures have evolved over time. For many years, the reactive approach formed the cornerstone of organizational security practices because threats were once highly noisy and the ability of systems to generate alerts was sufficient to detect most attacks. However, as threat actors became quieter and techniques grew more complex, the alert itself ceased to be a reliable guide. The reactive model turned into a reflex that responds only when a threat becomes visible; modern attacks, by contrast, now progress invisibly. Table 11.1 examines the differences in approach.

Dimension/ Perspective	Reactive Security Approach	Proactive Threat Hunting Approach	Notes from a Hunting Perspective
Trigger	An alarm, alert, or event log arrives; action is taken afterward.	No alert is awaited; a hypothesis is formed first, and then active searching is performed on data.	In the reactive model, the threat is seen when it reveals itself; in hunting, the threat is sought before it becomes visible.
Timing	Typically activated after the incident has occurred.	Searches for traces in the early stages of the attack chain, targeting invisible phases.	Time advantage is hunting's greatest strength; it enables early detection.
Focus	Individual alerts, IOCs, signatures, rule matches.	Behavioral patterns, TTPs, anomaly- and context-driven analysis.	Hunting pursues the question "Why did this happen?" rather than "What happened?".
Threat model	Noisy, signature-based, classic attack scenarios.	Low-noise implants, fileless attacks, abuse of legitimate tools (Living Off the Land Binaries [LOLBins], OAuth, etc.).	Modern APTs and complex campaigns often become visible only through hunting.
Telemetry usage	Logs corresponding to existing alerts are reviewed.	Telemetry is actively explored; gaps and blind spots are deliberately sought.	Proactive hunting also functions as a telemetry health test.
Core question asked	"What is this alert? Is it an incident or not?"	"Why is there no alert? What might be hidden behind this behavior?"	The type of question directly changes the team's mindset.
Human role	Alert triage, playbook execution, procedural handling.	Hypothesis development, anomaly sensing, behavior interpretation, context building.	Human intuition becomes the primary sensor on the hunting side.
Role of intuition/ experience	Relatively low; the process relies mostly on rules and playbooks.	Very high; the analyst's experience, curiosity, and skepticism play a critical role.	Good hunters draw from years of accumulated intuition, not reactive processes alone.

Table 11.1: Reactive Security Versus Proactive Threat Hunting Comparison

Dimension/ Perspective	Reactive Security Approach	Proactive Threat Hunting Approach	Notes from a Hunting Perspective
Success metrics	"How many alerts were closed? How many incidents were resolved?"	"How many blind spots were found? How many new detections were created? How many hypotheses were tested?"	The output of hunting isn't just incident count, but the maturity of the defense.
Approach to invisible threats	Generally, outside the radar; ignored because no alert is generated.	Specifically targeted; silent areas are placed at the center of investigation.	Alert fatigue can only be broken this way.
Error profile	Constrained by the balance of false positives/ false negatives.	Incorrect hypotheses, missing data, and interpretation errors—but learning oriented.	In hunting, even a false alarm produces knowledge; in reactive models, it's seen only as a burden.
Operational impact	Post-incident response, damage reduction, recovery (containment, eradication, recovery).	Pre-incident detection, attack surface reduction, providing input to detection engineering.	Hunting is a nourishing layer that improves the quality of incident response processes over time.
Cultural reflection	"We look when an alert arrives" culture.	"We look, question, and remain curious even if no alert arrives" culture.	Organizational security culture becomes more inquisitive and proactive with hunting.
Typical tools used	SIEM dashboards, ticketing systems, playbook/SOAR workflows.	SIEM + EDR + raw log queries, script-based analysis, custom query sets, pivot-driven investigation.	Even when the same tools are used, the mindset is completely different.
Risk	The chance of noticing the attacker before they shout is low.	The chance of catching the attacker by their traces increases even if they never shout.	Hunting is essential, especially to reduce the risk of long-term, silent compromise.

Table 11.1: Reactive Security Versus Proactive Threat Hunting Comparison (Cont.)

Reactive security is a model that waits for an event to be triggered. In this approach, an action begins only after the system generates an alert; in other words, the starting point of defense is the signal produced by a threat. Proactive hunting, however, treats silence as data. Instead of waiting for alerts, the hunter questions what appears normal: Why is there no alert? Why does behavior remain perfectly consistent? Why does a user never generate anomalies? These questions create a time advantage, allowing threats to be identified before they become fully visible.

This distinction becomes clearer when comparing signature-based and behavior-focused approaches. Reactive security often relies on indicators: Did an IOC match? Is an IP malicious? Is a file hash known? Yet many modern attacks leave no clear indicators at all. They rely on legitimate tools, mimic normal user activity, and blend into expected traffic patterns. Proactive hunting therefore shifts the focus from indicators to behavior. It asks whether an action aligns with what is natural: Did access occur faster than expected? Is there a subtle irregularity in timing? Does a process chain move with unnatural precision? Even when intent is concealed, behavior still leaves signals.

The difference also extends to telemetry. In reactive models, telemetry is primarily used to confirm alerts. In hunting, telemetry is explored proactively. Analysts examine completeness, consistency, and context, identifying blind spots as well as threats. In this sense, telemetry is both the evidence and the object of investigation.

Another key distinction lies in the human role. Reactive workflows emphasize procedure: review alerts, follow playbooks, and resolve cases. Proactive hunting requires interpretation. The analyst acts as an investigator who forms hypotheses, connects context, and interprets weak signals. This makes hunting both a technical and an analytical discipline.

Success is measured differently as well. Reactive security focuses on the number of alerts processed or incidents resolved. Hunting evaluates progress through improved visibility, stronger detection logic, and the discovery of previously unseen patterns. Even when no attack is found, the process contributes to a deeper understanding of the environment.

Perhaps the most important difference appears in the treatment of silent threats. Reactive systems rarely detect what does not generate alerts. Hunting, by contrast, begins precisely where visibility ends. Low-noise techniques such as credential abuse, cloud-based manipulation, and fileless execution often become visible only through behavioral analysis.

These differences reflect a broader shift in security culture. Reactive security follows a *respond-when-alerted* mindset. Proactive hunting adopts a different perspective: investigate *continuously*, even in the absence of alerts. This shift transforms defense from a passive function into an active practice.

11.1.4 Hypothesis-Driven Approach

The hypothesis-driven approach is more than a technique used in cyber defense, but rather a way of thinking. This approach prevents security teams from becoming fixated on what happened and instead directs them toward a far more difficult, yet far more valuable question: "If an attacker were present here, what would they be forced to do?" This question marks the true starting point of modern threat hunting because a hypothesis creates a defensive reflex in which intent—not data—stands at the center.

We'll see how the hypothesis-driven approach forms a key part of cybersecurity in the following sections and explore a few examples to demonstrate.

The Role of the Hypothesis-Driven Approach in Cyber Defense

Reactive security models, by their nature, look to the past. An event that generates an alert is already the result of a behavior that has occurred. The hypothesis-driven approach, by contrast, is forward-looking. It mentally constructs an attack path that hasn't yet occurred but is logically possible. For this reason, a hypothesis is a journey into the attacker's mental map. When a hunter asks, "If someone wanted to escalate privileges in this environment, which doors could they pass through without forcing them?" defense evolves from passive waiting into active inquiry.

To operationalize this mindset, the hypothesis-driven approach can be translated into a set of guiding questions that shape how telemetry is interpreted and investigated. Rather than introducing new tools, this model reframes how existing data is approached, turning raw signals into structured lines of inquiry. As illustrated in Table 11.2, each hypothesis type corresponds to a specific analytical question, a relevant telemetry domain, and a distinct advantage over purely reactive analysis. This mapping demonstrates how abstract thinking patterns evolve into actionable hunting practices within real environments.

Hypothesis-Driven Thought	The Core Question the Hunter Asks	Telemetry/Data Type Examined	How It Appears in a Reactive Approach	Hunting Value
Absence of alerts is data.	"Why are there no alerts here?"	Baseline behaviors, long-term silence.	Mostly ignored.	Starting point of silent threats.
Flawless behavior is suspicious.	"Why does this account never make mistakes?"	Login success patterns, token lifetimes.	Considered clean.	Indicator of credential abuse.
The chain matters, not the single event.	"What else is this event connected to?"	Process trees, timing, parent-child relationships.	Examined in isolation.	The behavioral story emerges.
What is unseen is a defensive weakness.	"If we can't see this, where are we missing it?"	Log coverage, audit policy, telemetry gaps.	Not questioned.	Identification of blind spots.
A hypothesis can be falsified.	"Is there data that disproves this hypothesis?"	Negative evidence.	Often not examined.	Learning and maturation.

Table 11.2: Hypothesis-Driven Thinking: Mental Model → Operational Mapping

Hypothesis-Driven Thought	The Core Question the Hunter Asks	Telemetry/Data Type Examined	How It Appears in a Reactive Approach	Hunting Value
The question, not the tool, is central.	"Which data is required to answer this question?"	Targeted queries.	Tool-centric execution.	Time and energy efficiency.
Intelligence must be translated.	"How would this TTP appear here?"	Environment-specific logs.	CTI remains theoretical.	CTI becomes operational.

Table 11.2: Hypothesis-Driven Thinking: Mental Model → Operational Mapping (Cont.)

Let's dive deeper into Table 11.2 to understand the main drivers of your hypotheses:

- **Absence of alerts is data.**
  Accepting the absence of alerts as data is one of the most difficult yet most powerful reflexes of proactive hunting. Reactive security treats an alert as a starting point; if there's no alert, investigation usually doesn't begin. The hypothesis-driven approach does the opposite: it treats the absence of alerts as a signal gap. The critical idea here is that some attacks are designed specifically not to generate alerts, therefore the absence of alerts doesn't mean there's no threat; it can simply mean that the threat is well hidden. The question this row gives to the hunter is "Why are there no alerts at all?" The moment this question is asked, attention naturally shifts toward baselines: normal user behaviors, normal service rhythms, and normal outbound network patterns. Long-term silence, especially silence in places where activity is expected, turns into a kind of compass in the hunter's hands. The place the defense can't see is where the attacker walks most comfortably.
- **Flawless behavior is suspicious.**
  This aligns with the logic of modern identity attacks and silent persistence. Reactive security is usually sensitive to loud indicators such as failed logins, locked accounts, or MFA errors. A mature attacker, however, aims for the opposite: not making mistakes. An account that produces no anomalies for months, shows no failed logins, and exhibits no suspicious actions may appear reassuring. Yet the hypothesis asks: "Why does this account never make mistakes?" This question moves the hunter's focus from failures to the context of successes. Login success patterns, token lifetimes, session continuity, device fingerprints, and the same account appearing across different geographies or client types become valuable areas of analysis. What appears clean may in fact represent an extremely controlled use. The hunting value of this row lies in its ability to expose the quietest form of credential abuse attacks: The attacker hides by appearing flawless.
- **The chain matters, not the single event.**
  This elevates hunting from the level of reading log lines to the level of constructing a behavioral narrative. Reactive processes often examine individual alerts: a single login, a

single process, or a single connection. The hypothesis-driven approach treats a single event as part of a story and asks, "What else is this event connected to?" This question leads to process trees, parent-child relationships, timing alignment, and additional artifacts occurring within the same time window. For example, "`powershell.exe` executed" may be ordinary on its own, but when it's triggered by an unusual parent, executed with abnormal speed, aligned with a specific network egress within the same minute, and combined with unexpected file access under the same user context, a behavioral chain emerges. The true power of this row is that the hunter reassembles the traces that the attacker deliberately scattered to remain hidden.

- **What is unseen is a defensive weakness.**
  This emphasizes that hunting is about both finding threats and identifying blind spots in the defense. Here, the hypothesis turns toward the defense itself, not the attacker: "If we can't see this, where are we missing it?" This question directs attention to log coverage, audit policy, telemetry pipelines, agent health, and SIEM normalization. Sometimes, the reason an attack can't be found isn't that the attacker is exceptionally stealthy, but that the organization is seeing incorrectly or incompletely. For example, certain Event IDs never appearing, DNS telemetry being mismatched due to Network Address Translation (NAT) or proxy behavior, or EDR producing partial data on some hosts are issues typically not questioned in reactive workflows because there's no alert. For the hunter, however, these gaps are the attacker's safe corridors. The hunting value of this row is a direct increase in maturity: Finding blind spots builds the foundation for detecting future attacks earlier.
- **A hypothesis can be falsified.**
  This represents the most professional aspect of hunting: the discipline of disproving oneself. Forming a hypothesis is easy; testing it is difficult. But actively searching for data that would invalidate it shows true maturity. When the question becomes "Is there data that disproves this hypothesis?" the hunter seeks supporting evidence *and* counterevidence. Negative evidence becomes critical here, such as an expected behavior never occurring, a certain pattern appearing only in a specific segment, or the same signal being explainable by legitimate automation. Reactive workflows often stop once a reasonable explanation is found; hunting, by contrast, produces more robust conclusions through falsification. The hunting value of this row is both a reduction in false positives and an increase in report credibility. Even if a hypothesis is proven wrong, the result isn't "We wasted time," but rather "We've drawn clearer boundaries of the environment."
- **The question, not the tool, is central.**
  This breaks a common trap in modern SOCs: tool dependency. Teams sometimes start with "Is there something in the EDR?", then move to the SIEM, and eventually get lost among dashboards. The hypothesis-driven approach works in reverse: question first, data second. When the question "Which data is required to answer this?" is asked, the direction of the hunt becomes clear. Targeted queries, narrowed time windows, host- or account-focused pivots, and analyses built on minimal yet meaningful fields emerge.

The tool-centric nature of reactive approaches often generates noise; the hypothesis-driven approach reduces it. The hunting value of this row is operational efficiency: producing more accurate results with fewer resources and reducing team fatigue.

- **Intelligence must be translated.**
  This statement directly connects this section to CTI. CTI often states, "This group did this." But if that information isn't translated into the local environment, it remains on a bookshelf. The hypothesis is the engine of that translation: the question "How would this TTP appear here?" leads the hunter to environment-specific logs. The same technique can leave different traces in different organizations. For example, a TTP may be visible through Sysmon in one environment, while in another, it may only appear in proxy logs or cloud audit logs. In reactive processes, CTI often remains theoretical; the hypothesis-driven approach operationalizes it: TTP → visibility assumption → data source selection → testing → detection output. The hunting value of this row is the transformation of CTI from "information" into "action."

Example 1: Detecting Silent Token Abuse Within Flawless Successes

A hypothesis is sometimes born not from the presence of an alert, but from its absence. This is especially evident in cloud identity attacks. An attacker doesn't shout by repeatedly making incorrect attempts after obtaining a user's password; on the contrary, they infiltrate through successful sessions. Moreover, they do this without breaking MFA, using methods that appear legitimate.

Let's start with our hypothesis: "If an account is being used through refresh token or session token abuse, logins will appear 'successful,' but the natural rhythm will be disrupted in terms of device, client type, and session continuity."

This hypothesis breaks you away from classic failed-login searches. Because there's no noise here, the hunter must examine the context of success.

The analysis relies on the following telemetry and data sources:

- IdP sign-in logs (Microsoft Entra ID, Okta, etc.)
- Client app/user agent/device ID fields
- Session continuity (different clients for the same user within short intervals)
- Conditional Access results and risk fields (if available)
- Session patterns indicating token refresh behavior

This form of token abuse manifests in practice as follows:

- While the user typically accesses from a single device during the day, an undefined client or a different client app appears intermittently on the same day.
- Sessions are too perfect: No MFA challenges and no failures occur, yet session continuity is unusual.
- A protocol or client that the user normally never uses (e.g., a legacy mail client) suddenly appears.

The hunter's critical question is "Why are these successes so flawless, and why do they appear with so many different faces?"

If the hypothesis is confirmed, the outcome goes beyond simply stating "the account is compromised":

- Restricting risky clients
- Tightening conditional access policies
- Reviewing token lifetime and session management
- Implementing defensive improvements such as admin consent and application permission reviews

Even if the hypothesis is falsified (e.g., due to a new device onboarding process), the organization's "normal" profile becomes clearer. This sharpens the next hunt.

> **Note**
>
> This example operationalizes two rows of the table simultaneously: "Flawless behavior is suspicious" and "Intelligence must be translated" because CTI says "token-based abuse," and you translate it into your own log fields.

Example 2: Detecting Fileless/Low-Noise Execution Through the Chain, Not the Single Event

On the endpoint side, the greatest trap is that the analyst looks at a single event, says "there's nothing here," and moves on. Yet most modern attacks aren't caught in a single line; they are caught as a chain. The hypothesis-driven approach comes into play here and connects the gaps between logs.

Let's start with our hypothesis: "If an attacker is performing fileless or low-noise execution, individual events may appear benign; however, when the process tree, timing, and memory access chain are examined together, an unnatural flow emerges."

This hypothesis pushes you away from a single PowerShell line and toward examining how PowerShell came into existence.

The analysis relies on the following endpoint telemetry and data sources:

- Process creation (Sysmon Event ID 1/EDR process telemetry)
- Parent-child relationships and command-line arguments
- Script block logging (if available) or module load traces
- Process access/memory access telemetry (Sysmon Event ID 10, EDR memory events)
- Timing: parent to child spawn time, sequence

This type of fileless or low-noise execution manifests in practice as follows:

- A process that normally never invokes PowerShell triggers PowerShell.
- The parent to child transition occurs at a speed that isn't human, but mechanical (excessively fast, excessively deterministic).
- The PowerShell command line appears short, but the modules loaded in the background are suspicious (e.g., a nonautomation module set).
- Within the same time window, memory access or "touching" behavior toward another process is observed.

The hunter's critical question is "Does this process chain represent a natural user action or an automated execution pipeline?"

If the hypothesis is confirmed, a second output emerges that is as important as attack detection itself: rule design. That is, instead of "PowerShell executed," you can now build detection logic based on behavioral constructs such as "PowerShell spawned under this parent, at this speed, with these module loads, and with these memory touches."

Even if the hypothesis is falsified, you clearly map the organization's automation sources (System Center Configuration Manager [SCCM], remote monitoring and management [RMM], script runners, scheduled tasks, etc.). This significantly reduces noise in future hunts.

> **Note**
> This example represents the real-world application of the table rows "The chain matters, not the single event" and "The question, not the tool, is central."

Example 3: Detecting a Beacon via Timing and Entropy Through Absence of Alerts

On the network side, the absence of alerts is even more deceptive. Attackers no longer have to use malicious domains. They hide behind CDNs, leverage popular services, and blend their traffic into normal HTTPS flows. As a result, reactive systems remain silent when they see no bad IOC. The hunter, however, examines the shape of the silence.

Let's start with our hypothesis: "If a C2 beacon is hiding with low noise, the content may appear legitimate; however, it will diverge from the rhythm of 'human' traffic in terms of timing, packet size distribution, and destination diversity."

This hypothesis moves you away from domain blacklists and toward rhythm analysis.

The analysis relies on the following network telemetry and data sources:

- NetFlow/IP Flow Information Export (IPFIX) or proxy logs
- Time series: request intervals, jitter distribution
- TLS metadata (SNI, JA3/JA4 if available), HTTP header patterns (if available)
- Destination diversity: the same host repeatedly communicating with the same target
- Data volume: regular, small outbound requests

This type of low-noise C2 beaconing manifests in practice as follows:

- Traffic appears random, but actually exhibits controlled randomness: It pretends to have jitter, yet the distribution is unnatural.
- Traffic becomes more stable during nighttime hours (the opposite of human traffic).
- The same endpoint communicates with a single destination using small packets over a long period; even if the content is legitimate, the rhythm beats like a pulse.

The hunter's critical question is "Is the randomness of this traffic natural, or is it engineered randomness?"

If the hypothesis is confirmed, you detect C2 without relying on IOCs. This is one of the most valuable detection types in modern defense because you capture the attacker's behavior rather than a copyable indicator.

Even if the hypothesis is falsified, you gain a deeper understanding of the organization's normal network rhythms: Legitimate periodic flows such as backups, patching, telemetry uploads, and software as a service (SaaS) synchronization are cataloged. This significantly reduces false positives in future hunts.

Note

This example represents the network-side counterpart of the table rows "Absence of alerts is data," "A hypothesis can be falsified," and "Treat behavior as a chain."

11.1.5 Filling Information Gaps: The Role of the Hunter

Cyber defense is built on what we can see: logs, alerts, and telemetry pipelines. Yet this visibility is never complete. Modern attacks often succeed not by triggering alarms, but by operating within the gaps of what isn't observed. This is where the true role of the threat hunter emerges: not to monitor what is visible, but to question why what should be visible is missing.

In a typical SOC environment, analysts work from the same dashboards and datasets. The difference lies not in what is seen, but in what is questioned. While most focus on what is happening, the hunter asks, "What should be happening but isn't?" In many cases, the absence of expected signals is more meaningful than the presence of obvious ones.

Information gaps rarely present themselves as clear deficiencies. A log may never arrive, an Event ID may lack context, or behavior may appear technically normal yet inconsistent with operational reality. Hunting, therefore, isn't limited to detecting anomalies; it's the discipline of measuring the difference between expected and observed states.

Consider an AD environment where authentication and privilege events appear complete. On the surface, everything functions as expected. Yet the hunter asks, "Is there truly no lateral movement?" In large environments, the complete absence of such activity is unlikely. This question shifts the focus from searching for an attacker to identifying where visibility may be lacking, whether in missing Server Message Block (SMB) telemetry, unseen NTLM fallbacks, or untracked service account movements.

The hunter's role extends beyond identifying these gaps to include explaining them. It's not enough to say, "There's no data." The hunter must determine why, whether due to missing log collection, incomplete telemetry pipelines, or broken normalization and correlation. In doing so, the hunter becomes a bridge between detection and architecture, turning uncertainty into actionable improvement.

Information gaps also emerge through subtle behavioral shifts. A service account may continue to access the same systems at the same times, yet begin interacting with different resources or performing new types of actions. No alert is triggered, and no explicit violation occurs. Still, the following question arises: "Is this behavior truly unchanged, or is it evolving under the appearance of normality?" Here, the gap isn't technical, but contextual.

Comparison is one of the hunter's most effective tools. Users are compared with users, systems with systems, and present behavior with historical baselines. A single event may seem insignificant in isolation, but gains meaning when viewed over time. While reactive systems capture moments, hunting focuses on change.

Working with information gaps inherently involves uncertainty. There are no clear indicators, only partial signals and unanswered questions. Rather than avoiding this ambiguity, the hunter operates within it. Progress in cyber defense often begins in these gray areas, where visibility is incomplete and assumptions must be challenged.

In cloud environments, for example, audit logs are abundant but noisy. A technically valid API call may stand out not because it's malicious, but because it has never been observed before in that context. This novelty forms an information gap. The hunter doesn't respond with an alert, but with a question: "Why now, why here, and why with this identity?" The answer may reveal an attack or a legitimate change, but in both cases, visibility improves.

Ultimately, the hunter's role is to find threats and refine understanding. As information gaps are identified and addressed, the defense shifts from assumption to evidence. Recognizing that nothing is happening is replaced with noting that nothing is happening for a verifiable reason. This transformation quietly strengthens the organization's security posture.

In the following sections, we examine the tools and techniques used to identify and close information gaps, along with practical examples that demonstrate how this process unfolds in real environments.

The Hunter's Advanced Toolset for Filling Information Gaps

The hunter's job is often described as finding threats, but in practice, the real work frequently becomes something else: making information gaps visible and designing the methods to close those gaps. This isn't done by randomly browsing logs, but through specific techniques. Let's walk through the technical toolkit that produces the highest value in practice and is systematically used by mature teams:

- **Baseline engineering and establishing the mathematics of "normal"**
  "Normal" may be perceived intuitively, but it must be managed with evidence. The purpose of a baseline isn't to derive a single average, but to capture the range of normal behavior, its seasonality, role-based differences, and system-specific rhythms. In practice, this requires defining normal behavior as a structured, multidimensional model rather than a single static reference point:
  - Multilayered baselines: User (human), service account (nonhuman), server role (DB/APP/DC), segment (prod/dev), and location (branch/headquarters) are modeled separately.
  - Seasonality: Patch nights, backup windows, payroll days, and month-end closures generate anomalies; the baseline labels these periods.
  - Distribution-focused thresholds: Instead of a single threshold, percentiles (P95/P99), interquartile range (IQR), and maximum tolerable downtime (MAD) are used to define the boundaries of normal.
  - Silence baseline: Situations where an expected event doesn't occur are included in the baseline (e.g., DNS log volume for specific servers, Sysmon event generation cadence).

  This technique turns the absence of alerts from good news into a measurable anomaly.
- **Telemetry gap hunting: Blind spot detection**
  Before catching the attacker, you must ensure that the eyes meant to catch them are open. Blind spot hunting dismantles the assumption that "logs are coming." In practice, this involves systematically validating whether telemetry is present, complete, consistent, and reliable across the environment:
  - Coverage validation: Which hosts lack agents, which hosts have policy drift, and which segments experience log loss?
  - Field integrity testing: Logs arrive, but are critical fields empty (`client_ip`, `process_guid`, `parent_image`, `device_id`, etc.)?
  - Timestamp consistency: Clock drift destroys correlation; NTP skew, indexing delays, and forwarder latency are measured.
  - Normalization loss: Parsers/normalizers may "swallow" fields; raw log versus normalized log differences are sampled.
  - Signal generation testing: Controlled mini-scenarios (test users, test hosts) verify whether expected events actually occur.

  Even if no attack is found, the hunter produces a critical outcome here: guaranteed visibility.
- **Chained correlation and behavioral storytelling**
  A single signal is often benign. Hunting isn't about reading individual events, but about turning events into a story. This requires linking events across processes, systems, and time to reconstruct a coherent behavioral sequence via the following:

- Process tree + timing: This involves analyzing parent-child process relationships together with execution timing to determine whether the sequence and speed of process creation reflect normal user behavior or automated execution patterns (e.g., multiple processes spawning within 1–3 seconds in a deterministic chain).
- Cross-domain pivoting: This refers to correlating related activity across multiple telemetry domains—such as endpoint, identity, network, and cloud audit logs—by linking events through shared attributes of user identity, device, and time window.
- Artifact chains: This approach reconstructs the sequence of attacker activity by connecting related artifacts, such as process execution, module loading, outbound network communication, credential access, and persistence mechanisms, into a single behavioral flow.
- Weak signal fusion: Individually low-risk signals combine to produce high risk.

This technique breaks the attacker's strategy of dispersing traces: Even if traces are scattered, the story recombines.

- **Perfection analysis: Examining the context of successes**

  Reactive defense focuses on failed attempts. Hunters sometimes pursue excessively clean successes. This approach focuses on evaluating successful activity in relation to expected patterns, rather than accepting it at face value:

  - Success-only anomalies: Although there are no failures, no MFA challenges, and no risk events, the usage profile changes.
  - Session continuity inconsistencies: This includes analyzing session continuity factors such as device fingerprint, client application, token refresh behavior, and session duration to identify inconsistencies with expected usage patterns.
  - Geographic/ISP inconsistencies: This refers to not just impossible travel, but same-day internet service provider (ISP)/Autonomous System Number (ASN) changes.
  - Nonhuman account discipline: Service accounts behave like humans (after-hours activity, interactive login attempts).

  This technique is especially valuable in credential abuse, token misuse, and silent persistence scenarios.

- **Negative evidence and the discipline of falsification**

  The maturity test of hypothesis-driven hunting is whether you can disprove your own hypothesis. Seeking evidence is easy; seeking counterevidence is professional. This means actively seeking disconfirming evidence and testing the hypothesis against alternative explanations:

  - Counterfactual search: If this were true, what else should also be visible—and is it actually present?
  - Control groups: Do similar signals appear in systems with the same role or in the same segment?

- Alternative explanation catalog: Systematically eliminating legitimate causes such as automation, patching, backups, RMM, SCCM, and scheduled tasks.
- Negative outcome reporting: This requires more than reporting that "We found nothing," but reporting instead that "We falsified it for this reason and defined this boundary."

A falsification culture reduces false positives and prevents hunting from becoming an intuition story.

- **Time-series and rhythm analysis**
  Modern attacks are caught not by bad domains but by rhythm—especially for beaconing, low-volume exfiltration, and scheduled persistence. This involves analyzing temporal patterns to distinguish natural human behavior from automated or engineered activity:
  - Jitter profiles: Intervals appear random but they have artificial distributions.
  - Intraday rhythm breaks: Human traffic is volatile during the day and quieter at night; C2 traffic is often the opposite—stable at night.
  - Periodic small outbound traffic: Small-sized, regular flows have low destination diversity.
  - Signature over threshold: This refers to the combination of interval distribution + packet size distribution + destination loyalty.

  This approach is a cornerstone of IOC-less hunting.
- **Risk scoring and prioritization: Hunting triage**
  What makes hunting sustainable isn't looking at everything, but looking at the right thing first. In practice, this requires structuring investigations based on impact, likelihood, and data quality to focus effort where it matters most:
  - Asset criticality weighting: This involves prioritizing investigations based on the criticality of assets, where systems such as DCs, Exchange servers, certificate authorities (CAs), hypervisors, and jump hosts are assigned higher weight due to their potential to amplify impact if compromised.
  - Identity risk weighting: This requires assigning higher investigative priority to identities with elevated or sensitive privileges (such as administrators, service accounts, break-glass accounts, and application registration owners) because misuse of these identities can lead to disproportionate access and control across the environment.
  - Behavioral density: This refers to evaluating the concentration and rarity of observed behaviors, where low-frequency but high-impact actions are treated as higher priority, as they are more likely to indicate meaningful deviations from normal activity rather than benign noise.
  - Evidence quality: This involves assessing the reliability and completeness of the underlying data (such as raw log integrity, field population accuracy, and timestamp consistency) to ensure that investigative conclusions are based on trustworthy and coherent telemetry.

Good triage reduces team burnout and prevents hunting from becoming an endless investigation.

- **Transformation into detection engineering: Turning hunts into rules**
  The lasting organizational value of hunting lies in making findings reproducible. This step translates one-time findings into repeatable detection logic that can be operationalized at scale:
  - From hunt to analytic: Convert behavioral constructs into queries/correlation rules.
  - Scope and noise management: This involves defining the operational scope of the detection while systematically reducing noise by applying allowlists for known automation, introducing role-based exceptions, and optimizing thresholds to balance detection sensitivity with false positive control.
  - Test set creation: This requires building representative test datasets using both positive and negative samples derived from historical logs, enabling validation of detection logic through controlled testing and regression analysis.
  - Lifecycle management: This entails continuously monitoring detection performance over time by tracking rule drift, reassessing effectiveness as telemetry evolves, and recalibrating detection logic to maintain accuracy and relevance.

  Without this step, hunting remains a momentary success; with this step, it becomes institutional muscle.
- **Cross-source consistency checks**
  Even if an attacker leaves no trace in one place, they usually leave one elsewhere. The hunter looks for consistency across sources. This involves comparing related data across different telemetry sources to identify inconsistencies and hidden gaps:
  - EDR versus Windows Security: Process events present but no logon—or vice versa?
  - Proxy versus DNS: Outbound traffic in proxy logs but no DNS (DoH, hardcoded IPs, resolver bypass)?
  - Cloud audit versus on-premise: Heavy cloud activity while the same user is absent on-premises?
  - Ticket/change records: Behavior present but no corresponding change record?

  Inconsistency is often the fastest indicator of an information gap.
- **Hunt operations: Hunt cards, versioning, and repeatability**
  Advanced hunting grows through operational discipline, not through individual heroics. This requires formalizing hunting activities into structured, repeatable processes that can be tracked, shared, and improved over time:
  - Hunt card standard: This requires documenting each hunt in a consistent structure that captures the hypothesis, relevant data sources, queries used, findings obtained, falsification steps performed, final analytical decision, and resulting recommendations.

- Query versioning: This involves maintaining queries in a version-controlled repository—such as Git or an internal wiki—together with clear documentation explaining why the query exists, what behavioral problem it's intended to address, and how it has evolved over time.
- Playbook creation: This means converting successful hunts into repeatable workflows.
- Metrics: Instead of measuring the number of findings, measure the number of blind spots found, new analytics created, mean time to remediate (MTTR) impact, data quality improvements.

Without this discipline, knowledge accumulates but never becomes institutional memory; as teams rotate, the knowledge disappears.

Table 11.3 connects that underlying sense of unease to an operational method. The point that elevates a hunter to an advanced level is that the process begins with a feeling of "Something doesn't sit right," but it doesn't end there; that feeling is transformed into a measurable question and a testable hypothesis.

Information Gap Type	Hunter's Hypothesis (Testable)	Evidence/ Telemetry (What to Examine)	Mini-Test/ Falsification Path	Gap Closure Action
"Expected event doesn't occur at all." (silence)	"Is this event truly not occurring, or are we simply not seeing it?"	Log volume trends, agent health, Security/Sysmon/ EDR event cadence, index latency	Generate the event via a controlled (authorized) scenario and verify end-to-end visibility.	Fix audit policy, complete agent deployment, remediate pipeline/forwarder issues.
"Event exists but lacks context." (fields empty/ missing)	"The event arrives, but correlation breaks because critical fields are lost."	Field completeness rates (e.g., `client_ip`/ `device_id`/ `parent_image`/ `process_guid`), parser outputs	Compare raw logs versus normalized logs; perform parser drift testing.	Update normalization/parsing, correct field mappings, standardize the data model.
"Perfect success." (identity producing no anomalies)	"Successes aren't natural; the session surface may be changing."	Sign-in logs, device fingerprints, client app, token/session continuity, Conditional Access results.	Control group comparison with similar-role users; eliminate alternative explanations (new device/ onboarding).	Tighten Conditional Access, review token lifecycles, add risk-based controls.

Table 11.3: Information Gap → Testable Hypothesis → Telemetry Evidence → Closure Action Map

Information Gap Type	Hunter's Hypothesis (Testable)	Evidence/ Telemetry (What to Examine)	Mini-Test/ Falsification Path	Gap Closure Action
"Single events look benign; the chain is suspicious."	"Benign-looking events form an attack narrative when combined in sequence."	Process tree, spawn rates, module loads, network egress, time-window correlation.	Seek disconfirming evidence: Is the same chain legitimate automation?	Create behavior-based rules, add allowlist/role-based exceptions, run detection regression tests.
"Sources are inconsistent with each other." (e.g., EDR present, no logon)	"An artifact unseen in one source should appear in another; inconsistency indicates a blind spot or bypass."	Consistency across EDR, security, proxy, DNS, and cloud audit logs.	Pivot within the same time window; author expected match rules.	Align data sources, cover bypasses (DoH/hardcoded IP), enable missing log sources.
"Change observed but no change record exists."	"If this behavior is legitimate, a change record should exist; absence increases suspicion."	ITSM/change/ticket records, configuration management database (CMDB), deployment logs.	Correlate event timing with change windows.	Integrate processes, bind change windows to telemetry, design alerts for unauthorized changes.
"Normal has shifted unnoticed." (drift)	"Baseline drift occurred; what we consider normal today was anomalous yesterday."	Long-term trends, P95/P99, MAD/IQR, seasonality tags.	Temporal slicing: month/week comparisons; control segment analysis.	Version baselines, add season/business-cycle tags, recalibrate thresholds.

Table 11.3: Information Gap → Testable Hypothesis → Telemetry Evidence → Closure Action Map (Cont.)

The table is the map of that transformation in the following ways:

- **Names the type of gap**
  One of the hardest challenges in an SOC is to correctly categorize a vague sense of discomfort. Is there silence? Is context missing? Are sources inconsistent? Without this classification, every investigation turns into an energy-consuming "log wandering" spread across a wide surface.
- **Makes the hypothesis testable**
  This is what removes hunting from the realm of subjectivity. Not "It felt suspicious to me," but rather: "If this is true, then this data should also look like this." For example, in

the AD case, the claim of "no lateral movement" only becomes meaningful through a controlled mini-test and end-to-end visibility validation.

- **Clarifies where the evidence resides**
  Not every hypothesis can be tested with every type of telemetry. Searching EDR screens for evidence of perfect identity behavior is often futile; the real evidence lies in the context of sign-in logs. For network rhythm, the opposite applies: Instead of hunting for IOCs, time series and destination affinity are what speak.
- **Makes the most critical output visible: Closure action**
  The hunter's role is not only to find but also to prescribe the improvement that will close the gap. For this reason, the table anchors every row to a clear line of look here → test like this → fix this. This is the part that turns hunting into institutional muscle.

Example 1: The Absence of Lateral Movement Gap in Active Directory

In a corporate AD environment, authentication logs, Kerberos requests, and privilege escalation events are flowing into the SIEM. Dashboards look clean. Ticket volume is low. Moreover, the EDR isn't generating any serious alerts. On paper, the environment appears mature.

The hunter's discomfort begins at this point: In a large environment, especially one with hundreds or thousands of servers and endpoints, the conclusion that there's no sign of lateral movement at all is statistically weak. In other words, this may indicate that there's no attack or that we're simply not seeing it.

Let's start with our hypothesis: "If lateral movement is truly occurring (or could occur) in this environment, at least some traces should appear in SMB/NTLM session artifacts, remote service creation, admin share access, or remote process execution patterns. If there are no traces at all, then either the behavior doesn't exist—or the telemetry is blind."

This hypothesis doesn't immediately search for the attacker; it first tests the defense's ability to see.

The hunter proceeds down the following technical path:

1. **Defining the list of expected normal traces**
   Before attempting to detect real lateral movement, the hunter defines what normal operations should produce:
   - On servers managed remotely, SMB sessions are formed at regular intervals.
   - Certain service accounts periodically connect to specific servers.
   - Specific shares are accessed during administrative operations.
2. **Quantifying the gap**
   Saying "there is no gap" is an intuition. The hunter turns it into metrics:
   - Distribution of SMB session/Logon Type 3 events by server count and user count over the past 30 days

  - NTLM usage rate (even if Kerberos is dominant, it's not expected to be zero)
  - Connection density from management servers to target servers

3. **Validating the telemetry gap**
   The critical point here is that logs capable of showing lateral movement may be reaching the SIEM, but with missing fields—or may not be arriving at all from certain segments. The hunter specifically checks the following:
   - Are DCs visible while security logs from member servers are missing?
   - Is there audit policy drift on endpoints?
   - Is Sysmon/EDR agent health consistent?
   - Are timestamps skewed (breaking correlation)?
4. **Performing a controlled mini-test (evidence generation).**
   The cleanest way to validate the claim that "We're not seeing it" is to generate expected traces through a permitted test scenario:
   - A legitimate SMB access from a management server to a target server
   - A legitimate remote management session (e.g., Windows Remote Management [WinRM])

The question then becomes clear: "Did this activity appear in the SIEM with the correct fields?" If the test activity doesn't appear in the SIEM, the problem isn't the attacker; it's a log collection, audit, or pipeline issue. This is the concrete manifestation of an information gap.

If the test activity appears, but real-world distribution is unreasonably low, then two possibilities become stronger:

- Operational management channels may be using a different path (e.g., an RMM tool), and you should search elsewhere.
- Lateral movement may have shifted to another protocol (e.g., WMI/WinRM/HTTP-based), making SMB-based expectations incorrect.

This example reinforces the information gap concept in two ways:

- Shows that the conclusion "There's no lateral movement at all" isn't evidence—it may be a blind spot
- Demonstrates that the hunter's output isn't limited to suspicious activity, but includes a remediation plan that guarantees visibility (audit policy, log source coverage, normalization, time synchronization, agent health)

In this scenario, the hunter strengthens the defense without ever catching an attacker. That's the real gain.

Example 2: Endpoint + Network Rhythm Gap

For an endpoint, the EDR isn't generating alerts. There are no malicious domains in the SIEM, as well as no malicious hashes or user complaints. In other words, the reactive workflow says it's clean. However, long-term network telemetry in the hunter's hands whispers something else: The same host consistently sends very small amounts of data outbound at night. Moreover, the destination appears legitimate (possibly behind a CDN or a popular service). There's no IOC, but there's a rhythm.

Let's start with our hypothesis: "If a low-noise implant/C2 exists on this host, the content may appear legitimate, but the timing distribution, destination affinity, and data size pattern will be incompatible with the natural rhythm of human traffic."

This hypothesis doesn't say to look for a bad domain; it says to look for an unnatural rhythm.

The hunter proceeds down the following technical path:

1. **Comparing the rhythm against a baseline**
   First, the hunter establishes what normal night traffic looks like for that host: patching, updates, backups, telemetry uploads, and SaaS sync. These are also periodic. The hunter avoids the periodic = malicious mistake and instead classifies periodic traffic in the following ways:
   - Legitimate periodic flows usually show destination diversity or target specific business services.
   - Packet size distributions are wide (some large, some small).
   - Timing variability is tied to operational processes.
2. **Searching for signs of designed randomness**
   C2 traffic often uses jitter. But natural jitter differs from designed jitter. The hunter looks for signals such as these:
   - Intervals that appear random but are constrained within a narrow distribution band
   - Requests lacking the typical irregularity of human behavior
   - Traffic becoming more stable during hours when humans are inactive
3. **Linking the chain on the endpoint side**
   Network rhythm alone isn't a verdict. Following the "not a single signal, but a chain" principle, the hunter pivots to the endpoint:
   - Is there a scheduled task running during these hours?
   - Is a specific process being triggered regularly in the same time window?
   - What is the parent process and command-line pattern?
   - Is there a link between the process opening network connections and user interaction?

4. **Seeking negative evidence (falsification)**
   Could this traffic be generated by a corporate agent (EDR/monitoring agent, update agent, inventory agent, RMM)? The hunter falsifies this with evidence not with allowlists:
   - The same agent should produce similar rhythms on other hosts.
   - Destinations should be similar across the organization.
   - Process signatures, installation paths, and version distributions should be consistent.

   If this rhythm exists on only one host, the agent explanation weakens.

If the network rhythm is destination-loyal, low-volume, stable at night, with artificial timing intervals and an endpoint-side process or scheduled task is regularly triggered in that window, this constitutes a high-value finding without relying on IOCs.

Even if it turns out to be a legitimate agent, the hunter still gains two benefits:

- The organization's normal periodic traffic is cataloged (reducing future false positives).
- The telemetry signature of such agents becomes clearer, allowing anomalies to be separated more quickly in the future.

This example demonstrates how a hunter creates analytical space through behavior and rhythm precisely where an information gap is masked as "no IOC." While a reactive system remains silent because it can't point to anything malicious, the hunter examines the shape of the silence and formulates questions. This is critically important in modern defense because silent attacks exist exactly in this space.

11.2 Intelligence-Driven Hunting Methodologies

In threat hunting, up to a certain point, everything progresses through intuition, experience, and deep familiarity with the environment. However, this isn't sufficient on its own because intuition provides direction, while intelligence extends range. This is precisely where intelligence-driven hunting methodologies come into play: They sharpen the hunter's perspective, mature the questions being asked, and force consideration of not only how something appears here, but how it manifests elsewhere in the world.

In threat hunting, intuition and experience help identify what doesn't feel right within an environment. However, intuition alone isn't enough. While it provides direction, intelligence provides context and extends the hunter's perspective beyond the local environment. Intelligence-driven hunting is often misunderstood as simply ingesting IOCs into detection systems. In reality, intelligence isn't just data, but context. An indicator gains value only when understood within the broader narrative of a campaign—its purpose, techniques, and intent. This shift enables hunters to move from identifying signals to understanding behavior.

Without intelligence, hunting becomes inward-focused. Attackers, however, continuously adapt across different environments. Intelligence allows hunters to anticipate how these evolving techniques might appear before they are encountered locally, providing a proactive advantage.

The key isn't asking if we've been attacked but asking the more specific question, "If we were attacked, how would current adversary behaviors manifest in our environment?" This transforms hunting from reactive analysis into hypothesis-driven exploration. For example, if intelligence suggests abuse of token lifetimes rather than MFA bypass, the hunter examines session behavior within their own environment instead of searching for specific indicators. In this way, intelligence becomes a guide for forming testable hypotheses.

Ultimately, intelligence-driven methodologies help hunters focus on what is plausible, reduce uncertainty, and understand not only where to look, but why.

The methodologies that follow in this section will show, step-by-step, how intelligence transforms into hunting:

- How CTI knowledge is systematically converted into testable hunting hypotheses
- How adversary TTPs are translated into environment-specific telemetry and observable patterns
- How intelligence from global campaigns is mapped to local behavioral signals within the environment
- Most importantly, how intelligence is operationalized into repeatable hunting workflows and detection logic

True power doesn't lie in reading more reports, but in knowing which question a report should lead you to ask. Intelligence-driven hunting is the methodology of those who know how to ask those questions.

11.2.1 Data Flow from CTI to the Operational Hunter

The flow of data from CTI to operational hunting isn't a straight line such as "Take information and write a rule." On the contrary, it's a layered, filtered, and often feedback-driven process. This is because the nature of CTI is general, while the nature of hunting is specific. CTI looks at the world; the hunter looks at their own environment. There's no direct translation between these two perspectives; interpretation is required. And the one who performs this interpretation isn't a tool, but a human.

The most common mistake in the field is that a CTI report is opened, IOCs are extracted and loaded into the SIEM, and it's declared that "Intelligence has been integrated." Technically, something has been done, but has operational value been produced? Most of the time, the answer is no because it's not the attacker's IP that persists; it's their behavior. IPs change, domains change, and hashes change, but the attacker's way of thinking, their timing preferences, and their risk appetite evolve much more slowly. This is exactly what the operational hunter is looking for.

In the following sections, we'll break down the flow of data transformation to understand how CTI knowledge is applied to operational threat hunting. We'll then explore three examples of this data flow in action.

Exploring the Data Transformation

Data flowing from CTI to the hunter gains meaning not in its raw form, but in its digested form. This digestion process passes through three fundamental transformations.

The first step is *decontextualization*, which is the separation of CTI information from its original context. This may sound counterintuitive because intelligence is already context, but that context belongs to someone else. When you read a sentence in a report such as "APT X executes memory-resident payloads via PowerShell," this information may not have a one-to-one correspondence in your environment. PowerShell may already be heavily used in your organization. So, what do you do with this information? This is where the hunter's mindset comes into play.

The question is reframed: "Why might this attacker have chosen PowerShell?" The answer is usually the same: silence, legitimacy, invisibility. This answer is far more valuable than any IOC.

The second step is *adaptation*, that is, to adapt this abstract behavior to the environment. The operational hunter takes the technique described in CTI and asks their own telemetry: "How would this technique leave traces here?" The same TTP can produce completely different signals in two different environments. In one organization, PowerShell `ScriptBlock` logging is enabled; in another, it's disabled. In one, the EDR produces memory telemetry; in another, it doesn't. A CTI report tells you what was done; the hunter searches for the answer to how it would appear in their own environment. At this stage, technical details become critical, including Event IDs, fields, time windows, and role-based accounts.

At this point, the tone shifts slightly because bringing CTI into operations often begins with disappointment. The trace you expect isn't there. The field you're looking for is empty, and logs are missing. This moment is the most valuable moment in the CTI-hunting pipeline because it reveals an information gap that reveals this statement, "If this attack had happened, we would not have been able to see it." This sentence isn't a failure; it's a defense insight worth its weight in gold. The hunter doesn't turn this insight into a report, but into action: Logging is enabled, policies are updated, and telemetry is expanded. CTI hasn't found a threat here; but it has strengthened the defense.

The third step is to convert CTI knowledge into a *hypothesis*. Without this step, operational hunting doesn't exist. CTI says, "This group did this." A hypothesis says, "If this behavior had occurred in our environment, we should have observed these anomalies." The difference is small, but critical. The first describes the past; the second tests a possibility. The operational hunter never converts CTI directly into a query; first, it's converted into a question.

For example, if a report states that attackers performed silent lateral movement using service accounts, the hunter's hypothesis isn't "Is there this IOC on service accounts?" Instead, it becomes the following: "In our environment, at what times do service accounts connect, to which systems, and for what purposes? Are there movements outside this pattern that don't generate alarms?" This question turns CTI into a living hunting scenario.

The flow of data from CTI to the operational hunter isn't one-directional. This is very important. In mature hunting teams, the flow is bidirectional. The hunter notices something in the field: a particular behavior resembles a TTP described in reports, but doesn't match it exactly. At this point, CTI reenters the loop. New reports are reviewed, variants are examined, and campaign evolutions are tracked. Sometimes the hunter's finding generates CTI's next question. When this loop is established, the organization moves closer to becoming not only a consumer of intelligence, but a producer of it.

There's also a human dimension to this process. CTI reports are cold, objective, and itemized. Hunting, on the other hand, is lonely and silent. Sometimes nothing emerges for hours or even days. CTI gives meaning to this silence: "Look, what you're searching for right now has appeared like this elsewhere in the world." This sentence is a small but powerful source of motivation for the hunter. It reinforces the sense that the hunt isn't random, but part of a larger picture.

To summarize with a simple metaphor: CTI tells the hunter at what depth the fish are swimming. But when to cast the line, how long to wait, and how the water is flowing that day are things the hunter knows. The operational hunter doesn't memorize CTI; they blend it with intuition, and this blend builds defense not from rules, but from understanding.

Ultimately, the flow of data from CTI to the operational hunter both an integration problem and a thinking problem. Tools matter, but the real value lies in treating CTI not as static information, but as a source of questions. When approached this way, intelligence becomes a practical guide for hunting, turning reports into direction and making the process truly intelligence driven.

To make this transformation concrete, Table 11.4 illustrates how different layers of CTI can be systematically translated into operational hunting actions, showing how intelligence is converted into questions, mapped to the environment, and ultimately turned into actionable outputs.

CTI Layer	Information Presented in CTI	Question Transformed by the Hunter	Operational Mapping	Hunting Output
Strategic CTI	General objectives and motivation of the threat actor.	"Why would this actor target an environment like ours?"	Focus on critical assets and business processes.	Hunt prioritization, scope definition.

Table 11.4: Transformation Chain from CTI Knowledge to Operational Hunting Action

CTI Layer	Information Presented in CTI	Question Transformed by the Hunter	Operational Mapping	Hunting Output
Operational CTI	Campaign narratives, employed TTP sets.	"How would this campaign appear in our architecture?"	Environment-specific behavioral hypotheses.	Hypothesis set.
Tactical CTI	Techniques, procedures, tool classes used.	"What traces would this technique leave in our environment?"	Log source + field mapping.	Targeted data collection.
Technical CTI	IOCs, artifacts, low-level details.	"How can I detect the same behavior without IOCs?"	Behavior-based hunting queries.	IOC-independent detection.
Meta-CTI	Variants, evolution, evasion techniques.	"If we're not seeing this, why?"	Telemetry gap analysis.	Blind spot identification.
Feedback	Deviations observed in the field.	"What did we observe that CTI doesn't describe?"	CTI assumption updates.	Organizational learning.

Table 11.4: Transformation Chain from CTI Knowledge to Operational Hunting Action (Cont.)

Let's move through Table 11.4 row by row to see how intelligence is reshaped within the hunter's mind:

- **Strategic CTI: Defining the direction of the hunt**
  Strategic CTI is generally the most abstract layer. It describes geopolitical motivations, sectoral targets, and long-term objectives. Reactive teams store this information as knowledge. The hunter doesn't stop there. The hunter asks the following question: "Why would this actor choose us?"

  This question draws the boundaries of the hunt because not every threat is equally meaningful for every organization. Strategic CTI also teaches the hunter where not to begin, which saves time. It's the quietest yet most critical gain of the hunt.

- **Operational CTI: Bringing the story into the environment**
  Operational CTI tells a campaign story. It has a timeline and an attack chain. But that chain unfolded in a different infrastructure. The hunter doesn't copy it; the hunter adapts it.

The question, "If this campaign had occurred here, where would we feel it first?", directly generates a hypothesis that serves as the steering wheel, not the fuel, for hunting. At this stage, CTI is no longer read; it's rewritten using the language of the environment and the reality of the organization.

- **Tactical CTI: Bringing it down to telemetry**
  Tactical CTI is where hunters make the most mistakes because techniques are listed, and they are tempting. But not every technique is visible in every environment.

  A skilled hunter performs the following transformation: technique → expected trace → visibility assumption.

  At this stage, CTI moves away from saying "This technique was used" and arrives at "This Event ID/this field/this timing should have existed." If it doesn't, the absence is more likely a visibility or telemetry gap than an indication of attacker absence.

- **Technical CTI: Transition from IOCs to behavior**
  IOCs are fast but short-lived. The hunter knows this. Therefore, the IOC is seen not as a target, but as a clue. The "What did the attacker have to do to produce this IOC?" question transforms the IOC into behavior.

  The hunting outputs produced at this stage aren't dependent on IOCs, yet they have the power to detect the same attack across different variants. This is the part that expands organizational defense.

- **Meta-CTI: The courage to find the blind spot**
  Meta-CTI is usually covered under evasion techniques and is often glossed over by most teams. For the hunter, however, this is a gold mine.

  The "If we're not seeing this, is it truly not happening, or are we simply not seeing it?" question tests the defense more than it searches for a threat. The outcome at this point may not be an attack, but the realization that "We're blind here" is far more valuable than an alert.

- **Feedback: From the hunt back to intelligence**
  This is the most mature point. The hunter observes a deviation in the field. It doesn't exactly match what is described in CTI. Many teams dismiss this by saying, "Then it must be wrong." A mature team does something else: "What is this deviation telling us?"

  Sometimes this is the evolution of the actor, an environmental difference, or a reality that CTI hasn't yet documented. At this point, hunting moves closer to a role that doesn't merely consume intelligence, but feeds it.

Example 1: Cloud Persistence Exists in CTI, Silent Privilege Appears in the Environment

A CTI report describes how a threat group achieves long-term persistence in cloud environments by abusing app registrations and broad OAuth permissions. The techniques are explicit in the report: `Application.ReadWrite.All`, `Directory.Read.All`, tenant-wide consent, and token abuse. A reactive approach typically asks, "Is there a known malicious app ID?" If not, the case is closed.

The hunter starts from a different point. They take the CTI and form the following hypothesis: "If this technique were used in our tenant, the attacker would prefer not to create a new application, but to expand the permissions of an existing, rarely used application to avoid generating high noise." This hypothesis shifts the hunter away from IOC searching and toward permission-change timing.

To validate this hypothesis, the hunter examines audit telemetry within Microsoft Entra ID, specifically focusing on Microsoft Graph activity and directory audit logs, where application permission changes, consent operations, and identity actions are recorded:

- Which applications in the tenant have had their permission sets changed in the past 90 days?
- At what times were these changes made?
- Was the identity making the change a human or an automation?
- Do these applications actually use these permissions, or do they just "carry" them?

As a result, an application is identified: one that has existed for years, is rarely used, yet holds permissions that haven't been exercised in recent weeks. There's no alert. There's no IOC. But the CTI → hypothesis → context chain has functioned. In this example, CTI didn't "show" the attack, but the hunter knew where to look thanks to CTI. This is where the operational value lies.

Example 2: LOTL Is Described in CTI, But Timing Speaks in the Environment

A CTI report states that attackers rely exclusively on built-in system tools to avoid leaving signatures: PowerShell, WMI, `schtasks`, and `rundll32`. There are no IOCs; only behavior is described. For reactive security, this is a dead end because these tools exist everywhere. The hunter interprets CTI differently: "The attacker can be caught not by the tool, but by timing."

The hypothesis is that "If this actor were present in our environment, they would be forced to use these tools with a rhythm that differs from human behavior."

Hunting begins with this hypothesis. Individual commands aren't examined. Instead, the following are analyzed:

- At what times do PowerShell executions concentrate?
- Do these executions overlap with periods of user interaction?
- How regularly do similar commands repeat for the same user?
- Is there a sense of automation in parent-child relationships?

At some point, a pattern emerges: PowerShell commands executed outside business hours, every night at similar minutes, with similar durations. The content is benign, but the rhythm is artificial. CTI stated that "PowerShell is being used." The hunter translated CTI into "How is PowerShell being used?"

Example 3: Service Account Abuse Is Written in CTI, But Behavioral Drift Is Detected in the Environment

A report explains that attackers favor service accounts. The reason is clear: no MFA, low noise, and behaviors that are rarely scrutinized. The reactive view asks, "Are there failed logins on this account?" If not, it's deemed clean.

The hunter, however, derives the following question from CTI: "Is this service account always doing the same job, or has it evolved into something else while appearing the same? " This question drives the hunter toward behavior analysis rather than privilege analysis. The following are compared:

- The systems this account accessed over the past six months
- The purpose of those accesses (read versus configuration)
- Access times
- Network egress activity performed by the same account

The result reveals the following: the account still accesses the same places, but it now touches more objects. The privileges haven't changed, and there's no alert, but the intent has changed.

CTI stated that "Service accounts are targeted." The hunter translated CTI into: "When a service account changes its behavior, the quietest attack begins." This distinction is a marker of operational hunting maturity.

11.2.2 Adversary Modeling: APT, Ransomware, and Insider Threat Profiles

Observations in the field have repeatedly shown the following: The same technical trace—same PowerShell invocation, same SMB session, same token renewal pattern—can carry three entirely different meanings for three different actors. Yet the intent behind them is completely different. Adversary modeling is the art of distinguishing that intent. The hunter's strongest tool isn't exposure, but the ability to differentiate.

In this section, we'll examine three primary adversary profiles (APT, ransomware, and insider threat) not only from the perspective of what they do, but through how they think, where they become impatient, and which mistakes they make. Technical traces gain meaning only when combined with this mental framework.

APT Profile: Patience, Consistency, and Invisibility

The first rule of understanding an APT is that an APT doesn't rush because time works in its favor. Its objective is clear: not access and noise, but persistence and stability.

When working against an APT profile, the hunting approach fundamentally changes. Because here, anomalies often don't exist. Sometimes things are even excessively normal. APTs particularly favor the following:

- Legitimate administrative tools (PowerShell, WMI, WinRM)
- Low-frequency outbound network traffic
- Reuse of existing privileges rather than privilege escalation
- Service accounts and automation identities
- Behaviors that don't trigger alerts but demonstrate continuity

The following are applied techniques for APT-focused threat hunting:

- **Long time window analysis**
  An APT isn't caught overnight. A 7-day window is often insufficient. Time series of 30–90 days are required. In the field, the most valuable analyses doesn't begin with "What happened on a single day?" but with "What didn't change over three months?"
- **Consistency hunting**
  APT behavior is highly consistent: the same hours, similar intervals, and similar paths. Here, the hunter doesn't look for deviation, but for excessive regularity. Human behavior fluctuates; APT behavior resembles mathematics.
- **Privilege nonuse analysis**
  An APT often doesn't use the privileges it possesses because using them creates visibility. For this reason, the question "Why did this account not do this?" is just as valuable in APT hunting as "Why did it do this?"

In APT hunting, sometimes nothing emerges for days. This silence is frustrating. But the right silence is more valuable than the wrong noise. Here, patience isn't a virtue; it's a necessity.

Ransomware Profile: Speed, Propagation, and Impact

The ransomware actor is the opposite of an APT. Time is the enemy. The goal isn't to hide, but to create the widest possible impact in the shortest amount of time. In this profile, adversary modeling equips the hunter with a different reflex. The ransomware operator does the following:

- Rapidly turns initial access into a chain of actions
- Aggressively uses lateral movement
- Doesn't hesitate to escalate privileges
- Targets defensive layers such as backups, antivirus (AV), EDR, and logging
- Doesn't avoid noise, but feeds on impact

The following are applied techniques for ransomware-focused threat hunting:

- **Propagation dynamics analysis**
  Here, momentum matters more than isolated events. A single account accessing many systems in a short period or operations normally spread over weeks being compressed into hours are critical signals.
- **Defense-blinding hunting**
  Before the attack, ransomware blinds the defense. Stopped backup services, disabled EDR agents, and sudden changes in logging services are hunted at this stage.
- **Final preparation artifacts**
  Test accesses before encryption, small file writes, short-lived access to shares often don't generate alerts, but they are loud in terms of timing.

An APT is like a watchmaker: slow and flawless. Ransomware is like a sledgehammer: fast and destructive. Even if they leave the same hammer mark, the way they strike is different.

Insider Threat Profile: Context, Emotion, and the Gray Zone

Insider threat is technically the most difficult profile because here, the adversary doesn't come from the outside—they are already inside. The most critical characteristic of this profile is that behaviors may be technically legitimate, but contextually wrong. Insider threat hunting carries the highest cognitive load due to the following:

- Privileges are real.
- Access is legitimate.
- Tools are normal.
- But intent has changed.

The following are applied techniques for insider-focused hunting:

- **Role-behavior alignment analysis**
  What matters isn't what a user does, but how aligned that action is with their role. The same query may be normal for a DBA, but not for a finance user.
- **Behavioral drift detection**
  An insider threat doesn't erupt suddenly. It drifts slowly. First, working hours change, then the accessed objects, and then data volume. The hunter overlays these small shifts over time.
- **First-time hunting**
  Actions performed for the first time in a user's history are extremely valuable. First export, first bulk download, and first administrative API call aren't IOCs, but indicators of intent.

Insider hunting is uncomfortable because technical correctness and ethical boundaries stand side by side. A good hunter must learn to protect both the defense and human dignity.

Distinguishing Thinking Across Profiles

The greatest mistake is anchoring a technique to a single adversary profile. For example, the use of PowerShell shouldn't be attributed to a single adversary profile, as the same technique may be employed by APT actors, leveraged in ransomware operations for lateral movement or payload execution, or even observed in insider activity—making context, intent, and behavioral patterns essential for accurate interpretation. What makes the difference is context + timing + intent.

Adversary modeling can be summarized in the following sentence: The same trace doesn't convey the same meaning when it's left with a different intent:

- In APT scenarios, consistency is the alert.
- In ransomware scenarios, speed is the alert.
- In insider scenarios, context is the alert.

Adversary modeling doesn't teach the hunter a new technique. It teaches where, for what purpose, and against whom to apply the techniques they already know.

The most important gain of this section is that hunting ceases to be technical matching and becomes a mental confrontation. In that confrontation, the winner isn't the one with the most tools, but the one who understands the adversary best.

Key Considerations

Adversary modeling isn't a body of knowledge that is learned once and then shelved. On the contrary, it's a mental calibration that is readjusted in every hunt and reshaped in every environment. For this reason, errors most often arise not from not knowing, but from failing to question what one believes they know. The following represent critical points of attention where practitioners most frequently stumble in the field, yet when mastered, dramatically elevate the quality of hunting:

- **Avoiding profile memorization**
  One of the most dangerous fallacies is that "This behavior resembles an APT; therefore, it's an APT." An APT profile isn't a signature. It's a collection of tendencies.

  In real environments, actors don't operate in pure forms. An APT may borrow ransomware tactics. Ransomware operators may attempt to remain quiet like an APT. Insider behavior may resemble an external threat. Profile boundaries are blurred in practice. What the hunter must do here isn't to affix the label early, but to defer labeling until the end. First, the behavior must be understood; only then should one assess which profile it most closely aligns with.

  The golden rule is that a profile isn't the starting point of a hypothesis; it's the result of validation.

- **Not ignoring the gap between technical trace and intent**
  A technical trace doesn't always reveal intent. But intent often leaks through the spaces between technical traces, for example:

- The same credential access technique
- The same administrative share access
- The same PowerShell invocation

All of these may appear across three different profiles. What differentiates them is timing, frequency, risk tolerance, and the next step to take.

If the hunter renders judgment based on a single technique, adversary modeling collapses because the adversary isn't a single step; it's a sequence of steps. Therefore, the most critical skill in adversary modeling isn't asking "What did I see first?" but rather "What happened after what I first saw?"

- **Miscalibrating time perception**

  Just because a behavior is fast doesn't necessarily mean it's ransomware. Being slow doesn't necessarily mean the behavior is an APT either. When time is detached from context, it becomes misleading. The hunter must ask this key question: "Is the speed of the behavior abnormal relative to the normal pace of this environment?" Consider the following:

  - In automation-heavy environments, rapid propagation may be normal.
  - In structures with dense DevOps pipelines, serialized access may be legitimate.

  For this reason, absolute judgments such as "too fast" or "too slow" lead to serious errors in adversary modeling. The correct reference point isn't global threats, but local baselines.

- **The trap of overtechnicalization in insider threats**

  The most common mistake in insider threat hunting is reducing the issue solely to technical traces. Yet in insider threats, technology is often a cover, not an accomplice. An insider does the following:

  - Uses their own account
  - Operates with legitimate tools
  - Accesses resources within granted privileges

  Here, the technical activity is normal; it's the shift in intent that's abnormal. The hunter need not ask "Is this behavior technically possible?" but if they fail to ask the following question in insider modeling, they become blind: "Why has this behavior become possible for this individual now?" This question places the human factor at the center of the hunt. And at this point, technical analysis alone is insufficient; context, process knowledge, and organizational awareness are required.

- **Treating one's own bias as part of the model**

  The most invisible risk in adversary modeling is the hunter's own mental shortcuts. The following color the hunter's lens:

  - The most recently read CTI report
  - The most recent major incident
  - The last APT case that was detected

Every new behavior begins to pass through this filter. This is particularly dangerous for experienced analysts because intuition can silently transform into bias. For this reason, in mature teams, adversary modeling isn't a single-person judgment. It's tested through multiple perspectives. The question "What else could this be?" is asked deliberately.

The reflex question is "What could invalidate my own model?"

- **The operational cost of early profile fixation**
  If a behavior is labeled as APT too early, the following happens:
  - Alert severity is unnecessarily elevated.
  - Management panic may arise.
  - The team focuses on the wrong area.

  If it's labeled as ransomware too early, the following occurs:
  - Incident response processes become overly aggressive.
  - Operations may be disrupted.
  - Incorrect containment decisions may be made.

  If it's labeled as an insider threat too early, the following happens:
  - Legal and ethical risks emerge.
  - Trust relationships are damaged.

  For this reason, haste in adversary modeling produces technical errors and organizational risk. One of the hunter's most important skills is resisting the pressure for certainty. Sometimes, saying "We don't yet know" is the most correct decision.
- **Treating the model as static: The quietest collapse**
  Adversary profiles change over time. APTs accelerate. Ransomware actors become quieter. Insider behaviors become increasingly digital. A model that was correct yesterday may be misleading today. For this reason, adversary modeling is a living document. Every hunt, every incident, and every failed hypothesis sharpens the model a little more. A model that isn't updated silently produces false confidence. The hunter's task isn't to defend the model, but to constantly challenge it.

The considerations in adversary modeling don't comprise a checklist; instead, they are difficult questions that hunters must ask themselves:

- Am I rushing?
- Am I defaulting to labels?
- Am I missing context?
- Am I following my own bias?

A good hunter is one who can model their own mind before modeling the adversary. True mastery lies not in seeing technical traces, but in refusing to believe them too early.

11.2.3 CTI Enrichment

When CTI enrichment is mentioned, many still default to a familiar question: "Has a new IOC arrived?" This perspective is incomplete. IOCs are inherently short-lived: IPs change, domains rotate, and hashes evolve. What persists far longer is attacker behavior. A useful distinction is that IOCs help you catch activity; behaviors help you understand it.

The value of intelligence, therefore, lies not in expanding lists, but in refining perspective. CTI enrichment isn't about adding more data, but about structuring and layering information so it becomes meaningful. More data doesn't automatically produce more insight; in many cases, it creates noise. Effective enrichment reduces ambiguity by sharpening context rather than increasing volume.

An IOC-centric approach is appealing because it offers clarity—something either matches or it doesn't. However, modern attackers adapt precisely to evade such detection. Infrastructure is rotated, tools are modified, and observable indicators are deliberately minimized or manipulated. In this context, IOCs should be treated not as definitive answers, but as entry points to a deeper question: "What behavior produced this indicator?" Without understanding the underlying process, detection remains superficial.

Behavior, unlike isolated events, unfolds over time. It reflects patterns, including timing, repetition, and consistency under similar conditions. Subtle signals such as unusually consistent sessions, small but regular outbound connections, or deviations from expected operational rhythms may not trigger alerts individually. Yet when correlated, they reveal intent. Enrichment is the process of bringing these signals together in a way that reduces uncertainty and narrows possible interpretations.

This process isn't purely technical; it's cognitive. Enrichment challenges the assumption that absence of evidence equals absence of activity. In some cases, the lack of expected signals indicates a visibility gap rather than a clean state. Recognizing this distinction is critical. The goal isn't simply to detect what is present, but to understand what may be missing.

Ultimately, CTI enrichment connects external intelligence with internal context. Intelligence doesn't dictate what is malicious; it frames how certain behaviors may appear under specific conditions. The hunter then translates this into the local environment by asking: "If this behavior occurred here, what traces would it leave?"

In this sense, CTI enrichment isn't about accumulating data, but about improving perception. Instead of producing more indicators and longer lists, it produces clearer understanding and stronger analytical judgment.

In the following sections, we'll examine the key tools you need to enrich your CTI data, walk through enrichment techniques, examine enrichment scenarios, and discuss an in-depth example.

The Building Blocks of Enrichment: IOCs, TTPs, and Behavior

To understand CTI enrichment, we need to revisit some concepts that should be very familiar to you by this point in the book: IOCs, TTPs, and behavior. IOCs are signs. They aren't the path. An IP address tells you to "Look here," but it doesn't answer the question "why." The same IP can carry different intentions at different times. A domain may be malicious today and become part of a legitimate CDN tomorrow. Hashes change, infrastructures rotate, and signatures evolve.

In the field, IOCs are like footprints on the ground. You see the footprint, but by itself, it doesn't tell you who passed, why they passed, or where they were going. Sometimes it can even lead you in the wrong direction. This is where CTI enrichment comes into play and reframes the question: "What did the attacker have to do to leave this trace?" This question elevates the IOC and changes it from being a purely technical artifact into the outcome of a behavior.

One step above IOCs are tactics, techniques, and procedures (TTPs). These form the skeleton of intelligence. They have shape, but not yet a soul. Statements such as "Credential dumping was performed," "Living-Off-the-Land was used," or "Cloud Persistence was established" don't, by themselves, translate into operations.

TTPs are like X-ray images. They reveal structure, but not movement. Hunting, however, is the art of capturing movement. Here, enrichment performs the following transformation: TTP leads to these questions:

- Why this TTP?
- Why does an attacker choose an LSASS dump on a system without EDR instead of using Mimikatz?
- Why do they infiltrate an existing automation rather than creating a new scheduled task?
- Why do they expand the permissions of an old, forgotten application instead of registering a new cloud app?

These "why" questions remove the TTP from being a technical heading and turn it into a preference. Where there's preference, there's mindset, and where there's mindset, there's behavior.

Behavior is the true target of CTI enrichment. Behavior isn't the attacker's signature; it's their character. Behavior isn't a single event but is woven over time. It emerges through rhythm, speaks in intervals, and sometimes reveals itself not through what is done, but through what isn't done.

None of these—a service account accessing the same systems at the same hours for months, PowerShell usage that always occurs outside business hours but never on holidays, or a cloud token that is continuously refreshed but produces almost no API calls—is malicious on its own. They don't generate alerts or shout on an SIEM dashboard. But when placed side by side, they whisper something. CTI enrichment exists to be able to hear that whisper.

At this point, the tone changes because you're no longer asking "What happened?" but "How did it happen?" Sometimes you even move to "It didn't happen, but if it had, what would it look like?" This is the threshold where intelligence turns into hunting.

A healthy CTI enrichment process progresses as a chain:

- IOC: By which technique was it produced?
- TTP: How would this technique appear here?
- Behavior: When, how often, and in what context would this appearance emerge?

Each step in this chain involves abstraction. But this abstraction doesn't produce uncertainty; on the contrary, it creates clarity because each level makes the one beneath it more meaningful.

Many times in the field, the IOCs come back clean. There are no matches. The reactive view relaxes at this point. The hunter becomes uneasy because if the behavior described in the CTI report leaves no trace in the environment, there are two possibilities: Either there's no attack, or we're blind. CTI enrichment is the courage to put this second possibility on the table.

The Quiet Cost of Behavioral Hunting

Behavior-focused hunting isn't easy. It requires patience, time, and sometimes the maturity to accept finding nothing at all. But in return, it offers something critical: freedom from misleading certainty.

When an IOC matches, a sense of relief arrives. You say, "I caught something." When you catch behavior, a different feeling arrives: discomfort. Because now you know this is only one instance and there may be others. This discomfort is unpleasant, but necessary. Good hunters don't flee from it; they use it like a compass.

Applied Techniques

This is precisely where CTI enrichment is translated into practice. Saying "We look at behavior" is easy; explaining how that behavior is actually observed is difficult. The techniques employed are modes of thinking and analysis developed to overcome this difficulty. They are perspectives rather than tools, and habits rather than rules.

Now, let's dive in:

- **Extending the time horizon**

  The first, and perhaps most fundamental, technique is extending the time horizon. Many security operations are naturally confined to short time windows: the past 24 hours, the past 7 days, at most 30 days. Behavior, however, isn't captured in haste. A hunter performing CTI enrichment deliberately extends time: 60 days, 90 days, sometimes more. This extension isn't for collecting more data, but for perceiving rhythm. What is noise in a short window becomes a pattern in a long one. What distinguishes whether a movement is singular or habitual is precisely this temporal depth.
- **Comparative norm construction**

  Immediately following this is the technique of comparative norm construction. Behavior is neither good nor bad in isolation; it gains meaning relative to a reference. For this reason, the hunter first defines what is normal. But this definition isn't universal. The same behavior may be routine for one team and alarming for another. Here, an expected behavior profile is derived for the same role, the same system, and the same time frame. Then the narrative provided by CTI is placed alongside this profile. The objective isn't to find difference, but to sense misalignment. When alignment breaks, it doesn't produce an alarm; it produces a question.

- **Negative space analysis**
  Another critical technique is negative space analysis. This is one of the most frequently overlooked approaches, yet it yields some of the most valuable outcomes. Negative space means thinking not through what exists, but through what doesn't. CTI may describe a behavior, yet in your environment the natural side effects of that behavior aren't visible. There are no logs, no traces, no echoes. In such cases, it's easy to say, "Then, there's no attack." The hunter performing enrichment looks from a different angle: "If this had happened, what should we have seen?" If we didn't see it, the problem isn't the attack, but visibility. This technique expands defense because it requires accepting absence. Acceptance is a difficult virtue, but defense matures precisely here.
- **Chain-breaking approach**
  Another commonly used technique on the path to behavior is the chain-breaking approach. Attack narratives are often presented as a chain: access, persistence, movement, impact. Reactive systems try to observe the entire chain. The hunter deliberately breaks the chain. They focus on a single link and ask, "For this link to exist here, what must have happened beforehand?" This approach reduces dependency on IOCs. It transforms one segment of the chain into behavior and builds a perception capable of detecting that segment across different variants. Thus, even if the attacker alters the chain, as long as the intent remains the same, it continues to be visible.
- **Timing analysis**
  Timing analysis is also indispensable in CTI enrichment. *When* a behavior occurs is as important as the behavior itself. Human behavior is irregular; attacker behavior is often excessively regular. Activities that hover around the edges of business hours, operations that deliberately avoid holidays, and accesses performed at similar minutes each time may appear technically benign. But timing speaks without shouting. The hunter takes the CTI narrative and tests timing with the following question: "At which hours would this intent hide most comfortably in this environment?" The answer is often more revealing than the behavior itself.
- **Role-context separation**
  Another important technique is role-context separation. The same action carries different meanings for different roles. Enrichment deliberately reads technical action together with role information, not apart from it. An API call, an export operation, and a privilege query are technical details on their own, but when "for this person, at this moment, at this frequency" is added, they become behavior. CTI provides guidance here regarding

which roles attackers favor, which identities they use, and which roles are less likely to be scrutinized. The hunter takes this information and looks within their own environment for areas with a high likelihood of remaining silent.

- **Deliberate slowing**
  A technique frequently used in CTI enrichment but rarely discussed explicitly is deliberate slowing. Speed is often presented as a virtue in security. In behavioral hunting, however, speed frequently produces errors. The hunter deliberately lets certain hypotheses sit. They don't rush to conclusions. They allow a behavior to manifest itself for a few more days. This waiting isn't passivity; it's observation because some behaviors reveal themselves only through repetition. This technique is particularly decisive in APT profiles. It requires listening to silence rather than breaking it too early.
- **Falsify assumptions**
  Finally, perhaps the most human technique in CTI enrichment is the reflex to attempt to falsify one's own assumption. For every behavioral hypothesis produced, the hunter directs the following question at themselves: "What could be an innocent explanation for this?" This question doesn't weaken defense; on the contrary, the question strengthens it. As easy explanations are eliminated, what remains is a more robust meaning. CTI here functions not as a validation tool, but as a test set. The knowledge that "This has appeared like this elsewhere in the world" prevents the hunter from blindly defending their own conclusion.

The common thread among these techniques is that none is miraculous on its own. They derive their strength from the relationships they form with one another: Time is extended, norms are constructed, negative space is questioned, chains are broken, timing is listened to, role context is added, and finally assumptions are deliberately challenged. This process is exhausting and, at times, frustrating. But in return, it provides something invaluable: awareness independent of alerts.

Example: Enrichment Scenarios

The following examples illustrate how CTI enrichment is applied in practice by transforming intelligence into behavioral hypotheses and observable signals across different environments. Rather than relying on static indicators, they demonstrate how patterns, timing, and context can be used to uncover otherwise invisible activity and translate CTI into actionable hunting logic.

- **Windows Server 2022: Detecting silent account attempts through Kerberos preauthentication anomalies**
  In this scenario, CTI describes a credential access technique in the form of password spraying, without providing specific indicators of compromise. Instead of relying on identifiable artifacts such as IP addresses or user agents, the hunter focuses on the behavioral signal reflected in the rhythm of failed authentications.

  Within a Windows Server 2022 AD environment, this behavior typically manifests as low-noise authentication attempts, where attackers perform one or two login attempts per user and distribute them over time to avoid account lockouts. As a result, individual Event ID 4625 entries may appear normal in isolation, and their significance becomes visible only when evaluated within a defined time window.

To validate this hypothesis, the hunter analyzes security logs on DCs, focusing on Event ID 4625 (Failed Logon Attempts), Event ID 4771 (Kerberos Pre-Authentication Failures), and Event ID 4768 (TGT Requests). Relevant patterns include multiple user accounts being targeted from the same source IP within short intervals, repeated identical failure codes or status values, and consistent logon type patterns.

From a hunting perspective, the approach shifts from detecting a high number of attempts against a single user to identifying a pattern of limited attempts distributed across many users. This can be operationalized by calculating the ratio of distinct user accounts per source IP within a 15–30 minute window. Under normal conditions, authentication issues such as helpdesk or virtual private network (VPN) errors tend to cluster around a single user, whereas password spraying results in a broader distribution across multiple accounts.

- **Linux: IOC-less LOLBins persistence through subtle `cron` and `systemd` drift**
  In this case, CTI describes persistence and defense evasion techniques that rely on legitimate system tools, without producing identifiable IOCs. The key signal isn't a binary artifact, but the unexpected placement and timing of persistence mechanisms.

  In Linux environments, attackers may avoid deploying new binaries and instead establish persistence by making minimal changes, such as adding a single line to a `cron` job, creating a low-frequency `systemd` timer or service, or slightly modifying an existing script. These actions are designed to appear benign, often using short and commonly accepted commands.

  The hunter evaluates telemetry related to configuration changes, including files such as `/etc/crontab`, `/etc/cron./`, user-specific `crontabs`, and `systemd` service or timer definitions under `/etc/systemd/system/` and user directories. In production environments, these files typically change infrequently and are expected to align with documented change management processes.

  Rather than asking whether a new file exists, the hunter examines how frequently these files change for a given host class, which user or process initiated the change, and where auditing data is available. The strongest signal emerges from the combination of a minor configuration change, long-term periodic outbound network activity (e.g., DNS or HTTP communication), and consistently low resource usage.

- **Windows Server 2022 and Linux: Detecting lateral movement through identity and token usage patterns**
  Here, CTI indicates that lateral movement is performed using legitimate identities, without bypassing MFA, by exploiting token or session lifetimes. The distinguishing signal isn't a specific network artifact, but the unusually consistent and structured use of identities over time.

In hybrid environments, this behavior appears as a shift in access patterns. On Windows systems, a service or administrative account that normally accesses a limited number of servers may begin interacting with a broader and more diverse set of targets. On Linux systems, a user or Secure Shell (SSH) key that typically connects to a small number of hosts may suddenly expand its reach, while the executed commands remain minimal and inconspicuous.

To investigate this behavior, the hunter analyzes telemetry such as Windows Event ID 4624 (Successful Logons), including fields such as `LogonType`, `WorkstationName`, and `SourceNetworkAddress`, as well as Event ID 4769 (TGS Requests) to evaluate service access diversity. On Linux systems, relevant sources include `auth.log` or secure logs for SSH activity, privileged access management (PAM) records, and `journald` entries related to SSH sessions and `sudo` usage.

A key indicator is the same identity accessing an unusually large number of distinct systems within a short period without triggering alerts. The hunting approach focuses on measuring this expansion by comparing the number of unique hosts accessed over time against historical baselines. Finally, the temporal rhythm of activity is examined, as human-driven behavior tends to be irregular, whereas automated attack patterns are typically consistent and structured.

Example: Nighttime Signature of Service Accounts + First-Seen Anomalies Using Python

This example in Figure 11.1 deliberately disables the reflex question of "Has a new IOC arrived?" in CTI enrichment and instead centers on the following question: "How does intent manifest as rhythm in this environment?" The four layers we see in the output (`SUMMARY → TOP HOURLY RHYTHMS → FIRST-SEEN → QUALITY`) represent the operationalized form of this approach.

This output is generated by a Python-based telemetry analysis pipeline that ingests authentication and event logs, normalizes key fields (e.g., principal, host, timestamp, and source IP), and then aggregates behavior over time windows. The pipeline groups events by identity and hour to derive rhythm patterns, calculates diversity metrics such as unique hosts and IPs, and tracks first-seen combinations to highlight novel behavioral interactions. In addition, it measures ingestion latency and data consistency to ensure that observed patterns reflect actual behavior rather than pipeline artifacts.

From a practical standpoint, this approach can be implemented using common data processing libraries (e.g., pandas) on exported SIEM/EDR logs, or directly integrated into a streaming pipeline. The key requirement isn't the specific tool, but the ability to model time-based behavior, track identity-context relationships, and preserve sufficient historical depth to distinguish normal rhythm from anomalous activity.

```
C:\Users\admin\PycharmProjects\Books\Telemetry Builder>python main.py --demo 5000

--- SUMMARY ---
Events (deduped): 5000
Principals: 6
Output CSV: C:\Users\admin\PycharmProjects\Books\Telemetry Builder\outputs
Parquet dataset: C:\Users\admin\PycharmProjects\Books\Telemetry Builder\data\events

--- TOP HOURLY RHYTHMS (count desc) ---
 principal  hour  count  unique_hosts  unique_ips
svc_backup     5    112             6           6
svc_backup     2    108             6           6
svc_backup     3    107             6           6
svc_backup     4    107             6           6
   svc_sql     3    107             6           6
   svc_sql     4    103             6           6
svc_backup     0    101             6           6
   svc_sql     5     98             6           6
   svc_sql     0     96             6           6
svc_backup     1     89             6           6

--- FIRST-SEEN (latest 10) ---
                  event_time_utc  principal      host     action       src_ip        first_seen_type
2025-10-14 00:02:06.319903+00:00    svc_sql      DC01 login_fail 10.10.14.99 principal_host_action
2025-10-05 05:58:59.319903+00:00        bob  SERVER02 login_fail 10.10.12.50 principal_host_action
2025-10-03 21:29:33.319903+00:00      alice  WS10-001 login_fail 10.10.12.50 principal_host_action
2025-10-01 09:12:58.319903+00:00 app_reg_01    JUMP01 login_fail  10.10.10.5 principal_host_action
2025-09-30 18:39:50.319903+00:00    svc_sql  WS10-001 login_fail 10.10.13.77 principal_host_action
2025-09-30 14:01:53.319903+00:00    svc_sql    JUMP01 login_fail  10.10.10.5 principal_host_action
2025-09-30 01:10:29.319903+00:00 helpdesk01  WS10-001 login_fail 10.10.11.20 principal_host_action
2025-09-29 14:38:33.319903+00:00        bob  WS10-014 login_fail 10.10.14.99 principal_host_action
2025-09-28 22:24:51.319903+00:00 app_reg_01  SERVER02 login_fail 10.10.12.50 principal_host_action
2025-09-28 20:10:00.319903+00:00      alice    JUMP01 login_fail 10.10.10.12 principal_host_action

--- QUALITY (ingest delay p50/p90, negatives) ---
delay_p50=910.0s  delay_p90=1640.1000000000004s  negative_delay=0
```

Figure 11.1: Telemetry Rhythm Analysis Output Highlighting Nighttime Service Account Activity and First-Seen Behavioral Signals

Let's start by looking at the `SUMMARY` section output:

- `Events (deduped): 5000`
- `Principals: 6`
- `Output` written in `CSV` and `Parquet` formats

The critical point here is that this system doesn't operate with a "I saw two logs, and I made a judgment" approach. Instead, it works with a behavior memory built on an extended time window. The `Principals` value in the summary reflects the number of distinct identities included in this behavioral scope, providing context for how patterns are distributed across users and service accounts rather than being tied to a single entity. Likewise, the `Output` lines indicate that the processed dataset has been written in structured formats (CSV and Parquet), enabling the same behavioral analysis to be reused, validated, and extended beyond a single execution. Five thousand events, when examined individually, may appear as noise; however, when combined with the bucket + `FIRST-SEEN` + `QUALITY` layers, they cease to be fog and become a readable rhythm map.

The `TOP HOURLY RHYTHMS` section highlights how activity is distributed across time and identities, allowing us to observe recurring patterns and concentration points. From this view, several key observations emerge:

- The svc_backup and svc_sql accounts stand out clearly in the 00–05 time band.
- Each row shows unique_hosts=6 and unique_ips=6.

The following examples illustrate specific instances of this pattern in practice:

- svc_backup hour=5 count=112 unique_hosts=6 unique_ips=6
- svc_sql hour=3 count=107 unique_hosts=6 unique_ips=6

The TOP HOURLY RHYTHMS output (principal-hour distribution table shown in Figure 11.1) conveys the following:

- **Timing is a signature.**
  Human behavior is usually irregular; automation or scheduled work is "flat" and repetitive. The concentration of svc_* accounts during nighttime hours doesn't constitute a single alert; it produces a behavioral signature.
- **Host and IP diversity generates a separate question.**
  As shown, unique_hosts=6 and unique_ips=6 mean the following:
  - This service account isn't operating on a single server; it spans multiple hosts.
  - Source IP diversity appears high. (In this demo, diversity is deliberately generated; in real environments, this metric varies by hour and by user role.)

  Here, CTI enrichment doesn't say "Find an IOC"; instead, it prompts the following:
  - Is this level of regular nighttime activity normal?
  - If it's normal: "This is our automation rhythm."
  - If it's not normal: "Who might be imitating this rhythm?"
- **First-Seen combinations reveal behavioral novelty.**
  The First-Seen (latest 10) list on the screen produces records such as these:
  - The svc_sql account generating a login_fail on DC01 (principal_host_action)
  - The alice account generating a login_fail on WS10-001
  - The app_reg_01 account generating a login_fail on JUMP01
  - The helpdesk01 account generating a login_fail on WS10-001

  The critical point here is that First-Seen states that this combination has been observed for the first time without requiring any IOC. In other words, rather than producing an alarm, it produces a hypothesis. If an application/service-like identity such as app_reg_01 generates a login_fail on a JUMP host for the first time:
  - Why did this identity interact with JUMP?
  - Does this identity normally access JUMP?
  - Is there role-context alignment?

  Here we see the transformation described in the text applied directly:
  - IOC: "This IP is malicious."
  - Behavior: "This identity performed this action on this host for the first time → why?"

This directly aligns with the core statement in the text: "Good CTI doesn't say 'this is bad'; it says 'Under these conditions, intent appears like this.'" The first-seen output provides the raw material required to be able to say "Intent appears like this."

Finally, let's examine the `QUALITY` section of the screen, which contains the following lines:

- `delay_p50` = 910.0s
- `delay_p90` = 1640.1s
- `negative_delay` = 0

We can see that ingestion delay has been measured, which means the impact of data arriving late can be controlled. A `negative_delay` value of 0 means there's no inversion between `event_time` and `ingest_time`, and so no obvious out-of-order issue is observed. Therefore, the produced rhythm is based on the intrinsic temporal patterns of the events rather than on a pipeline error.

Even though this is a demo, it demonstrates that the architectural approach has been correctly established.

11.2.4 Target Prioritization Through Threat Landscape Analysis

After a certain threshold is crossed in CTI enrichment, a new reality settles at the table: Threats are infinite, and time is limited. At some point, the hunter stops asking "Can I find it?" and starts asking "Where should I begin?" Threat landscape analysis is precisely the backbone that bears this burden.

Real-World Example

When I first began to take this topic seriously, I felt a certain resistance. Hunting is often guided by intuition, and mapping or prioritization can feel restrictive. However, during a week when I was handling multiple critical alerts at once, I realized that the real target had slipped past me. That moment made one thing clear: prioritization isn't the enemy of hunting; it's insurance. Threat landscape analysis isn't, as many assume, a list of which APTs exist. You can make such a list, hang it on the wall, and feel a brief sense of relief—but it adds nothing to your defense because the name of the attacker rarely explains the behavior moving through your systems. Landscape analysis does something different: It views threat not as a label, but as a pressure field, like weather. The name of the rain is irrelevant; what matters is whether you have an umbrella, where the roof is leaking, and which direction the wind is blowing.

At this point, the tone shifts slightly. This section demands cold reality rather than romantic intuition. The idea of "let's hunt everything" is a pleasant fantasy but fantasy collides with budget and human hours in the real world. Landscape analysis tells the hunter, "You can't catch everything. But if you look in the right place, you'll see the highest impact." This sentence can be comforting because it reduces the burden but also unsettling because it forces you to choose—and choice carries responsibility.

What we call prioritization is, in fact, the superposition of three distinct questions: Who are we similar to, what do we value, and where are we exposed? Without the convergence of these three, the notion of a "target" remains abstract. How the sector is targeted, from which surface your organization appears visible, and what meaning your assets carry for an attacker all must meet within the same sentence. Otherwise, you read CTI, and reports grow thicker, yet the hunt keeps circling the same places: where the noise is.

In the following sections, we'll discuss the key steps and considerations that lead to successful prioritization through threat landscape analysis. We'll also cover the techniques you need and apply core concepts through a few examples.

The Flow of Threat Landscape Analysis

When prioritizing targets through threat landscape analysis, begin with a simple question: "If someone were to target us today, why would they do it?" This "why" carries more than a technical inquiry because "why" simultaneously reveals the attacker's motivation and the value of the target. If it's financial, behavior flows one way; if espionage or insider leakage, it flows two other ways. The same technical trace settles into a completely different map under a different intent. Prioritization begins here—not by assuming intent, but by ordering the probabilities of intent.

The most useful distinction in the field is the difference between general risk and risk specific to us. What everyone is talking about may not be your greatest risk. Sometimes the opposite is true: something no one discusses, yet highly feasible in your environment, is your greatest risk. This is why landscape analysis favors what is feasible, not what is popular. And that is exactly what attackers favor as well: feasibility. The target is not the most brilliant technique, but the one that works with the least friction. The hunter's task is to see where friction is low and harden that space.

The second step in prioritization isn't labeling an asset as "critical," but defining the asset through flow. This is necessary because the attacker doesn't attack a single server; they attack the life connected to that server. For example, a DC is the heart of identity flow, a file server is both file storage and the memory of the business, and an email system is both communication and the carrier of trust relationships. Therefore, target prioritization doesn't end with looking at an asset inventory and saying "These are important." The real work lies in understanding which behaviors these assets make possible.

Consider this simple metaphor: Assets are rooms, and behaviors are corridors. Attackers don't search for rooms; they search for corridors because corridors connect rooms. Landscape analysis shows where corridors are narrow and where they are wide. The widest corridor is usually identity and access infrastructure, then management channels, backup and monitoring layers, and finally data. This order isn't identical in every organization, but it's surprisingly similar in most organizations.

The third step is to stop reading the attack surface through open ports and to start reading it through behavioral entry points. For example, one organization may not use VPNs at all, yet rely heavily on Remote Desktop Protocol (RDP) for administration. Another may have RDP closed, but rely on established and legitimate administrative channels such as PowerShell remoting, WinRM, or SCCM. In one case, the primary risk originates from the internet; in the other, it propagates from within. Threat landscape analysis captures this distinction because the weakest link is sometimes not a technical vulnerability, but an operational habit—such as using the same service account everywhere, or having the same administrative workstation touch everything. This is a form of operational convenience that no one objects to, but that attackers find highly appealing.

There's a subtle emotional micro-expression here: Sometimes people feel discomfort when they begin to see the habits of their own environment because those habits have kept operations running for years, becoming a comfort zone. But in security, comfort is often the attacker's comfort as well. Accepting this reality isn't a technical realization; it's a cultural confrontation. This is one of the reasons why prioritization is difficult.

What matures target prioritization through threat landscape analysis is bringing *impact* and *likelihood* into the same decision frame. Most teams understand impact: If a critical system goes down, the business stops. Likelihood, however, is often misread: the feeling of "It won't happen to us" is the most expensive illusion. Likelihood is related both to how much an attacker wants you and to how much convenience you provide to the attacker. Even if an actor isn't specifically targeting you, an exposed surface can turn you into a target. For this reason, prioritization takes the attack path seriously, not the threat actor.

CTI reenters the picture here, but in a different role. Now CTI doesn't explain what happened; it provides directions for "Where they would walk." You take this set of directions and draw it onto your own organizational map. The question "Where is the equivalent of this in our environment?" is no longer theoretical. It becomes concrete. Is it the identity layer, management channels, cloud permissions, backups, or log visibility? You must choose one because you can't simulate attacks on all of them at once, hunt all of them simultaneously, or give depth to all of them at the same time.

The key takeaway from this section is simple: Hunt scenarios don't need to be elegant—they need to be well-placed. A technically advanced hunt may produce no results if it targets the wrong area, while a simple, well-formed question can reveal meaningful signals. In practice, asking the right question is often more valuable than applying complex detection logic. Threat landscape analysis provides both direction and discipline. It enables the hunter to justify where to focus while also defining where not to spend effort. When combined, hunting becomes a deliberate and structured activity rather than reactive investigation. In essence, CTI enrichment develops an understanding of behavior, while threat landscape analysis defines priority. Effective hunting requires both.

Key Considerations

As target prioritization matures through threat landscape analysis, a new reality becomes clear: the primary risk isn't chasing the wrong thing, but treating the right thing with the wrong level of seriousness. This section focuses on fine-tuning points that rarely even appear as marginal notes in most documents, yet repeatedly cause real damage in the field. It doesn't shy away from technical depth; it steps directly into it. Mistakes made at this stage don't generate noise—they quietly weaken the defense.

Let's dive in to these key points to consider:

- **Treating priority as a static label**
  Labels such as "critical," "high," or "medium" are comforting, but real-world threats don't respect these labels. Priority isn't a permanent adjective written onto an object; it's a shifting weight that changes over time. A surface that appears low risk today may, with a single configuration change tomorrow and become the most attractive entry point. For this reason, mature teams don't view prioritization as an annual exercise, but as a continuously fine-tuned balance. Otherwise, the organization becomes perfectly defended against yesterday's threats and vulnerable to today's.
- **Stripping CTI of its context**
  Reading a sentence such as "Sector X is being targeted" in a CTI report doesn't automatically mean we're a target. Likewise, saying "We're not a target" is an equally dangerous illusion. The real question is where do the attacker's needs intersect with your weaknesses? This intersection is far more valuable than any sector label. This is experienced repeatedly in the field: It's not the APT groups everyone talks about that cause the most damage, but the actors no one names—yet who adapt exceptionally well. Adaptability is more lethal than popularity.
- **Over-personalizing the threat actor**
  Statements such as "This APT would never come after us" or "This ransomware targets mostly small businesses" are often driven not by data, but by a psychological need for reassurance. Attackers aren't romantic when selecting targets. They decide based on suitability and friction. If an environment enables a technique at low cost, the actor profile instantly becomes secondary. Therefore, target prioritization should ask "Who would be comfortable here?" before asking "Who would come here?"
- **Organizational roots of prioritization blindness**
  Some assets are technically critical but politically untouchable. Some systems are extremely risky but excluded from hunting due to the fear that the business would stop. This is where danger lies. Attackers intuitively exploit organizational hesitation. The least scrutinized areas are often the least questioned. A mature hunting approach aims to make these political blind spots visible through technical language—not by assigning blame, but by concretizing the cost of risk.
- **Underestimating cascading effects**
  Target prioritization is often performed over individual assets: "Is this server critical?" Yet attacks rarely remain confined to a single asset. The real impact emerges in the flows

between assets. An identity, a management channel, and an automation account may each appear low impact in isolation, but together they can produce catastrophe. Advanced prioritization therefore focuses not on assets themselves, but on the transitions between them. The question "If this is compromised, what is the cheapest next move?" is invaluable here.

- **Limiting the threat surface to only what is explicitly exposed**
  Open ports, internet-facing services, external IPs are visible and therefore attract disproportionate attention. Yet a significant portion of modern attacks don't pass through what is open, but through what is overly trusted. Internal automations, continuous integration/continuous delivery (CI/CD) pipelines, monitoring agents, and backup access are often assumed to be safe. For an attacker, however, anything that is assumed to be safe is a potential accelerator. During prioritization, what is most trusted should be examined alongside what is most vulnerable.
- **Measuring priority solely by technical impact**
  Technical impact matters, but it's not sufficient on its own. What does the failure of a system mean for the business? Which processes are poisoned when a behavior remains invisible? Some attacks don't halt systems but distort decision-making mechanisms. Log manipulation, for example, gradually erodes trust. Such effects rarely appear on classic "critical system" lists. But for teams that truly understand the threat landscape, these silent impacts carry high priority because loss of trust is often detected later than loss of data.
- **Feeding prioritization with past success stories**
  The phrase "We caught this last time" quietly creates a new form of blindness. Attackers don't observe your successes; they observe your habits. Assuming that a technique has been abandoned simply because it was hunted once is naive. The more likely scenario is that the technique remains the same, but the context changes. Prioritization should account not for past victories, but for future friction.
- **Misreading the time dimension**
  Prioritization often focuses on what is most risky, but sometimes the right question is "What will happen first?" A scenario with massive impact but low probability shouldn't be weighed equally against one with moderate impact but imminent likelihood. Making this distinction requires viewing the threat landscape not only as a set of threats but also as a calendar. Nontechnical factors such as patch cycles, business calendars, holiday periods, and personnel changes significantly influence attack timing.
- **Carrying the emotional weight of prioritization decisions**
  These decisions also involve saying, "We won't look here for now," and sometimes, something happens precisely there. This is unavoidable. Mature teams don't interpret this as personal failure. Prioritization isn't the art of being flawless; it's the discipline of taking conscious risks. What matters isn't covering everything, but being aware of what you're not covering.

Target prioritization through threat landscape analysis is a way of aligning defense with organizational reality. Effective hunting requires recognizing habits, assumptions, and blind spots—and focusing attention where it matters most.

Techniques and Software Used

When threat landscape analysis turns into target prioritization, the techniques and software employed become an extension of a way of thinking. At this point, the issue isn't which product you use, but which question you can ask, and at what depth. Dozens of teams may collect the same data; what makes the difference is the perspective from which you break it apart. The following techniques and uses of software operate quietly in the field, often unnamed, yet decisive in their outcomes:

- **Time-series normalization**
  Most security platforms present events in real time or within short windows. Target prioritization, however, requires the expansion of time. The approach used here is not merely counting events, but extracting the time-distributed frequency of the same behavior. This technique operates on a SIEM or a data lake, but its logic is independent of the tool. Daily, weekly, and monthly rhythms are derived and then segmented by role, asset, and network segment. In this way, behaviors that are rare but always occur in the same form become visible. Such behaviors are generally low-noise but high-meaning.
- **Contextual enrichment chain**
  This refers to matching data from a single source with at least three different contexts: identity context, asset context, and time context. For example, an access event is read not only with an IP and a user, but together with the user's role, the function of the accessed system, and the normality of the time window in which the access occurred. This enrichment doesn't require classic IOCs because the real value lies not in the event itself, but in the inconsistency of the context in which the event is situated.

  At this stage, the software that comes into play is typically SIEM, extended detection and response (XDR), or data lake architectures. However, in advanced teams, these platforms are used as hypothesis-testing environments, not report-producing tools. Queries written for prioritization don't have to generate alerts. On the contrary, most queries are written to "find nothing." This is a deliberate choice because target prioritization seeks to rank risks *and* blind spots. The statement "We see nothing here," when said in the right place, is one of the most valuable outputs.
- **Transition cost analysis**
  This technique attempts to measure how much effort an attacker must expend to move from one asset to another. The software used here typically includes Identity And Access Management (IAM) logs, network flow records, and management-channel telemetry. But the core focus is on the following question: "From this point on, what is the attacker's cheapest next step?" This analysis shifts prioritization from individual systems to transitions between systems. The answers to these questions reorder the target list:

  - Which identities can touch multiple segments?
  - Which automations silently cross boundaries?
  - Which management channels are normally invisible yet critical?

- **Negative signal correlation**
  Most software captures what is present. Advanced prioritization also puts what should be present but isn't on the table. For example, the absence of expected log types in a given system class may not be an indicator of an attack, but it's a definitive indicator of a defensive weakness. This technique is used together with the health metrics of SIEM or log management systems. But the real value comes from reading it through a threat perspective: "If an attack came through this surface, where should we have seen it, and why can't we see it?"

- **Role-based behavior profiling**
  In this approach, software treats users not as individual entities, but as behavior clusters. Normal behavior ranges for accounts with the same role are established; then movements that fall outside this range yet don't generate alerts are flagged. This technique dramatically improves target prioritization, especially for service accounts, automation identities, and administrative roles because these accounts are both high-value and low-visibility for attackers. The software used is often simple; what truly makes the difference is the comparison logic.

- **Operational friction measurement**
  Friction refers to how many steps are required to perform an action, how many controls are encountered, and how much trace is left behind. Software doesn't measure this directly; instead, inferences are drawn by combining signals such as proxy logs, MFA triggers, and authorization approval processes. From a prioritization standpoint, low-friction paths are always more critical. Because attackers choose the quiet and short path, not the complex one.

- **Reusable hypothesis libraries**
  At this point, the software used is usually simple documentation tools or platforms operating with a code-repository logic, such as internal wikis (e.g., Confluence), version-controlled repositories (e.g., Git-based platforms such as GitHub or GitLab), or knowledge management tools (e.g., Notion). But the content is extremely valuable. Every threat landscape analysis produces new hypotheses. These hypotheses aren't lost; they are classified, tagged, and reused. Thus, target prioritization ceases to be an exercise performed from scratch each time. An institutional behavioral memory is formed. This memory is more valuable than even the most expensive security products.

- **Metrification of prioritization decisions**
  These metrics aren't Common Vulnerability Scoring System (CVSS)–like scores. Rather, they measure things such as these:

- How long did this hunting scenario remain open?
- How many different system classes does it cover? How many different threat paths does it affect?
- What was the visibility gain?

The software here produces dashboards, but the true goal isn't reporting; it's monitoring decision quality. Prioritization errors become visible in this way.

- **Conscious simplification**

  Finally, above all of these techniques, there's an approach that isn't limited to software: conscious simplification. Advanced teams don't use every piece of data or collect every signal. They deliberately leave some things out. This isn't a deficiency; it's mastery. Because threat landscape analysis isn't the art of increasing complexity, but the discipline of intensifying meaning. No matter how powerful the software used may be, without this discipline, target prioritization turns into nothing more than a polished table.

Example 1: Active Directory Certificate Services Invisible Privilege Path (Windows Server 2022)

In some environments, DCs are already on the table because they are critical. But the real accelerator is often not the DC itself, but the alternative gateway to identity. Active Directory Certificate Services (AD CS) running on Windows Server 2022, if not properly managed, can turn into a corridor that allows an attacker to gain privileges without generating password noise. The threat here isn't a single vulnerability; it's the combination of misconfiguration and certificate processes that appear normal.

To operationalize this scenario, the following elements outline how the CTI context is translated into concrete questions, telemetry, and prioritization signals within the environment:

- **Threat landscape context: Why does it become a priority?**

  If the identity infrastructure is being targeted, the attacker shifts from passwords to proofs of identity. AD CS, when improperly designed, doesn't generate noise like password attempts; on the contrary, it flows as a legitimate operation. The impact is chained: one certificate → one identity → many systems.

- **The question to ask**

  The question that translates a CTI-derived TTP into the environment is "In our environment, do certificates operate only for devices and services, or can incorrect templates allow someone to carry an identity 'on behalf of another'?"

- **What is monitored?**

  Certificate Services logs and related Windows event records are monitored, specifically the following:

  - Certificate issuance events: Who requested it, with which template, with which subject/SAN?
  - Templates usage: The sudden appearance of templates that are normally rarely used, or the use of templates that are incompatible with the user's role.

  - Traces of AD changes (template/CA configuration changes): Are template permissions, enrollment permissions, or manager approval/`autoenroll` behaviors changing?
  - The identity usage side: The same identity accessing unusual systems shortly after certificate issuance (e.g., an increase in TGS requests).
- **The detail that makes it a priority**
  There's almost no IOC here. Even if you search for IPs or hashes, you usually find nothing. The prioritization signal is that the normal rhythm of certificate flows is disrupted, yet the SIEM doesn't raise an alarm. Therefore, the hunting output isn't "capture" but "narrowing the corridor" of which templates are truly necessary, who should be allowed to enroll, which templates should remain device-only.
- **Techniques/software**
  The following techniques and software would be used for this example in practice:
  - Windows advanced auditing + CA/AD CS logs (and preferably a centralized SIEM)
  - Identity-focused visibility (e.g., identity anomaly approaches similar to Microsoft Defender for Identity) and AD audits that track template changes
  - First-seen logic: "Has this template/subject combination been seen before?"

This example ties target prioritization to the following reality: Identity corridors are the widest corridors. If AD CS is misconfigured, that corridor can suddenly turn into a highway.

Example 2: systemd Silent Persistence

On the Ubuntu side, the most dangerous surface is sometimes not an internet-facing service. Sometimes it's the internal management layer that everyone trusts: `systemd`. Especially actors who need persistence may choose transient (temporary) units, timers, or user-level `systemd` units instead of classic `cron` entries because this method can do the following:

- Proceed without leaving files or with very few files
- Become visible for a short time and then blend back into normal
- Bypass traditional `cron` checklists

To translate this technique into a practical hunting approach, the following elements define how it should be interpreted, questioned, and prioritized within the environment:

- **Threat landscape context (Why does it become a priority?)**
  On Linux servers, `systemd` activity is often normalized as "There's already a lot of admin work anyway," which is exactly what the attacker likes: over-normalized management traffic.
- **The question to ask**
  The question that translates a CTI-derived TTP into the environment is "In our environment, does persistence leave traces as a file, or does it flow as a unit? Who can do this, and when?"

- **What is monitored?**

  To make this behavior observable in practice, the hunter focuses on the following telemetry sources and analytical signals:

  - `systemd` event flow via `journald`: Unit started/stopped rhythms in which units started regularly during night hours can't be explained by workload.
  - `systemd` unit inventory: Are there new units/timers under `/etc/systemd/system/` and user units (`~/.config/systemd/user/`) that are unexpected for the role?
  - Process–network correlation: Very low-frequency but continuous outbound (DNS/HTTP) activity, that is, low CPU, low noise, but high intent.
  - If `auditd` is present (extremely valuable): Who modified `systemd` unit files, and through which process? This is where role-context separation truly shines.

- **The detail that makes it a priority**

  Saying "There's a `systemd` unit" by itself means nothing. The prioritization signal follows:

  - Role-context mismatch: Why does this type of unit exist on this host class?
  - Timing patterns: Automation-like, not human.
  - Low-noise persistence: Small, steady actions over weeks.

- **Techniques/software**

  The following techniques and software would be used for this example in practice:

  - Centralized collection of `journald/syslog`
  - `auditd` (especially on critical server classes)
  - An inventory/query layer such as `osquery` (unit files, timer lists, first-seen approach)
  - If necessary, extended Berkeley Packet Filter (eBPF)–based runtime observation (Falco-like) with the goal: "Which process generated which network egress?"

This example ties target prioritization to the following principle: It's not the open port, but the over-trusted management layer that produces risk.

Example 3: Lateral Movement via Corridor Prioritization

This example demonstrates how lateral movement can be analyzed through a corridor prioritization approach, where the goal isn't simply to identify compromised assets, but to model the most efficient path an attacker would take through the environment. Instead of focusing on isolated events, we construct a graph of identity and system relationships and evaluate transitions based on cost factors such as effort, risk, noise, and probability of success.

The output shown in Figure 11.2 represents this model as a simplified attack graph, where nodes correspond to identities and systems, and edges represent possible transitions between them. The image highlights key elements such as the lowest-cost attack path, dominant corridors, edge-level cost breakdowns, and automatically generated hunting hypotheses, all of which are derived from the same underlying structure.

This output is generated using a Python-based analysis pipeline that ingests authentication and access telemetry, models relationships between identities and hosts, and assigns weighted scores to each transition. By applying a cost function and graph traversal logic (e.g., shortest-path analysis), the system identifies the most likely attacker routes and surfaces the points where visibility is lowest and impact is highest.

```
C:\Users\admin\PycharmProjects\Books\lateral_cost_engine>python main.py

=== GRAPH SUMMARY ===
Nodes: 4 | Edges: 3
Node list: svc_backup, jumpbox01, backup01, dc01

=== CHEAPEST ATTACK PATHS (k-shortest) ===

[1] Path: svc_backup -> jumpbox01 -> backup01 -> dc01
    TotalCost: 4.742
    Insight : Dominant corridor: backup01 -> dc01 (priv=domain_admin, auth=cert; low-visibility corridor; l
    Edge breakdown:
      - svc_backup -> jumpbox01 | w=2.456 | eff=0.4 risk=0.9 noise=0.4 | p=0.9 | priv=admin, auth=kerberos; l
      - jumpbox01 -> backup01 | w=1.368 | eff=0.2 risk=0.6 noise=0.2 | p=0.95 | priv=root, auth=ssh-key; low-
      - backup01 -> dc01 | w=0.918 | eff=0.2 risk=0.3 noise=0.1 | p=0.85 | priv=domain_admin, auth=cert; low-v
    Auto-generated hunting hypotheses:
      * [H1] FIRST-SEEN: Observe a first-time access pattern 'backup01' -> 'dc01' during the night bucket (auth=cert,
      * [H1] FIRST-SEEN: Observe a first-time access pattern 'jumpbox01' -> 'backup01' during the night bucket (auth=
      * [H1] FIRST-SEEN: Observe a first-time access pattern 'svc_backup' -> 'jumpbox01' during the night bucket (auth=
      * [H2] CORRIDOR RHYTHM: If 'backup01' contacts 'dc01' repeatedly off-hours, validate whether it maps to approved
ng.
      * [H2] CORRIDOR RHYTHM: If 'jumpbox01' contacts 'backup01' repeatedly off-hours, validate whether it maps to appr
staging.
      * [H2] CORRIDOR RHYTHM: If 'svc_backup' contacts 'jumpbox01' repeatedly off-hours, validate whether it maps to ap
l staging.
      * [H3] PRIVILEGE INTENT: Any 'backup01' -> 'dc01' movement with privilege=domain_admin should be reviewed for 'wh
      * [H3] PRIVILEGE INTENT: Any 'jumpbox01' -> 'backup01' movement with privilege=root should be reviewed for 'why n
      * [H3] PRIVILEGE INTENT: Any 'svc_backup' -> 'jumpbox01' movement with privilege=admin should be reviewed for 'wh

=== CHOKEPOINT (CORRIDOR) ANALYSIS ===
  - jumpbox01 -> backup01 | centrality=0.333 | weight=1.368 | priv=root, auth=ssh-key; low-noise
  - svc_backup -> jumpbox01 | centrality=0.25 | weight=2.456 | priv=admin, auth=kerberos; low-noise
  - backup01 -> dc01 | centrality=0.25 | weight=0.918 | priv=domain_admin, auth=cert; low-visibility corridor; low

=== REPORT ===
JSON written to: report.json
```

Figure 11.2: Corridor-Prioritized Lateral Movement Analysis Showing the Lowest-Friction Path from svc_backup to the Domain Controller

Let's understand this output, starting with the `GRAPH SUMMARY`; that is, the environment's attacker map. You can see the following code:

- `Nodes: 4 | Edges: 3`
- `Node list: svc_backup, jumpbox01, backup01, dc01`

This topology isn't a classic asset list; it's a model that derives value through transitions:

- **`svc_backup`**
  An identity/service account (initial carrier).
- **`jumpbox01`**
  Management jump point (the gate of the corridor).

- **backup01**
  A highly trusted system (overprivileged in many environments).
- **dc01**
  The crown jewel (identity/impact center).

The critical point here is that instead of talking about critical servers, we're talking about critical transitions.

Next is the `CHEAPEST ATTACK PATH`: the route the attacker can "walk most comfortably." Here's the output:

- `Path: svc_backup -> jumpbox01 -> backup01 -> dc01`
- `TotalCost: 4.742`

What does this mean? According to the cost model used in this analysis—where transitions are evaluated based on effort, risk, noise, and probability of success—expressed as (effort + risk + noise)/probability of success—this route represents the lowest-friction path to reach the DC.

"Cheapest" doesn't mean only technical ease; it also implies the combination of the fewest traces and the highest likelihood of success.

In the `Dominant corridor`, we can see why `backup01 -> dc01` is the most dangerous corridor. Here's the output:

- Dominant corridor: `backup01 -> dc01`
- Label: `priv=domain_admin, auth=cert; low-visibility corridor; low-noise`

This finding is extremely valuable because it says two things at the same time:

- **Privilege level is very high.**
  A `domain_admin`-level transition creates cascading impact in a single move.
- **Visibility/noise is low.**
  `low-visibility` + `low-noise` means this corridor can flow without shouting in EDR/SIEM. This is exactly what an attacker likes: high impact + low friction.

This brings us to the `Edge breakdown`: In this context, an edge represents a transition between two nodes (e.g., from one identity or system to another) within the attack graph, and each edge carries a calculated cost based on factors such as effort, risk, noise, and probability of success. The following format is used to describe how this cost is composed for each transition:

```
w (weight) | eff | risk | noise | p (success_prob) | note
```

We now examine each transition in detail:

- **svc_backup -> jumpbox01**
  This transition represents the initial pivot point in the attack path, where an identity moves from its origin into the managed access layer:

  - `w=2.456` (the most expensive step on the path)
  - `priv=admin`, `auth=kerberos; low-noise`

  This step carries the majority of the cost. This usually implies one of two things: either the initial pivot step is more visible/costly, or there's greater control/friction here (policy, access, segmentation). Despite this, the fact that it's `low-noise` suggests that the cost is driven more by the risk/effort components than by noise.

- **`jumpbox01 -> backup01`**
  This transition reflects movement within trusted infrastructure, where administrative access is commonly normalized:
  - `w=1.368`
  - `priv=root`, `auth=ssh-key; low-noise`

  Root access from the `jumpbox` to the backup system via SSH key is, in the real world, often a corridor that masquerades as normal operations. Being `low-noise` means it has very little potential to generate anomalies.

- **`backup01 -> dc01`**
  This transition captures the final escalation step, where high-trust systems enable direct access to critical identity infrastructure:
  - `w=0.918` (the cheapest yet the most critical)
  - `priv=domain_admin`, `auth=cert; low-visibility corridor; low-noise`

  The danger here isn't that the cost is low; it's that while the cost remains low, the privilege level is maximal. Certificate-based authentication (`auth=cert`) appears as a legitimate flow in many environments and doesn't generate password noise. For this reason, the corridor turns into a "silent highway."

Next, we find the `Auto-generated hunting hypotheses`, which are outputs that directly turn into hunting scenarios. There are three hypothesis sets:

- **`[H1] FIRST-SEEN`**
  Example: Investigate if `'backup01' -> 'dc01'` is seen for the first time in the night bucket.

  The first-seen approach is IOC-independent: even if the attacker arrives with a new method, it can still be detected. The emphasis on the night bucket enables intent reading through operational rhythm.

- **`[H2] CORRIDOR RHYTHM`**
  Example: Repeated off-hours contact: Automation or staging?

  This is hunting for behavioral rhythm rather than a single event. Especially on the backup/DC line, regular but low-frequency contacts may align with staging behavior.

- **`[H3] PRIVILEGE INTENT`**
  Example: In transitions performed with `domain_admin/root/admin` privileges, ask "Why now?" and "Why this corridor?"

This hypothesis standardizes analyst reflexes rather than enforcing a technical control:

- Why now?
- Why this transition?
- Which workflow does this transition normally belong to?

This is an excellent design because it automatically answers the question "Why are we hunting this?" in team communication.

Finally, we have the `CHOKEPOINT (CORRIDOR) ANALYSIS`, which distinguishes between topological bottleneck and operational criticality. In this context, a *topological bottleneck* refers to a transition that is structurally central in the graph (meaning many potential paths pass through it), while *operational criticality* refers to a transition that may not be the most traversed, but carries the highest real-world risk due to its privilege level, low visibility, or potential impact.

In the output, the highest centrality is `jumpbox01 -> backup01 (0.333)`. Due to graph structure, centrality highlights the corridor most frequently traversed; in a small graph, this is expected. However, your output makes a valuable distinction:

- Topological chokepoint: `jumpbox01 -> backup01`
- Operational dominant corridor: `backup01 -> dc01` (high privilege + low visibility)

Reading these together reflects a professional approach: Tightening the chokepoint cuts lateral spread, making the dominant corridor visible exposes the silent highway mentioned earlier.

Here are some practical takeaways from this output for clearer detection:

- **Most critical risk**
  The `backup01 -> dc01` path producing `domain_admin` behavior in a silent, low-visibility manner.
- **Best first hardening point**
  Making transitions through `jumpbox01 -> backup01` (the chokepoint) more difficult and/or more observable.
- **Best first hunting rule**
  First-seen and repeated rhythm of `backup01 -> dc01` in the night bucket.

11.3 Summary

This chapter introduced a graph-based approach to detection engineering, modeling adversary movement as transitions across systems rather than isolated events. By focusing on paths and relationships, we showed how seemingly normal activities can reveal meaningful patterns when analyzed in sequence and context.

We distinguished between structural chokepoints and high-risk operational corridors, emphasizing that the most dangerous paths aren't always the most visible. By combining topology with context such as privilege and behavior, we identified where detection and monitoring should be prioritized.

These insights were translated into practical outcomes, enabling more focused hunting and stronger defensive positioning. In the next (and final) chapter, we'll build on this foundation by incorporating temporal dynamics and anomaly scoring to further enhance detection accuracy.

Chapter 12

Automation and Threat Intelligence Platforms

Automation doesn't exclude humans; it relieves them of burden. The analyst's task isn't to carry data, but to extract meaning from it. Automation clarifies this distinction by delegating noise to machines and entrusting intuition to humans. A balanced cyber threat intelligence operation walks precisely along this line.

When first establishing serious engagement with threat intelligence, a clear expectation typically forms that accurate information will arrive, and we'll act accordingly. At the beginning, this expectation is met for a time. A few indicators of compromise (IOCs), a few reports, and a few alerts, and the system appears to function. However, as time passes, the picture changes. Data multiplies, sources expand, and the threat surface grows. At that point, it becomes apparent that the problem is no longer a lack of information but an excess.

In modern security operations center (SOC) environments, the primary struggle isn't finding the threat, but filtering it. Manual cyber threat intelligence (CTI) processes quietly falter here. The human mind is simply not designed to continuously and consistently track a river in which thousands of indicators flow simultaneously. After a while, analysis gives way to reflexes, and reflexes give way to assumptions. At this point, automation becomes a necessity.

Viewing automation solely as a means of gaining speed is an incomplete perspective. The real gain is rhythm. Automated workflows provide CTI processes with a continuity that humans alone can't sustain. The time of day, the day of the week, or the analyst's fatigue ceases to be a determining factor. The system approaches every piece of data with the same level of attention; it processes information without emotion, yet fairly. This may appear cold, but it's an operationally vital advantage.

Threat intelligence platforms (TIPs) represent a mode of organization that embodies the effort to centralize fragmented information, enrich it with context, and render it communicable with operational systems. In a sense, TIPs represent the memory of the SOC. People change, shifts end, and teams are renewed; yet the intelligence memory remains in place and continues to grow.

The primary focus of this chapter isn't so much how automation is built, but how it's sustained. Open-source platforms, standardized data models, and machine-to-machine integrations gain meaning only when combined with sound operational reasoning. Otherwise, automation risks turning into a mechanism that produces errors more quickly. Such a system may create a sense of confidence, but it doesn't produce security. To understand how to effectively implement automation, we'll discuss the limitations and benefits of automation in CTI in the following sections, followed by a look through the major TIPs and concluding with a discussion of threat information sharing through MISP Threat Sharing (MISP).

Ultimately, this chapter seeks to answer a question far more fundamental than which platform is better. How does CTI survive within a growing organization? The answer is clear, though not simple. Survival lies in maturing through automation.

12.1 Introduction to CTI Automation

There's a common moment experienced by everyone who works with CTI. At first, everything is under control. Incoming threat reports are read, IOCs are noted, and a few manual correlations are performed. The system seems to speak, and you listen to it. For a while, this relationship continues. Then one day, data begins to flow simultaneously from 3 different sources, then 5, and then 10. At that moment, you realize that threat intelligence is quietly slipping out of your hands.

The problem isn't that information arrives. The problem is that information never stops arriving.

Manual CTI processes are meaningful for limited volumes. The human mind is strong at building context, but fragile when it comes to continuity. It's not possible to track thousands of indicators with the same quality, the same attention, and the same speed. At a certain point, analysis is delayed, priorities become blurred, and the most dangerous thing happens: normalization begins. What is abnormal starts to appear ordinary. CTI automation comes into play at this breaking point—not as an accelerator, but as a stabilizer. Automation quietly takes over the tasks that humans shouldn't be doing—repetitive, rule-based activities with limited context. In doing so, the analyst returns to where real value is created: interpreting, doubting, and questioning.

CTI automation is also a memory mechanism. Humans can forget. Shifts change, teams disperse, and priorities drift. Automation doesn't forget. It remembers when an indicator was first observed, in what context it gained meaning, and which systems it was associated with. This continuity transforms threat intelligence from a momentary reflex into an institutional capability.

Of course, automation doesn't solve everything. Poorly designed automation accelerates mistakes as well. For this reason, CTI automation can't be approached with a "set and forget" mindset. It must be continuously reviewed, nourished, and adjusted. Humans still remain at the center, but now as those providing direction instead of carrying the load.

The CTI automation approach discussed in this chapter isn't a competition between humans and machines, but a deliberate division of labor. Noise is left to machines; meaning is left to humans. Mature CTI operations emerge at the point where this distinction is successfully achieved, and automation is the first threshold of that maturity.

Now let's take a closer look at the limitations of manual CTI and how automation optimizes the CTI lifecycle.

12.1.1 The Limitations of Manual CTI Processes

At first glance, manual CTI processes appear reassuring. Because they pass through human hands, they create a sense of being "controlled." Every IOC is reviewed by an analyst, every report is read, and every decision is made consciously—at least in theory. In practice, however, the picture is far more complex. Threat intelligence doesn't flow at the pace of humans, but at the pace of the internet.

In real life, this means that during a single shift, IOCs arrive from three different sources: one is an open-source report, one a commercial feed, and one an internal incident record. The analyst reads them all, takes notes, considers some important, and sets a few aside to look at later. Then an alert fires. Priorities change. The IOCs that weren't revisited quietly remain in the background. Is this negligence? Most of the time, no. This is the natural limitation of a manual process.

Time is the greatest enemy of manual CTI. When threat intelligence is delayed, it loses its value—even if it's correct. Is an IP address still meaningful when it's examined hours later? Does a malicious hash truly count as useful if it's detected after the campaign has ended? These questions are asked repeatedly in SOC environments, but are rarely voiced aloud because the answer is uncomfortable.

Human error is inevitable here. Fatigue, loss of focus, and shift intensity aren't weaknesses, but natural consequences of being human. Manual CTI processes, however, fail to account for this reality. They expect the same analyst to produce the same quality at the end of the night as in the first hours of the day. They expect it, but don't receive it. The issue is then labeled as "analyst performance," yet the problem lies in the process itself.

Inconsistency is an even more insidious problem. The same IOC may be interpreted differently by two different analysts. One labels it critical, the other assigns it low priority. Which one is wrong? Perhaps neither. But from a system perspective, this situation produces ambiguity. Manual CTI struggles to generate standards because decisions depend on context, experience, and the analyst's mental state at that moment. In large-scale SOCs, this inconsistency gradually turns into a loss of visibility.

In a real SOC scenario, this situation typically unfolds as follows: A domain that arrived months earlier was marked as "benign" at the time. Weeks later, the same domain appears again as part of a different campaign. However, manual records are fragmented. Who reviewed it? Based on what criteria was the decision made? No one remembers. The system doesn't remember either. This is the most fragile point of manual CTI: memory.

Manual processes are also not scalable. A workflow that functions with 5 analysts becomes complicated with 10 and unmanageable with 20. Having more people doesn't always mean more capacity. Sometimes, they simply mean a greater need for coordination. However, CTI operates through flow, not coordination.

This doesn't mean that manual CTI is wrong. What's wrong is assuming that it's sufficient on its own. Manual CTI generates context and cultivates intuition, but it can't provide continuity. For this reason, in large-scale and high-tempo SOCs, manual processes remain a foundation—but they can't serve as the backbone. The backbone is built through automation. Manual CTI becomes a conscious layer operating on top of it.

As long as these limitations aren't acknowledged, CTI processes appear busy but ineffective. Once they are accepted, a transformation begins. This transformation is the psychological threshold of the transition to automation.

At this stage, a subtle but deeply impactful threshold emerges in most SOCs. No one yet says out loud, "We need to move to automation," but everyone begins to feel the same thing: Even though the system appears to be working, it's quietly wearing itself down. Even if the number of alerts doesn't increase, the cognitive load does. Analysts don't make fewer mistakes; they simply notice them later. This delay creates a silent erosion within the nature of CTI. It's usually here that the automation discussion begins. But this discussion doesn't emerge in a technical meeting; it takes root in everyday remarks: "Have we looked at this before?" "Who tagged this IOC?" "Why does this alert feel familiar, but I can't remember?"

The very presence of these questions is an indication that the limits of manual CTI have been reached because the problem is no longer understanding a single threat, but preserving threat memory. The human mind can generate context, but it can't preserve that context with the same clarity over years. The value automation offers here is continuity, before speed.

The transition to automation often creates not a sense of relief, but a mild unease because it feels like control is changing hands. Where once one could say "I reviewed this," the phrase now becomes "the system processed it." This shift isn't easily accepted, especially by experienced analysts because automation takes over not only tasks, but habits as well. Over time, however, one realization becomes clear: Automation doesn't reduce control; it changes the form of control. The analyst no longer chases individual IOCs; instead, they observe patterns. Rather than focusing on isolated events, they monitor clusters of behavior. The question "Is this IP malicious?" is replaced by "What does this behavior mean in this environment?" This transition marks the point at which CTI matures.

There's an important balance here. When automation begins to do everything on its own, it produces more harm than benefit. Blind thresholds, correlations detached from context, and incorrectly prioritized playbooks are all possible. For this reason, mature CTI automation doesn't remove humans from the loop; it elevates them. The analyst's role is no longer to process data, but to guide the system.

In real life, this translates into the following: When a threat is blocked through automation, most of the time, no one notices. There's no alert, no one panics, and often a report isn't even written. But something didn't happen. And in security, sometimes the greatest success is that nothing happens. These silent successes are rarely achieved through manual processes because manual systems demand visible action. Automation, on the other hand, produces invisible gains.

After this stage, the CTI operation becomes two-layered. At the bottom is the machine: it doesn't tire, forget, or lose its rhythm. At the top is the human: one who questions, builds context, and breaks rules when necessary. When these layers are properly aligned, CTI ceases to be a reactive activity and becomes a proactive defensive capability.

More often than not, this transformation is completed not through a major project announcement, but through a small acceptance: "We can no longer do this manually."

12.1.2 The Role of Automation in the CTI Lifecycle

On paper, the CTI lifecycle always appears orderly: collect, enrich, correlate, distribute, and receive feedback. When read, it almost conveys a mechanical sense of flow. However, once this cycle is applied in the field, it becomes clear how fragile it actually is. Each step depends on the previous one functioning correctly and on time. When a single link in the chain slows down, the entire structure falters. The role of automation here isn't so much to accelerate the cycle as to keep it in motion.

Real-World Example

When I first realized automation's true role, the problem wasn't the misclassification of an IOC. The problem was that a correctly classified IOC never reached anyone. It had been collected and enriched. But somewhere along the way, it had stalled. CTI had turned into information that no longer moved. This is where the true meaning of automation becomes clear: keeping information alive.

In the following sections, we'll break down the role of automation in CTI and look at a few examples.

Automation in Each Phase of the Lifecycle

To better understand how automation sustains the continuity of this otherwise fragile process, it's essential to examine how it functions within each phase of the CTI lifecycle. Its contributions—though sometimes subtle—collectively ensure that intelligence remains active, connected, and operational. Let's walk through the phases:

- **Collection**
  During the collection phase, automation scans a scope far broader than a human could ever perceive. Humans must be selective with open sources, commercial feeds, and internal systems; automation does not. It collects everything, and then filters. This

distinction is critical. In manual processes, filtering usually comes first, followed by collection. In automation, collection comes first, and interpretation follows. This approach significantly reduces the likelihood of missed threats.

- **Enrichment**
  In the enrichment phase, the value of automation is quieter but deeper. Historical behaviors associated with an IP address, related domains, and previous campaigns can't realistically be recalled manually every time. Automation functions here like a reflexive memory. The moment a human says, "This seems familiar," the system can show why it's familiar. This reduces the analyst's burden and improves the quality of decisions.
- **Correlation**
  In the correlation phase, automation doesn't attempt to imitate human intuition; instead automation complements intuition. Humans recognize patterns, but they can't see them all at once. Automation, even with limited context, is scalable. It patiently tracks recurring behaviors, timelines, and clusters of relationships across hundreds of events. The important point here is that automation doesn't say, "This threat is definitively X." It says, "These elements may be meaningful together." The decision still belongs to the human.
- **Distribution**
  The distribution phase is often the most underestimated part of the CTI lifecycle. Yet information that's delayed at this stage can nullify all prior effort. Automation delivers intelligence to the right place, in the right format, and at the right time—to security information and event management (SIEM) systems; security orchestration, automation, and response (SOAR) platforms; and security controls. Humans no longer act as intermediaries; they become architects. They define in advance who should receive what and when, and then observe the system applying these rules consistently.
- **Feedback**
  Feedback is the stage that determines the maturity of automation. In manual CTI, feedback is often verbal, fragmented, and delayed. In automated environments, feedback becomes measurable. Information on which IOCs were effective, which generated noise, and which had no impact at all flows back to the beginning of the lifecycle. The system begins to recognize itself. At this point, CTI ceases to be a static process and becomes a learning mechanism.

Real-World Example

Personally, this is where I felt the greatest difference. Once automation was introduced, CTI stopped being something we were constantly trying to keep up with. It became calmer and more controlled. Analysts ran around less, but thought more. Noise decreased, while the signal became clearer. This was a gain achieved not through speed, but through balance.

Automation doesn't make the CTI lifecycle perfect, but it makes it continuous. In security, continuity often matters more than perfection. As long as the cycle doesn't stop, intelligence doesn't die. That's why automation isn't CTI's accelerator, but its life support.

The practical impact of automation becomes more evident when each phase of the CTI lifecycle is examined in contrast with manual processes. As summarized in Table 12.1, automation not only expands the scale at which each stage operates but also introduces consistency, continuity, and measurable outcomes that are difficult to achieve manually. The comparison highlights how tasks that are traditionally dependent on individual effort and timing are transformed into structured, repeatable processes, ultimately enabling CTI to function as an integrated and operational capability rather than a fragmented analytical activity.

CTI Lifecycle Stage	Situation in Manual Processes	Contribution Provided by Automation	Operational Impact (Real-World Implication)
Collection	The analyst follows specific sources and is often forced to be selective.	Continuously and uninterruptedly scans a broad pool of sources.	The rate of missed IOCs decreases; "later-discovered" threats are reduced.
Enrichment	The analyst relies on memory and previous notes.	Automatically adds historical incidents, relationships, and context.	The feeling of "this seems familiar" is transformed into concrete data.
Correlation	Manual relationships can be established among a limited number of events.	Establishes time- and behavior-based correlations across hundreds of events.	Campaigns and patterns become visible instead of isolated events.
Prioritization	This is based on the analyst's experience and momentary intuition.	Performs consistent ranking based on risk, recurrence, and context criteria.	Critical threats are prevented from being lost within noise.
Dissemination	Information is often delivered late or incompletely.	Delivers to the right system, in the right format, at the right time.	CTI ceases to be mere information and becomes a direct defensive input.
Feedback	Feedback is verbal, fragmented, and often delayed.	Produces measurable feedback that is fed back into the cycle.	Over time, the system learns what works.

Table 12.1: The Role of Automation in the CTI Lifecycle and Comparison with Manual Processes

CTI Lifecycle Stage	Situation in Manual Processes	Contribution Provided by Automation	Operational Impact (Real-World Implication)
Continuity	This is dependent on shifts, individuals, and workload.	Operates independently of humans, rhythmically, and continuously.	CTI operations become structure-driven rather than person-dependent.
Institutional memory	This is dependent on notes, personal files, and experience.	Establishes persistent, traceable, and retrievable memory.	The question "Haven't we seen this before?" doesn't remain unanswered.

Table 12.1: The Role of Automation in the CTI Lifecycle and Comparison with Manual Processes (Cont.)

Example 1: Automation Revealing the Silent Blind Spot

CTI automation is most often implemented to "see more." Yet in practice, it produces another, unexpected effect: It makes previously unnoticed blind spots visible. At first, this paradox is difficult to grasp. While automation introduces new data, it simultaneously begins to challenge old assumptions.

In a SOC environment, after automation is put into operation, there may be a feeling for weeks that nothing is happening. Alert counts drop, IOCs are rapidly filtered out, and playbooks run smoothly. This creates a sense of satisfaction in many teams, until someone asks, "Is this environment really this clean, or are we no longer seeing some things at all?"

In manual CTI processes, this question is rarely asked. Manual processes are already incomplete, and that incompleteness has been normalized. Automation, by producing consistency, makes deviations far more visible. Months of receiving no signal at all from a specific network segment constitutes an alert in an automated environment. In manual setups, such silence often goes unnoticed.

At this point, automation assumes an unexpected role: It prompts scrutiny not of the threats, but of the detection architecture itself. Sometimes the problem isn't the threat, but the sensor. Automation exposes structural gaps such as a mispositioned log source, a missing telemetry stream, or faulty normalization because the system's rhythm has been disrupted. Human intuition enters the picture here and recognizes, "This silence isn't natural."

This effect is rarely discussed in the literature because automation is almost always described in terms of "better detection." Yet in mature CTI environments, the true value of automation lies in making us aware of where nothing is being seen. This awareness is nearly impossible to achieve through manual processes alone.

Example 2: Automation Transforming Analyst Behavior

One of the least discussed effects of CTI automation is how it changes the way analysts think. Technical outputs are measured, and metrics are reported; but the cognitive transformation often remains invisible. An analyst working with manual CTI spends most of the day trying to keep up. When one list is finished, another arrives. Priorities shift constantly. Over time, this tempo makes the analyst's decisions increasingly short-term. The question "Is this IOC relevant now?" turns into "Can I close this today?" This isn't laziness; it's a survival reflex.

When automation comes into play, this reflex gradually dissolves. The analyst no longer has to touch every IOC individually. The system has taken over repetitive decisions. In the space that opens up, the analyst's mind shifts elsewhere: to patterns.

At first, this change is difficult to notice. But over time, the language analysts use begins to change. Instead of asking "Is this IP malicious?", they begin to ask, "What does this behavior represent in this environment?" This marks a critical threshold. CTI ceases to be a reactive activity focused on producing alerts and begins to generate environmental awareness. The analyst no longer focuses on isolated threats, but on the overall health of the environment. This moves CTI from the periphery of the SOC to its center.

Such a transformation rarely occurs through manual processes. Manual structures constantly pull the analyst back into operational details. Automation, by contrast, retains the details within the system and leaves thinking to the human. For this reason, automation reshapes not only processes, but professional reflexes as well.

This effect rarely appears in reports, but it's evident in mature SOCs: analysts who rush less, question more, react less impulsively, and yet make more accurate decisions.

12.2 Overview of Threat Intelligence Platforms

As CTI matures, the first reality that becomes apparent is that the main problem of threat intelligence isn't a lack of information, but the fragmentation of information. Reports exist, but they are in different places. IOCs exist, but they are detached from context. Analyses have been conducted, but they are forgotten over time. Everything exists; yet it's not together. This is exactly where TIPs begin to gain meaning: at the center of this fragmentation.

When first engaging with a TIP, the expectation is usually simple: "Let's gather everything in one place." This expectation soon proves insufficient because TIPs aren't just repositories. A well-designed TIP doesn't store intelligence; it puts it into circulation. The platform takes information out of a passive archive and turns it into an operational component. This difference may look small on paper, but it's very evident in the field. Realizing that an IOC's history often matters more than the IOC itself made this clear. The same indicator can carry different meanings depending on time and context. In manual processes, this context fragments, whereas TIPs preserve it, providing both technical clarity and cognitive relief.

The value of TIPs becomes apparent as they grow. At small volumes, manual notes and spreadsheets can seem sufficient. But as threat intelligence expands, relationships become complex. An actor appears in multiple campaigns. A piece of malware repeatedly confronts us through different infrastructures. At this point, the human mind struggles to build the connections. TIPs serve as a kind of map here. They may not show the way, but they clarify where you stand.

To understand a TIP correctly, it should be seen not as a "tool," but as a way of working. Because what matters isn't the platform itself, but how you use it. The same TIP can be nothing more than an IOC repository for one team, while for another it can become an organizational threat memory. The difference isn't technology; it's in the approach.

Here, the tone shifts slightly because TIPs aren't magical. A TIP established with the wrong expectations will quickly cause disappointment. The idea that it will "connect everything" and "explain everything" isn't realistic. TIPs don't make decisions or stop threats. TIPs only provide a ground on which the right questions can be asked. Without this ground, CTI remains an effort based on intuition.

In real life, the strongest impact of TIPs is often invisible. It shows itself between incidents, not during one. Traces of a campaign seen months ago quietly surface in a new alert. The analyst says, "We've seen this before," but this time doesn't have to remember; the system provides the reminder. These small moments are the clearest indicators of why TIPs are critical.

We'll dive deeper into the role of TIPs and their architectural structure in the following sections.

12.2.1 What Is a TIP and What Does It Do?

At a certain point, everyone working with CTI realizes that the problem isn't the absence of threat data, but where it resides and how it moves. Reports arrive, feeds flow, and IOCs are produced; yet this information often remains fragmented, stored in files, emails, or individual memory. TIPs exist to eliminate this fragmentation and ensure that intelligence remains accessible and actionable.

In its simplest form, a TIP centralizes intelligence. However, centralization alone isn't its defining value. A TIP enables data from different sources to be understood within a shared context. It evaluates indicators not in isolation, but alongside historical events, related infrastructure, and operational impact. For this reason, a TIP a system of relationships, in addition to its role as a repository. This distinction becomes clear in practice. An indicator that once triggered only a vague sense of familiarity can, within a TIP, be immediately contextualized—when it was observed, which campaign it was associated with, and what outcomes it produced. The system doesn't replace the analyst's thinking; it supports their thinking by making context visible and persistent.

TIPs don't stop threats, generate alerts, or execute response actions. Their role is more foundational: They make intelligence fluid. Information isn't confined to where it's collected; it's distributed to where it's needed (across SIEMs, SOAR platforms, and security controls) while preserving its context. In this sense, a TIP functions less as a static center and more as a structured distribution mechanism.

Another critical function of TIPs is standardization. By providing a shared structure for data and context, they enable different analysts and teams to interpret threats consistently. This reduces ambiguity in decision-making and strengthens both speed and confidence in operational processes.

Over time, the presence of a TIP shifts how analysts work. Less effort is spent searching for information, and more is devoted to analysis and interpretation. Intelligence becomes cumulative rather than fragmented, and knowledge is embedded into the system rather than remaining dependent on individuals. This reduces the fragility associated with personnel changes and operational variability. However, the effectiveness of a TIP depends on how it's used. Poorly structured data, inconsistent tagging, and weak correlation practices can quickly degrade its value. A TIP requires discipline: clear definitions, consistent enrichment, and deliberate relationship building. Without these, it risks becoming a storage space rather than an intelligence platform.

In mature environments, TIPs also serve as a bridge between technical analysis and organizational understanding. They transform raw indicators into structured insights—highlighting patterns, relationships, and trends—without losing technical depth. This increases the operational relevance of CTI across the organization.

Ultimately, TIPs define how intelligence is maintained, interpreted, and operationalized. They enable CTI to move from isolated observations toward a continuous and structured capability. While they don't replace the human element, they provide the foundation upon which meaningful analysis and informed decision-making can take place.

TIPs don't begin to produce value the moment they are deployed. In fact, in their early stages, they often create the impression of failing to meet expectations. The reason is usually not the inadequacy of the platform, but its incorrect positioning. Viewing a TIP as a central system that will solve everything is one of the greatest mistakes that can be made from the outset. A TIP doesn't provide answers; it enables analysts to ask the right questions. Until this distinction is accepted, the expected impact won't emerge, no matter how powerful the platform may be.

The following are key points to consider regarding automation with a TIP:

- **Data discipline**
  TIPs are fed by data, but not all data is useful. IOCs that are ingested blindly quickly disrupt the signal-to-noise balance. Especially when low-confidence indicators, indicators with weak context, or repetitive indicators are taken into the system without proper tagging, a TIP ceases to be an intelligence platform and turns into a complex list. What matters at this point isn't how much data is present, but which data is there and for what reason.

- **Correlation**
  The strength of TIPs lies not in individual IOCs, but in relationships. However, not every relationship is meaningful. Linking two indicators that aren't part of the same campaign solely because of technical similarity can lead to false inferences. For this reason, correlation rules should be defined deliberately, not hastily. Human intuition is still decisive here. Automation establishes relationships, but humans decide which ones should be preserved.
- **Time dimension**
  One of the most common mistakes in TIPs is carrying the past directly into the present. Threat intelligence, however, is context-dependent. An infrastructure that was critical months ago may be completely irrelevant today. Therefore, mechanisms must be established that allow indicators to "age" over time. Otherwise, a TIP turns into a historical archive that doesn't reflect the current situation. This quietly erodes analysts' trust in the platform.
- **Integrations**
  Integrating a TIP with SIEM, SOAR, and security devices isn't a goal in itself, but a means. Not every integrated piece of data should turn into an automatic action. Some intelligence is for awareness only. Some should be monitored. Some truly require intervention. Integrations built without making this distinction produce noise rather than speed. Fast but incorrect actions are riskier than taking no action at all.

12.2.2 Core Components and Architectural Structure

The moment you begin to truly understand a TIP usually comes with the following question: "Why is this platform designed this way?" In other words, why is it layered instead of being a single, monolithic structure? Why does each function reside in a separate component? The answer to this question is less technical than operational. Because threat intelligence isn't a single linear flow, but a constantly expanding, looping, and sometimes overlapping field of movement, the architecture must be built in a way that can carry this complexity.

Let's dive into this architecture, related points to consider, and a few helpful examples.

Understanding the Architecture of Threat Intelligent Platforms

The first layer, *data collection*, is often the most visible yet the most misunderstood component. The task here seems simple: pulling data from feeds, reports, internal systems, and so on. But what becomes apparent in practice is that the collection layer defines not only what is taken in but also what isn't. The choices made at this layer shape the character of the platform. A collection scope that is too narrow creates blind spots. An excessively broad collection amplifies noise. The first balance is established here.

Here's an example: Hundreds of IOCs were coming from the same source but produced no operational impact at all. The problem wasn't the IOCs. The problem was that the collection layer was operating without context. This is where the second layer comes into play: normalization and enrichment.

Normalization is often seen as a tedious technical detail. In reality, it's the backbone of a TIP. It ensures that data coming from different sources speaks the same language. An IP address is an IP address everywhere, but if one arrives as a string, another as an object, and another as a relationship, analysis becomes impossible. Normalization removes this complexity. Then, enrichment follows, which is where a TIP begins to demonstrate its real power.

Enrichment is the process of adding context to data. But *context* doesn't always mean more information—sometimes it means less of the right information. An IP address may carry dozens of labels, for example, but which of them is meaningful today? The enrichment layer seeks to answer this question. It turns the analyst's sense of familiarity into concrete data. This transformation is quiet, but it significantly affects decision quality.

Then, the *correlation engine*takes the stage. Here, the tone shifts slightly. Because correlation, when done incorrectly, is the component that causes the most harm. Connecting everything to everything else is easy. What is difficult is distinguishing what is meaningful together. The correlation engine doesn't attempt to replicate human intuition, but instead supports it. The engine observes timelines, recurring behaviors, and shared infrastructures patiently and says, "There may be a pattern here." It doesn't make the decision; it just points.

In a real SOC environment, this is felt as follows: Events that appear insignificant when viewed individually begin to tell a story when considered together. In manual processes, this story is often missed because the pieces reside in different places. The correlation engine brings these pieces to the same table. It doesn't write the story, but makes it readable.

The final layer, *API and integration*, is where the TIP communicates with the outside world. The greatest mistake made here is viewing integration as an "automatic action" mechanism. In reality, integration is a communication contract. What information goes where, when, and with what context is defined here. The data sent to a SIEM shouldn't be the same as the data sent to a SOAR. Intelligence delivered to security devices should be different from the intelligence presented to analysts. When this distinction isn't made, a TIP doesn't increase speed; it creates confusion.

When you consider this architectural structure as a whole, you see that a TIP is a flow architecture. Data enters, changes form, is correlated, and is carried to meaningful destinations. Each layer depends on the previous one, but doesn't repeat it. When this layered structure breaks down, a TIP either becomes too quiet or too noisy.

So, is this architecture truly necessary, or is it overengineering? In practice, the answer is clear. As threat intelligence grows, these separations become but a necessity, not a luxury. At small volumes, everything can be done in one place. At large volumes, however, everything must have its own place. Otherwise, the system collapses.

Ultimately, TIP architecture reflects the way CTI thinks. The clearer, more balanced, and more disciplined it is, the more reliable the intelligence it produces becomes. Architecture isn't just a technical preference; it's also a quiet expression of how you view threats.

From this point onward, architecture should be evaluated not only by how it works but also by how it can fail. Because in TIP architectures, the real test isn't in ideal flows, but under stress, with incomplete data and conflicting signals. A mature architecture is one that can quietly remain standing in these moments.

Key Considerations

TIP architecture may appear flawless on paper. The layers are clear, the flow is smooth, and the components are well defined. However, the real risks don't emerge when the architecture is drawn, but when it begins to be used. For this reason, the issues that require attention revolve less around the question of how it's built and more around how it's sustained.

The following are key points to consider:

- **Blurred boundaries between layers**
  One of the most common mistakes observed in practice is the correlation engine beginning to perform enrichment tasks, or the enrichment layer starting to generate correlation logic. When these boundaries become blurred, the architecture may seem to gain speed, but control is quickly lost. Each layer must answer only its own question. Collection asks "What has arrived?", enrichment asks "What does this mean?", and correlation asks "Are these meaningful together?" When these questions are mixed, the intelligence produced becomes blurred as well.
- **The reflex to fight noise early**
  Many teams, shortly after deploying a TIP, encounter the complaint that "We're seeing too much." The mistake made at this point is trying to suppress noise at the correlation engine. In reality, noise is usually generated in the collection and enrichment layers. No matter how well you correlate data whose source is flawed, you can't make it meaningful. Fighting noise doesn't begin at the top of the architecture, but at its base.
- **The time dimension**
  TIPs naturally tend to accumulate: data remains, relationships remain, and tags remain. However, in real environments, threats aren't permanent. The past importance of an IOC doesn't automatically determine its present meaning. For this reason, the architecture must include mechanisms that allow information to age. Indicators that lose weight over time should quietly recede into the background. Otherwise, a TIP ceases to be a living system and turns into a frozen past.
- **Integration drift**
  Another point that requires attention is integrations drifting beyond their intended purpose. The API layer of TIPs is powerful, and this power can easily be misused. Sending everything everywhere may seem attractive at first. However, this approach quickly dulls defensive mechanisms. A SIEM treating every TIP output as an alert, a SOAR turning every relationship into an action, or security devices blocking every IOC all serve to blunt the

system's reflexes. Integrations must be selective. Architecture should be shaped not by "Can I do this?" but by "Should I do this?"

- **Human intervention**
  The position of human intervention must also be consciously defined. In advanced TIP architectures, analysts don't work with individual IOCs; they shape system behavior. If this transition isn't clearly established, one of two extremes emerges: Either analysts constantly try to manually correct the system, or they become completely passive. Both situations are dangerous. A healthy structure is one in which analysts act as rule setters, and the system applies those rules consistently.
- **Tagging and taxonomy**
  No matter how solid the architecture may be, without tagging discipline, a TIP gradually loses meaning. Labeling the same type of threat with different names leads to blurred relationships. This situation is often noticed late because the system appears to be functioning. However, at a certain point analysts begin to derive different conclusions from the same data. This isn't a sign that the architecture has failed, but that discipline has collapsed.
- **Definition of normal behavior**
  Another issue that deserves attention is the definition of normal behavior. TIPs are often built to catch "what is bad." Mature architectures, however, first know what normal looks like. Silence isn't always good; sometimes it's blindness. The architecture must also be able to interpret situations in which nothing is happening. Without this capability, a TIP produces only reactions, not awareness.
- **Independence from humans**
  Finally, when evaluating TIP architecture, we must determine if this structure can produce correct decisions even when humans aren't present? If the system depends on the experience of specific individuals, the architecture isn't yet complete. A mature TIP must be able to produce consistent behavior independently of individuals. Human contribution remains valuable, but it shouldn't be decisive. Architecture is successful to the extent that it can transfer experience from individuals into the structure.

In summary, the key considerations lie less in technical details and more in behavioral discipline. When a TIP is implemented correctly, it operates quietly, reducing noise and clarifying decisions. When implemented poorly, effort increases while progress declines. The difference is shaped not by technology, but by how the architecture is understood and applied.

Example 1: When the Correlation Engine Produces a False Correctness

In an SOC environment, the TIP architecture appeared technically flawless. The collection layer was rich, enrichment was strong, and the correlation engine was running dozens of rules. For months, the system was reported as highly successful. Alert volumes had decreased, and incidents were being closed more quickly. The problem was noticed at the following point: Detections of real attacks had also decreased.

At first, this was interpreted as a positive outcome. However, a subsequent retrospective analysis revealed that infrastructure belonging to a specific threat actor had been quietly operating within the environment for months. The TIP had seen this infrastructure and had enriched it. The correlation engine had even linked these IOCs to different events. Yet it had never elevated them to a "critical" level. Why?

The correlation engine was misinterpreting the time dimension. The same IP address had appeared across different campaigns over the course of months, but each time, it was assessed as "isolated and low-frequency." The correlation engine perceived distribution, not intensity. Sparse but consistent repetitions over time weren't considered meaningful by the system.

This architectural flaw isn't immediately apparent because the system appears to be functioning, producing no alerts and generating no noise. But in reality, it produces a false sense of calm.

The solution wasn't to add a new rule. The solution was an architectural change—a separate evaluation path was added to the correlation engine for "entities that are low-intensity but recurring over time." This path was completely separated from classical correlation and placed behavioral continuity at its core.

This example shows that even if a layered architecture is built correctly, if time awareness is designed incorrectly, a TIP produces a calmness that appears correct but is dangerous. This silence is riskier than noise.

Example 2: When the Integration Layer Weakens Security

In another environment, the TIP architecture was built with the goal of full automation. The TIP → SOAR → security devices pipeline was operating almost continuously. When an IOC crossed certain thresholds, it was automatically blocked, and firewall and endpoint detection and response (EDR) rules were updated. In the first weeks, this setup worked perfectly. Then, something unexpected happened: The behavior of the security devices began to degrade.

EDR systems started responding with delays due to excessive rule loads, firewall tables became bloated, and some connections began to drop randomly. More critically, analysts could no longer track what the system was blocking and why. The decisions may have been correct, but they weren't traceable.

The problem wasn't with the TIP. The problem was a misunderstanding of the architectural role of the integration layer. The TIP had begun to behave like an authority. The SOAR was executing actions without questioning. Security devices, in turn, were reacting without context.

The corrective action taken at this point was to slow down the integrations. However, this didn't mean reducing performance. The integration layer was split into two paths:

- Context-carrying integrations (those feeding analysts, reports, and investigations)
- Action-producing integrations (those feeding devices)

Only time-resilient, recurring, and feedback-validated IOCs were allowed to remain on the action path. The rest were made visible only. The result was that fewer automatic actions were taken, but the overall security level increased because the system was no longer producing reflexes—it was producing awareness.

This example clearly demonstrates that when the integration layer is mispositioned in TIP architecture, the system operates faster but becomes more fragile. In advanced architectures, speed isn't the priority; controlled flow is.

12.3 Using MISP for Community-Based Threat Sharing

At a certain point, you realize that alone, you can see many things, but you can't see everything. In the early days, this gap appears to be a technical problem. If we collected more logs, wrote more rules, and pulled more IOCs, it seemed the picture would be completed. But it didn't work because the problem wasn't a lack of data. The problem was loneliness. While threats were moving together, the defense side was trying to think alone. Community-based threat sharing begins to make sense at this point. This approach isn't "taking other people's data"; it's more subtle than that. It's the effort to reinterpret what others have seen within our own context. This effort isn't sustainable without the right foundation, which is where MISP aligns.

At first, it's easy to see MISP merely as an IOC sharing platform. There are IPs, hashes, domains, and the list goes on. But after using it for a while, you realize that MISP isn't actually a list tool; it's a collective memory mechanism. Each event isn't just an indicator; it's a timestamp, a fragment of context, and a silent transfer of experience.

There's an important distinction here. Community-based sharing is an echo that gains meaning at the right time, not a rapidly fed stream. MISP organizes this echo. It doesn't eliminate noise entirely—and it shouldn't—but it makes visible what repeats within the noise, what is new, and what simply passes by.

At this point, the tone becomes a bit more serious. Community sharing isn't a romantic idea of "Let's unite." When misconfigured, when incorrect trust relationships are established, or when consumed without context, MISP can turn into a source of misdirection rather than a defensive tool. Not every piece of shared information is valuable, not every community is trustworthy, and not every data point may fit your environment. But when it's set up correctly, then MISP makes possible something a single organization could never create: distributed intuition. A weak signal that doesn't yet look "meaningful" in one place may have already triggered alarms elsewhere. You don't see it in your own systems, but you notice its projection in MISP.

In this section, we'll discuss why community-based sharing is more than automation alone, cover the boundaries within which MISP meets this need, and most importantly, talk about where it should stand. Perhaps most importantly, this section will pursue when MISP should speak and when it should remain silent because real threat intelligence begins with knowing what shouldn't be shared.

12.3.1 Understanding the Role of MISP in CTI

Automation and MISP are often confused, but in simple terms, automation produces speed, and MISP produces awareness. The difference between them is like the difference between an alarm sounding and understanding why the alarm sounded. The first requires reflex; the second requires judgment.

One of MISP's strongest aspects is that it doesn't treat shared information as correct on its own. An IOC within MISP is never independent of context. Tags, relationships, timelines, and source trustworthiness are all quietly woven around that indicator. For this reason, data in MISP doesn't behave like a raw alert dropped into a SIEM. It's more like a case file with marginal notes added to it. Here, a small but critical mental shift occurs. Analysts asking "Is this IOC malicious?" is no longer sufficient. The real question becomes "What does this IOC look like in my world?"

This question removes the defense side from being a passive consumer. When used correctly, MISP becomes less a system that feeds you and more a structure that challenges you. Every piece of shared information carries an implicit expectation: "What do you think about this?" At this point, MISP's community dimension ceases to be a technical detail and becomes an operational reality. Trust relationships begin here. What matters isn't who you share data with, but with whom you speak the same language. Being in the same sector isn't enough. One must share similar threat perceptions, comparable operational reflexes, and similar risk appetites. Otherwise, data flows, but meaning does not.

One of the most frequent mistakes in the field is feeding MISP with a "let's pull everything in" mindset. At first, it's attractive. Numbers increase, dashboards fill up, and metrics climb. But soon, a quiet fatigue begins within the team because nothing truly feels urgent. Everything is somewhat important and somewhat uncertain. At that moment, MISP ceases to be a collective memory and becomes a collective burden. Yet MISP's strength lies in selectivity: few but meaningful, rare but resonant.

There are some key points to consider when working with MISP, which we'll discuss next, alongside some examples that demonstrate the role MISP plays in your CTI process.

Key Considerations

The greatest misconception about MISP is that the points requiring attention are technical. In practice, however, the real fractures occur at mental thresholds, not in configuration files. Most teams that misuse MISP don't lack knowledge, but they're overconfident. Phrases such as "We can filter the data," "We can resolve the context," or "Our analysts are experienced" are frequently heard. Yet MISP is a system that amplifies the blind spots of experience as much as it rewards experience itself.

Keep the following points in mind when using MISP:

- **Balance the intent to share and the reflex to consume.**
  Many organizations position MISP as a source that pulls data from outside but produces no data internally. In the short term, this is comfortable. You take no risks. You don't share incorrect information. But without realizing it, you lose something: the community's perception of you. MISP communities don't technically block passive consumers; they filter them intuitively. A node that doesn't contribute gradually falls out of context. The data it receives is still "correct," but it no longer speaks to you.
- **Maintain a perception of time.**
  Every piece of data in MISP is time-bound, yet this dependency is often assumed to be superficial. Timestamps are read, first-seen/last-seen fields are checked, and then passed over. But the real temporal impact isn't here. The real issue is the threat phase at which an IOC is shared. Data shared while an attack is still in its propagation phase doesn't carry the same weight as data shared after the incident has been closed and turned into a report. Technically, they are the same; operationally, they are not. Most teams using MISP miss this distinction and build a pool filled with truths that arrived too late. Here, a small but unsettling reality emerges: MISP sometimes forces you to be right at the wrong time.
- **Beware of tag inflation.**
  Tagging is one of MISP's most powerful mechanisms. But when power is used without restraint, it distorts the sense of direction. Adding dozens of tags to every IOC doesn't enrich the data; on the contrary, it destabilizes the data. The analyst begins to struggle with the chaos of tags surrounding the indicator, not the indicator itself. Over time, a particular feeling emerges: "Everything is related to everything." This feeling is toxic to threat hunting because hunting requires discrimination and clarity. It requires few but distinct signals.
- **Don't assume trust is static.**
  Trusting a MISP community once doesn't mean you can trust it forever. Organizations change, teams reorganize, and priorities shift. Data that is carefully shared today may become a by-product of an automated pipeline tomorrow. For this reason, trust in MISP isn't a setting, but a continuously observed behavior. Who shares how often? Has the tone of the shares changed? Has the depth of context diminished? These aren't technical metrics, but a good MISP user notices them.
- **Don't disturb MISP's silence.**
  Not every indicator needs to speak. There are some IOCs that shouldn't be shared. Data that isn't yet sufficiently understood, lacks context, or carries a high risk of misinterpretation isn't a contribution to the community; it's a burden. Mature teams in the field know this: Sometimes not sharing is the most correct form of sharing. This isn't cowardice, but respect for collective memory.

Here, the tone hardens slightly. When misused, MISP doesn't strengthen defense; it distorts the perception of defense. The team becomes unable to distinguish what is truly important from what is statistically frequent. The number of alarms doesn't decrease, but the density of meaning erodes. At this point, MISP fails quietly. No one says, the system has collapsed, but no one listens to it anymore either.

- **Lower your ego threshold.**
  Finally, there's a detail that is rarely written in books but often noticed in the field: The ego threshold of the analyst working with MISP must be lowered because MISP confronts you with what others saw before you did. This is uncomfortable but also instructive. Teams that can say, "We missed this," benefit from MISP. Those who say, "We already knew this," gradually silence MISP over time.

Example 1: An IOC Quietly Settling into the Wrong Place

Once, a domain shared in MISP caught my attention. Everything looked "correct." The source was reliable, the tags were consistent, and the timestamps were coherent. Moreover, the same domain had been reported by three different organizations. Under normal circumstances, this is a signal that should be taken seriously. Yet something didn't sit right with me.

The domain was visible in our environment, yes, but only along a specific proxy path, within a specific user group, and only during business hours. This behavior didn't align with the shared attack scenario. The context in MISP suggested "command and control," whereas on our side everything appeared overly orderly. I didn't touch it for a while. I neither escalated an alert nor suppressed the IOC. I simply observed.

A week later, I realized that the domain was part of a new content delivery network (CDN) routing implemented by a third-party software as a service (SaaS) provider. In the organization that first reported it, this SaaS had appeared as a side artifact during an attack. Others, however, had echoed it without checking the context. Technically, there was no incorrect data, but operationally, there was an incorrect echo. This example taught me the following: An IOC that multiplies within MISP doesn't always grow stronger. Sometimes, it just becomes louder.

Example 2: A Hunt Saved by Unshared Data

During an internal hunting exercise, the behavior of a particular service account caught my attention because it was very short-lived and low-volume, but it had regular connections during nighttime hours. There was no signature and no known IOC. Technically, it was too early to label it "suspicious," yet too symmetrical to ignore. At that moment, someone on the team asked, "Should we put this into MISP?" We didn't because we didn't yet know what it was. We only knew what it wasn't.

Two weeks later, the same behavior appeared again in a different segment—this time with a slight variation. At that point, the picture was complete. What we were seeing was a low-noise lateral discovery technique that hadn't yet entered the literature. The sharing was done at that moment—but not hastily. It was shared with its context, its boundaries, and especially with notes on what we weren't certain about. This sharing enabled another organization to recognize the same pattern early. If we had shared it on the first day, it would most likely have been mislabeled and subsequently disregarded. It became valuable not because we withheld it, but because we shared it at the right time.

Sometimes MISP works not by speaking, but by remaining silent.

Example 3: The Quiet Change of a Trusted Source

We had received data from a MISP community for a long time. Their shares were clean, and their context was strong. Over time, we had begun to trust them almost automatically. Incoming IOCs were checked, but rarely debated. One day, I noticed that the tone of the shares had changed. There were more signs of automation. The tags were more generic, and the descriptions had become shorter. At first, I thought, "They have scaled up." But then I noticed a small inconsistency in the description of an IOC. It wasn't a technical error; it was a conceptual disconnect.

Upon investigation, we learned that the team at that organization had changed. Experienced analysts had left, and sharing processes had become semiautomated. The platform and the source name were the same, but the mindset had changed. From that point on, we began to read data from that source through a different lens. We didn't cut it off entirely, but we reduced its echo weight.

This example demonstrated that, in MISP, trust isn't a certificate. It's a habit, and habits can break.

12.3.2 The Fundamental Structure of the MISP Architecture

When trying to understand MISP, most people make the same mistake: They look at the screens first. They examine the interface, click through buttons, and navigate the menus. The events, attributes, tags, and so on all appears to be in their proper places, but after a while, a question inevitably arises: "Why was this structure designed this way?" Without answering that question, it's not truly possible to begin using MISP effectively. MISP is an open-source and community-driven platform, but more importantly than that, it establishes a deliberate balance between centralization and distribution. This balance is felt at every layer of the architecture. A structure that seems simple at first glance, appears as a sum of intentional design choices upon closer inspection. On first serious examination of the MISP architecture, the following analogy came to mind: This isn't a data repository; it's an intersection.

In the following sections, we'll examine the different components of the MISP architecture.

Central Server

At the heart of MISP lies a central server. This server stores events, attributes, relationships, and metadata. However, this center doesn't behave like an all-knowing brain. Rather, it resembles a structure that retains memory while leaving interpretation to the user. This design is intentional because in threat intelligence, there's no centralized "truth." The same data can produce different outcomes in different environments. MISP reflects this reality in its architecture. The center stores the data, and meaning is produced at the edge.

In real-world practice, you see this very clearly. Consider two teams using the same MISP instance: one in the financial sector, the other in manufacturing. They pull the same event, yet only one generates an alert. The other makes a note and moves on. The system doesn't object because the architecture is designed not to object.

Responsibility

There's a central data repository, but there's no central authority. At first, this feels reassuring, but then it becomes slightly unsettling because this architecture removes responsibility from the platform and places it on the user. When an incorrect event is shared, the system doesn't halt. No error message appears. No red warning light turns on. Yet that incorrect information finds an echo elsewhere, and the architecture quietly leaves the consequences to you. For this reason, working with the MISP architecture requires both technical skill and a professional stance.

Event Structure

In MISP, the concept of an *event* differs from the notion of an *incident* found in many platforms. An event doesn't have to represent a single moment of attack. Sometimes it encompasses a campaign, a reconnaissance process, or merely a suspicious observation. The critical question is "What does this event represent?"

The MISP architecture doesn't want this question to have a single answer. Therefore, events become context capsules, composed of attributes, relationships, temporal information, and tags. Data that appears ordinary when viewed in isolation produces meaning when it stands together.

In practice, this is experienced as follows: A single IP address is insignificant. But when that same IP is associated within the same event, within a specific time window, for a specific purpose, and with a specific actor, it suddenly gains weight. MISP's architecture doesn't force this change in weight but does make it possible. Attributes are the smallest yet most sensitive components of the MISP architecture. Each attribute is an atom of information. It may be an IP, a hash, or a URL. But more important than the atom itself is the context in which it resides. Here lies an architectural subtlety that many overlook: Attributes aren't considered "true" on their own. Every attribute is bound to an event. In other words, there's no free-floating truth in MISP. This is a design choice that's highly prone to misunderstanding, yet extremely valuable.

In real-world use, its benefit becomes clear in this way: If the same hash appears in two different events, it doesn't automatically carry the same meaning. The architecture doesn't allow it. Meaning remains confined within the boundaries of the event. This is one of the fundamental differences that distinguishes MISP from simple IOC lists. Let's examine a few examples.

Example 1: Silent Privilege Mapping

In a Windows Server 2022 environment, Active Directory Certificate Services (AD CS) was enabled. The system had been operating smoothly for years. Certificate requests were regular, and event logs were quiet. At first glance, everything appeared normal. However, the event opened on the MISP side didn't represent a single "attack," but rather a reconnaissance process that had taken shape over time.

The event contained the following attributes:

- Failed certificate template requests made on behalf of different user accounts
- Requests originating from the same source IP but attempting different template names
- Requests observed only outside of business hours
- A recurring but low-frequency pattern of Windows Event ID 4886 and Event ID 4887 records

When examined individually, these attributes didn't generate alerts. Certificate errors appeared to be misconfigurations or user mistakes. However, when all of them were correlated within the same event, within the same time window, and with the same behavioral intent, the picture became clear. This event didn't represent a moment of attack; it represented an initial reconnaissance activity silently probing the privilege levels of certificate templates. Had the same errors been scattered across different events, this context would never have emerged.

Thanks to MISP's event structure, the following occurred:

- Attributes ceased to be perceived as mere errors.
- Behavior evolved into an intent-bearing pattern.
- The preparation phase became visible before any exploitation occurred.

This example clearly demonstrates how attributes lose their meaning without an event.

Example 2: Lateral Reconnaissance Emerging Over Time

The second example occurred in a file infrastructure running on Windows Server 2022 with Server Message Block (SMB) and the Distributed File System (DFS). There was neither an exploit nor a known piece of malware involved. The attributes covered by the event were as follows:

- Short-lived but repeated SMB logon attempts originating from different clients
- Each attempt resulting only in read-only access
- Queries targeting listed but not accessed shares within the DFS namespace
- The same client attempting different servers on different days

None of these attributes was suspicious on its own. SMB sessions were normal. DFS queries were routine. Read-only access was already authorized. However, when considered together within a single event and along a timeline, they acquired a different meaning. The event represented that an actor was quietly seeking the answer to the question of what existed where within the file infrastructure. There were no write attempts and no privilege escalation—only mapping. Here, the MISP architecture played a critical role:

- The same client's activities across different days were retained within a single event.
- Relationships between attributes were established contextually rather than automatically.
- This behavior transitioned from mere noise into a lateral reconnaissance pattern.

If these data points had been treated only as logs or IOC lists, each would have been considered normal in isolation. Through the event as a context container, intent that formed over time became visible.

Sharing and Synchronization

One of the most frequently criticized yet most deliberately designed components of the MISP architecture is its synchronization model. Some ask, "Why doesn't everything flow everywhere instantly?" The answer is simple: because it shouldn't. Sharing in MISP is controlled. There are push and pull mechanisms, but these mechanisms are intentionally designed with a brake applied. Filters narrow the scope of shared data. Not every event goes everywhere, and it shouldn't.

In practice, the value of this is felt very clearly. You share an event, but only with specific communities because you know that visibility doesn't equal understanding. The architecture supports this selectivity. At this point, the tone softens slightly because this design choice is less technical than it is ethical. The MISP architecture keeps user roles superficially simple—admin, user, sync user, and so on. Yet this simplicity is deceptive. Roles answer both who sees what and who influences what.

An incorrectly configured role distribution can render MISP ineffective in a short time. Grant too much authority, and chaos emerges. Restrict too much, and sharing withers. The architecture continuously forces you to remain in balance here.

Real-World Example

With good intentions, we enabled everyone to tag everything. The first week was excellent. In the second week, tags multiplied. By the third week, no one could find anything. The architecture hadn't failed. We had drawn the boundary incorrectly. Wondering why the MISP architecture is so "cautious" and why it seems to slow you down rather than accelerate you? Simply put, in threat intelligence, the most expensive mistakes are fast ones.

The core structure of MISP is designed to protect you from these mistakes. Every layer, every relationship, every synchronization choice whispers the same thing: "Think once more." And perhaps for this reason, the MISP architecture is a written form of a way of thinking.

As you spend more time with the MISP architecture, you begin to notice something: This structure doesn't force you to find the truth, but rather to recognize the falsehood early. This is a subtle yet important distinction. Most security platforms demand speed from the user: faster decisions, faster actions, and faster closure. Through its architecture, MISP asks "Are you sure?" This question is written nowhere in the system, yet it's felt at every layer.

Correlation Logic: Not Forcing a Link, But Placing Side by Side

One of the less discussed yet highly decisive aspects of the MISP architecture is its approach to correlation. Here, correlation doesn't operate automatically on a matched equals related logic. It's possible to establish relationships between attributes, but these relationships don't claim certainty. They are more akin to "seating them at the same table." This is a deliberate choice.

In reality, threat data rarely speaks clearly. A hash, an IP, or a domain doesn't shout on its own. What matters is what sits beside it. By design, MISP leaves this seating arrangement to you. It establishes the relationship but doesn't impose the conclusion.

In practice, the same attribute can acquire different meanings within different sets of relationships. The architecture is deliberately rigid in order not to break this flexibility, while keeping its distance from automated inference because a single incorrect automated conclusion can overshadow even a correct manual observation. In MISP, temporal information doesn't merely establish a before and after relationship. The timeline is the memory layer of the architecture.

A mere logging mindset doesn't record when an event was first seen, when it was updated, and when a particular attribute was added. The architecture treats time as part of the narrative. This makes it possible to read retrospectively how a threat evolved, what was noticed first, and what was recognized later. This makes a difference especially in retrospective analyses. A detail that went unnoticed at the time can gain meaning when reviewed along the timeline.

At some point, analysts ask themselves, "What would I have done if I had seen this data today?" This is what the MISP architecture exists to make possible.

> **An Architecture at Arm's Length from Automation**
>
> MISP isn't opposed to automation, but it doesn't sanctify it either. APIs exist, integrations exist, and data flows are possible, yet the architecture doesn't allow automation to have the final word. Every piece of automatically ingested data still resides within an event, still requires context, and is still expected to pass under human review. This becomes especially evident in SIEM and SOAR integrations. MISP feeds other systems,

but it doesn't think on their behalf. The architecture draws this boundary very clearly. For this reason, organizations that attempt to turn MISP into a fully automated IOC production pipeline quickly run into a wall because the MISP architecture accepts one fundamental premise: Thinking can't be delegated.

12.3.3 Feed and Event Management

At a certain point while using MISP, a reality becomes apparent: The real issue isn't where you obtain the data from, but where you stop it. The concept of a feed often appears innocent. "External IOC source," "automatic ingestion," and "enrichment" all sound productive, as if more feeds necessarily meant stronger defense. It's often assumed that adding more sources increases visibility. Over time, however, we find that when feeds aren't managed correctly, they don't become channels of information; they become generators of habit. Something that flows continuously begins to be accepted without being questioned.

MISP's approach to feeds is deliberately cautious. Feeds don't enter MISP as "truth"; they acquire meaning only when they are placed inside an event. This seemingly small distinction is, in fact, one of the architecture's hardest boundaries. MISP leaves us alone with a question: "Are you merely ingesting this data, or are you actually taking ownership of it?"

Feed Does Not Equal an Event

The moment you truly grasp the difference between a feed and an event, your perspective on MISP changes irreversibly. Until that moment, you have—consciously or unconsciously—been using MISP as a kind of list manager. The list grows, dashboards fill, and numbers increase. The system appears to be working. Yet somewhere, something feels missing. That missing element is context.

Feeds are, by nature, fast. They want to flow. They originate somewhere and enter somewhere else. A hash is seen, a domain is reported, or an IP is flagged, and a feed transmits what it sees, not what it thinks. For this reason, what feeds produce isn't knowledge, but probability. "This might be malicious," "This caused an issue somewhere before," and "This deserves attention," are all vague half-full sentences. A feed never says "why"; it simply says "Here it is" and moves on.

An event, by contrast, is slow—deliberately slow. An event implies not only that something was observed, but that it was noticed. When you open an event, you're effectively making a claim: "This is worth remembering." But for how long, with what meaning, and within which boundaries hasn't yet been defined. An event isn't a finalized judgment; it's more like an open-ended question. This is why an event produces context, time, relationships, and, most importantly, hesitation.

A hash coming from a feed is merely a possibility for MISP—nothing more, nothing less. On its own, it's neither true nor false. But when that hash is placed within an event, it's no longer alone. Time joins it: "When did we see it?" Place joins it: "Where did we see it?" Other behaviors sit beside it: "What else was it associated with?" At that point, the hash ceases to be a technical indicator and becomes a small part of a story. The question is no longer "Is this hash malicious?", but "What does this hash represent in this story?"

For this reason, there's no independent technical discipline called feed management in MISP. There can't be because a feed isn't something that can be managed on its own. A feed gains meaning only within events. Feeds serve events; events don't serve feeds. The moment this balance is disrupted, MISP continues to appear operational, but it shuts down cognitively. No one notices. The system produces no errors, and logs continue to flow, but the questioning voice inside the analyst falls silent.

Real-World Example

I saw how destructive this can be very clearly in a financial environment. An aggressive external feed was automatically pushing thousands of domains into MISP. From a technical standpoint, it was a flawless setup. Synchronization was correct, tags were in place, and the source appeared reliable. But after a few months, the analysts' language began to change. When a domain was discussed, they said, "This is already in MISP." The subtext was obvious: "There's no need to think."

That was the moment the system was finished. Because MISP's reason for existence isn't to take over thinking. On the contrary, it exists to force thinking. The reflex "If it's in MISP, it's already bad" turns MISP from a collective memory into collective rote. No one asks "Why is this here?" anymore. Questions such as "Who added this?", "In what context was it shared?", or "What does this look like in our environment?" quietly disappear. What remains is only a tag. And when a tag is detached from context, it's nothing but noise.

This is where MISP's value emerges. It doesn't tell you "This is bad"; It tells you "This is here," and that's far more uncomfortable because "bad" provides clarity; "here" assigns responsibility. What you do with something that is here is your decision. You may ignore it. You may monitor it. You may wait. But you can't pass over it without thinking. This is MISP's silent power. It makes you work with questions, not answers.

When the distinction between feed and event is missed, MISP continues to function for a while as if nothing is wrong. Dashboards fill, IOC counts rise, and integrations keep flowing. But the system no longer produces context. It just transports data. From that point on, MISP is no longer an intelligence platform; it's an expensive list manager.

Perhaps the most painful part is that the collapse isn't loud. No one says, "MISP has failed." But no one truly listens to it anymore either because when MISP falls into the category of "already bad," not a single meaningful question remains to be asked.

Internal Feeds: The Most Difficult Sources to Manage

External feeds are noisy but familiar. The real challenge lies with internally generated feeds (telemetry-derived lists). Domain lists derived from proxy logs, hashes coming from EDR, and IPs filtered from firewall deny records are the organization's own shadow, and people are less suspicious of their own shadow.

The most common mistake when managing internal feeds in MISP is to automatically regard them as "high trust." Yet in the field, internal data that has been stripped of context can become more dangerous than external data. Because it looks familiar, the analyst's reflex weakens. For this reason, in mature environments, even internal feeds don't directly become events. They are first placed in a holding area. There, they are observed quietly: Do they repeat? Do they coincide with other behaviors? Do they take shape over time?

Becoming an event isn't a right; it's a decision.

Event Management

When you open an event in MISP, you are, in fact, placing your reflexes on the table because the security world has taught you the same lesson for years: an alarm sounds, you run. You investigate, decide, and close. Everything that is closed gives a sense of completion. That file is closed, and you move on to the next one. In MISP, however, this habit is almost reversed. Opening an event isn't about finishing a case; it's about keeping it alive. This feels strange at first. Humans are uncomfortable with things left open. Uncertainty settles like a burden. But MISP's event management teaches you how to carry that burden in the right place.

An event isn't a work order opened to be closed, nor should it be thought of as a simple incident record. An event is more like a file to be read (and sometimes it's not even a single file); it's a narrative that can grow, expand, and branch out. Moreover, this narrative seeks an answer to "What does this look like?", not a chronological list of what happened. Threats often don't try to reveal themselves directly; they often want to remain unseen. What you see isn't the threat itself, but its shadow. MISP's event management tries to hold that shadow as a film strip formed over time, not as a single snapshot.

For this reason, a good event is incomplete on the first day. This incompleteness isn't inexperience. It's not a mistake at all. It's a deliberate gap because the data you have on day one is usually not sufficient; it's just interesting. You see an IP, a user behavior, a certificate request, and a DNS pattern that all may point somewhere, but you can't yet name it. Classical tools push you to two extremes here: Either you say "not suspicious" and move on, or you say "suspicious" and raise the alarm. MISP opens a third space in between: "I don't yet understand this, but I don't want to forget it." This sentence seems simple, but in the field, it marks a team's maturity threshold. Being able to say this requires avoiding both panic and arrogance.

What we call event management is, in essence, learning to live with this gap. Living with a gap isn't easy. Especially in operational teams, the expectation of results applies constant pressure. Amid questions like "What came out of this?", "What are we doing?", or "Can we close it?", saying "We're monitoring for now" can sometimes be perceived as weakness. Yet some threats exploit this impatience. They want you to name them hastily because an incorrect name kills a correct observation. This is why MISP's event management advances with rhythm, not speed. Instead of forcing you to make fast decisions, MISP forces you to asks the right question at the right time.

Real-World Example

I felt this most clearly in the field with the behavior of a particular service account on the internal network. On Monday nights, when no one was around, short-lived LDAP queries appeared. There was no brute force, no burst of failed logons, and no trace of privilege escalation, just a regular, small, and overly "neat" movement. We could have automatically reached for a SIEM rule. Writing an alert called "service account performing LDAP queries on Monday night" would have been easy. But the practical outcome of such an alert in the field is usually that it fires a few times in the first week, everyone gets used to it in the second week, and by the third week, it becomes "not worth looking at." In other words, the alert itself turns into noise. We didn't do that.

We opened an event, but wrote almost nothing. This requires discipline. Humans want to fill empty space. The reflex to write a description, add a tag, or say something kicks in. We added only three things: a time window, the account identity, and a behavioral note, plus our small internal discomfort—that "wait a minute" feeling. Then, we waited. Days passed. The event remained silent. In most systems, such silence means false positive. In MISP, silence sometimes means "the film hasn't started yet."

Later, another team noticed a similar pattern in a different segment that had the same service account, a different target, and a slight variation: The queried objects had changed, the target OU was different, the access rhythm was the same, but the intent was slightly more explicit. That was the moment the event grew. That growth wasn't the addition of data; it was the formation of meaning. Notice that there was no feed here. There was no external "bad IOC" list and no signal that perfectly matched any threat report. Yet there was an event taking shape. Sometimes what we call an incident isn't an IOC, but choreography—a repeating, consistent, and relocating form of behavior.

At this point, MISP's event management teaches something that classical tools can't: "Catching" an incident doesn't mean seeing it once—it means getting to know the incident over time. Knowing something doesn't happen through data collection alone; it happens through memory. The event is the container of that memory. You don't place only the finding there; you also place the "suspicion it left in you that day." Weeks later, when you return, you need to remember that suspicion. And human memory alone isn't reliable for this. This is why MISP ceases to be merely a technical platform and becomes an operational memory mechanism.

Here, a small but critical transformation occurs: You begin to see an event not as a completed case, but as a maturing hypothesis. On the first day, the hypothesis is bare. Then it slowly gets dressed as an attribute arrives, a relationship is added, a timeline gains meaning. Then one day, another signal appears somewhere, and that hypothesis suddenly stands up. At that moment, you understand why you didn't want to forget something you didn't yet understand because what you didn't forget was a small trace that caught you in the right place at the right time.

MISP's event management also builds an internal team language. The question "Should we close this?" gradually turns into "Where does this belong?" Instead of "Is this IOC malicious?", you start asking "What does this behavior look like in our environment?" This change may appear small on paper, but in practice it alters the backbone of defense because defense no longer just reacts—it constructs context. A team that can construct context doesn't drown in noise. It carries noise, but doesn't surrender to it.

In the end, a quiet calm settles inside because you accept that you don't have to solve everything immediately. This acceptance isn't laziness. It's a methodology. MISP's event management teaches that sometimes the most correct action isn't to tag something immediately, but to wait long enough to formulate the right sentence at the right time. Saying "I don't yet understand this, but I don't want to forget it" is the patience muscle of defense. When that muscle develops, many threats become visible before they fully grow.

In MISP, every feed has a lifespan. But this lifespan isn't technically defined; it's felt through human intuition. You notice when the tone of a feed changes. More generic tags, fewer explanations, more signs of automation, and so on indicate that the feed is still running, but it's no longer speaking. It's merely making noise. Mature teams don't generate alarms when shutting down a feed. They quietly pull it back because in MISP, dramatic decisions are usually wrong decisions. Feed management isn't loud management; it's fine-tuning. Feed and event management may appear to be the technical backbone of MISP. But in reality, this section is a silent negotiation the analyst conducts with themselves with these questions:

- How much will I leave to automation?
- How much will I hold by trusting my own intuition?
- When will I speak, and when will I remain silent?

MISP deliberately doesn't answer these questions, but it leaves you alone with them. And perhaps that is why being a good event manager in MISP requires not so much being a good tool user, but being a good listener. Some threats don't shout, but in the right place, within the right event, and at the right time, they whisper.

After working with MISP for a while, analysts might begin—often without realizing it—to ask the following question: "Am I managing this system, or is the system training me?" Because what we call feed and event management, while appearing to be a technical process, gradually turns into a habit that shapes the analyst's mindset. When you encounter a new piece of data, does your hand automatically move to "add," or do you pause and take a step back? This reflex is the thin line that separates teams that use MISP from teams that are used by MISP.

At a certain point, your perspective on feeds begins to change. You no longer see them as "nutrition." They are more like air currents. They blow continuously, change direction, sometimes cool you down, and sometimes irritate you. But they are never meaningful on their own. Meaning emerges only when you decide when to open the window. In MISP, that window is the event. The feed enters through it. But not everything that enters needs to remain in the room. Some things simply pass through, some leave a trace, and some alter the atmosphere of the room even after the window has been closed.

At this stage, event management ceases to be merely a technical act of record keeping. A form of editorial work begins. You decide which data remains in the text, which becomes a footnote, and which is removed entirely. And you make these decisions without rigid rules at hand. What guides you is experience, intuition, and the faint regret of things once missed. MISP operates within this emotional space because threat intelligence isn't as mathematical as it's often assumed to be. It's closer to language. The same word can carry entirely different meanings in different sentences.

The quiet boundary of feeds emerges here. The fact that a feed is still technically correct doesn't mean it's operationally useful. Recognizing this requires crossing a difficult threshold because people are reluctant to abandon sources they have grown accustomed to. "We've been pulling data from here for years" is a phrase frequently heard in MISP environments. But the threat landscape isn't static, as actors change, motivations shift, and techniques evolve. When the language of a feed remains unchanged, it may in fact be falling behind. MISP won't tell you this through a report or generate an alert. It will simply let you feel it. One day, you realize that the data coming from that feed no longer grows events—it merely fills them. At that moment, you decide whether to cut it off, pull it back, or reduce its weight. These decisions aren't made loudly because in MISP, the most correct decisions are often the least spoken ones.

As incident management progresses, the language within the team begins to change as well. Conversations are no longer about "how many IOCs we added." Instead, the question becomes, "What does this event tell us?" Analysts begin to ask, "Does this behavior feel familiar?" This is a sign that defensive reflexes are maturing because we stop speaking in numbers and start speaking in narratives. Narratives are far more durable in defense. An analyst may forget thousands of hashes, but a well-written, properly cultivated event can be remembered even years later. The statement that "We've seen something like this before somewhere" is born from this place.

MISP's feed and event management may appear to slow you down. And, at times, it genuinely does. But this slowness doesn't come from inertia; it comes from awareness. Automation produces speed, but speed isn't always an advantage. When you move quickly in the wrong direction, you only get lost sooner. This is why MISP is deliberately reserved when it comes to speed. It intentionally places friction in your path, saying "Don't turn this into an automatic event," "Don't share this directly," or "Let this wait a little." None of these instructions are written on the screen, yet the entire architecture whispers them.

Over time, you come to realize that a well-managed event in MISP doesn't always contain a large volume of data. Sometimes it tells a great deal with very little. It has gaps, has question marks, and may even contain explicitly written statements such as "We're not certain." This isn't a weakness. On the contrary, these are honest traces left for collective memory. When another team encounters that event, they don't look only at the conclusion; they look at the process as well. They ask, "How did they think about this?" This is where community-based intelligence finds its meaning. Not only data is shared, but a way of thinking.

Perhaps the most critical point is that feed and event management temper the analyst's ego because MISP frequently shows you that what you believe to be new may already be old to someone else. A detail that escaped your attention may have been noticed long ago elsewhere. For this reason, teams that mature within MISP gradually stop talking about what they already knew and start saying, "We looked at this before, but it looks different now."

In the end, the analyst working with MISP comes to accept that not everything needs to speak. Some feeds gain value by remaining silent, some events by waiting, and some data by not being shared at all. MISP doesn't push you to constantly produce; it forces you to stand in the right place. This is a discipline that is rarely found in the modern defensive landscape.

As illustrated in Table 12.2, feeds primarily function as high-volume carriers of signals, while events transform those signals into contextualized narratives that can be interpreted and acted upon. This comparison highlights a fundamental shift: from receiving data to assigning meaning. Understanding this difference is critical, as relying solely on feeds can accelerate activity without improving insight, whereas event-driven analysis enables more deliberate and context-aware decision making.

Dimension	Feed (Stream)	Event (Context Capsule)	Practical Decision/Note
Primary purpose	Carry IOC/IOA signals.	Transforms signals into a narrative.	Feeds bring data; events produce meaning.
Underlying assumption	"It may have caused an issue somewhere."	"What does this look like in our world?"	Feeds carry probability; events require judgment.
Temporal behavior	Fast, continuous, high-volume.	Slow, selective, evolving.	Speed can amplify early mistakes.
Sense of correctness	Appears "correct" based on source reputation.	Remains "pending" until tested through context.	Correct data can be a burden at the wrong time.
Noise generation	High (especially aggregator feeds).	Controlled (through capsule boundaries).	What reduces noise isn't filtering, but context.

Table 12.2: Operational Distinction Between Feeds and Events

Dimension	Feed (Stream)	Event (Context Capsule)	Practical Decision/Note
Update mechanism	Automatically refreshed, pulls along.	Manually/semiautomatically matures.	Event growth isn't addition, but increase in meaning.
Risk	Produces rote thinking ("If it's in MISP, it's bad").	Produces hesitation ("Why is this here?").	Rote thinking accelerates defense but blinds it.
Integration role	Tends toward raw feeding into SIEM/SOAR.	Context-aware triggering for SIEM/SOAR.	Raw feeds mean action pipelines increase the false-positive cost.
Success metric	IOC count, coverage.	Captured patterns, recurring choreography.	The goal isn't many IOCs, but a few with resonance.
Operational ownership	A sense of "received."	A sense of "being read."	Feeds are received; events are owned.

Table 12.2: Operational Distinction Between Feeds and Events (Cont.)

12.3.4 Attributes, Tagging, and Taxonomies

When examining MISP, attention naturally focuses on the event—what is included and what is left out. Over time, however, it becomes clear that attributes and tags shape how that content is understood. Without this context, even accurate data can be misinterpreted, raising a simple but critical question: Are we truly understanding the data, or merely observing it?

Attribute, tag, and taxonomy structures enable intelligence to be classified in terms of context and trust level. In the field, however, raw threat data often resembles the following: a handful of sand in your hand is real, yet its individual grains provide no direction. Tagging and taxonomy are like bringing a magnet close to that sand, causing the iron filings to separate so that a pattern emerges. The sand remains the same sand, but now it speaks.

The following sections will look at event attributes, tagging, and taxonomies in more detail, including examples and important considerations.

Attributes

In MISP, an attribute is an "information atom": an IP, a domain, a URL, an email address, a hash, a file name, a registry key, a user account, a certificate fingerprint, a mutex, and the list goes on. But the critical distinction is that an attribute doesn't "shout" like a SIEM alert. It simply stands there. And it unsettles you. When it does, ask these questions:

- What does this attribute say on its own?
- What does it become when it sits inside the event?
- What does this correspond to in my environment?

The same IOC can take on entirely different characters in two different worlds. Sometimes a domain that screams "command and control" elsewhere is, in your environment, merely a new CDN route of a corporate SaaS service. At other times, a certificate request that looks like nothing at all turns out to be a silent reconnaissance choreography.

Tagging

Tagging is one of MISP's most powerful muscles and also one of the most easily abused. A tag doesn't explain the data; it shapes how the data is read. An incorrect tag can poison even correct data.

Good tagging typically serves three purposes:

- Sharing boundaries: Who should see it/who should not?
- Operational priority: Now, later, or observe
- Trust and context: Is the source reliable? How strong is the evidence?

For this reason, in mature teams, tags aren't decorative. They function more like traffic signs. And as traffic signs multiply, roads don't necessarily become safer; sometimes they become more confusing. When tag inflation begins, everything is labeled as everything else. The result? Nothing is clear.

Taxonomy

Taxonomy is the "grammar" of tags. In MISP, taxonomies typically exist under specific namespaces (e.g., `tlp:`, `pap:`, `misp:`, `ecsirt:`, `kill-chain:`). Using a taxonomy allows your team to speak using the same words. The following commonly used taxonomies illustrate how this shared language is structured and applied in practice:

- **Traffic Light Protocol (TLP)**
  Commonly used to indicate information-sharing restrictions (e.g., `tlp:green`, `tlp:amber`).
- **Permissible Actions Protocol (PAP)**
  Preferred for defining what actions may be taken with the information (e.g., whether it may be automated).
- **MISP taxonomies**
  Used to provide structure across areas such as operational classification, incident type, sector, trust level, and analysis status.

What Are We Classifying?

One of the most frequently confused issues in the field is that confidence and severity aren't the same thing:

- You may have very high confidence in an IOC, yet its impact may be low.
- You may be looking at something highly critical, yet the evidence in hand may be weak (low confidence).

In MISP, tagging and taxonomies make this distinction visible. Consider it this way: A photograph may be clear but show something insignificant, or unclear yet capture something critical. The question isn't clarity alone, but which truly warrants action.

Three Different Examples

In this section, we'll go through three examples related to threat sharing with MISP. The purpose of these examples isn't to show how MISP is installed, but to demonstrate how operational decisions change when attributes, tags, and taxonomies come together.

Note

Real environments may vary; the logic and mechanics illustrated in these examples are generic.

Let's walk through our examples:

- **Windows Server 2022/AD CS: certificate template probing behavior**

  Active Directory Certificate Services (AD CS) is enabled on Windows Server 2022. Logs appear calm. Yet at night, something "small" happens. The initial reaction is often a shrug: "It's probably just a certificate error."

  Within this event, several attributes can be identified to capture and contextualize the observed probing behavior:

  - `ip-src`: Source IP sending the template requests
  - `user-account`: Requests attempted under different user names
  - `windows-eventlog`: AD CS–related event IDs (especially template request/error flows)
  - `certificate-template`: Template names being attempted
  - `datetime`: First-seen/last-seen window (out-of-hours rhythm)

  What looks like a mere "error" when viewed individually turns into something else when assembled within an event: privilege mapping. Probing template names is essentially a silent way of asking, "Which template can be obtained by whom, from where, and under which conditions?"

  The tagging/taxonomy approach follows:

  - Sharing boundary: `tlp:amber` (as an example, because it may contain internal context)
  - Operational state: An analysis flow such as `misp:analysis-state="ongoing"` (depending on the taxonomy set in use)
  - Confidence: Evidence strength for source accuracy (e.g., high if logs are validated, medium if based on a single observation)

This works because MISP doesn't reduce this behavior to a single IOC. Even without a hash, it preserves the choreography. When you tag that choreography, a clear sentence emerges: "This isn't an error; this is probing."

The internal reaction is usually a small one: "Alright... this is something." It doesn't shout, but it shifts position.

- **Windows Server 2022/SMB + DFS: a silent map of lateral reconnaissance**
  File servers, a DFS namespace, routine SMB sessions, and the background noise of normal operations comprise a favorite place for attackers.

  Within this event, several attributes can be identified to capture and contextualize the observed probing behavior:

  - `hostname`: Targeted server names
  - `ip-src/ip-dst`: Client-to-server flows
  - `user-account`: Account(s) performing the attempts
  - `share-name`: Shares being listed or accessed
  - `datetime`: Repetition window spread across days
  - (If available) details such as `logon-type`, `auth-method` (via EDR/SIEM enrichment)

  Viewed individually, this looks like an authorized user browsing shares. But when examined within an event and along a timeline, a different picture appears: a read-oriented, nonwriting, probing hand. The question "What exists here?" is being asked across DFS. There are no writes, no errors, and no noise. That silence is unsettling.

  The tagging/taxonomy approach follows:

  - Marking the behavior as a listing/reconnaissance pattern (using the relevant taxonomy classes)
  - Splitting confidence into two dimensions:
  - Confidence in the data (Is the log source reliable?)
  - Confidence in the interpretation (How strong is the intent assessment?)

  The operational outcome is that on the SIEM side, the decision is often not block immediately, but rather observe and correlate. A premature automated action could disrupt production. Here, MISP acts as a brake pedal by asking, "Are you sure?"

- **Ubuntu/DNS + proxy feed: two lives of the same domain**
  Scene: A domain repeatedly appears via an Ubuntu-based security node (e.g., DNS resolver, proxy, Zeek/Suricata sensor, or log collector). The same domain is also flagged in external feeds. A teammate reacts reflexively: "It's in MISP, so it's bad." However, the same word doesn't mean the same thing in every sentence.

  Within this event, several attributes can be identified to capture and contextualize the observed probing behavior:

  - `domain`: The suspicious-looking domain

- `dns-question/dns-answer`: Resolution patterns (`NXDOMAIN`? Short time to live [TTL]? Rapidly changing answers?)
- `http-host/url`: Access patterns via the proxy
- `user-agent`: Or client fingerprint (if available)
- `first-seen/last-seen`: Business hours or nighttime activity
- `source`: Distinction between feed source and local observation source

The tagging/taxonomy approach follows:

- Tagging that keeps feed-derived data "pending": "Externally flagged; local validation pending"
- Sharing boundary: Potentially a more open TLP (because the domain is already public), but more restrictive if internal context is added
- Most importantly: Expressing the two possible lives of the same domain separately:
- "May be command-and-control"
- "May be a benign CDN or telemetry-related infrastructure component"

What do you do in practice? Instead of branding the domain as simply "bad," you keep it within an event, tied to behavioral context:

- DNS response rhythm on the Ubuntu side
- Proxy access targets
- User group and time window

Sometimes the conclusion is that the domain isn't inherently malicious; it appeared as a side artifact during an attack. At other times, the opposite is true: The domain behaves like a benign service while quietly hiding behind dynamic infrastructure. What makes the distinction isn't the tags themselves, but the discipline carried by the tags.

Here's a simple rule to follow: If a tag doesn't influence a decision, it shouldn't exist. A tag must affect sharing, priority, confidence, or correlation; otherwise, it only adds noise. A second consideration is how the tag will be interpreted by the next analyst. MISP's value lies not only in the present but in how data is understood over time. Consistent tagging preserves this continuity, while disorganized tagging undermines it.

Ultimately, tagging and taxonomies are less about technical structure and more about disciplined communication. Their purpose isn't to maximize volume, but to maintain clarity because effective defense depends on precise, meaningful signals.

Key Considerations

Up to this point, we've discussed how attributes, tags, and taxonomies function. Now we arrive at a more difficult place: how they fail. In the field, MISP is rarely harmed by a lack of information; it's more often damaged by misplaced comfort. The system is up, synchronization works, and dashboards are full. Yet something is missing, and that missing element is usually not technical, but mental. At this point, the first question to ask is "Does this structure still force me to ask questions?" If not, then we've gone wrong somewhere.

Let's dive into the important points to consider:

- **Allowing tags to replace decisions**

  One of the most dangerous thresholds is that the tag becomes the decision itself:

  - `tlp:red` indicates urgent
  - `misp:confidence-high` indicates true
  - Advanced persistent threat (APT) tag indicates critical

  At first, this reflex seems to save time. Then it quietly does something else: it removes judgment from the equation. A tag should be an input to a decision, not its output. If an analyst sees a tag and closes the file, MISP is no longer producing memory; it has begun producing rote responses.

  A small but strict boundary to use is "If seeing a tag doesn't trigger a 'why?' inside me, that tag is too powerful." A strong tag guides, but an excessively strong tag suppresses thinking.

- **Treating taxonomy as universal truth**

  Taxonomies provide language, but they don't provide truth. The same taxonomy can take on different meanings in different organizations. A behavior considered high risk in a financial environment may be routine on a production line. Traffic patterns that are normal in a cloud-native architecture can be red flags in a closed network.

  The danger lies in reading taxonomy as an environment-independent reality. In teams that fall into this trap, you often hear statements like these:

  - "That's what the taxonomy says."
  - "That's what the standard defines."
  - "Everyone labels it this way."

  But MISP's silent assumption is that every environment is different. Taxonomy gives you the language. Constructing the sentence is still your responsibility.

- **Confusing confidence with comfort**

  A high-confidence tag creates a sense of relief in many teams: "Alright, this is solid." And right there, a subtle danger appears: comfort. Confidence measures correctness, not importance. Yet in practice, these two are often conflated. High-confidence but low-impact data occupies more space than it deserves simply because it feels reliable. Meanwhile, low-confidence but high-potential behavior is pushed aside with "We're not sure." And yet threats often enter through exactly this door: the place you're not sure about.

  For this reason, in mature environments, a confidence tag is never read in isolation. It's always accompanied by questions such as these:

  - Which decision does this data influence?
  - What is the cost if this turns out to be wrong?
  - What is the gain if this turns out to be right?

Confidence is a compass, but a compass doesn't tell you the destination.

- **The quiet beginning of tag inflation**

  Tag inflation doesn't happen overnight, nor is it a noisy failure. It begins quietly:

  - First week: "Let's be a bit more descriptive."
  - Second week: "Let's add this as well—no harm done."
  - Third week: "Just in case. . . ."

  And suddenly, a single attribute carries dozens of tags. All of them are correct, yet none of them are distinctive. At this point, MISP ceases to be a decision-support system and turns into a tag collection.

  There's a clear sign of this in the field: When analysts open an event and look first at the tag list rather than at the attributes, inflation has begun because the mind is doing this: "Someone has already thought this through." But MISP's value lies in the opposite: "Someone thought about this—but I must think again."

- **Sharing courage instead of sharing discipline**

  Another common trap in community-based work is the belief that the more we share, the more valuable we are. No. Sometimes the opposite is true. Sharing isn't a matter of volume, but of timing. Correct data shared at the wrong time becomes a burden for the community. An attribute whose context hasn't yet settled can trigger false alarms elsewhere—and once that happens, no one listens to it again.

  Here, a difficult but necessary virtue comes into play: the ability to remain silent.

  Ask this question: "If I share this information, will it improve someone else's decision or make it harder?" If the answer is unclear, wait. Waiting isn't laziness; it's respect for collective memory.

- **Forgetting the human behind the tags**

  In MISP, behind every tag and every attribute, there's a human being who had an observation, a suspicion, or a quiet "Wait a minute" feeling. But as the system grows, people fade from view. Sources are reduced to "feeds" and analyses to "tags." At this point, the greatest risk is the erasure of intent.

  If, when reading an event, you can no longer ask these questions, then MISP may appear to be functioning, but communication has ended:

  - What might the author have seen?
  - At which point did they hesitate?
  - Where were they unsure?

  A good MISP record carries both conclusions and uncertainty.

- **The final threshold: "Is this system training me?"**

  Perhaps this is the most critical question of all. After working with MISP for a while, two paths diverge: Either you manage the system, or the system trains you.

A system that trains you slows you down. It makes you stop and look. It forces questions and creates discomfort. If MISP no longer unsettles you, no longer makes you pause, or no longer prompts you to ask "Am I sure?", then something has been lost.

Because intelligence isn't a comfortable space, it produces controlled unease. The most expensive mistake in MISP isn't a wrong tag, it's reaching a point where tags no longer make you feel anything. At that moment, the system is still running, but you've stopped thinking with it.

12.4 Summary

This chapter explored how CTI evolves from a fragmented, human-dependent activity into a structured and sustainable capability through the use of automation and TIPs. We began by examining the limitations of manual CTI processes, highlighting that the core challenge is no longer the absence of data, but the inability to maintain continuity, consistency, and memory at scale. Automation was therefore positioned not as a tool for speed, but as a mechanism for rhythm—ensuring that intelligence remains active, connected, and operational over time.

We then examined how automation supports each phase of the CTI lifecycle, from collection to feedback, and how it transforms both processes and analyst behavior. Beyond improving efficiency, automation introduces stability and visibility, making both threats and blind spots observable. In doing so, it shifts the analyst's role from processing individual indicators to interpreting patterns and guiding the system itself.

The discussion then moved to TIPs, emphasizing that their value lies not in centralizing data, but in preserving context and enabling relationships. TIPs were framed as organizational memory systems—structures that allow intelligence to persist beyond individuals and evolve into a continuous capability. Their architecture, when properly designed, reflects technical decisions, operational discipline, balancing data flow, enrichment, correlation, and integration.

Finally, we explored the role of the MISP in community-based threat sharing. MISP was presented as a collective memory and decision-support environment. The distinction between feeds and events, the importance of context over volume, and the disciplined use of attributes, tags, and taxonomies were all shown to be critical in maintaining clarity and preventing cognitive overload. Throughout, a consistent theme emerged: Effective threat intelligence depends on how technology is interpreted, constrained, and applied.

Ultimately, this chapter demonstrated that mature CTI isn't defined by how much data is collected, but by how meaning is constructed and sustained. Automation, TIPs, and platforms such as MISP provide the structure, but it's disciplined thinking, contextual awareness, and deliberate decision-making that turn intelligence into a reliable defensive capability.

Appendix A
Bibliography

Bianco, D. J. (2015). *The Threat Hunting Reference Model (Part 1: Measuring Hunting Maturity). Sqrrl. www.threathunting.net/files/The%20Threat%20Hunting%20Reference%20Model%20Part%201_%20Measuring%20Hunting%20Maturity%20_%20Sqrrl.pdf*

Caltagirone, S., Pendergast, A., & Betz, C. (2013). *The Diamond Model of Intrusion Analysis.* Center for Cyber Threat Intelligence and Threat Research. *www.activeresponse.org/wp-content/uploads/2013/07/diamond.pdf*

CERT-EU. (2022). TLP Version 2.0 Primer. *https://cert.europa.eu/blog/tlp-version-2-primer*

CIRCL/MISP. (n.d.). *User Guide of the MISP Intelligence Sharing Platform. www.circl.lu/doc/misp/*

CISA. (2021). *Known Exploited Vulnerabilities (KEV) Catalog. www.cisa.gov/known-exploited-vulnerabilities-catalog*

CISA. (2022). *STIX Best Practices Guide (v1.0.0).* OASIS. *www.cisa.gov/resources-tools/resources/oasis-stix-21-best-practice-guide-v100*

Claise, B., Trammell, B., & Aitken, P. (2013). *RFC 7011: Specification of the IP Flow Information Export (IPFIX) Protocol for the Exchange of Flow Information.* IETF. *www.rfc-editor.org/rfc/rfc7011*

ENISA. (2024). *ENISA Threat Landscape 2024. https://www.enisa.europa.eu/sites/default/files/2024-11/ENISA%20Threat%20Landscape%202024_0.pdf*

ENISA. (2025). *ENISA Threat Landscape: Finance Sector (January 2023–June 2024 Incidents). www.enisa.europa.eu/publications/enisa-threat-landscape-finance-sector*

European Union. (2022). *Directive (EU) 2022/2555 (NIS 2 Directive). https://eur-lex.europa.eu/legal-content/EN/TXT/?uri=CELEX%3A32022L2555*

European Union. (2022). *Regulation (EU) 2022/2554 (Digital Operational Resilience Act – DORA). https://eur-lex.europa.eu/eli/reg/2022/2554/oj/eng*

Europol. (2024). *Internet Organised Crime Threat Assessment (IOCTA) 2024. www.europol.europa.eu/publication-events/main-reports/internet-organised-crime-threat-assessment-iocta-2024*

FIRST. (2022). *Traffic Light Protocol (TLP): FIRST Standards Definitions and Usage Guidance – Version 2.0. www.first.org/tlp/*

Hutchins, E. M., Cloppert, M. J., & Amin, R. M. (2011). *Intelligence-Driven Computer Network Defense Informed by Analysis of Adversary Campaigns and Intrusion Kill Chains.* Lockheed

Martin. *www.lockheedmartin.com/content/dam/lockheed-martin/rms/documents/cyber/LM-White-Paper-Intel-Driven-Defense.pdf*

ISO/IEC. (2023). *ISO/IEC 27035-1:2023 – Information Security Incident Management—Part 1: Principles and Process. www.iso.org/standard/78973.html*

Microsoft. (n.d.). *Audit Process Creation: Advanced Audit Policy Configuration.* Microsoft Learn. *https://learn.microsoft.com/en-us/windows-server/identity/ad-ds/plan/security-best-practices/advanced-audit-policy-configuration*

Microsoft. (2025). *Command Line Process Auditing (Event ID 4688 Configuration Guidance).* Microsoft Learn. *https://learn.microsoft.com/en-gb/windows-server/identity/ad-ds/manage/component-updates/command-line-process-auditing*

Microsoft. (n.d.). *Security Auditing Event 4688(S): A New Process Has Been Created (Windows 10).* Microsoft Learn. *https://learn.microsoft.com/en-us/previous-versions/windows/it-pro/windows-10/security/threat-protection/auditing/event-4688*

Microsoft. (n.d.). Sysmon (System Monitor): Official Documentation. Microsoft Learn. *https://learn.microsoft.com/en-us/sysinternals/downloads/sysmon*

MISP Project. (n.d.). *MISP Core Software Repository.* GitHub. *https://github.com/MISP/MISP*

MISP Project. (n.d.). *MISP Documentation (misp-book).* GitHub. *https://github.com/MISP/misp-book*

MITRE. (n.d.). *MITRE ATT&CK® Knowledge Base. https://attack.mitre.org*

NIST. (2012). *SP 800-61 Revision 2: Computer Security Incident Handling Guide. https://csrc.nist.gov/pubs/sp/800/61/r2/final*

NIST. (2024). *NIST SP 1299: NIST Cybersecurity Framework 2.0 – Resource and Overview Guide. https://csrc.nist.gov/pubs/sp/1299/final*

NIST. (2024). *The NIST Cybersecurity Framework (CSF) 2.0 (NIST CSWP 29). https://nvlpubs.nist.gov/nistpubs/CSWP/NIST.CSWP.29.pdf*

NIST. (2024–2025). *Publications Supporting the Integration of CSF 2.0 and SP 800-61 Revision 3. https://csrc.nist.gov/Projects/incident-response*

NIST. (2025). *SP 800-61 Revision 3: Incident Response Recommendations and Considerations for Cybersecurity Risk Management. https://csrc.nist.gov/pubs/sp/800/61/r3/final*

OASIS. (2021). *STIX™ Version 2.1 (OASIS Standard). https://docs.oasis-open.org/cti/stix/v2.1/stix-v2.1.html*

OASIS. (2021). *TAXII™ Version 2.1 (OASIS Standard). https://docs.oasis-open.org/cti/taxii/v2.1/taxii-v2.1.html*

OASIS CTI Technical Committee. (n.d.). *STIX/TAXII Documentation and Resources. https://oasis-open.github.io/cti-documentation/*

Paas, F. G. W. C., & Van Merriënboer, J. J. G. (1994). *Variability of worked examples and transfer of geometrical problem-solving skills: A cognitive-load approach.* Journal of Educational Psychology, 86(1), 122–133. *https://doi.org/10.1037/0022-0663.86.1.122*

Sweller, J. (1988). *Cognitive load during problem solving: Effects on learning.* Cognitive Science, 12(2), 257–285. *https://doi.org/10.1207/s15516709cog1202_4*

ThreatHunting.net. (n.d.). *A Framework for Cyber Threat Hunting. www.threathunting.net/files/framework-for-threat-hunting-whitepaper.pdf*

Tor Project. (n.d.). *Tor Rendezvous Specification - Version 3. https://gitlab.torproject.org/tpo/core/torspec/-/raw/main/attic/text_formats/rend-spec-v3.txt*

Tor Project. (2021). *V3 Onion Services Usage. https://blog.torproject.org/v3-onion-services-usage/*

Verizon. (2025). *2025 Data Breach Investigations Report.www.verizon.com/business/resources/reports/dbir/*

Volatility Foundation. (n.d.). *The Volatility Framework: Project Overview. https://volatility-foundation.org/the-volatility-framework/*

Volatility Foundation. (n.d.). *Volatility 3 Documentation. https://volatility3.readthedocs.io/en/latest/*

Volatility Foundation. (n.d.). *Volatility 3 Source Repository. https://github.com/volatilityfoundation/volatility3*

Wireshark Foundation. (n.d.). *Wireshark User's Guide. www.wireshark.org/docs/wsug_html_chunked/*

Appendix B
The Author

Haydar Yener Arıcı is a senior systems and cybersecurity specialist with more than 23 years of experience in IT infrastructure, system administration, digital forensics, and open-source intelligence (OSINT). Throughout his career, he has conducted extensive work in critical areas such as the design, operation, and security of enterprise IT infrastructures; digital evidence analysis; and the establishment and development of corporate cybersecurity processes.

Haydar holds one bachelor's degree and three master's degrees, combining his academic background with extensive field experience to develop projects at both the technical and strategic levels. He is the author of six books published in Turkey covering Windows Server administration, enterprise IT security, and digital evidence management. These works serve as important reference resources for system administrators, cybersecurity professionals, and digital forensics specialists, and they are widely used within the industry.

For many years, Haydar served as a Microsoft Certified Trainer (MCT), during which he delivered advanced training to numerous IT professionals in system administration, infrastructure security, and enterprise security architectures.

He has also collaborated with universities and academic institutions, providing training, consultancy, and academic contributions in areas such as forensic readiness, cybersecurity policy development, and enterprise system hardening.

Haydar continues his work by combining technical expertise with academic knowledge and field experience, focusing on enterprise security architectures, digital forensics, and advanced system security.

Index

A

B

C

D

E

F

G

H

I

P

Q

R

S

T

U

V

W

X

Y

Z